The Regulation of Turkish Network Industries

Muzaffer Eroğlu · Matthias Finger
Editors

The Regulation of Turkish Network Industries

Springer

Editors
Muzaffer Eroğlu
Faculty of Law
Boğaziçi University
Istanbul, Turkey

Matthias Finger
European University Institute
Florence, Italy

Istanbul Technical University
İstanbul, Turkey

ISBN 978-3-030-81719-0 ISBN 978-3-030-81720-6 (eBook)
https://doi.org/10.1007/978-3-030-81720-6

This Springer imprint is published by the registered company Springer Nature Switzerland AG
The registered company address is: Gewerbestrasse 11, 6330 Cham, Switzerland

Foreword

It is my honor to introduce this edited volume on the governance and regulation of Turkey's network industries. This is the first comprehensive and systematic overview of all the network industries, ranging from telecommunications to energy, water, postal services, and the different transport modes. Each of the 24 chapters competently highlights the achievements, practices, and challenges of regulating the various aspects of Turkish network industries. The introduction puts all this into a historical perspective, which makes these achievements even more impressive.

All chapters have been written by Turkish scholars, working in Turkey, some in the regulatory authorities and some in academia. I am pleased to see that such expertise exists in Turkey, and I congratulate the editors for having identified and coordinated all these experts in one single volume.

Even though the book applies the highest academic standards, it is practice-oriented and ultimately aims to improve what is already quite a remarkable quality of regulating network industries. I am sure it will inspire many junior researchers to engage in such an interdisciplinary intellectual endeavor with concrete, practical outcomes for the citizens, the economy, and the country.

As the president of the Istanbul Centre for Regulation board, I can testify that this book will not remain an isolated contribution. It is an integral part of a much broader initiative to foster Turkish expertise on the different network industries and their cross-cutting issues, such as regulation, governance, and financing.

It is precisely to that effect that in 2018 we established the Istanbul Centre for Regulation (IC4R), based at Istanbul Technical University. By way of our quarterly newsletter—*Network Industries Quarterly Turkey*—we are already reaching out to managers and policymakers in these different network industries. In the medium term and long term, we aim to become the center of expertise in matters of de- and reregulating infrastructures, not only in Turkey but also in the region.

In this sense, this edited volume is undoubtedly a first significant step.

Ankara, Turkey

Dr. Alparslan Bayraktar
President of the Board, Istanbul Centre for Regulation
Deputy Minister of Energy and Natural Resources, Republic of Turkey

Contents

Editors and Contributors

About the Editors

Muzaffer Eroğlu is an assistant professor at University of Boğaziçi Faculty of Law; he lectures on the company law, energy law, and competition law. He holds an LLB from Ankara University, an LLM from the University of Kent, and a Ph.D. from Queen Mary University of London. He is a member of the Ankara Bar. His academic research focuses on company law, corporate governance, competition law, and energy law. He published a book *Multinational Enterprises and Tort Liabilities*. He has several publications in the field of company law, competition law, and energy law in both English and Turkish. He also carried out researches as a visiting academic at the University of Oslo, Max Planck Institutes for Comparative and International Private Law in Hamburg, and European Institute of Florence.

Matthias Finger is a Professor Emeritus in charge of Digital Governance and Regulation at the Center for Digital Trust (C4DT) at Ecole Polytechnique Fédérale in Lausanne, Switzerland (EPFL). He holds a Ph.D. in political science from the University of Geneva. He has been an Assistant Professor at Syracuse University (New York), an Associate Professor at Columbia University (New York), a Professor of Management of Public Enterprises at the Swiss Federal Institute of Public Administration, and between 2002 and 2020 the Swiss Post Chair in Management of Network Industries at EPFL. Since 2010, he is a part-time professor at the European University Institute (EUI) in Florence, Italy, where he directs the Florence School of Regulation's Transport Area (FSR-T). Since 2017, he is also a professor at the Faculty of Management at Istanbul Technical University (ITÜ), directing the Istanbul Center for Regulation (IC4R).

Contributors

Eylem Apaydin Kocaeli University Law School, Kocaeli, Turkey

Şahin Ardıyok Balcıoğlu Selçuk Ardıyok Keki Attorney Partnership, Istanbul, Turkey

Remzi Özge Aritürk Turkish Competition Authority, Ankara, Turkey

Armanç Canbeyli Balcıoğlu Selçuk Ardıyok Keki Attorney Partnership, Istanbul, Turkey

Bağdagül Kaya Caner Head of Group at the Energy Market Regulatory Authority of Turkey (EMRA), Ankara, Turkey

Volkan Recai Cetin Presidency of Turkey, Strategy and Budgeting Office, Ankara, Turkey;
The World Bank, Washington, USA

Faruk Cirit Presidency of Turkey, Strategy and Budgeting Office, Ankara, Turkey

Uğur Emek Başkent University, Ankara, Turkey

Muzaffer Eroğlu Boğaziçi University Faculty of Law, Istanbul, Turkey

Ömer Faruk Erol School of Law, Ibn Haldun University, Istanbul, Turkey

Soner Esmer Iskenderun Technical University, İskenderun, Turkey;
Dokuz Eylul University, İzmir, Turkey

Matthias Finger Istanbul Technical University, İstanbul, Turkey

Bülent Gökdemir Bilkent University, Ankara, Turkey

Mehmet Suat Kayikçi Partner At LBF Law & Consultancy, İstanbul, Turkey

Burak Kepkep Polat Enerji Sanayi ve Ticaret AŞ, Toronto, ON, Canada

Burak Keskin Faculty of Economics and Administrative Sciences, Department of Business Administration, Cankiri Karatekin University, Cankiri, Turkey

Emin Köksal Bahçeşehir University, Istanbul, Turkey

Mehmet Kürkçü Energy Market Regulatory Authority of Turkey, Ankara, Turkey

Taner Şahin Law School, Ondokuz Mayıs University, Samsun, Turkey

Engin Saygin Faculty of Law, Department of Administrative Law, Ankara Yıldırım Beyazıt University, Ankara, Turkey

Evren Sesli Balcıoğlu Selçuk Ardıyok Keki Attorney Partnership, Istanbul, Turkey

Ali Osman Solak Department of Public Finance, Faculty of Economics and Administrative Sciences, Bolu Abant Izzet Baysal University, Bolu, Turkey

Nilgün Görer Tamer Department of City and Regional Planning, Faculty of Architecture, Gazi University, Ankara, Turkey

Refik Tiryaki Energy Market Regulatory Authority of Turkey (EMRA), Çankaya/Ankara, Turkey

Umut Alkım Tuncer EPFL ME, Ras Al Khaimah, UAE

Mehmet Yalili Energy Market Regulatory Authority of Turkey (EMRA), Çankaya/Ankara, Turkey

Kazım Yeni Iskenderun Technical University, İskenderun, Turkey

Barış Yüksel BASEAK, Istanbul, Turkey

Engin Zeki Ecole Polytechnique Fédérale de Lausanne, Lausanne, Switzerland

Abbreviation

3G	IMT-2000/Universal Mobile Telecommunications System
4.5G	IMT-Advanced
AAC	Airspace Architecture Study
ACC	Area Control Center
ADR	European Agreement Concerning the International Carriage of Dangerous Goods by Road
ADSL	Asymmetric Digital Subscriber Line
AIPH	International Association of Ports and Harbors
AIS	Aeronautical Information Services
ALÇESU	Alaçatı-Çeşme Su İşletmeleri San. Tic. A.Ş.
ALDAŞ	Antalya Altyapı Yönetim ve Danışmanlık Hizmetleri Sanayi ve Ticaret A.Ş.
AMAN	Arrival Management
ANSPs	Air Navigation Service Providers
ANTSU	Antalya Su İşletmeleri A.Ş.
ASAT	Antalya Metropolitan Municipality Directorate General for Water and Wastewater Management
ASKİ	Ankara SKİ
ATAK	Mediterranean and Aegean Tourism Infrastructure and Coastal Management
ATC	Air Traffic Control
ATM	Air Traffic Management
Avea	Avea İletişim Hizmetleri A.Ş.
Aycell	Aycell Haberleşme ve Pazarlama Hizmetleri A.Ş.
BCOs	Beneficial Cargo Owners
BLT	Build–Lease–Transfer
BO	Build and Operate
BOO	Build–Own–Operate
BOT	Build–Operate and Transfer
BOTAŞ	Petroleum Pipeline Corporation (Boru Hatları ile Petrol Taşıma A.Ş.)

BOTs	Built Operate and Transfer Agreements
BTK	Information and Communication Technologies Authority
BULATSA	Bulgarian Air Traffic Services Authority
BUSKİ	Bursa Water and Sewerage Administration
Cable TV	Cable TV network
ÇALBİR	Çeşme and Alaçatı Environmental Protection Infrastructure Facilities Construction and Operations Union
CAPEX	Capital expenditures
CAS	Country Assistance Strategy
ÇEAŞ	Çukurova Elektrik A.Ş.
CGE	Compagnie Generale des Eaux
CIRT	Cyber Incident Response Team
CMA	Capital Market Authority
CNG	Compressed natural gas
CNS	Communication, navigation, and surveillance
CoS	Council of State
CRM	Capacity Remuneration Mechanism
CTP	Continuous Trade Platform
DAMs	Day-ahead markets
DBOT	Design–Build–Operate–Transfer
DDK	State Supervisory Council
DEA	Data Envelopment Analysis
DGCA	Directorate General of Civil Aviation
DGH	Directorate General of Highways
DGS	Directorate of General Security
DHMI	General Directorate of States Airport Authority (Devlet Hava Meydanlari Isletmesi)
DPC	Driver Professional Competency
DSİ	Directorate General of State Hydraulic Works
DSO	Distribution System Operator
DT	Deutsche Telekom
EASA	European Aviation Safety Agency
EBT	Electronic Bulletin Tables
EC	European Commission
ECA	Electronic Communications Act No. 5809
ECAC	European Civil Aviation Conference
ECJ	European Court of Justice
ECL	Electronic Communication Law
EDS	Electronic Control Systems
EEC	European Economic Community
EGO	General Directorate of Ankara Electricity, Gas and Bus Enterprises
EHABS	Electronic Communication Infrastructure Information System

EIA	Environmental Impact Assessment
EİEİ	Electrical Power Resources Survey and Development Administration (Elektrik İşleri Etüt İdaresi)
EIF	Environmental Impact Factor
EML	Electricity Market Law No. 4628
EMRA	Energy Market Regulatory Authority
EMSSD	Electricity Energy Market and Supply Security Strategy Document
EPDK	Enerji Piyasası Düzenleme Kurumu
EPİAŞ	Energy Exchange Istanbul (Enerji Piyasaları İşletme Anonim Şirketi)
ESHOT	General Directorate of Electricity, Water, Gas and Trolleybus
EU	European Union
EÜAŞ	Electricity Generation Company (Elektrik Üretim Anonim Şirketi)
EUROCONTROL	European Organization for the Safety of Air Navigation
FABs	Functional Airspace Blocks
FDPS	Flight Data Processing Systems
FERC	Federal Energy Regulatory Commission
FIRs	Flight Information Regions
FSRU	Floating Storage and Regasification Unit
FTTB/H	Fiber-to-the-Building/Home
FUA	Flexible Use of Airspace
GAT	General Air Traffic
GATS	General Agreement on Trade in Services
GDHR	General Directorate of Highway Regulation
GDSAA	General Directorate of State Airports Authority
GDTSR	General Directorate of Transportation Services Regulation
GFC	Global Financial Crisis
GOP	Grid Operation Principles
GRF	Daily Reference Price
GRT	Gross tonnage
GSM	Global System for Mobile Communications
GTT	General transaction terms
HGS	Fast Passing System (Hızlı Gecis Sistemi)
HHI	Herfindahl–Hirschman Index
HKEP	Watershed Protection Action Plans
HPP	Hydroelectric power plant
HST	High-speed trains
ICAO	International Civil Aviation Organization
ICC	Infrastructure Coordination Centers
ICS	International Chamber of Shipping
ICT	Information and Communication Technologies

ICTA	Information and Communication Technologies Authority
ICTB	Information and Communication Technologies Board
IEA	International Energy Agency
IEC	International Electrotechnical Commission
İETT	Istanbul Electricity, Tramway and Tunnel Enterprises
IGA	Istanbul Grand Airport
İGDAŞ	İstanbul Gaz Dağıtım A.Ş.
ILO	International Labor Organization
IMM	Istanbul Metropolitan Municipality
IMO	International Maritime Organization
İOAŞ	İstanbul Otobüs A.Ş.
IPSAS	International Public Sector Accounting Standards
IPSASB	International Public Sector Accounting Standards Board
ISGOTT	International Safety Guide for Oil Tankers and Terminals
ISI	Import Substitution Industrialization
İSKİ	Istanbul SKİ
ISO	Independent System Operator
ISO	International Organization of Standardization
ISPS	International Ship and Port Facility Security
ISPs	Internet Service Providers
İŞ-TİM	İŞ-TIM Telekomünikasyon Hizmetleri A.Ş.
ITCA	Information Technologies and Communication Agency/Authority
ITO	Independent Transmission Operator
ITS	Intelligent Transport Systems Applications
JAA	Joint Aviation Authorities
JICA	Japanese International Cooperation Agency
Kayseri A.Ş.	Kayseri ve Civarı Elektrik A.Ş.
Kepez A.Ş.	Kepez Elektrik Ticaret A.Ş.
KGM	General Directorate of Highways
KGS	Card Passing System
KID	Lease-Operate-Transfer
Law No. 6461	The Act regarding the Liberalization of Railway Transportation No. 6461
LLU	Local-Loop Unbundling
LNG	Liquefied natural gas
LOI	Ladder of Investment
LPG	Liquefied petroleum gas
LSSIP	Local Single Sky Implementation
MARPOL	International Convention for the Prevention of Pollution from Ships
MENR	Ministry of Energy and Natural Resources

MESKİ	Mersin Water and Sewerage Administration
METU	Middle East Technical University
MM	Mid-level Manager
MNC	Mobile network codes
MNOs	Mobile network operators
MoC	Memorandum of Cooperation
MoEU	Ministry of Environment and Urbanization
MoI	Ministry of Interior
MOT	Ministry of Transport and Infrastructure
MoTF	Ministry of Treasury and Finance
MoTI	Ministry of Transport and Infrastructure
MTA	Mineral Research and Exploration Institute
MTOW	Maximum Take-off Weight
MTR	Mobile termination rates
MVNOs	Mobile virtual network operators
NC	National Carbonizing
NCAs	National Competition Agencies
NewPENS	New Pan-European Network Service
NGANs	Next-Generation Access Networks
NGML	Natural Gas Market Law
NRA	National Regulatory Authority
NRAs	National Regulatory Agencies
NRF	New Regulatory Framework
OAT	Operational Air Traffic
OCIMF	Oil Companies International Marine Forum
OCM	On-the-Day Commodity Market
OGS	Automatic Passing System
ÖHO	Private bus operators (Özel Halk Otobüsleri)
ÖİB	Directorate of Privatization Administration
OPEX	Operational expenditures
OTC	Over-the-Counter
OTT	Over-the-Top
PANEP	Pan-European Partner
PFI	Private Finance Initiative
PFSO	Port Facility Security Officer
PGC	Portfolio Generation Companies
POP	Points-of-Presence
PoSB	Presidency of Strategy and Budget
PPA	Power Purchase Agreements
PPP	Public–Private Partnership
PPPs	Public–Private Partnerships
Presidential Decree No. 4	Presidential Decree About the Organization of Institutions Related to Ministries and Other Institutions and Organizations
PSC	Public Sector Comparator

PSD	Electricity Energy Sector Reform and Privatization Strategy Document
PSTN	Public Switched Telephone Network
PTA	Public Transport Authority
PTSQES	Public Transport Service Quality Evaluation System
PTT	Turkish Postal Administration (Posta Telegraf ve Telefon İdaresi)
RE	Renewable energy
REL	Renewable Energy Law
RESs	Renewable energy sources
RF OFF	Off-net fees
RF ON	On-net fees
RFS	Radio frequency spectrums
RK	Rekabet Kurumu
RMS	Regulating and Metering Station
RSO	Recognized Security Organizations
RTL	Road Transport Law
RTR	Road Transport Regulations
SAR	Search and rescue
SDGs	Sustainable Development Goals
SDPS	Surveillance data processing system
SeMS	Aviation Organizations Security Management System
SES	Single European Sky
SESAR	Single European Sky ATM Research
SESAR-JU	SESAR-Joint Undertaking
SHGM	Directorate General of Civil Aviation
SHT-17.2	Civil Aviation Security Training and Certification Instruction
SHT-17.3	Instruction
SHT-17.6	Air Cargo and Mail Security Instructions
SHT-6AF	Instruction on Auditing of Commercial Air Enterprises in Financial and Fiscal Matters
SHT-HES	Airport Safety Standards Instruction
SHT-OLAY	Instruction for Reporting of Civil Aviation Safety Incidents
SHT-RAMP	Instruction for Safety Assessment for Domestic and Foreign Aircrafts
SHT-SMS	Instruction on the Implementation of the Safety Management System in Commercial Air Transport Companies, Flight Training and Maintenance Organizations
SHT-SMS/HAD	Instruction for the Implementation of the Safety Management System at Airports
SHY-6502	Regulation on Reporting and Evaluation of Safety Events Related to Air Traffic Services

SHY-6A	Commercial Air Operations Regulation (Regulation on Commercial Air Enterprises)
SHY-GÖZETİM	Regulation on Safety Surveillance in Air Traffic Management
SHY-İPC	Regulation on Administrative Fines to be imposed by the Directorate General of Civil Aviation
SHY-RAMP	Regulation on Safety Assessments of Domestic and Foreign Aircraft
SHY-SMS	Regulation on Safety Management System in Civil Aviation
SHY-TD01	Regulation on Duties, Authorities, Responsibilities, Working Procedures and Principles of Technical Auditors of DGCA
SHY-YDTGMK	Regulation on Authorized Audit, Technical Audit and Consultancy Institutions
SHY-YOLCU	Regulation on the Rights of Passengers Traveling by Airline
SKİs	Water and Sewerage Administrations (Su ve Kanal İdareleri)
SM	Senior Manager
SMEs	Small and medium-sized enterprises
SMP	Significant Market Power
SNPs	Social Network Providers
SOEs	State-owned enterprises
SOLAS	Safety of Life at Sea
SPK	Sermaye Piyasası Kurumu
SPO	State Planning Organization
SPV	Special-purpose vehicle
STAs	Standardized Transport Agreements
SUA	System Usage Agreement
SUEN	Turkish Water Institute
SWIM	System-Wide Information Management
TA	Telecommunications Authority
TANAP	Turkish Anatolian Pipeline Project
TCA	Turkish Competition Authority
TCB	Turkish Competition Board
TCC	Transport Coordination Centers
TCC	Turkish Commercial Code
TCDD	Turkish State Railways
TCO	Turkish Code of Obligations
TCoC	Turkish Commercial Code
TCPs	Telephone Call Providers
TDI	Turkish Maritime Facilities

TEAŞ	Turkish Electricity Generation and Transmission Company (Türkiye Elektrik Üretim İletim Anonim Şirketi)
TEDAŞ	Turkish Electricity Distribution Company (Türkiye Elektrik Dağıtım Anonim Şirketi)
TEDES	Traffic Electronic Control System
TEİAŞ	Turkish Electricity Transmission Company (Türkiye Elektrik İletim Anonim Şirketi)
TEK	Turkish Electricity Authority (Türkiye Elektrik Kurumu)
TELKODER	Turkish Competitive Telecom Operators Association
Telsim	Telsim Mobil Telekomünikasyon Hizmetleri A.Ş.
TETAŞ	Turkish Electricity Trade and Contracting Corporation (Türkiye Elektrik Ticaret ve Taahhüt Anonim Şirketi)
The Board	Energy Market Regulatory Board
The Regulation	Electricity Market Consumer Services Regulation
The TCoC	Turkish Commercial Code
THY	Turkish Airlines
TKI	Turkish Coal Enterprise
TMAs	Terminal maneuvering areas
TOM	Touch-on-Memory
ToOR	Transfer of Operating Rights
TOR	Transfer of Operating Rights
TPA	Third-party access
TPAO	Turkish Petroleum Corporation
TPCs	Transition Period Contracts
TPP	Thermal power plant
TSO	Transmission system operator
TSR	Turkish State Railways
TT	Turk Telekom
TTAŞ	Türk Telekomünikasyon A.Ş.
TTL	Telegraph and Telephone Law No. 406
TUBİTAK	Scientific and Technological Research Council of Turkey
Türk Telekom	Türk Telekomünikasyon A.Ş.
Turkcell	Turkcell İletişim Hizmetleri A.Ş.
Türksat	Türksat A.Ş.
TURKSTAT	Turkish Statistical Institute
UAB	Ministry of Transportation and Infrastructure
U-ETDS	Transportation Electronic Tracking and Control System
UkSATSE	Ukranian State Air Traffic Service Enterprise
UNDP	United Nations Development Programme
UNFCCC	United Nations Framework Convention on Climate Change
US	United States

USAŞ	Aircraft Service Corporation (Ucak Service Anonim Sirketi)
USDC	Unit Service and Depreciation Charge
USO	Universal Service Obligation
USP	Universal Service Provider
VAT	Value-added tax
VfM	Value for money
VoIP	Voice over Internet Protocol
VTP	Virtual Transfer Point
WACC	Weighted average cost of capital
WHO	World Health Organization
YEKA	Renewable energy zone
YEK-G	Organized Renewable Energy Certificate
YEKDEM	Renewable energy resources support mechanism

Network Industries in Turkey: A Historical Approach

Muzaffer Eroğlu and Matthias Finger

Abstract The aim of this introductory chapter is to put the development, the governance, and both the regulation and deregulation of the different Turkish network industries into a historical perspective. The chapter distinguishes between five phases: the (late) Ottoman empire period; the phase of the early Republic (1923–1950); the first steps of a liberalizing economy yet with still state-owned enterprises (1950–1980); the liberalization period, with a strong US inspiration (1980–2000); and the last 20 years, inspired by EU-style reregulation. Finally, we summarize the main arguments of the 24 chapters contained in this volume.

Keywords Turkey · Network industries · Liberalization · Competition · Regulation

1 Introduction

Network industries constitute the "foundational infrastructures" of a country (Finger, 2020; Foundational Economy Collective, 2018). Such foundational infrastructures are even more important in the case of an emerging economy, such as Turkey, as they lay the groundwork for societal development, just as they did in Europe and the USA over 100 years ago. This requires substantial investments, planning, and institutional capacity more generally. On the other hand, emerging economies do not have to deal with legacy infrastructure systems, the rehabilitation of network infrastructure, and the adaptation to the needs of the twenty-first century, which may be even more costly. Of course, Turkey is not starting from scratch in matters of infrastructures, as some were already highly developed in the Ottoman times, such as postal services and railways (Christensen, 2017). However, postal services are going through an

M. Eroğlu (✉)
Boğaziçi University, Faculty of Law, Istanbul, Turkey
e-mail: muzaffer.eroglu@boun.edu.tr

M. Finger
Istanbul Technical University, İstanbul, Turkey
e-mail: matthias.finger@yandex.com

M. Eroğlu and M. Finger (eds.), *The Regulation of Turkish Network Industries*,
https://doi.org/10.1007/978-3-030-81720-6_1

existential crisis because of digitalization, and railway infrastructures need more than simple rehabilitation today. For the most part, however, infrastructure development in Turkey has taken place in recent decades, notably in energy, road, aviation, and telecommunications infrastructures.

However, this book is less about the infrastructures per se than their evolving governance as a result of their liberalization since the 1980s, more concretely about their deregulation and reregulation over time. In this context, one talks about network industries, rather than infrastructures. Most of the writing about these transforming network industries has been shaped by economic theory, typically of either American (neo-liberal economics) or European (institutional economics) origin (see Finger & Jaag, 2015). Even though an impressive number of Turkish scholars has already written about the network industries, their governance, their privatization, and their regulation—and this edited volume testifies to that fact—most of them refer to the US and European literature.

This is notably the case of *The Political Economy of Regulationin Turkey* (Cetin & Oguz, 2011), a book that, ten years ago, pursued the same ambition as ours. Unfortunately, Cetin and Oguz's edited volume has remained the only systematic attempt to study the deregulation and reregulation of the Turkish network industries. That book came slightly too early in this transformation process and was therefore motivated by normative, theoretical, and generally US-inspired liberalization theory. As we will see, most of the transformations of the network industries in Turkey turned out to be inspired by the much more pragmatic EU approach to deregulating and reregulating the different infrastructure sectors, owing of course to Turkey's hopes, in the beginning of the twenty-first century, to join the EU.

Of course, we do not pretend that infrastructures are different in the USA, in Europe or in Turkey, for that matter. On the contrary, infrastructures worldwide rely on the same technologies, just as their construction relies on the same infrastructure firms (most of which are European), even though firms from emerging countries (China, India, and Turkey) are catching up. However, just as there are large institutional and political differences in the governance of the network industries between the USA and Europe, it is highly likely that emerging countries also have unique ways of developing and especially governing infrastructures.

With this book, we want to offer to a broader audience an understanding of where network industries in Turkey are situated. How have they evolved, especially over the past decades? What does their ownership look like? How are they regulated and how are they governed more generally? With the exception of the above-mentioned, but now dated book by Cetin and Oguz (2011), this has not been tried so far, mostly due to the highly fragmented nature of Turkey's academic and research community.

Moreover, the present book covers all the relevant network industries in Turkey, from communications and energy to transport and water. We also address cross-cutting issues such as privatization, public–private partnerships (PPPs), competition law, and regulatory authorities, among others. We have asked the contributing authors to provide a systematic overview of these different network industries, both from a historical and from a multidisciplinary perspective, thus indicating how they were developed, often deregulated and, like everywhere else, rapidly reregulated. When reading through the different chapters, the reader will rapidly realize that there is a strong influence of the European approach to deregulating and reregulating the different network industries. This can be explained by geographical and cultural proximity to Europe, but also because, in the early twenty-first century, there was a deliberate political agenda to integrate the common European (infrastructure) market.

Nevertheless, the US's much more neo-liberal influence in terms of liberalizing the network industries in Turkey has been strong, especially between 1980 and 2000, the heyday of globalization. Still the combination of the Ottoman origins, the strong role of the state during the first 60 years of the Republic, US-inspired neo-liberalism between 1980 and 2000, a strong EU influence at the turn of the twenty-first century, and a much more autonomous approach over the past 10 years, has created a unique approach to the governance of the different network industries in Turkey. It is this approach that we wish to highlight in our book. It is also what we will briefly present in this introductory chapter.

Our approach to the network industries in Turkey is factual rather than political and pragmatic rather than ideological. While there are things to criticize, and the authors in this book to not refrain from criticism, the achievements of Turkey in matters of infrastructure development, but also in matters of their governance, are impressive, not only in themselves, but also if compared internationally. Our book aims to replicate this ambition at the academic level, that is, to create a community of researchers studying the Turkish network industries from a multidisciplinary perspective. This has not been tried before, even though there is substantial intellectual expertise within the country. Therefore, we have deliberately given the voice to these authors who, despite having typically obtained their master's and Ph.D.s from abroad, have returned to the country and written "from within," that is, based on a solid understanding of the Turkish network industries in all their dimensions: historical, technological, economic, social, legal, and political.

In this introductory chapter, we have chosen a historical approach: while some network industries such as ports, railroads, electricity, postal services, domestic water, and roads can be traced back to the Ottoman Empire, others are new, such as natural gas, energy more generally, Internet infrastructure, and mobile telecommunication. In order to understand the historical origins of governing the infrastructures, therefore, we will first recall Ottoman-era regulations. However, it is important to remember that modern Turkey has its roots in the Republican time, which started in 1923.

We can divide this Republican period into four phases (Akyıldız & Eroğlu, 2004). The first phase was the early Republican period between 1923 and 1950, during which the state was the major player in operating network industries by way of state-owned enterprises (SOEs). Consequently, there were few regulations. From 1950 onwards, most economic activities were opened to the private sector, and thus, liberal economy principles were introduced to economic regulation. However, the network industries remained for the most part exempt from this initiative. Turkey's economy expanded rapidly, but this rapid pace triggered economic and political crises. From the 1980s until around the turn of the Millennium, in the context of economic globalization, there was a push to introduce more liberal policies into network industries as well.

Consequently, this period saw the introduction of modern regulatory approaches to network industries and thus the emergence of independent regulatory authorities. Especially during the 1990s, Turkey developed new, more liberal approaches to the network industries. During this period, many laws and regulations were changed in order to become EU-compliant. However, many political crises during these 20 years negatively affected the efficient application of this EU regulatory approach.

The last century ended for Turkey with a terrible economic crisis, and thus, since 2000, Turkey has adopted much more comprehensive economic policies, resulting in substantial changes in most of the network industries. This is especially the case in the telecommunications, energy, and transportation sectors, as we will see in the chapters of this book.

2 Infrastructures During the Ottoman Empire Period

Infrastructure development, on a massive scale, really only started with the Industrial Revolution, most prominently with railways, even though postal or rather courier services go back to the Middle Ages. This was no different for the Ottoman Empire, so we start with the late nineteenth and early twentieth centuries. The Ottoman Empire also sought to modernize its economy and to develop more generally in order to match developments in Europe. Despite reformist political movements during the eighteenth and nineteenth centuries, modernizing the Ottoman Empire turned out to be a challenge for the Ottomans but also for the other contemporary empires, namely the Austro-Hungarian and Russian Empires. All three ended during or soon after the First World War (Pamuk, 2004). Consequently, during Ottoman times, infrastructure development was painfully sluggish.

Existing network industries were mostly governed by way of complicated traditional laws. Transport infrastructures were typically operated by the state and there were few regulations for users (Halaçoğlu, 2002). Roads were publicly owned and mostly free to use. However, the road network had not been developed much. Similarly, most of the ports were operated by the state and there were tariffs for users but not much regulation to deal with other issues such as access and security (Halaçoğlu, 2002).

Since the state was in debt and the Empire's financial capacity was mostly inadequate in order to build new infrastructures toward the late nineteenth and early twentieth century, public infrastructures were opened up to private investors and operators. Accordingly, the state entered into concession agreements with private companies (mostly from Europe) to build and operate important network industries. For example, there were concession agreements to build several roads in different regions of the Empire during the second half of the nineteenth century (Çetin, 2013) Also, even though the state had built some railroads, most of the railroads were built by foreign companies based on concession agreements (Christensen, 2017; Schoenberg, 1977). Similarly, most of the port and related business was transferred to private owners by concession agreements during the nineteenth century.

Postal and telegraph networks were operated by the state and, like everywhere else, merged at the end of the nineteenth century into a state-owned institution (Halaçoğlu, 2002). However, telephony was introduced in the Ottoman Empire much later than to other Western States. In 1911, telecom operations were outsourced to an English–American company based on a concession agreement (Kubilay, 2017).

The first serious use of electricity started in 1911 in İstanbul, when electricity was first produced and sold based on concessions agreements (Erol, 2007). At that time, the state signed a concession agreement for 50 years for electricity generation and distribution with a Germany/Hungarian company (Özdemir, 2016).

To summarize, until the middle of the nineteenth century, it was mainly the Ottoman State that built and regulated the infrastructures, yet regulation was not very developed, nor mature. Some networks, such as water, were also built and operated by local authorities without central government control and unified rules. As of the middle of the nineteenth century, privately built and operated infrastructures based on concession agreements that were very similar to modern-day built operate and transfer agreements (BOTs) started to emerge alongside state-owned infrastructures. These concession agreements became the main regulatory tool for many industries, including electricity, telephony, railroads, and roads (Örsten Esirgen, 2010). Each of these agreements was separate from each other, which led to highly complex operations of the infrastructure systems (Atam, 2012). Furthermore, these agreements lacked government oversight, leading to the abuse of power by the private companies involved, as much as it hindered infrastructure developments in many areas.

3 The Period of the Early Republic (1923–1950)

The establishment of Republic in 1923 after the devastating First World War and the Independence War was very painful, as the state had lost most of its economic power. As in most other countries affected by the First World War, there was an urgent need to build infrastructure and the economy more generally. However, Turkey did not have many private businesses left that could develop and operate large-scale infrastructures; also, the economic conditions generally were not suitable for foreign investment. Consequently, the state engaged in huge infrastructure developments in almost all the industries (Takim & Yilmaz, 2010), while still relying, at least in the early Republican years, on some concession agreements signed with foreign companies (Erol, 2007).

Over time, however, the state was increasingly centralized and after the 1929 global economic crisis, and ensuing Great Depression, there were not many foreign companies willing to invest. The dominant economic model of these time was Keynesian, with the state playing a big role. Thus, from the 1930s onwards, most of the network industries were nationalized, and concession agreements were often terminated by paying compensation. For example, by 1938, the entire electricity networks and infrastructures were state-owned (Erol, 2007). Similarly, the telephone company was also nationalized (Üçer, 2021).

However, these institutions were not even SOEs; they were entirely integrated within the central state's administration. For example, the building operations of the railroad and roads were planned by the central state, and central authorities such as Turkish Rail Road Operation (Utikad, 2012) and Turkish Road Operation were established. Similarly, postal services, telegraph, and telephony were operated under the Posta Telegraf ve Telefon İdaresi (PTT), which was integrated into the central government (Halaçoğlu, 2002). Turkey did not stand out as an exception. Similar institutional arrangements could be observed all around the world at that time.

As for the ports, there were some concession agreements in the early Republican times, but later, most of the new facilities were built by the state institutions and during the 1930s the state took control of the almost all ports in Turkey, which subsequently continued as public services.[1] As for transport, during the early Republican times, the state invested mostly in railroads and later shifted its attention to the main roads (Çetin et al., 2011). PTT, an administrative entity of the state, built new telephone lines and extended postal services to every corner of the Republic (Halaçoğlu, 2002).

The air transport infrastructure started during this period thanks to the initiative of the government, which created Turkish Airlines.[2] The General Directorate of State Airports Authority (DHMİ) another administrative entity that still exists today, was also established at that time to extend airport capacity to many cities as well as to operate air traffic control.[3]

In short, during the early Republican times, almost all the network industries were operated as entities within the central administration. A few were operated by local municipalities, such as domestic water, some electricity distribution, as well as ports. Most of them were operated as a public service; that is, without pursuing any profit motive. Again, the same model can be observed, at that time, in most Western countries (except the USA). Accordingly, "regulation" of these network operators was mostly done by the central government based on codes and bylaws. In other words, the different network industries' sectors were treated in the same way as all the other public services such as education and health care. They were all part of the central government system and the state was responsible for providing services, regardless of whether it was a hospital or a road. If conflicts arose, they were addressed in the administrative courts, based on administrative law. Private law applied only to those few services run by private service providers, such as coach transport.

---

[1] https://www.tdi.gov.tr/tarihce/.

[2] https://www.turkishairlines.com/tr-tr/basin-odasi/hakkimizda/hikayemiz/.

[3] https://www.dhmi.gov.tr/Sayfalar/hakkimizda.aspx.

4 State-Owned Network Industries as a Foundation for a Liberalizing Economy (1950–1980)

After the Second World War, Turkey transitioned into a modern democracy with free elections and political parties (Pamuk, 2014). In parallel, the economy was opened up to foreign investors and foreign enterprises; this was also the beginning of private Turkish enterprises. In short, the liberal economic model applied to most of the economy, but not yet to the network industries, which remained state-owned. Even more so, before this period, the last concession agreements with private firms operating in the network industries were either terminated or transferred to the state. This was the main reason why network industries were considered to serve as a basis for the liberal economy to prosper.

Infrastructures had to be built (by the government) in order to meet the new demands triggered by the by the expansion of a now more liberal economy. For example, transport shifted from railroad to motorways. Consequently, Turkish railroads, like those all over Europe, became loss-making. During this period, a consolidated Central Motorway Authority extended motorways and main roads to many cities in Turkey (Çetin et al., 2011). Port capacity was expanded, with the state building several new ports. The same evolution took place for airports, with many cities being newly served.

In electricity, generation, transmission, distribution, and retail were consolidated into a single state-owned organization named Turkish Electricity Institution (TEK) in 1970.[4] This new industrial behemoth built new generation capacities and extended the electricity grid throughout the country, now serving the majority of the Turkish population (Erol, 2007). As for communications, postal and telephone services were also improved to match demand.

Almost all of these network industries were state-owned and operated. As in many other especially Southern European countries, they were considered to be public services and thus operated under administrative law.

5 Liberal Economy with Liberalized Network Industries (1980–2000)

A long economic and political crisis period during the 1970s ended with a military coup in 1980 (Kibritçioğlu, 2001). The newly installed government came with a liberal agenda, leading to the liberalization of the economy, including in the different network industries (Pamuk, 2014). In the early 1980s, the vast majority of the network industries were still operated by the state. However, with the new government, the active participation of the private sector in infrastructure developments became conceivable for the first time since the creation of the Republic in

---

[4] https://www.tedas.gov.tr/#!tedas_hakkimizda.

1923 and was gradually promoted. For example, during this period, some electricity generation capacity was privatized, and some BOT contracts were signed (Aslan & Yavaş, 2012).

On the other hand, the state was still the main player in network industries in the 1980s: telecom infrastructures were extended to cover all of Turkey by the state-owned PTT, leading to a tremendous increase in users. The Internet became available for the first time thanks to publicly owned telecom lines. Postal coverage was also extended throughout the country (Halaçoğlu, 2002). In the energy sector, natural gas was imported to Turkey through state-built natural gas pipelines and then distributed locally by municipality-owned companies (Yardımcı, 2011).

Transport systems, with the exception of railways, were modernized thanks to public investments into and construction of roads, ports, and airports. Some of the existing transport infrastructures, as well as the right to build new ones, were also transferred to private companies under the BOTs' contracts.

All of these developments during the 1980s resulted in more developed, but also more complex network industries. This was also a result of the state, for the first time, allowing the private sector to develop entire networks. For example, in the early 1990s, mobile telephone operations rights were licensed to two companies in order to roll out a mobile telephony network. These two companies (Turkcell and Telsim) engaged in huge infrastructure investment and created one of the most modern mobile communication networks (Kar, 2018). Similarly, most of the new natural gas distribution capacity, other than municipality-owned distribution companies, was also licensed to private companies and natural gas network started to spread all over Turkey (EPDK, 2019).

In addition, the first liberalization efforts along an EU approach also started during the 1990s. For example, the state monopoly institution in electricity (TEK) was divided up into four companies separating generation, transmission, wholesale, and distribution (Doğru, 2010). This, in turn, opened up new opportunities for further liberalization and even privatization. As in most other countries worldwide, PTT was also separated into a telecom operator (Türk Telekom) and a postal operator. Interestingly, the postal operator has retained the name PTT until today.

The first independent regulatory authority set up in Turkey was the Sermaye Piyasası Kurumu (SPK) (Capital Market Authority (CMA)) in 1982. In 1997—quite late in international terms—Rekabet Kurumu (RK) the Turkish Competition Authority (TCA) followed. These were later followed by other independent sector-specific regulators.

In our opinion, the biggest problem pertaining to the governance of the network industries in Turkey was the lack of clear guidelines for privately owned companies operating in some of the infrastructures. There were movements from administrative to private law principles, but the law was not properly enacted in order to guide these transformations. Also, during this period, this complex structure and lack of regulatory oversight resulted in many problems. For example, the privatization of some of the electricity distribution and retail operations turned out to be disastrous during the 1990s as there were no regulatory tools and no authority to oversee the newly created market (Aslan & Yavaş, 2012). There were huge problems between the privatized companies and other companies belonging to the states as there were no regulatory authority to solve conflicts. Moreover, privatized companies started to behave abusively toward their customers due to lack of regulatory oversight.

Even though this period saw many private investments to some network enterprises, the failure to regulate properly created problems in almost all the network industries. For example, mobile operators were involved in anticompetitive behavior regarding access until the competition authority became actively involved (Koktürk, 2012). Even though this period saw the emergence of independent Internet services providers (ISPs), many of them exited the market, mostly because of access problems created by the dominant telecom operators. Here, we must also mention problems during this period in the broader political environment, which resulted first in the crisis and subsequently in the collapse of the banking system. In fact, it is this banking crisis that has facilitated a more comprehensive regulatory approach to all the network industries in Turkey since the 2000s.

6 EU-Style Regulation (2000–2020)

After the major economic crisis of 1999, Turkey applied IMF-backed liberalization and privatization policies, accompanied by EU-style regulation of the different network industries. Consequently, Turkey went through a huge codification process. For example, in the energy sector, Turkey enacted laws to regulate electricity, natural gas, petrol and other related markets, such as the electricity market law in 2001, as well as the gas market law in the same year. In parallel, EMRA—the Electricity Market Regulatory Authority (Enerji Piyasası Düzenleme Kurumu, EPDK)—was established to oversee these markets.

A year earlier, in 2000, the information and communications technology law was changed and a corresponding regulatory authority was established, which was renamed in 2008 as the Information and Communication Technologies Authority (ICTA) (BTK). Furthermore, variouscommunication laws were created in order to regulate landlines, fiber, the Internet, and mobile communications. Most of these changes were directly inspired by EU regulations. Moreover, the postal market was liberalized only in the 2013 Postal Services Act and ICTA's powers were enlarged to also include the postal market.

Similarly, railways were deregulated and the railway market was partially liberalized, following a new railway law. TCDD, the state-owned railway company, was unbundled along the EU model, that is, it was separated into a monopolistic infrastructure company on one hand and a competitive transport operator on the other hand, thus opening the way to potential new entrants.

This period also saw major privatization efforts. For example, in 2001, almost the entire electricity generation capacity was still owned by the state. In 2020, however, the state only owned around 20 percent of generation, while energy trading companies are now all private. A similar situation can also be observed in communication networks. Here, the biggest event was the privatization of the vertically integrated state-owned telecommunication operator: as of early 2020, all of Turkey's landlines, the fiber cables, Internet services providers, and the mobile telephony operators were owned privately. However, in September 2020, the state-owned Turkish wealth fund took a controlling share in Turkcell, the biggest mobile operator.

Another important development during these past 20 years was the substantial rise of PPPs and other BOT contracts, a movement called "contractual liberalization." Most of the transport networks, including airports, ports, motorways, which were operated by the state in the past, were contractually liberalized as of 2000, if not entirely privatized. Most of the existing airport operations were contracted out to private operators and most new developments were awarded by way of BOT contracts, including Istanbul Airport, one of the largest in the world. Similarly, some of the important motorways and bridges were also built based on BOTs, such as Osman Gazi Bridge, one of the longest bridges in the world, as well as the third bridge over the Bosphorus, connections roads, and the underwater tunnel connecting the Asian side of İstanbul to its European side.

As a result, infrastructure development in Turkey accelerated during the past 20 years, but network industries became also more complex, both in terms of operations and regulation, due to the coexistence among public, private, and contractual operations. It is mainly this complexity that has led to significant regulatory dynamics, most of it triggered by the independent regulatory authorities, which, by so doing, managed to increase their power. In the sectors where independent regulatory authorities do not exist, new laws were introduced to oversee the market. In rail transport, for example, the law was changed in 2013 to open the market to competition and establish a system to oversee the market.

Even though market opening and competition was the main agenda, privatizations sometimes pursued revenue-maximizing objectives. Many of these efforts were stopped by a very active TCA, but some were not. For example, the competition authority clearly contributed to making the electricity market more competitive. On the other hand, the privatization of the already dominant Turk Telecom resulted in a very concentrated market. As a result, the competition authority had to interfere several times and punish Turkish Telecom for abusing its dominant position (Koktürk, 2012).

Therefore, we can characterize the past 20 years as the period of the emergence of EU-style regulation in Turkey, namely because of the unbundling of the different network industries. Such unbundling, as in the EU, leads to a separation of the remaining infrastructure monopoly from the commercial activities, and subsequently, to access regulation to the monopolistic networks. In addition, tariffs for the monopolistic infrastructures must be regulated and, not surprisingly, many ensuing problems of imperfect markets call for further regulation (Finger, 2020).

7 Where Do We Stand Today?

The result of this evolution is a complex but also interesting picture of the Turkish network industries. This book aims to shed light on this complex reality: while market and competition is king in some of the infrastructures, in others, the state is still the dominant player and liberalization progresses slowly. In electricity, over the past 20 years, the market has been almost completely liberalized and privatized: the state still owns 20 percent of generation capacity, operates high-voltage transmission (TEİAŞ), as well as Energy Exchange (EPİAŞ), while most generation and all retail is privately owned. Moreover, during the past TEN years, generation capacity has been diversified and generation from renewables now accounts for a significant share. The market operates smoothly. There are still some problems, of course, but the regulatory authority seems capable of overcoming these hurdles.

On the other hand, natural gas market liberalization has not progressed much since the early efforts. Most import capacity is still managed by the vertically integrated state monopoly company BOTAŞ. Also, the import market, and therefore also the wholesale market, are still not competitive, thus leading to less competition in the retail market. There have been several efforts at regulatory intervention, but they have not had much success so far.

In the area of telecommunications, and even though regulation has so far failed to create competitive landlines and ADSL Internet, recent regulatory intervention has resulted in a more competitive mobile communication and fiber Internet market. Also, the deregulation of the postal sector has resulted in a highly competitive postal and delivery market, fortunately just before the e-commerce boom hit and customers could benefit from it.

In the transport sector, the state's reliance on PPPs, BOTs, and other operating agreements led to success, such as in airport and port developments, but also led to failures, as in some cases of roads and bridges projects, where the expected level of users was not reached, at least not until today. Some infrastructures continue to be operated by the public authorities, such as in the case of domestic water distribution, and not open to competition.

As for institutional policies, we observe regulatory efforts in almost all the network industries. The basic model is, along the EU approach, to establish an independent regulatory authority, such as the one in the energy (EMRA) and telecommunications sectors (ICTA). The other network industries are also subject to extensive regulation, even without an independent regulatory authority. Typically, there is a government body that oversees operations, such as TCDD in the case of railways and DHMI and SHGM for air transport.

8 Outline of the Main Chapters

The aim of our book is to introduce readers to all the network industries in Turkey and identify their main regulatory dimensions. It is structured into four parts: energy and water (seven chapters); transport and logistics (nine chapters); communications (three chapters); and issues that cut across all the sectors, namely privatization, PPPs, regulatory authorities, and the articulation between sector-specific and competition regulation (four chapters).

Refik Tiryaki and Mehmet Yalılı (Chapter "Regulation of Electricity Generation in Turkey") examine the regulation of electricity generation in Turkey. Regulation electricity generation has become highly sophisticated in recent years, both licensed and unlicensed. Moreover, there is a separate system for renewable electricity generation covering both support schemes and renewable energy zones. We see high-level private sector participation in electricity generation both for non-renewables and renewables and also both for licensed and unlicensed electricity generation. The chapter concludes that regulation for electricity generation in Turkey is sophisticated and mature, with the sector displaying a high level of competition and with a remarkably diverse generation portfolio.

Organized electricity trading is an important dimension of a liberalized electricity market, and Turkey has recently achieved great success in the electricity trading. Taner Şahin (Chapter "Regulation of the Turkish Wholesale Electricity Market: A General Overview") examines these recent developments and offers his view on electricity trading. He states that establishing the electricity trading company EPİAŞ was an important step for achieving success in day-ahead and other markets. He concludes that the Turkish electricity trading mechanism is fully compatible with similar systems in the EU and in other countries.

Eylem Apaydin and Muzaffer Eroğlu (Chapter "Regulation of Trade and Supply Contracts in the Turkish Electricity Market") examine the current situation in electricity supply contracts, as well as the way these contracts are regulated. Turkey improves its supply-side liberalization yearly by reducing compulsory last resort contract limits. Today, the market is open to 97 percent of the electricity consumed in Turkey. Moreover, there is a new tariff system that forces large electricity consumers in particular to transfer to free contracts. Turkey also introduced detailed contract regulation, which contains more protection for lower-consuming consumers compared to higher-consuming ones. A negative point is that EMRA sometimes abuses its power of last resort in tariff regulation, notably by deciding non-competitive tariffs, which delays market liberalization.

Turkey started its liberalization of the natural gas market together with the electricity market. However, the process was not as smooth as it was in electricity. Bağdagül Kaya Caner (Chapter "Regulation of the Natural Gas Grid") examines the regulations covering the natural gas grid, which was first introduced in 2001, covering many issues, such as non-discriminatory third-party access, tariffs, transparency, as well as the role of the regulator. The current law has also come to cover LNG terminal, storage facilities, and balancing mechanism. The biggest problem is that the unbundling of BOTAŞ has not been done so far until now. However, according to the author, this vertically integrated structure of BOTAŞ has not impeded the utilization of the transmission network by third parties.

Suat Kayıkçı and Mehmet Kürkçü (Chapter "Regulatory Aspects of Natural Gas Distribution in Turkey") examine the regulation of gas distribution to the clients, including retail. This is done by way of tendered licenses by EMRA for particular geographical regions. The Natural Gas Market Law and corresponding bylaws contain many detailed regulations, including access, tariff, and consumer protections. As a result, after the introduction of this new law, gas distribution networks have dramatically expanded to provide gas to all 81 provinces of the country. As of 2020, only the İstanbul region remains operated by a municipality-owned company, while the rest of the country is served by private operators.

The most problematic area of the natural gas sector in Turkey remains the liberalization of supply contracts. Burak Kepkep (Chapter "Regulation of Natural Gas Contracts") examines the regulation of these contracts and shows that, compared to the Turkish electricity market, whose liberalization started at the same time, the liberalization of the natural gas market has failed so far. He argues that the primary reasons for this are the lack of evolution of natural gas supply agreements due to Turkey's dependency on foreign sources, as well as BOTAŞ's quasi-monopolistic position in the market.

One of the least liberalized networks in Turkey is domestic water supply. Nilgün Görer Tamer (Chapter " Regulation of Domestic Water in the Framework of Provincial Greater City Municipality Water and Sewerage Administrations") examines the regulation of domestic water within the context of the Provincial Greater Municipality Water and Sewerage Administration. Since 1986, the model of independently operated water utilities (Su ve Kanal İdareleri-SKİs-) has been established and domestic water has been turned into a commodity. Even though the SKİs are not privately owned companies, they are organizations with an independent budget and an independent public legal entity. Today, SKIs are strong actors of water management that provide rural and urban water services in the 30 provinces. The author concludes that political discussions regarding water supply will lead to reconsidering water supply as public services.

Air transport is one of the most developed sectors in Turkey, displaying tremendous growth during recent years. Within this context, Burak Keskin (Chapter "Airport Regulations in Turkey: An Efficiency Review") shows how Turkey has privatized and liberalized airport operations in recent years, leading to almost all large airports being operated by private companies. The author highlights a complex and sophisticated regulatory system, covering slot allocation, airport charges, passenger rights, ground handling, and services.

Zeki Engin (Chapter "Air Traffic Management (ATM) Regulation in Turkey: What is the next step?") examines Turkey's Air Traffic Management (ATM) within the context of an ever more competitive air transport environment. He shows how Turkey's ATM is implementing a local Single-Sky approach. He concludes that the next step for Turkish ATM will be to align with the latest developments in the region, including the EU, by adopting new technologies as well as by adapting its regulations to an increasingly fast-paced technology-driven ATM landscape.

Another important aspect of the air transport industry is the liberalization of air carriers. Ömer Faruk Erol (Chapter "Regulation of Air Carriers") examines the deregulation and reregulation of air carriers in Turkey. The most important development was the opening of the domestic market to competitors in 2003. The author also examines the relationship between air carriers and the administration in charge of regulating their activities, both in safety and economic matters.

Şahin Ardıyok and Evren Sesli (Chapter "Regulation of Rail Transport") focus on the process of harmonizing Turkish rail liberalization policies and regulations with the EU. The Act regarding the Liberalization of Railway Transportation entered into force in 2013, which led to Turkey opening up the market for both freight and passenger train services. TCDD was restructured into an infrastructure operator under the Ministry of Transportation, while a new state-owned company for train operations was established under the name of TCDD Taşımacılık A.Ş. However, the authors also show that, because of the need for significant investments into railway operations, there has almost been no market entry yet.

Volkan Recai Çetin and Faruk Cirit (Chapter "Regulation of Motorways and Main Land Roads") examine the regulations regarding motorways and main land roads. Recently, the country has witnessed a paradigm shift, as the government has announced medium- and long-term plans to upgrade the quality of the national road network. While new road investments and maintenance of existing roads have traditionally been financed by the national budget, the government has for the last ten years utilized the PPPs model as a means of creating an alternative financing method for major motorway projects. These developments have led to a rapid increase in freight and passenger transport volumes. In parallel, the government has implemented new regulations to protect road infrastructures, raise efficiency, and increase traffic safety.

Rapid urbanization in Turkey has created important problems regarding urban transport networks as of recently. Umut Alkım Tuncer (Chapter "Regulation of Urban Transport Infrastructures and Services—The Case of Istanbul") examines the urban transport regulation through the example of Istanbul. He identifies the main the actors of a complicated urban transportation system and reviews the current market structure in urban transport. The author concludes that urban transportation is undergoing rapid technological changes owing to new emerging transport modes and technological innovations (digitalization and automatization), all of which have led to unprecedented regulatory challenges.

In Turkey, coach transport is the most common mode of intercity passenger transportation. Interestingly, coach services have been offered by the private sector under free-market conditions since 1926, without much government interference. Ali Osman Solak (Chapter "Regulation of Turkish Intercity Coach Industry") examines coach transport regulation in Turkey, especially in light of a new law enacted in 2003. Thanks to this new law and corresponding regulations, the government hopes to better control the coach industry, reduce idle capacity, increase efficiency, and professionalize coach-operating companies, as well as to make them financially viable. Solak concludes that the number of operators in the market has decreased in recent years, both because of these new regulations and because of increased intermodal competition.

Global trade growth affects Turkish ports and there is a growing interest from national and international investors in Turkish ports. Kazım Yeni and Soner Esmer (Chapter "Regulation of Postal Sector in Turkey") examine the regulation of ports in Turkey, covering, among other things, current regulations regarding the various aspects of port management. They show how the privatization of ports, which started in 1997, has led to the public sector largely abandoning port management, and now focusing on rule-setting and regulation. There are now approximately 180 port facilities in Turkey, 80 of which are large-scale, mainly operated by the private sector.

Volkan Recai Çetin (Chapter "Regulation of Ports in Turkey") examines postal sector regulation. Parcel and express markets have grown rapidly in Turkey as a result of e-commerce and digitalization. This transformation has raised the importance of both deregulation and effective reregulation in order to ensure sustainable universal service provision, at the same time creating a competitive environment. The sector is mainly regulated by the Postal Services Law, which was enacted in 2013. Çetin concludes that postal liberalization has not been completed in Turkey and that the incumbent universal service provider (PTT) is granted quite a large reserved area, along with the resources levied on its competitors.

Fixed telecom lines have historically comprised the Turkish telecommunications network, but technological developments in the telecommunications sector have been exponential. Bülent Gökdemir (Chapter "Regulation of Fixed Telecommunication Services") examines this evolution in detail. He shows that Turkey's regulatory approach to the telecommunications market has been largely inspired by the EU, promoting retail competition thanks to effective network regulations. After nearly two decades of liberalization, both retail and infrastructure competition have made significant improvements. The market power of the incumbent monopoly, Türk Telekom, has been eroding. Fixed voice traffic has been shifting rapidly to mobile, as well as VoIP. The author concludes that Türk Telekom no longer has sufficient market power to determine prices.

Recent growth in Internet demand has put broadband Internet networks at the center of the debate. Emin Köksal (Chapter "Regulation of Fiber and the Internet") examines the regulatory issues in the Turkish broadband market, focusing mainly on fiber, namely by discussing the current regulatory issues related to fiber deployment. In order to promote the deployment of investments into next-generation networks, ICTA, the independent telecommunications regulator has announced a regulatory holiday, thus indicating its preference for facility-based competition. In addition, the Ministry of Transportation has interfered into the sector by imposing access obligations to the existing fiber infrastructure owners. The author concludes that there is a lack of vision in both demand-side and supply-side regulatory policies and suggests a more comprehensive regulatory approach.

One of the success stories of telecommunications in Turkey has certainly been the development of mobile telephony networks. Barış Yüksel (Chapter "Regulation of Mobile Communications Markets in Turkey") explores the emergence and the development of the mobile communications market in Turkey, focusing on the effects that the legal framework has had on the overall market structure and its performance. He indicates that the modern legal framework and the effective use of *ex-ante* regulations by ICTA repaired some of the initial problems created by the early policy mistakes. However, the structure of the mobile communications market, which comprises only three players and an ever more powerful market leader, is far from being optimal. Yüksel concludes that ICTA's failure to keep up with the EU and to prioritize OTT services may have already created problems in the markets for OTT services.

Remzi Özge Arıtürk's contribution (Chapter "Privatization of Turkey's Network Industries") is the first of four cross-cutting chapters. In it, the author addresses the privatization history and record in Turkey, covering electricity and telecommunications. In particular, he examines the privatization process that were accepted by the regulators. He also reviews contribution of regulation, technological development, and other relevant governmental policies on privatization when creating competitive markets. Arıtürk concludes that regulation subsequent to privatization should not be used as an indirect public interference into the market, but should merely serve as a liberalization tool.

In recent years, there has been extensive use of PPP contracts in Turkey's network industries. Uğur Emek (Chapter "Public–Private Partnerships in Turkey") examines these PPPs and discusses their current problems. He argues that Turkey has a fragmented legal and institutional structure for handling PPP contracts; there is also no clear PPP policy. He shows that the (government) contracting agencies pay little attention to the preparation of the feasibility studies and concludes that although regulations require the procuring agencies to prepare well-designed feasibility studies, in practice, they are unable to do this, given the lack of specific methodologies for the appraisal of financial, economic, environmental, and fiscal implications.

Along the EU regulatory approach, one of the main institutional changes in Turkey was the establishment of independent regulatory authorities in Turkey. Engin Saygın (Chapter "Regulatory Authorities in Turkey") examines the authorities that have power and duties in the Turkish network industries, especially the EMRA and the ICTA. He identifies the power and duties of these authorities, as well as their impact. He concludes that even though the relevant laws of regulatory authorities clearly determine their administrative and financial autonomy, there is a need to strengthen "regulatory governance" in Turkey.

The relationship between sector-specific and competition regulators plays an important role in terms of harmonized enforcement and an effective free-market economy. Şahin Ardıyok and Armanç Canbeyli (Chapter "Relationship Between Competition Law and Sector-Specific Regulations") examine this relationship in the case of Turkey. The chapter focuses on the interaction between these two types of regulators, while setting out the structural and practical distinctions between their respective functions. The authors distinguish the technical regulations from the competition regulations and discuss the risk of double sanctioning, as well as exploitation of jurisdictional overlaps via forum shopping.

9 Outlook

As can be seen from the short summaries of all the chapters, our book offers a solid and comprehensive overview of the way the different network industries are regulated in Turkey today. The country has already had impressive infrastructure developments in the past and continues to have plans for the future. There does not seem to be a lack of political will, nor real financial constraints to these plans, even though questions are often raised as to financial and ecological sustainability of some of these infrastructure projects (such as airports and highways).

Today, many of the country's infrastructure projects are PPPs. This raises the question of whether the regulation and the governance of the Turkish network industries will be up to the task of ensuring continued private sector investment and long-term involvement in the different Turkish network industries. Of course, this question must be answered sector by sector. So far, the country has developed many of its regulatory institutions in line with EU network industry regulation. This, along with the fact that Turkey is an emerging economy, seems to explain the attractiveness and the successful infrastructure developments in Turkey. In our view, Turkey is well advised to continue on this track.

However, good network industry governance, it is not only about emulating EU regulatory policies and practices. It is also about the relationship with Europe, or, more specifically, the EU. It should be in the EU's interest to have Turkey join the EU, given its relatively advanced state of the economy, its domestic market, but even more its growth potential. But when it comes to the network industries, another factor for the EU to consider is Turkey's geographical position as both an energy and a transport hub in between Europe and Asia. Therefore, EU membership or not, Turkey and the EU will become ever more integrated, at least in matters of infrastructure networks, and therefore also in matters of network industry regulation and governance.

References

Akyıldız, H., & Eroğlu Ö. (2004). Türkiye Cumhuriyeti Dönemi Uygulanan İktisat Politikaları. *Süleyman Demirel Üniversitesi, İktisadi ve İdari Bilimler Fakültesi, 9*(1), 43–62. https://dergipark.org.tr/tr/pub/sduiibfd/issue/20843/22343

Aslan, E., & Yavaş, Ö. (2012). Çukurova Elektrik A.Ş. ve Kepez ve Antalya Havalisi Elektrik Santralleri A.Ş. İmtiyaz Sözleşmelerin Feshi. *İdare Hukuku ve İlimleri Dergisi, 15*(1), 117–166. https://dergipark.org.tr/tr/pub/iuihid/issue/1264/14838

Atam, Ş. (2012). Osmanlı Devleti'nde Bir Yol Çıkmazı: İzmit-Ankara Şosesi İnşa ve İşletme İmtiyazı. *Çankırı Karatekin Üniversitesi Sosyal Bilimler Enstitüsü Dergisi, 3*(1), 119–138. https://dergipark.org.tr/en/pub/jiss/issue/25909/273059

Çetin, E. (2013). Tanzimat Döneminde Yol Yapım Çalışmalarına Bir Örnek: İzmir-Birunâbâd Şosesi İnşa ve İşletme İmtiyazı. *Ankara Üniversitesi Dil ve Tarih-Coğrafya Fakültesi Dergisi, 53*(2), 145–164. http://dtcfdergisi.ankara.edu.tr/index.php/dtcf/article/view/259/460

Cetin, T., & Oguz, F. (Eds.). (2011). *The political economy of regulations in Turkey*. Springer.

Çetin B., Bakış S., & Saroğlu, S. (2011). Türkiye'de Karayollarının Gelişimine Tarihsel Bir Bakış. *Çankırı Karatekin Üniversitesi İktisadi ve İdari Bilimler Fakültesi Dergisi, 1*(1), 123–150. https://dergipark.org.tr/tr/pub/ckuiibfd/issue/32888/365355

Christensen, P. (2017). *Germany and the ottoman railways. Art, empire and infrastructure*. Yale University Press.

Doğru, C. (2010). Türkiye'de Elektrik Piyasasının Yeniden Yapılanması Sürecine Bir Bakış. *Sosyal Bilimler Metinleri*, 2010(1), 1–30. https://dergipark.org.tr/tr/pub/sbm/issue/48003/610194

EPDK. (2019). *Doğal Gaz Piyasası Sektör Raporu*. https://www.epdk.gov.tr/Detay/Icerik/3-0-94/dogal-gazyillik-sektor-raporu

Erol, E. (2007). *Türkiye'de Elektrik Enerjisinin Tarihi Gelişimi 1902–2000, İstanbul Üniversitesi Sosyal Bilimler Enstitüsü, Yayımlanmamış Doktora Tezi*. https://tez.yok.gov.tr/UlusalTezMerkezi/tezSorguSonucYeni.jsp

Finger, M. (2020). *Network industries. A research overview*. Routledge.

Finger, M., & Jaag, C. (Eds.). (2015). *The Routledge companion do network industries*. Routledge.

Foundational Economy Collective. (2018). *Foundational economy*. Manchester University Press.

Halaçoğlu, Y. (2002). *Osmanlılarda Ulaşım ve Haberleşme*. Ankara: PTT Genel Müdürlüğü Yayınları.

Kar, M. (2018). Türkiye Mobil Telekomünikasyon Piyasasının Analizi ve Mobil Numara Taşınabilirliği Düzenlemesinin Etkilerinin Değerlendirmesi. *Social Sciences Studies Journal, 4*(26), 5658–5672. https://doi.org/10.26449/sssj.1028

Kibritçioğlu, A. (2001). Türkiye'de Ekonomik Krizler ve Hükûmetler; 1969–2001. *Yeni Türkiye Dergisi, Ekonomik Kriz Özel Sayısı, 1*(41), 174–182.

Köktürk, N. S. (2012). *Telekomünikasyon Sektöründe Dikey Bütünleşik Firma Davranışlarının Düzenlenmesi: Sektörel Düzenleme ve Rekabet Hukuku*, Rekabet Kurumu Uzmanlık Tezi. https://www.rekabet.gov.tr/Dosya/uzmanlik-tezleri/120-pdf

Kubilay, Ç. (2017). 1923–1939 Yılları Arasında Türkiye'de Telekomünikasyon Politikaları. *Ankara Üniversitesi İlef Dergisi, 4*(2), 37–66. https://doi.org/10.24955/ilef.357092.

Örsten Esirgen, S. (2010). *Osmanlı Devleti'nde Yabancılara Verilen Kamu Hizmeti İmtiyazları, Ankara Üniversitesi Sosyal Bilimler Enstitüsü, Yayımlanmamış Doktora Tezi*. https://tez.yok.gov.tr/UlusalTezMerkezi/tezSorguSonucYeni.jsp

Özdemir, N. (2016). İmparatorluktan Cumhuriyete Türkiye'de Elektriğin Tarihsel Gelişimi. *Osmanlı Medeniyeti Araştırmaları Dergisi, 2*(3), 17–32. https://doi.org/10.21021/osmed.89880

Pamuk, Ş. (2004). Institutional change and the longevity of the Ottoman empire, 1500–1800. *The Journal of Interdisciplinary History, 35*(2), 225–247. https://www.jstor.org/stable/3656813?origin=JSTOR-pdf&seq=1

Pamuk, Ş. (2014). *Türkiye'nin 200 Yıllık İktisadi Tarihi*. Türkiye: Türkiye İş Bankası Kültür Yayınları.

Schoenberg, P. E. (1977). The evolution of transport in Turkey (Eastern Thrace and Asia Minor) under Ottoman Rule, 1856–1918. *Middle Eastern Studies, 13*(3), 359–372.

Shaker, S. (1995). *State, society, and privatization in Turkey, 1979–1990*. Woodrow Wilson Center Press.

Takim, A., & Yilmaz, E. (2010). Economic policy during Ataturk's Era in Turkey (1923–1938). *African Journal of Business Management, 4*(4), 549–554. https://doi.org/10.5897/AJBM.9000233

Üçer, S. (2021). Istanbul telephone company from concession to nationalization: A study in business history (1909–1936). *Muhasebe ve Finans Tarihi Araştırmaları Dergisi*, (20), 63–90. https://dergipark.org.tr/en/pub/muftad/issue/60034/867643

UTİKAD, Dünden Bugüne Türkiye'nin Demiryolu Serüveni 2012. https://www.utikad.org.tr/Detay/Sektor-Haberleri/10478/dunden-bugune-turkiye

Yardımcı, O. (2011). Türkiye Doğal Gaz Piyasası: Geçmiş 25 Yıl, Gelecek 25 Yıl. *Ekonomi Bilimleri Dergisi, 3*(2), 157–166. https://dergipark.org.tr/tr/pub/ebd/issue/4858/66833

Muzaffer Eroğlu is an assistant professor at Boğaziçi University, Faculty of Law; he lectures on the Company Law, Energy Law and Competition Law. He holds an LLB from Ankara University, an LLM from the University of Kent, and a Ph.D. from Queen Mary, University of London. He is a member of the Ankara Bar. His academic research focuses on company law, corporate governance, competition law, and energy law. He published a book "Multinational Enterprises and Tort Liabilities." He has several publications in the field of company law, competition law, and energy law both in English and Turkish. He also carried out researches as a visiting academic at the University of Oslo, Max Planck Institutes for Comparative and International Private Law in Hamburg and European Institute of Florence.

Matthias Finger is a professor emeritus in charge of Digital Governance and Regulation at the Center for Digital Trust (C4DT) at Ecole Polytechnique Fédérale in Lausanne, Switzerland (EPFL). He holds a Ph.D. in Political Science from the University of Geneva. He has been an assistant professor at Syracuse University (New York), an associate professor at Columbia University (New York), a professor of Management of Public Enterprises at the Swiss Federal Institute of Public Administration, and between 2002 and 2020 the Swiss Post Chair in Management of Network Industries at EPFL. Since 2010, he is a part-time professor at the European University Institute (EUI) in Florence, Italy, where he directs the Florence School of Regulation's Transport Area (FSR-T). Since 2017, he is also a professor at the Faculty of Management at Istanbul Technical University (ITÜ), directing the Istanbul Center for Regulation (IC4R).

Regulation of Electricity Generation in Turkey

Refik Tiryaki and Mehmet Yalılı

Abstract The electricity generation sector in Turkey has experienced significant changes since the foundation of the Turkish Republic in 1923, from concession agreements with foreign companies to a mature market with enhanced private sector participation and high level of competition. By the time the first Electricity Market Law (Law No. 4628) was introduced and the independent energy market regulatory authority (EMRA) was established in 2001, the electricity generation sector had changed significantly by allowing the private companies to invest in the sector freely via licenses granted by EMRA. The most prominent attempts realized after this period and the number of private sector-owned power plants, relying on various kinds of resources, increased steadily. With the aim of organizing the market structure, market participants, market entry, monitoring, third-party access, and auditing more efficiently, the new Electricity Market Law (Law No. 6446) came into force in 2013 and the regulations prepared regarding all the subfields including the generation in the electricity sector raised the level of market maturity. Through this law, the electricity generation activity was separated into two models—licensed and unlicensed generation—while the former still holds the majority of total installations. Unlicensed electricity generation, on the other hand, encouraged the investors to build renewable-based power plants for their own consumption, by also giving them the right to sell the excess electricity that they generate. The last model for electricity generation is the renewable energy zone (YEKA) tendering model, which aims to establish large-scale power plants in shorter periods in regions where the resource potential is substantially high. The regulations and organization of the tenders for YEKA model are carried out by MENR. This chapter sheds light on the progress of the electricity generation sector in Turkey by explaining the models adopted for electricity generation.

Keywords Electricity · Generation · Licenced · Unlicensed · Privatisation

R. Tiryaki (✉) · M. Yalılı
Energy Market Regulatory Authority of Turkey (EMRA), Mustafa Kemal Mahallesi 2078. Sokak No:4 06510 , Çankaya/Ankara, Turkey
e-mail: rtiryaki@epdk.gov.tr

M. Yalılı
e-mail: myalili@epdk.gov.tr

M. Eroğlu and M. Finger (eds.), *The Regulation of Turkish Network Industries*,
https://doi.org/10.1007/978-3-030-81720-6_2

1 Introduction

The evolution of an electricity generation sector varies significantly depending on the availability and diversity of the resources, as well as the regulatory structure of each individual country. In Turkey, the electricity market has experienced a long and challenging period since the foundation of the Republic of Turkey in 1923. Increasing demand for electricity, associated with economic and population growth, has forced the government to take measures to meet the electrical energy required. Therefore, the electricity generation sector first encountered a state-owned period following the private sector participation, and finally more enhanced and market oriented one. The liberalization process and electricity market reform after 2001 introduced a well-designed and organized market structure with the establishment of an independent regulatory authority, by highly encouraging the private sector's participation in the electricity generation sector. Market growth has been rapid since private companies started investing in electricity generation. The regulatory framework changes frequently depending on the needs and advances in the market so in this chapter the current regulations and practices in the electricity generation sector have been covered in detail.

This chapter begins with an overview of Turkey's experience in electricity generation, from its beginnings towards the present time, dividing the overall period into two stages: state oriented and sector oriented. Such a division also offers a chance to observe how the generation sector evolved over time. Furthermore, the details of three methods of electricity generation activity in Turkey—the licensed and unlicensed mechanisms, as well as the YEKA model—will be explained in the following sections to provide a broad vision regarding how electricity generation market is regulated and operated in Turkey by also describing the role and rights of the market players. The concluding section summarizes the most crucial points mentioned in this chapter by emphasizing the regulations in force, related to the electricity generation and the current situation of the generation mix, as well as expectations for the future.

2 A Brief Historical Account of Electricity Generation in Turkey

Electricity demand has increased steadily during the reign of the Republic in light of economic development and population growth. The governments made significant efforts to satisfy this demand and increase the people's access to electricity. The institutional structure of electricity generation sector in Turkey will be explained in two subsequent periods—a state-oriented period and a sector-oriented period—obviously omitting the early private sector-driven years, based on sector policies pursued by governments alongside the planned economy and market economy policies.

2.1 The State-Oriented Period (1900s–1984)

Electricity generation started in Turkey with a 2 kW hydraulic turbine installed in 1902, which was shortly thereafter transformed into 60 kW power plant, at Tarsus, Mersin. In 1914, the Ottoman Electricity Corp. constructed a proper generation plant in Silahtarağa, Istanbul, a thermal generation facility with a capacity of 15 MW. Up until 1983, when its economic useful life ended and the doors of this power plant were closed, the electricity produced at this coal-fired power plant was used to meet the electricity requirement of Istanbul (Karagöl & Tür, 2017).

The total installed capacity and electricity generations were 33 MW and 45 GWh, respectively, while the population of the Republic of Turkey was around 12 million when it was founded in 1923 (Yılmaz & Uslu, 2007). The electricity generation was conferred upon foreign and newly established Turkish companies via concession contracts, during the Ottoman era and first decade of Republic. Meanwhile, utilization of natural resources was assigned to state entities following the Great Depression. Several institutions had been established to determine and utilize natural resources, such as the Electrical Power Resources Survey and Development Administration (Elektrik İşleri Etüt İdaresi-EİEİ in Turkish) and the Turkish Coal Enterprise (TKI in Turkish). The main aim of the administration was to determine the energy needs of Turkey and to conduct research and exploration regarding water and other types of natural resources. This institution operated until 2 November 2011, when it was closed and the relevant duties were transferred to the General Directorate of Renewable Energy (later incorporated into the General Directorate of Energy Affairs). In addition, with the establishment of the General Directorate of State Hydraulic Works in 1954, which aimed to enhance the usage of water and soil sources, the government concentrated primarily on the construction of hydroelectric power plants. As a result, the share of hydroelectric power in the total installed capacity rose to 35% over a 15-year period (Tutuş, 2006). Besides, in order to make the energy policies of Turkey, the Ministry of Energy and Natural Resources (MENR) was also established in 1963 (Alboyacı & Dursun, 2008).

Until 1970, a number of power plants and local distribution networks were constructed through the concession agreements signed with foreign firms. Both the government and private companies have built and operated the power plants within this period. In addition to the advancements in generation, the transmission network developed substantially during that time. Nevertheless, the presence of numerous entities in the electricity industry resulted in an unorganized structure.

The increasing need for organizing electricity generation, transmission, and distribution operations in Turkey necessitated the establishment of a national corporation in 1970. Therefore, a vertically integrated publicly owned entity named the Turkish Electricity Institution (Türkiye Elektrik Kurumu (TEK) in Turkish) was established in 1970 for the purpose of uniting the operations related to the electricity supply under a single publicly owned entity (Çetintaş & Bicil, 2015).

The global energy crisis during the 1970s and 1980s also affected Turkey, due to its external dependence on the fuel sources of thermal power plants. This led to an imbalance between the generation and consumption. Despite these drawbacks, Turkey's installed capacity reached 5118 MW thanks to the numerous large-scale hydroelectric power plants built within this period (Karagöl & Tür, 2017). In 1982, control of the municipal distribution facilities and the power plants operated by regional concession companies were transferred to TEK; by this means, the generation, transmission, and distribution of electricity were monopolized under this institution.

2.2 *The Sector-Oriented Period (1984–Present)*

At the start of the 1980s, the Turkish economy was in deep crisis due to global energy crises and political embargos. New economic policies, following the coup d'état in 1980 established faith in the market economy. Inadequate investment budget, both for the new and maintenance investments, was the prominent reason for energy sector reforms. The period starting from 1984 will be described under three subsections that represent (a) the years when the private sector started to be involved in the sector, (b) the period after the first electricity market law entered into force, and (c) the period after 2013, during which the market participation has largely become a competitive issue with the emergence of an extensively organized regulatory structure.

2.2.1 Private Sector Participation

Law No. 3096 introduced in 1984 removed the monopoly of the Turkish Electricity Institution (TEK) and constituted a legal basis for private sector participation in the electricity generation and distribution activities. In this regard, some investment models were developed and implemented, such as the build, operate, and transfer (BOT) contracts for new generation facilities, the transfer of operating rights (TOR) contracts for existing generation assets, and the auto producer system for industrial consumers to generate their own electricity. The BOT scheme was a concession agreement through which a private company would build and operate a power plant for max. 99 years (later amended to 49 years) and, after this period, transfer it to the government at no cost. TOR, on the other hand, was an agreement by which a company would operate or rehabilitate and operate publicly owned generation or distribution facilities for a given period of time (Atiyas et al., 2012). Later, in 1997, the build and operate (BO) scheme was also introduced through Law No. 4283. All of those models have attracted the attention of the private sector. However, these concession contracts could not show the expected impact on the energy investments despite its effect on the minimization of the investors' financial risks (Çetin, 2010).

In 1993, TEK was included in the government's privatization program and separated into two state-owned companies in order to operate more functionally, as the Turkish Electricity Generation and Transmission Company (later abrogated Türkiye

Elektrik Üretim İletim Anonim Şirketi (TEAŞ) in Turkish) and the Turkish Electricity Distribution Company (Türkiye Elektrik Dağıtım Anonim Şirketi (TEDAŞ) in Turkish). Through this division, TEAŞ was held responsible for the generation and transmission activities, whereas TEDAŞ was responsible for the distribution and retail sales operations. Later, in 2001, through Decree 2001/2036 of the Council of Ministers, TEAŞ was separated into three public enterprises as the Electricity Generation Company (Elektrik Üretim Anonim Şirketi (EÜAŞ) in Turkish) for carrying out electricity generation operation, the Turkish Electricity Transmission Company (Türkiye Elektrik İletim Anonim Şirketi (TEİAŞ) in Turkish) for transmission operations, and the Turkish Electricity Trade and Contracting Corporation (abrogated Türkiye Elektrik Ticaret ve Taahhüt Anonim Şirketi (TETAŞ) in Turkish) for wholesale of the electricity (Kemal, 2006).

Seeking to become a full member of the European Union and due to the Turkey's electricity sector's experience of a deep crisis in the late 1990s resulted in serious supply deficits, which required the implementation of supply restrictions and energy-saving measures (Özkıvrak, 2005), Turkey took the initiative to open its electricity market to competition in 2001, which marked a turning point for Turkey, with the design and legal framework of the new market model adapted from that of the European Union. Since 2001, there have been some major developments in the electricity market of Turkey.

2.2.2 *The Market Experience (2001–2013)*

During the liberalization process, Turkey enacted its first Electricity Market Law No. 4628 in 2001. This law was a milestone for the liberalization of the electricity market. The aim of the law was written in the first article, which was "to establish a financially strong, stable, transparent and competitive electricity market". Through this law, an autonomous regulatory body, the Energy Market Regulatory Authority (EMRA), was established by initiating a major electricity market reform program. With the separation of TEAŞ into three entities during this period, the transmission, generation, and wholesale operations were unbundled to form a competitive market.

The repealed Electricity Market Law No. 4628 included the following provisions:

- A licensing framework for market participants
- A centrally run wholesale electricity market primarily based on bilateral contracts between market participants
- Eligible consumer concept to ensure freedom for eligible consumers to choose their own suppliers
- Non-discriminatory pricing and tariff mechanisms
- Regulated third-party access to transmission and distribution grids in a non-discriminatory manner.

Participants needed to receive a license for the generation, transmission, distribution, retail, and wholesale as well as import and export activity (Bagdadioglu & Odyakmaz, 2009). The licensing concept introduced by the law, allowed the private sector companies to build and operate power plants. Before this regulation, the private sector could only build power plants through BO, BOT, and TOR contracts with the government or auto-production plants for self-consumption. Therefore, by this reform action, Turkey's electricity sector took a vital step to promote sufficient electricity capacity in a more sustainable and competitive manner (IEA 2010). Later, in 2013, the first Electricity Market Law was further revised with the Electricity Market Law No. 6446, which superseded the previous version.

The Turkish electricity sector has significantly changed in line with the EU harmonization process. Through this period, the electricity market underwent a liberalization process regarding electricity regulation, renewable energy, energy supply; the establishment of an energy market regulatory authority; energy price reform; the creation of a functional electricity market and large-scale private sector participation through privatization.

With the first Electricity Market Law (Law No. 4628) enacted in 2001, Turkey began to create a new market dynamic with the design and legal framework of a new market adapted from that of the European Union.

2.2.3 The Enhanced Market Experience (2013–Present)

With the new Electricity Market Law (Law No. 6446) E-mevzuat (2020e), which came into force in March 2013, the legislation concerning the market structure, market participants, market entry, monitoring, third-party access, and auditing was organized. The most important issue related to the generation process was the introduction of the pre-license concept during which all necessary permits/obligations must be completed and the project owner can start the construction of the power plant afterward.

Based on Law No. 6446, the Turkish electricity market is based on bilateral agreements complemented with the balancing and settlement market. The private sector may participate in all segments of the electricity market, except for transmission, by receiving relevant licenses from EMRA. Regulated third-party access to the network without discrimination is also in place under the supervision of EMRA.

The Turkish Electricity Transmission Corporation (Türkiye Elektrik İletim Anonim Şirketi (TEİAŞ) in Turkish) is the independent transmission system operator and has responsibility for ownership, operation, and maintenance of investments in the national grid. On the other hand, distribution companies are responsible for planning, construction, and operation of the distribution network in their assigned regions. The assigned suppliers, which were legally unbundled from distribution companies in 2014, are required to engage in retail business.

Before 2013, market operations had been carried out by TEİAŞ through a separate department (PMUM in Turkish). Afterward, according to Law No. 6446, the market operation activity was defined as "the operation of organized wholesale power markets and the financial settlement of the transactions made in these markets". This activity is nowadays carried out by an independent company named Energy Exchange Istanbul (Enerji Piyasaları İşletme Anonim Şirketi (EPİAŞ) in Turkish) under a license. EPİAŞ is responsible for operating the organized wholesale markets such as Day-Ahead and Intraday Market. The Power Futures Market and Organized Renewable Energy Certificate (YEK-G in Turkish) Market, that have recently started operation, are also operated by EPİAŞ. It then performs duties for the development of the Energy Exchange in Turkey. Other changes, in addition to the Electricity Market Law (Law No. 6446), are as follows:

- Unlicensed generation: No need for a 'generation license' if installed capacity is up to 1 MW for renewable energy sources. This limit was increased to 5 MW by the presidential decision in 2019.
- Extended deadlines for 50% discount on the transmission system utilization fee and stamp duty exemption.
- Wholesale licenses and retail licenses are removed and combined under the Supply License.

EÜAŞ was the state-owned entity in charge of the generation segment and TETAŞ was the state-owned entity in the wholesale segment. It was envisaged that, with the EÜAŞ privatizations and the expiration of TETAŞ contracts, the share of the state-owned enterprises would reduce their share in the market. This process is still ongoing. EÜAŞ and TETAŞ, on the other hand, were united under EÜAŞ in 2018.

In bringing this part to an end, it can be said that the electricity market reform has introduced significant changes to the Turkish electricity sector. Some of the conclusions derived from the reform phase that started in 2001 are the establishment of new institutions and redefined roles of previously existing ones; the introduction of well-operating wholesale markets (such as day-ahead, intraday, balancing, futures, and YEK-G markets); continuous decrease in limits for eligible customers for further competition (in 2021, 1200 kWh/year); the introduction of the Renewable Energy Law (Law No. 5346 2005); and unbundling of the distribution and retail sale operations. Through the extreme competition in electricity generation sector, especially in the renewables over the last decade, the total installed capacity of Turkey reached 98.5 GW in August 2021, around 53% of which pertained to renewables (TEİAŞ, 2021).

The following sections consider and explain in detail the electricity generation models implemented in Turkey (licensed, unlicensed, and YEKA models).

3 Licensed Electricity Generation Model

In Turkey, electricity generation can be carried out via three models. The first and most common one is the licensed generation, the second is unlicensed generation, and the third is the renewable energy zone (YEKA in Turkish) model. The YEKA projects are also licensed by EMRA, which allows the power plants to be built.

The majority of power plants are constructed and operated under licensed models as it allows the construction of average and large-scale projects in addition to the small-scale ones. Under the scope of the unlicensed model, however, small-scale renewable energy projects can be applied, with several exceptions.

The types of electricity generation facilities may be hydroelectric, wind, solar PV, solar CSP, thermal (coal-fired), thermal (natural gas), nuclear, geothermal, or biomass. According to the Electricity Market Law, companies must first receive a pre-license following the generation license to construct a power plant. The applications for pre-license and licenses are made to the EMRA, and the companies can be granted them if the application is found to be adequate by the board of EMRA. The application procedure differs for wind and solar projects from the other type of sources. The pre-license applications for wind and solar projects can only be received by EMRA when the regional connection capacities are announced by TEİAŞ. Based on the capacities determined by TEİAŞ, the board of EMRA decides the dates within which the companies can make the pre-license applications. According to the By-Law for Electricity Market Licensing (2013) E-mevzuat (2020a), for both wind and solar projects, on-site metering data of at least a one-year period that has been collected within the previous eight years are mandatory for the pre-license application. In the case of solar projects, however, six months of metering data must be on-site. Apart from that, for the wind and solar pre-license applications that are made for the same grid connection points, or if at least two of the project sites intersect or overlap, a capacity allocation auction is carried out to allocate the allowable connection capacities. The auction mechanism for wind or solar power pre-license applications is based on the reduction from the feed-in tariff (YEKDEM) prices corresponding to the resource type. The fixed YEKDEM prices are replaced by each individual bid and the winners are granted the right to sell the power over their individual bids for 10 years of operation following the commissioning of their power plants. Nonetheless, the rights to the local content tariffs that are described in the appendix of Renewable Energy Law (Law No. 5346) EMRA (2020d) are reserved in case of participation to YEKDEM, independent of the bid price. It is crucial to note here that, by the amendment to the renewable energy law on December 2, 2020, the companies who won the auctions held before December 2, 2020, with zero or negative prices will not have the right to benefit from local content tariffs.

All in all, the capacity allocation auctions are applied only for the pre-license applications for wind and solar power projects, while the applications for pre-licenses for other sources can be made at any time when the necessary application documents are completed.

3.1 Application for Pre-License and License

All the regulations for pre-license and licenses, procedures, and evaluation of the applications and issues related to the granting of the licenses are covered in detail in the By-Law for Electricity Market Licensing. The pre-licenses, licenses, and applications for the amendment thereof are received online by the EMRA Online Application Platform (https://basvuru.epdk.gov.tr). This module allows investors to submit their applications and have the necessary documents finalized step by step by signing the application via electronic signature. The companies must apply EMRA to assign their authorized employees in order to be able to make processes on the Online Application Platform prior to the pre-license application. The documents required for the pre-license and license applications are defined under the "List of Information and Documents to be Submitted in Applications for Pre-licenses and Licenses" regulation prepared by EMRA (Board Decisions No.10252-2 and 10252-3 dated June 10, 2021). The common pre-license application documents that are mandatory for each type of source are as follows:

- Pre-license application form (formed automatically by the EMRA online application platform)
- A copy of the articles of the company that is certified by the trade registrar
- Fact sheet of the generation utility (varies depending on the resource type and all the details related to the project are included)
- Fact sheet of the partnership structure of the company
- Fact sheet for the capital of the company (the capital of the company must be equal to at least 5% of the total investment cost of the utility for a single application, and this amount is applied as 1% for domestic coal or nuclear power plants; however, if the company has another pre-license/license application or holds another pre-license/license, the capital of the company must fulfill the minimum capital requirement considering all of its pre-license/license or applications. The unit investment costs based on the resource type can be seen in Table 1 (Board Decision No. 9397 dated June 18, 2020 EMRA (2020a)))
- 1/25,000 and 1/5,000 scale maps representing the location of the facility
- Single-line diagram
- Zoning status sheet

Table 1 Total unit investment costs depending on the resource type

Resource	Total unit investment cost (TRY/MWm)
Coal	1,500,000
Natural gas/LPG	1,000,000
Fuel oil/Naphtha	1,000,000
Hydro	2,000,000
Wind	2,500,000
Geothermal	2,100,000
Biomass	1,900,000
Solar	3,000,000
Nuclear	6,000,000
Process waste heat	700,000
Others	1,400,000

Source EMRA, Board Decision No. 9397 dated June 18, 2020

Table 2 Application fees for pre-license and generation license

The installed capacity of the power plant"P (MWm)"	Application fee (TRY/MWm)*
$0 < P \leq 10$ MW	10,600
$10 < P \leq 25$ MW	20,600
$25 < P \leq 50$ MW	30,900
$50 < P \leq 100$ MW	51,500
$100 < P \leq 250$ MW	103,000
$250 < P \leq 500$ MW	205,900
$500 < P \leq 1000$ MW	308,800
$P > 1000$ MW	515,000

Source EMRA (2020b), Board Decision No. 9841 dated December 17, 2020
*For the pre-license and generation license applications based on renewable energy and natural resources, 10% of the application fees shown in Table 2 must be paid

- Fact sheet testifying to the non-sensitive nature of the location pursuant to the By-Law for the Environmental Impact Assessment (EIA) Annex-5
- Declaration on non-forbidden nature of the location (the project should not be deployed on a fertile agricultural land)
- Letter of guarantee (MWm of the project × 10,000 TRY, the upper limit to that value is decided to be 5,000,000 TRY) (Board Decision No. 9397 dated June 18, 2020)
- Receipt indicating that pre-license application fee was paid (the amounts for 2021 can be seen in Table 2. Those fees are revised annually by the board decision of EMRA).

Apart from the documents described above, there are some additional documents that must be submitted by the applicants depending on the resource type. For hydroelectric power plants, an original or certified copy of the fact sheet on the eligibility to sign water usage agreement with the General Directorate of State Hydraulic Works is necessary. For wind power plants, at least a one-year-long, for solar power plants at least six months long on-site metering data report that has been collected within the preceding eight years is compulsory. Moreover, for the geothermal power plants, an original or certified copy of the fact sheet on the acquisition of the right to use the related resources is necessary. For biomass plants, an official document on the resource procurement for at least three years of the operation or compatible with the license duration must be submitted, and for nuclear power plants, the site license is mandatory. Finally, the environmental impact assessment report is necessary for the projects except for wind, solar, hydraulic, geothermal, biomass, or domestic mines.

Once EMRA receives the pre-license application, the documents are controlled, whether they comply with the requirements and relevant legislation in a preliminary evaluation phase. EMRA can request for additional information related to the project to the applicant or to other public institutions. If the application documents have been submitted properly, meaning that they fulfill all the requirements, the application is announced for the public opinion on the EMRA's website to determine any violation of civil rights. Afterward, the opinion of the TEİAŞ and/or relevant distribution company is sought for the connection of the project to the national grid. The applicants can be granted by pre-licenses through the decision of the Board of EMRA. On the other hand, a technical evaluation phase must be carried out for the applications based on wind and solar energy after the first evaluation phase is completed. The General Directorate of Energy Affairs of the Ministry of Energy and Natural Resources (MENR) carries out the technical evaluation and notifies the results to EMRA. The wind and solar-based projects can only be granted pre-licenses if they pass the technical evaluation stage successfully and if the project owner wins the capacity allocation auction. As mentioned before, the capacity allocation auctions for wind and solar power pre-license applications are performed when the total capacity of the applications exceeds the regional grid connection capacity announced by TEİAŞ or if the land of at least two of the projects intersects or overlaps. The "reverse auction method" is used in these auctions; namely, the lowest bid wins the capacity, and the current FIT price becoming the price ceiling. Negative bids are also permitted, for which a financial penalty, calculated based on the bid and monthly electricity generation, must be paid after the power plant is commissioned.

The duration of the pre-license depends on the type of source and the installed capacity of the project as shown in Table 3 (Board Decision No. 9394 dated June 18, 2020).

The selection of the project sites is the investors' responsibility. Building two power plants at the same location are prohibited under the current regulation. The only exception is that if anyone has the property of a land inside the licensed project site, he/she can build an unlicensed generation facility in that land. In addition, if the project site selected and submitted in the pre-license application falls within a project land regarding an international contract or a licensing process for natural gas storage, refinery, or oil storage is ongoing, the pre-license application is rejected. On the other hand, domestic coal, imported coal, and renewable energy projects are prioritized respectively where more than one application exists at the same site, and the order of priority for renewable energy projects are specified based on the resource type

Table 3 Pre-license periods. **a** For the hydroelectric dam projects, excluding the ones declared in Table-4 by the General Directorate of State Hydraulic Works and for the thermal power plants using domestic or imported coal, the pre-license period is 36 months. **b** For the projects, except those given in a, the pre-license periods are as shown below

The installed capacity of the power plant"P (MWm)"	Pre-license period (month)
$0 < P \leq 5$ MW	24
$5 < P \leq 50$ MW	30
50 MW < P	36

Source EMRA, Board Decision No. 9394 dated June 18, 2020

Table 4 Letter of guarantee amounts. **a** For the projects using domestic coal or for YEKA projects, the amount of letters of guarantee in the generation license application is calculated in the table below. The upper limit is identified as 30,000,000 TRY. **b** For the projects except for given in A, the amount of letter of guarantee in the generation license application is calculated in the table below. The upper limit is identified as 78,600,000 TRY

a

Percentage applied to the total planned investment cost (%)	The formula used for the calculation of letter of guarantee amount

b

The installed capacity of the power plant"P (MWm)"	Percentage applied to the total planned investment cost (%)	The formula used for the calculation of letter of guarantee amount
$0 < P \leq 10$ MW	3	P X UIC* X 0,03
$10 < P \leq 100$ MW	2	UIC X [0,3 + (P-10) X 0,02]
100 MW < P	1	UIC X [2,1 + (P-100) X 0,01]

Source EMRA, Board Decision No. 9397 dated June 18, 2020
*UIC (TL/MWm) refers to the total unit investment cost depending on the resource type given in Table 1
*UIC (TL/MWm) refers to the total unit investment cost depending on the resource type given in Table 1

as geothermal, hydraulic, wind and solar under the By-Law for Electricity Market Licensing.

After the company receives the pre-license, it has a limited time to complete the *"obligations to be fulfilled in the pre-license period"* specified in Article 17 of the By-Law for Electricity Market Licensing. These requirements are indicated in the list below.

- For projects based on *wind, solar, hydraulic, geothermal energy, biomass, or domestic mines*, the pre-license holder must apply to the relevant institution within 90 days following the start of the pre-license period to receive the necessary document in the context of the environmental impact assessment
- For *wind energy projects*, the pre-license holder must apply to the relevant institution within 180 days following the start of the pre-license period to receive the Technical Interaction Permission Report
- Acquisition of ownership/usufruct rights of the power plant site (expropriation procedures included)
- Approval of the zoning plans of the project site
- Approval of the project
- Acquisition of construction permit
- Environmental impact assessment report
- Obtaining the final forestry permits, if subjected

- For wind power plants, obtaining the permit for technical interaction with military and civilian air services and the positive opinion pursuant to the Regulation on Military Forbidden Zones and Security Areas
- For hydroelectric power plants, finalization of the Water Usage Agreement with the General Directorate of State Hydraulic Works
- Application for the system connection and usage agreements with TEİAŞ or the relevant distribution company
- For wind and solar power plants, finalization of contribution margin agreement with TEİAŞ
- For geothermal power plants, the acquisition of the right to use relevant resources
- For pre-licenses granted to YEKA projects, a suitability report to the By-Law for Renewable Energy Resource Zones showing that there is no problem for the investor to be granted by the generation license to be issued by the General Directorate of Energy Affairs.

Once the companies holding pre-licenses complete all the above-mentioned mandatory documents related to the pre-license obligations, they must apply to EMRA to receive generation license until the end of the pre-license period. Unless they make this application in time, the pre-license is terminated automatically when the duration of the pre-license expires. If the pre-license holders can not complete some or any of the obligatory documents until the end of the pre-license period due to force majeure, they can apply EMRA for the time extension. In this case, the Board of EMRA can either give a decision for the extension of the pre-license period or the termination of the pre-license depending on the causes of delay.

The necessary documents in the application for generation license, apart from the documents indicated above, are as follows:

- License application form (formed automatically by the EMRA online application platform)
- A copy of the articles of the company that is certified by the trade registrar
- Fact sheet of the partnership structure of the company
- Business deadline plan
- Receipt indicating that the license application fee was paid (the amounts can be seen in Table 2. They are the same as the pre-license application fees. For the generation license applications in the context of Article 5/3 of the By-Law for Electricity Market Licensing, the whole amount represented in Table 2 must be paid)
- Fact sheet for the capital of the company (the capital of the company must be equal to at least 20% of the total investment cost of the utility for a single application, and this amount is applied as 5% for domestic coal or nuclear power plants, as well as YEKA projects. However, if the company has another pre-license/license application or holds another pre-license/license, the capital of the company must fulfill the minimum capital requirement considering all of its pre-license/license or applications. The unit investment costs based on the resource type can be seen in Table 1 (Board Decision No. 9397 dated June 18, 2020))

- Letter of guarantee (the amount is calculated based on the formula given in Table 4 (Board Decision No. 9397 dated June 18, 2020). The letter of guarantee offered in the pre-license application is repaid when the generation license is granted.

After the application for the generation license was submitted to EMRA's online application platform, is evaluated and if the application is found to be appropriate, the license is granted by the decision of Board of EMRA. The generation license is granted for 10–49 years; this is the permitted duration including both the construction and operation of the power plant. However, the licenses for YEKA projects are issued for 30 years. The granting of a generation license means that the investors can now start the construction of their power plants. The construction periods are determined by the board decision of EMRA (No. 9394 dated June 18, 2020) and demonstrated on each license document. The construction periods depend on the resource type and the capacity of the project (seen in Table 5). In case of non-realization of the project within the specified construction period, the license is terminated unless the reason for non-deployment is caused by force majeure or a justified reason. In these situations, board of EMRA can prolong the period of construction and the generation license is amended accordingly. However, this amendment does not affect the license term.

After the construction processes are completed and the facility is ready to operate, the commissioning of the power plant is carried out by the General Directorate of Energy Affairs of the MENR. Once the preliminary operational tests and the controls are completed, the company can start to generate electricity in the power plant. Following the commissioning of the power plant, license holders must apply to EPİAŞ for registering their power plants to the organized electricity market system to be able to operate at the organized electricity markets such as day-ahead and intraday markets.

3.2 *Amendment of Pre-License and License*

The current regulation allows for the amendment of pre-licenses and generation licenses in several ways. The rules and procedure are indicated in Articles 18 and 24 of the By-Law for Electricity Market Licensing for the pre-licenses and licenses, respectively. The types of amendments for the pre-licenses and licenses permitted in the legislation are as follows:

- Prolongation of the pre-license or license periods
- Prolongation of the construction periods
- Renewal of the licenses
- Change of the mechanical or electrical capacity of the project
- Change of the details of the project (that is, number of units, coordinates of the units/turbines, annual amount of electricity generation)
- Change of the project site
- Change of the connection point of the project to the national grid

Table 5 Construction periods for licensed power plants **a** Thermal power plants: **b** Hydroelectric power plants: **c** Other renewable power plants:

a

Type of power plant	*The installed capacity of the power plant "P (MWm)"*	*Construction period (month)*
Simple cycle engine	$P \leq 10$	18
	$10 < P \leq 50$	21
	$50 < P \leq 100$	24
	$100 < P$	30
Simple cycle gas turbine	$P \leq 10$	21
	$10 < P \leq 50$	27
	$50 < P \leq 250$	30
	$250 < P \leq 500$	36
	$500 < P$	42
Cogeneration (Solid/liquid/gas fired power plants)	$P \leq 10$	24
	$10 < P \leq 50$	30
	$50 < P \leq 250$	36
	$250 < P \leq 500$	42
	$500 < P$	48
Combined cycle power plants (Liquid/gas fired)	$P \leq 50$	38
	$50 < P \leq 250$	42
	$250 < P \leq 500$	48
	$500 < P$	54
Conventional (liquid/gas fired)	$P \leq 50$	38
	$50 < P \leq 250$	42
	$250 < P \leq 500$	46
	$500 < P$	54
Fluidized bed, conventional, super critical PC (solid/fossil fuel based power plants)	$P \leq 50$	38
	$50 < P \leq 250$	44
	$250 < P \leq 500$	54
	$500 < P$	66
Nuclear power plants	–	Determined based on the values on the international agreement, otherwise the board of EMRA decides by considering the view of the MENR

(continued)

Table 5 (continued)

a

b

Type of power plant	*The installed capacity of the power plant "P (MWm)"*	*Construction period (month)*
Run-of-the-river station	$P \leq 50$	38
	$50 < P \leq 100$	44
	$100 < P$	60
Reservoir*	$V_g \leq 1{,}000{,}000$	42
	$1{,}000{,}000 < V_g \leq 5{,}000{,}000$	54
	$5{,}000{,}000 < V_g \leq 7{,}500{,}000$	60
	$7{,}500{,}000 < V_g \leq 10{,}000{,}000$	66
	$10{,}000{,}000 < V_g$	72

c

Type of power plant	*The installed capacity of the power plant "P (MWm)"*	*Construction period (month)*
Wind	$P \leq 10$	22
	$10 < P \leq 50$	30
	$50 < P \leq 100$	38
	$100 < P$	46
Geothermal	$P \leq 50$	38
	$50 < P$	46
Biomass/biogas	$P \leq 10$	24
	$10 < P \leq 50$	30
	$50 < P$	38
Solar/hydrogen	$P \leq 10$	22
	$10 < P \leq 50$	30
	$50 < P$	36
Tidal power	$P \leq 10$	18
	$10 < P \leq 50$	30
	$50 < P$	38

(*Source* EMRA (2020c), Board Decision No. 9394 dated June 18, 2020)
*For pumped-storage hydroelectric plants, the periods given for reservoir units are valid

- Change of the current postal address or name/kind of the company

The applications for the amendment of the pre-license or licenses are received on the EMRA's online application platform as pre-license/license applications. The necessary documents required for the amendment of pre-license or licenses are defined under the "List of Information and Documents to be Submitted in Amendment Applications for Pre-licenses and Licenses" regulation prepared by EMRA (Board Decision No. 10252-4 dated June 10, 2021). The final decision for the amendment applications are given by either board of EMRA or by the Department of the Electricity Market, the details of which are described in Articles 18 and 24 of the By-Law for Electricity Market Licensing. Obviously, it is not guaranteed that all applications for amendments are accepted; force majeure or a valid reason is required for many types of amendments to be accepted, such as the prolongation of the pre-license/license/construction periods, change of the project location, etc.

4 Unlicensed Electricity Generation Model

Another electricity generation model in Turkey is the unlicensed generation method, for which the establishment of a company and receiving a generation license is not mandatory. The categories and exemptions from receiving a generation license and procedures for unlicensed generation are explained in the following subsections.

4.1 Fundamentals of Unlicensed Electricity Generation

According to the current regulation in the Electricity Market Law (Law No. 6446), natural or legal persons are exempted from receiving a generation license and establishing a company if they construct;

- Renewable-based power plants up to 5 MW, yet up to consumer retail sale contract capacity.
- Micro-cogeneration facilities up to 100 kW
- Cogeneration facilities
- Generation facilities for municipal waste disposal and treatment facilities
- Emergency groups or isolated generation facilities.

The surplus electricity generated is utilized by the price of energy under the corresponding user group in the tariff tables, which are declared quarterly by EMRA. Information regarding the types of power plants that can be constructed as unlicensed models, the limits for installed capacity grid connection modes, and evaluation of the excess amount of energy produced in those power plants are explained below.

- For the emergency groups, there is no power limit and they have to be established as isolated (that is, off-grid)
- For electricity generation units that are off-grid and for which all the power generated is consumed for own consumption, there is no limit assigned for installed capacity

- The capacity of electricity generation units based on renewable energy sources can have a maximum of 5 MW installed capacity, and must be connected to the distribution network. The surplus electricity must be purchased by the incumbent supplier at the current energy price declared on the tariff list
- For electricity generation units based on renewable energy sources using all of the energy generated without connecting to the transmission or distribution system and the generation and consumption of those are at the same measurement point, the electricity shall not be given to the network; otherwise, it would be purchased free of charge
- Those micro-cogeneration or cogeneration facilities that meet the efficiency level determined by the MENR and fall under the category to be specified by the board of EMRA can be connected either to the transmission or distribution network, but the excess electricity shall not be given to the network; otherwise, it would be purchased free of charge
- Microgeneration facilities up to 100 kW can be connected to the distribution network and the surplus electricity must be purchased by the incumbent supplier over the current energy price declared on the tariff list
- The electricity generation units, either based on renewable energy sources or not, established for the purpose of disposing mud from the solid waste facilities and treatment facilities of municipalities can be connected to the distribution network and the surplus electricity must be purchased by the incumbent supplier at the current energy price declared on the tariff list
- By legal entities, more than half of the capital is directly or indirectly owned by the municipality, a hydroelectric power plant can be established on the water or waste water distribution lines operated by the municipality, on the condition that it is technically feasible and deemed appropriate by the General Directorate of State Hydraulic Works. This power plant shall be connected to the distribution network and the surplus electricity must be purchased by the incumbent supplier at the current energy price declared on the tariff list.

The new regulation on Unlicensed Electricity Generation in the Electricity Market entered into force upon its publication in the Official Gazette on 12 May 2019 and the regulations regarding the unlicensed electricity generation in the electricity market dated 2013 have been abrogated (E-mevzuat, 2020c).

The recently published secondary legislation on unlicensed electricity production reflects a policy preference towards self-consumption rather than electricity trading purposes. The new regulation allows all individuals and companies to establish rooftop solar electricity generation facilities instead of utility-scale ones, depending on the amount of consumption and 5 MW capacity limit.

The monthly settlement method (net-metering system) based on the hourly calculated data regarding the purchase of surplus electricity has also been included in the new regulation. If we consider the newly adapted applications for the unlicensed generation, unlicensed solar energy generation facilities can only be established as roof or façade. However public institutions and organizations, agricultural watering, waste water, and drinking water facilities' energy need may be subject to PV installment on area. An unlicensed generation plant must be associated with a consumption unit and both generation and consumption units must be located at the same point (except for public institutions and organizations). There is no obligation for the specified amount of consumption associated with a generation facility anymore. The completion periods from signing the connection agreement for the unlicensed generation facilities are as follows:

- For hydraulic power plants on medium voltage within three years
- For other power plants on medium voltage within two years
- For power plants on low voltage within one year.

There is also an opportunity to build rooftop solar up to 10 kW on the roofs and façade of buildings. These power plants can directly be connected to the network if the distribution system operator provides an affirmative opinion about the connection of the utility to the national grid. For projects up to 3 kW, the system operator directly gives the affirmative connection opinion (EMRA 2018). Net-metering can be applied to those plants, meaning that the excess electricity generated can be settled monthly and priced at current energy prices declared on the tariff list.

5 Renewable Energy Zone (YEKA) Model

According to Article 4 of Law No. 5346, MENR has the duty of identifying, using, and protecting the renewable energy resource zones in Turkey. Therefore, the Ministry of Energy and Natural Resources is responsible for specifying the RE Zones where renewable power plants can be constructed in a relatively short period of time and operate more efficiently.

The MENR accordingly prepared legislation (By-Law for Renewable Energy Resource Zones 2016 E-mevzuat (2020b)) published in the Official Gazette on 9 October 2016. Through this regulation, a new renewable energy investment model has been introduced to support renewable energy investments and encourage local manufacturing or local use of the power plant equipment. The primary scopes of this model are to shorten the period of construction of the renewable power plants, to build them more efficiently by specifying RE Zones on the public, treasury or private-owned sites, to encourage local manufacturing/or use of the domestically manufactured equipment, and to render the technology transfer by the research and development (R&D) activities.

5.1 Fundamentals of YEKA Tenders and Related Procedures

The renewable energy resource zone and the utilization rights for the grid connection capacity can be proposed to a company or a consortium by means of RE Zone tenders. Two kinds of tender schemes are "Allocation on the Condition of Local Manufacturing" and "Allocation on the Condition of Using Domestically Manufactured Equipment". The MENR can announce a tender by selecting either type of scheme for any kind of renewable energy source. The wind and solar power tenders are currently the most common types in Turkey due to the large amount of resource potential.

For the first scheme, the company or consortium that wins the tender and the right to use the YEKA region and the corresponding connection capacity must establish a factory to manufacture the plant equipment/components locally and a research and development center. In this center, R&D operations must be carried out for a certain period of time and be compatible with the mandatory conditions such as budget limit, number, and qualification of employees (for example, there could be a limit for the number of native engineers working in the factory). In the YEKA project, the equipment and components manufactured locally must be used. In the second scheme, the same rights are given to the winner of the tender committing the procurement of domestically manufactured equipment/components to be used in the power plant. The equipment/components must satisfy the local content ratio indicated in the terms of reference and be compatible with national or international standards.

YEKA tenders are based on awarding a power purchase agreement (PPA), which grants the winning company/consortium the right to sell electricity at a constant tender price. The MENR makes the plans and programs, the dates for the applications, and the terms of reference of the tenders. In these tenders, the MENR decides a price ceiling over TRY/kWh and the lowest bid wins the usage rights of the YEKA region, the allowable connection capacity, and the right to sign a PPA. Therefore, it can be said that, unlike the licensed projects, no additional local content tariff is provided to the winner of the tender. However, the duration of PPAs is commonly longer than the YEKDEM support (currently 10 years). For example, in the previous four YEKA tenders (two solar PV tenders and two onshore wind tenders), the period of purchase guarantee in the PPAs was defined as 15 years.

After the tender is completed and the winner company/consortium is determined, they are asked to submit an additional letter of guarantee for signing the power purchase agreement. Once this agreement is signed, the company/consortium has a specified period to make pre-license application to EMRA. The licensing procedure is performed by EMRA in accordance with the By-Law for Electricity Market Licensing, similar to the applications directly submitted to EMRA for licensed electricity generation.

The specifications and results of the previous YEKA tenders are summarized in Table 6 (The Presidency of the Republic of Turkey Investment Office 2019).

Apart from the tenders shown in the above table, the MENR has completed another YEKA PV tender for 1 GW solar power connection capacity in May 2021. In this context, 74 separate solar PV tenders have been carried out in 36 different provinces. The capacities of the projects are 10, 15, or 20 MW and the duration of the PPA

Table 6 Summary of the YEKA tenders performed in Turkey

Description of the 1st tender	Karapınar YEKA GES-1
Capacity	1 GW
Date of the tender	20/03/2017
Power production capacity (kWh)	1.7 billion
Price ceiling (US cents/kWh)	8
Winning bid (US cents/kWh)	6.99
Winner entity	Hanwha Q cells-kalyon energy consortium
Total investment amount (US $)	1.3 billion
Local content ratio	60% and R&D investment
PPA duration	15 years
Capacity commissioned (as of August 2021)	329.28 MWe
Description of the 2nd tender	Karapınar YEKA GES-1
Capacity	1 GW
Date of the tender	03/08/2017
Power production capacity (kWh)	3 billion
Price ceiling (US cents/kWh)	7
Winning bid (US cents/kWh)	3.48
Winner entity	Siemens-Kalyon-Türkerler consortium
Total investment amount (US $)	1.1 billion
Local content ratio	65% and R&D investment
PPA duration	15 years
Capacity commissioned (as of August 2021)	–
Description of the 3rd tender	Karapınar YEKA GES-1
Capacity	1 GW (250 MW/Balıkesir, 250 MW/Aydın, 250 MW/Muğla, 250 MW/Çanakkale
Date of the tender	30/05/2019
Power production capacity (kWh)	3 billion
Price ceiling (US cents/kWh)	5.5
Winning bids (US cents/kWh)	Balıkesir: 3.53 Aydın: 4.56 Muğla: 4 Çanakkale: 3.67
Winner entities	Balıkesir: Enercon Aydın: Enerjisa Muğla: Enercon Çanakkale: Enerjisa
Total investment amount (US $)	1 billion
PPA duration	15 years
Capacity commissioned (as of August 2021)	–

is 15 years. The winning bids changed between 0.289 to 0.185 TRY/kWh, showing that the tenders were quite competitive. After the project sites are determined by the winners, they are expected to apply to EMRA to receive pre-license. Moreover, MENR has recently announced another 2 GW onshore wind and 1 GW solar PV tender notices, for which the applications will be accepted in October 2021 and March 2022, respectively. Furthermore the MENR, announced a new YEKA PV tender for 1, 5 GW solar power connection capacity in September 2021 based on domestically manufactured equipment model. In this context, according to the tender notice, published in the Official Gazette on 30 September 2021, 76 separate solar PV tenders are planned to be carried out in 23 different tender zones, mosly provinces. The capacities of the projects are announced to be 10, 20, or 30 MW. The duration of each PPA will be determined to 23 GWh electricity generation from each allocated MW. The applications for those will be received by MENR on January 2022 and a notice for the dates of tenders will be made subsequently on the MENR's website. The price ceiling for PPA for each tender were specified as 0.4 TRY/kWh.

6 Conclusion

Electrical energy is one of the most essential energy sources used in transportation, industry, households, and almost every part of daily life. The demand for electricity, therefore, increases continuously depending on technological developments, population growth, and economic changes. Because of the need to meet increasing demand and issues related to climate change, the method of electricity generation, distribution of the generation mix, and the national energy policies play a key role for the countries.

In Turkey, the generation sector can be evaluated mainly under two stages. During the state-oriented period (the period between the foundation of the Republic and 1984), new power plants mostly based on hydraulic and thermal sources were built as a result of concession contracts and the efforts of General Directorate of State Hydraulic Works. The private sector's participation in the generation sector started with the law introduced in 1984 and a number of power plants were established via models such as BO, BOT, TOR, and the auto producer scheme. Meanwhile, the institutional adjustments and decentralizations made for the generation, transmission, distribution, wholesale, and supply of electricity established a more organized structure in the electricity market during this period. A milestone for the Turkish energy market occurred in 2001 by the introduction of the electricity market law (Law No. 4628), with which the independent energy market regulatory authority (EMRA) was also established for making the necessary regulations in the market, as well as monitoring and supervising the market. The licensing mechanism for the market participants in each field of the energy market encouraged the private sector participation and electricity generation sector started to develop rapidly since then, providing market maturity.

The present electricity market law (Law No. 6446) and renewable energy law (Law No. 5346) are the primary legislation showing the rules of electricity generation in Turkey. Moreover, the existing secondary legislation such as By-Laws for Electricity Market Licensing, Unlicensed Electricity Generation, and Renewable Energy Resource Zones, describe the details for all the processes to be conducted related to the licenses during the construction of power plants.

The electricity generation models adopted in Turkey were expressed in three subsections as licensed, unlicensed, and YEKA models. In the licensed model, investors must first receive a pre-license from EMRA, and until the end of pre-license period, they must obtain all the necessary permissions and prepare the documents necessary for starting the construction. After all the obligations have been completed, the investors can be granted a generation license by the decision of the board of EMRA and start to construct their power plants. The documents and processes associated with the licensing may change depending on the type of resource to be used in the power plant. For example, for wind and solar power, a tendering mechanism is applied for the capacity allocation and the pre-license applications can be received by EMRA only if the allowable grid connection capacities are announced by the Turkish TSO (TEİAŞ). The unlicensed electricity generation model has also attracted the attention of the private sector over the past decade and around 6200 small-scale solar PV plants have been established by this model. It is expected that the new regulation on unlicensed electricity generation will lead to rooftop solar also becoming prevalent in Turkey in the near future. Finally, the large-scale YEKA tenders for solar and wind power carried out by the MENR will help to increase the share of renewables in the generation mix, as well as improving the technologies with the obligation of local manufacturing of the equipment used in the power plants and establishment of the R&D centers.

In summary, the Turkish electricity generation sector has experienced a great and challenging era in the 2000s, with the enhanced regulatory structure and a high level of market participation. The trend and demand for renewable energy investments accelerated the improvement of the market. By this means, the total installed capacity reached 98.5 GW at the end of August 2021, more than half of which is based on renewables. Moreover, the first nuclear power plant, with 4.8 GW of capacity, is expected to start operation in 2023. All in all, it can be said that the energy policies for electricity generation, based on the use of national and domestic resources, the private sector's long-standing willingness to enter the sector, and the high level of competition in the market, will continue to give favorable results for the Turkish electricity generation sector.

References

Alboyacı, B., & Dursun, B. (2008). Electricity restructuring in turkey and the share of wind energy production. *Renewable Energy, 33*, 2499–2505. https://doi.org/10.1016/j.renene.2008.02.008

Atiyas, I., Çetin, T., & Gülen, G. (2012). *Reforming Turkish energy markets, political economy, regulation and competition in the search for energy policy*. E-book, ISBN 978-1-4614-0290-9 Springer Science+Business Media New York.https://doi.org/10.1007/978-1-4614-0290-9

Bagdadioglu, N., & Odyakmaz, N. (2009). Turkish electricity reform. *Utilities Policy, 17*, 144–152. https://doi.org/10.1016/j.jup.2008.02.001

Çetin, T. (2010). Institutional change in the Turkish energy markets. In Çetin, T., Yilmaz F. (Eds.), *Understanding the Process of Economic Change in Turkey* (275–318). Nova Science Publishers, Inc. ISBN: 978-1-60876-945-2.

Çetintaş, H., & Bicil, İM. (2015). Restructuring in the electricity markets and structural transformation in Turkish electricity market. *Optimum Journal of Economics and Management Sciences, 2*(2), 1–15.

E-mevzuat. (2020a). *By-law for Electricity market licensing*. https://www.mevzuat.gov.tr/mevzuat?MevzuatNo=18985&MevzuatTur=7&MevzuatTertip=5

E-mevzuat. (2020b). *By-law for renewable energy resource zones*. https://www.mevzuat.gov.tr/mevzuat?MevzuatNo=22923&MevzuatTur=7&MevzuatTertip=5

E-mevzuat. (2020c). *By-law for unlicensed electricity generation*. https://www.mevzuat.gov.tr/mevzuat?MevzuatNo=31502&MevzuatTur=7&MevzuatTertip=5

E-mevzuat. (2020d). *The law on the utilization of renewable energy sources for the purpose of generating electrical energy (Law No. 5346)*. https://www.mevzuat.gov.tr/mevzuat?MevzuatNo=6446&MevzuatTur=1&MevzuatTertip=5

E-mevzuat. (2020e). *The electricity market law (Law No. 6446)*. https://www.mevzuat.gov.tr/mevzuat?MevzuatNo=6446&MevzuatTur=1&MevzuatTertip=5

EMRA (2018). *The secondary legislation for the establishment of rooftop solar up to 10 kW*. https://www.epdk.gov.tr/Detay/Icerik/6-3-3063/usulveesaslar

EMRA (2020a). *Board decision on the total unit investment costs based on resource type and on the amounts for letter of guarantee*. https://www.epdk.gov.tr/Detay/Icerik/3-0-86/elektriklisans-islemleri

EMRA (2020b). *Board decision on the license fees for 2021*. https://www.epdk.gov.tr/Detay/Icerik/3-0-86/elektriklisans-islemleri

EMRA (2020c). *Board decision on the reference construction periods for the licenses based on resource type*. https://www.epdk.gov.tr/Detay/Icerik/3-0-86/elektriklisans-islemleri

International Energy Agency. (2010). Energy Policies of IEA Countries: Turkey 2009 Review. ISBN 978-92-64-06041-8. 9 rue de la Federation 75739 Paris Cedex 15, France.

Karagöl, E. T., & Tür, M. R. (2017). Türkiye'de Elektrik Enerjisi. *Siyaset, Ekonomi ve Teknoloji Araştırmaları Vakfı (SETA), SETA Yayınları 96*. https://www.researchgate.net/publication/321197687_Turkiye'de_Elektrik_Enerjisi

Kemal, H. (2006). *Türkiye'de Elektrik Enerjisi Gelişiminin Kısa Tarihçesi ve Genel Üretim Bilgileri*. http://www.emo.org.tr/ekler/0082ac261d74f5a_ek.pdf

Özkıvrak, Ö. (2005). Electricity Restructuring in Turkey. *Energy Policy, 33*, 1339–1350. https://doi.org/10.1016/j.enpol.2003.12.010

TEİAŞ. (2021). Ağustos*2021 Kurulu Güç Raporu*. https://www.teias.gov.tr/tr-TR/kurulu-guc-raporlari

The Presidency of the Republic of Turkey Investment Office. (2019). *Guide to investing in Turkish renewable energy sector*. https://www.invest.gov.tr/tr/library/publications/lists/investpublications/turkiye-yenilenebilir-enerji-sektorune-yatirim-rehberi.pdf

Tutuş, A. (2006). Türkiye'de Elektrik Enerjisinin Tarihsel Gelişimi ve Yeni Piyasa Düzeni İçerisinde Hidroelektrik Enerjinin Yeri. *TMMOB 1st Congress on Water Policies*, 318–330. http://www.imo.org.tr/resimler/ekutuphane/pdf/9136.pdf

Yılmaz, A. O., & Uslu, T. (2007). Energy policies of Turkey during the period 1923–2003. *Energy Policy, 35*, 258–264. https://doi.org/10.1016/j.enpol.2005.10.015

Refik Tiryaki energy expert, lawyer, mediator and Ph.D. Iur. He is an energy expert at Energy Market Regulatory Authority (EMRA) of Turkey. He started his career as a lawyer in 1998, after completing his LLB in Ankara University Law Faculty. Then, during his post graduate study at the same University,he had worked as a research assistant at Faculty of Law for 4 years. In 2002,he joined to the EMRA as an energy expert, during its foundation period. During his professional career he also completed his PhD with the thesis "Constitutional Regime of Economic Liberties" which is published in Turkish. In 2013, he was the Visiting Fellow at London School of Economics (LSE), during the academic year. His work of "The Evolution of Turkish Electricity Market Regulation" was awarded as the 'Best Paper' by the organization of Competition and Regulation in Network Industries, 6th Annual Conference, Brussels, 22 November 2013. In 2014, he had appointed as the Head of Group for Solar and Wind Energy Licenses at the Electricity Market Department. In 2017, he completed trainings, passed exams and became a Mediator. In 2019, he had appointed as the Head of Electricity Market Department. He has several publications in the field of administrative law and energy law. He has given post graduate lectures at Ozyeğin University and Hacettepe University on energy law.

Mehmet Yalılı is an energy expert working at the electricity market department of the Energy Market Regulatory Authority of Turkey (EMRA).He holds a Bachelor's and Master's degrees in Mechanical Engineering from Middle East Technical University (METU), as well as Bachelor's degrees in Business Administration and International Relations from Anadolu University. He currently studies for doctorate in Energy Systems Engineering at Gazi University. During his Master's study, he was a research assistant at the Mechanical Engineering Department at METU. His main areas of work and research include the regulation of the electricity market, market monitoring and surveillance, generation of electricity from renewable and conventional energy sources and energy efficiency practices.

Regulation of the Turkish Wholesale Electricity Market: A General Overview

Taner Şahin

Abstract It is undeniable that electricity is an indispensable element of modern life. In order to provide high-quality electricity services more reliably and at an affordable price, Turkey began to liberalise its electricity market decades ago in parallel with developments in European Union (EU) energy markets. This paper aims to overview the regulation of the wholesale electricity market in Turkey that emerged during the long reform processes, particularly in the last two decades, and to provide some suggestions, especially for capacity remuneration mechanism implemented in Turkey. In terms of structure, after briefly describing the necessity for regulation in liberalised electricity markets, the chapter deals with reforms in the Turkish electricity market. Finally, the chapter examines the existing structure of the Turkish electricity market in general.

Keywords Wholesale electricity · Markets · Balance · Price renewable incentives · Demand-side

1 Introduction

Those who read books, articles and reports on electricity markets are accustomed to reading statements about the necessity of electricity for modern life as a classic introduction. This is undeniable truth—indeed, a modern life without electricity is unthinkable. However, this proposition has never been more strongly tested than during the Covid-19 pandemic. The Covid-19 pandemic has triggered a large, deep and prolonged economic crisis at the global level.[1] According to the International Energy Agency (IEA), the Covid-19 pandemic has led to enormous economic challenges not experienced since the 1930s (IEA, 2020, p. 13). Governments around the world have taken extensive financial measures to deal with this massive economic

[1] The word 'unprecedented' is likely one of the most commonly used words to describe the destruction caused by Covid-19 in the fields of health, economics, education and social psychology.

T. Şahin (✉)
Law School, Ondokuz Mayıs University, Samsun, Turkey
e-mail: taner.sahin@omu.edu.tr

M. Eroğlu and M. Finger (eds.), *The Regulation of Turkish Network Industries*,
https://doi.org/10.1007/978-3-030-81720-6_3

downturn. According to an IEA report published in June 2020, '[s]o far, they [governments] have announced measures worth about USD 9 trillion, focussing primarily on emergency financial and economic relief to prevent an even deeper crisis' (IEA, 2020, p. 13). In addition to this economic chaos, a major humanitarian disaster has occurred that no country has escaped.[2] In this period of significant economic destruction and humanitarian crisis, the importance of electricity in modern life has become better understood. On this matter, the IEA has stated:

> The energy sector, particularly electricity, has played a critical role in the global response to the Covid-19 crisis. Uninterrupted energy supplies have enabled hospitals to provide care, food and other essentials to be delivered, and millions of people to work and study from home while maintaining social contact online. Without access to reliable and affordable electricity, the lockdowns introduced by governments to tackle the public health crisis would have resulted in far greater economic damage. (IEA, 2020, p. 13)

In order to liberalise electricity markets, which are integral to the general good of humanity, many reforms have been carried out over the past 40 years in many countries. Electricity trading in free market conditions became possible as a natural outcome of these liberalisation reforms. With this understanding, and in parallel to the developments in the EU, Turkey began to liberalise its electricity industry in the early 2000s in the context of historical difficulties and institutional constraints (Cetin & Oguz, 2007, p. 1761). Although there is a long way to go to ensure a fully competitive environment, Turkey has nevertheless achieved remarkable success in liberalising its electricity market, particularly its wholesale electricity market (Dilli & Nyman, 2015, p. 96).

This paper attempts to present an overview and provide some recommendations with regard to the regulation of the wholesale electricity market, particularly for Capacity Remuneration Mechanism (CRM), in Turkey. First, the question of why regulation is a necessity in liberalised electricity markets is briefly investigated. After this general framework is set forth, the process of the liberalisation of the electricity market in Turkey is addressed in its historical context. After examining day ahead, balancing, intraday and bilateral markets, renewable energy incentives and CRMs in Turkey are analysed. Finally, some policy suggestions are provided.

---

[2] Clair noted in his article about the implications of Covid-19 for the EU wholesale electricity market that the process derived from the Covid-19 pandemic is, first and foremost, a human tragedy (Clair, 2020). Furthermore, in a speech about the economic effects of Covid-19 on YouTube, Gürkaynak (2020) rightly stated that the crisis human beings are currently experiencing is primarily a health crisis and that health should be the first aspect to be addressed. He added that this is supported both morally and economically, since the economic problems will not disappear until the health crisis is eliminated (Gürkaynak, 2020).

2 Why is Regulation a Necessary Element of Liberalised Electricity Markets?

Electricity trading in free market conditions became possible as a natural outcome of liberalisation reforms in the last 40 years.[3] However, liberalised electricity markets are still subject to a certain amount of regulation, as liberalisation in itself does not mean that regulation disappears in electricity markets (Mäntysaari, 2015, p. 93). To give an example from popular culture, an episode of *House of Cards* (a phenomenal political television show) portrayed a conversation discussing how regulation is an inseparable part of liberalised electricity markets (Coles, 2014):

> **Frank Underwood [US Vice President]**: You are deliberately keeping electricity prices at an inflated rate, which is illegal.
>
> **Raymond Tusk [a rich and powerful businessman in the energy sector]**: No. I am responding to the free market.
>
> **Frank Underwood**: A regulated free market.

While this dialogue is fictional, it is also clear that the fiction in question is based on an undeniable truth about electricity markets. As a matter of fact, it is difficult to say that *any* market is literally free, not only electricity markets. On this issue, Rodrik (2011) succinctly notes:

> [...] *markets are not self-creating, self-regulating, self-stabilising or self-legitimatizing.*[4] Every well-functioning market economy blends state and market, laissez-faire and intervention. The precise mix depends on each nation's preferences, its international position, and its historical trajectory. But no country has figured out how to develop without placing substantial responsibilities on its public sector. (Rodrik, 2011, p. 22).

Rodrik clearly—and rightly—states here that state and market are in balance in a well-functioning market economy and that this balanced mix can be determined in the context of the specific conditions of each country. This balanced mix of state and market must be determined not only by the circumstances of countries but also within the specific conditions of each market. This is particularly true for electricity markets. Due to the unique characteristics of the electricity industry (imbalances, congestion management, ancillary services, and scheduling and dispatch), a well-designed regulatory infrastructure is an unavoidable requirement for competitive electricity trading (Hunt, 2002, pp. 123–126). Furthermore, the fact that electricity, as a basic factor of production, affects the prices of all products in the market requires that the electricity market have specific forms of organisation and regulation (The State Supervisory Council (DDK), 2010; as cited in Çakmak, 2011, p. 83).

---

[3] Liberalisation in electricity markets first began in Chile and, from there, spread globally. Since the historical background of liberalisation in electricity markets around the world is well beyond the scope of this chapter, it is not further analysed here. For further analyses concerning liberalisation reforms in the Chilean electricity market, *See,* (Raineri, 2006).

[4] Italicisation is Rodrik's.

3 Liberalisation in the Turkish Electricity Market: A Brief History of Continuous Reforms

By the 1980s, the Turkish electricity market—as in every country in the world—was dominated by a state monopoly, and all market segments were owned by a vertically integrated state company called the Turkish Electricity Authority (TEK) (Atiyas et al., 2012, p. 20). Several attempts were made in the 1980s and 1990s to engage the private sector in the Turkish electricity market. In this sense, the first legal framework (Law No. 3096, which came into effect in 1984) was designed to introduce two types of contracts to promote private investment: build–operate–transfer (BOT) and transfer of operating rights (TOR) (Atiyas et al., 2012, pp. 20–21). A BOT was a concession agreement that allowed a company to build and operate a new generation plant for 99 years (subsequently reduced to 49 years) and then transfer the plant to the state at no cost (Atiyas et al., 2012, p. 21). A TOR was a lease-type agreement through which a private company could operate and, if needed, rehabilitate existing generation and distribution facilities for a certain period of time (Atiyas et al., 2012, pp. 21–22). TOR contracts were also designed to allow companies to generate their own electricity as auto-producers (Atiyas & Dutz, 2004, p. 7).

In 1993, in order to prepare TEK for privatisation, TEK was separated into two state-owned companies: the Turkish Electricity Generation Transmission Corporation (TEAŞ) and the Turkish Electricity Distribution Company (TEDAS) (Atiyas & Dutz, 2004, p. 7). In 1994, Law No. 3996 and Implementing Decree 5907 were enacted to empower the Undersecretariat of Treasury to grant guarantees for BOT projects and provide tax exemptions (Atiyas & Dutz, 2004, p. 7). An additional piece of legislation (Law No. 4283) that became effective in 1997 replaced the concession regime with a licencing regime through build–own–operate (BOO) arrangements, again with Treasury guarantees (Atiyas & Dutz, 2004, p. 7).

All BOT, TOR and BOO contracts concluded between the private party and TEAŞ or TEDAŞ included take-or-pay clauses that provided fixed prices to buy a predefined amount of electricity quantities for 15 to 30 years (Atiyas et al., 2012, p. 22). These contracts later led to several serious concerns, such as rewarding some contracts without a competitive procedure, purchasing expensive electricity, and irregularities in the design and implementation of these contracts (Atiyas et al., 2012, p. 22; Cakarel & House, 2004). According to Atiyas et al., (2012, p. 22), at least, it can be said that that the government did not negotiate these contracts in a rigorous way and did not use its bargaining power (Atiyas et al., 2012, p. 22). Therefore, the state took most of the commercial risk while providing private companies with significant rewards, particularly in terms of treasury guarantees to cover take-or-pay requirements throughout the life of contracts (Atiyas et al., 2012, p. 22). The contingent liabilities stemming from these take-or-pay obligations further adversely affected the promptly deteriorating public finance of the 1990s (Atiyas et al., 2012, p. 22).

As can be understood from the above, these contracts have been subject to harsh criticism that was indeed well deserved due to the financial burden they brought to the country. However, Çakmak (2011, p. 85) argues that, on the contrary, these contracts arose as necessary financing models under the conditions of the day. According to Çakmak (2011, p. 85), Turkey was a significant political and economic risk for foreign investors prior to 2000; therefore, it had to provide such treasury guarantees in order to attract foreign direct investments. In addition, Çakmak (2011, p. 85) admits that these projects did not provide the expected benefit but argues that the reason for this was the inadequate time and opportunity provided for subsequent projects. Ultimately, according to Çakmak (2011, p. 85), the period between 1984 and 2001 was a period of trial and error for Turkey; hence, Turkey incurred the costs of this trial process.

3.1 Post-2001 Era

The restructuring process of the electricity market, which began in the first half of the 1980s and was generally carried out unsystematically, entered a new phase after 2001. After this date, Turkey launched systematic liberalisation reforms to (1) increase efficiency in the electricity industry, (2) fulfil the demands of international financial institutions supporting the Turkish economy in crisis, (3) increase compliance with EU laws on the path to full EU membership and (4) attract investments to meet the growing demand for electricity (Erdogdu, 2007, p. 986; OECD, 2002, p. 8; Özkývrak, 2005, p. 1340). In its report on the Turkish electricity market, the Competition Authority defined the period until 2001 as a ‘preparation process’ and stated that Electricity Market Law (EML) No. 4628 was the turning point for liberalisation in electricity markets (Soysa et al., 2015, p. 7). First, according to Article 1 of Law No. 4628, this legislation was put into force to establish a financially sound, stable and transparent electricity market operating in a competitive environment and to found an independent regulatory authority (Electricity Market Law No. 4628, 2001). Accordingly, the Electricity Market Regulatory Authority was established as the regulator of electricity markets to oversee these markets (Atiyas et al., 2012, p. 23). Following the adoption of the Gas Market Law in 2001 and the Petroleum Market Law in 2003, the Electricity Market Regulatory Authority was extended to the natural gas and oil markets and its name was changed to the Energy Market Regulatory Authority (EMRA) (Atiyas et al., 2012, p. 23). The main functions of EMRA include implementing a licencing regime; creating secondary legislation for energy markets, including electricity, gas and oil markets; regulating transmission and distribution segments as well as retail services for non-eligible consumers; and overseeing whether market players comply with rules and imposing penalties and fines in the events of non-compliance (Atiyas et al., 2012, p. 24). In addition to the establishment of EMRA, Provisional Article 1 of EML No. 4628 provided a legal framework for the unbundling of TEAŞ (Electricity Market Law No. 4628, 2001). TEAŞ was divided into three state-owned companies: (1) the Turkish Electricity Transmission Company (TEİAŞ); (2) the Electricity Generation Company (EÜAŞ); and (3) the Turkish Electricity Trading and Contracting Company (TETAŞ). The assets of EÜAŞ and TETAŞ would potentially be privatised; however, TEİAŞ would continue to have a public monopoly over the transmission network (Atiyas et al., 2012, p. 24). TETAŞ's main responsibility was to take over whole existing energy sale and purchases contracts signed by TEDAS and TEAŞ, including BOT, Built and Operate (BO)[5] and TOR agreements as well as export and import contracts (Atiyas et al., 2012, p. 24). Moreover, TETAŞ would initially buy electricity generated by EÜAŞ under the logic that TETAŞ would balance the expensive electricity purchased from BOT, BO and TOR contracts with the relatively cheap electricity purchased from EÜAŞ, then sell electricity to TEDAŞ at a single price (Atiyas et al., 2012, p. 24).

3.2 New Turkish Electricity Market Law No. 6446

In March 2013, another vital step was taken in the process of reforming the electricity market. The New EML No. 6446 came into force on 30 March 2013, making fundamental modifications to the previous market structure. Briefly, the law brought innovations in three areas: (1) preliminary licences; (2) supply licences for retail and wholesale activities; and (3) establishing a new market operator (Tunç et al., 2019). In addition to other crucial changes in licence regimes, Article 11/1 of the new law introduced the activity of 'market operation' and defined it as the operation of organised wholesale electricity markets and activities of financial settlements conducted in these markets, together with other related financial transactions (Electricity Market Law No. 6446, 2013). To conduct this activity, according to Article 11/2 of the same law, a new private legal entity called the Energy Market Operation Corporation (EPİAŞ) was established. According to this law, EPİAŞ was responsible for conducting organised wholesale electricity markets with the exception of balancing and ancillary services markets, which would continue to be administered by TEİAŞ (see Article 3/1/y and Article 11/5 of the Electricity Market Law No. 6446, 2013).

[5] For the sake of clarity, it is appropriate to explain that BO was a new model to enable private sector to participate in the Turkish electricity market. This model was introduced by Law No. 4283 entered into force in 1997. The model was defined as a model that enables private parties to build and operate power plants while keeping the ownership of these plants. According to the Article 1 of this Law, only thermal power plants could be built under this model. Hydroelectric, geothermal, nuclear power plants and other RESs are outside of the scope of this model. (Construction and Operation of Electrical Energy Generation Facilities and Energy Sales in the Build-Operate Model Law No. 4283, 1997).

4 Marketplaces for Electricity Trading in the Turkish Wholesale Electricity Market

Markets for electricity trading can be designed beginning years in advance up to the real (actual) time at which the electricity generated in the power plant is made available to consumers (Stoft, 2002, p. 203). Therefore, it could be said that there is more than one marketplace for electricity trading established in terms of different timeframes. In this regard, Batlle (2013, p. 356) notes that 'the overall trading timetable covers a number of timescales: months or years before a trade is to be implemented; 'gate closure'; real time when the transaction takes place; and post-transaction settlement'. Keeping in mind that these categories are determined based on time, the markets where electricity can be traded can be listed as follows: (1) long-term markets; (2) day-ahead markets (DAMs); (3) intraday markets; and (4) balancing (in the EU terminology) or real-time (in the US terminology) markets (Batlle, 2013, p. 356).

4.1 Spot Marketplaces in the Turkish Electricity Market

4.1.1 Day-Ahead Market

The DAM in Turkey's electricity market was first established in 2011 (Sahin, 2018, p. 247). This market is operated by EPİAŞ and its main goal is to provide a platform that offers trading opportunities for electricity that has not yet been contracted bilaterally and unmet demand (Acar et al., 2019, p. 1300). The working principle of the DAM is based on market participants' making bids and offers for each hour of the next day, which EPİAŞ uses to draw supply and demand curves for each hour of the next day (Acar et al., 2019, p. 1300). The intersection points of these supply and demand curves form the market clearing price for that hour (Acar et al., 2019, p. 1300). In addition, the market operator is responsible for informing all market participants about the market clearing price (Acar et al., 2019, p. 1300).

According to EMRA's latest annual report, the establishment of the DAM was considered the biggest attempt towards achieving the formation of targeted electricity market' (EMRA, 2020a, p. 50). Article 4/1/z/dd of the Electricity Market Balancing and Regulation defines the DAM as '[t]he organised wholesale electricity market established for purchase and sale transactions of electricity to be delivered in the day ahead on the basis of settlement period […]' (Electricity Market Balancing and Settlement Regulation, 2009). Article 7/3 of the same regulation sets the objectives of the DAM. Accordingly, the DAM run by the market operator should be operated in accordance with the following aims:

> (a) provide market participants the opportunity to balance their generation and/or consumption needs and contractual obligation in the day ahead, (b) determine reference price for electricity energy, (c) provide System Operator a balanced system in the day ahead, (d)

> provide System Operator the ability to manage congestions in the day ahead, (e) provide market participants the opportunity to buy and sell electricity for the following day, in addition to their bilateral contracts. (Electricity Market Balancing and Settlement Regulation, 2009)

In accordance with these aims, according to EPİAŞ, the four important contributions of the DAM can be summarised as follows: (1) the DAM enabled the demand side to adjust its consumption level in terms of electricity price; hence, the demand side had the opportunity to protect itself against price fluctuations; (2) it provided market participants with the opportunity to balance their own portfolios, which 'allowed participants to present more balanced structure to the market and decrease of imbalances of generation/consumption units within their respective portfolios'; (3) it enabled financial settlement on a daily rather than monthly basis, which allowed market participants to continue their investments without any cash flow concerns; and (4) it set up a collateral mechanism whereby market participants provided guarantees for the market operator (EPİAŞ, 2016a).

4.1.2 Balancing Power Market

Helman et al. (2008, p. 180) defined a balancing market (or real-time market) as a marketplace where deviations from the schedule determined through the DAM are priced. According to Batlle, these deviations may occur if the programme in the DAM is not suitable for a power plant or the power plant for some reason does not perform as expected (Batlle, 2013, p. 357). In accordance with these approaches, Article 4/1/ş of the Electricity Market Balancing and Settlement Regulation defines a balancing power market as:

> [t]he organised wholesale electricity market, which is operated by the System Operator and where the reserve capacity, obtained by the change in output power within 15 minutes, is sold or purchased, to serve the purpose of real-time balancing of demand and supply. (Electricity Market Balancing and Settlement Regulation, 2009)

Although DAMs and intraday markets provide the system operator with a market that is as balanced as possible, there may be deviations from this balance in real time for various reasons, as discussed above (EPİAŞ, 2016c). Therefore, balancing power markets provide the system operator with the available capacity that can be activated within at most 15 min (EPİAŞ, 2016c).

4.1.3 Intraday Market

As a new step in the process of creating a more transparent, efficient and financially strong electricity market that was integrated with EU electricity markets, the intraday electricity market was established on 1 July 2015 (EMRA, 2020a, p. 61). EPİAŞ, the market operator, defines an intraday market as a bridge between the DAM and the balancing market which primarily contributes to the balance and sustainability of the electricity market (EPİAŞ, 2016a). Further, according to EPİAŞ, an '[i]ntraday market is a continuous market. Orders can be given until 60 min before the physical delivery and can be updated, cancelled or rendered inactive' (EPİAŞ, 2016b). Thus, the intraday market provides market participants with a near–real-time chance for electricity trading and the opportunity to manage their short-term portfolios (EMRA, 2020a, p. 61).

4.2 Bilateral Market

Article 3/1/j of EML No. 6446, titled 'Definitions and Abbreviations', defines bilateral agreements as the '[c]ommercial agreements between real and legal persons on trading electricity and/or capacity which are not subject to the approval of the Board [Paragraph s of the same article defines 'the Board' as the EMRA], but to the private law provisions' (Electricity Market Law No. 6446, 2013). Therefore, it can be understood from these provisions that bilateral agreements in the Turkish electricity market are not subject to regulation. According to the most recent EMRA electricity market report, the structure of the electricity market in Turkey is mainly based on bilateral agreements, and other markets are complementary to this market (EMRA, 2020a, p. 63). The aforementioned report also states that most of the current bilateral agreements were inherited from TETAŞ (EMRA, 2020a, p. 63), which is an abrogated institution according to Provisional Article 22 (added to the EML No. 6446 by the Article 9 of the Decree Law No. 703, dated 2 July 2017). According to Provisional Article 22, TETAŞ and EÜAŞ were merged into EÜAŞ. Consequently, all contracts inherited from TETAŞ began to be executed by EÜAŞ. In this context, the following two pieces of information stand out in EÜAŞ's 2019 annual report: (1) EÜAŞ received a 20-year supply licence from EMRA due to its merger with TETAŞ, and (2) EÜAŞ purchased 16.9 billion kWh of electricity within the scope of the contracts (BO-BOT-TOR) taken over from TETAŞ (EÜAŞ, 2020, pp. 26–28).

4.3 Renewable Energy Incentives in Turkey

Supporting renewable energy sources (RESs) through public incentives is, as in other jurisdictions, inevitable. RESs are strategically important for Turkey for two crucial reasons: (1) Turkey needs to increase the share of RESs in electricity consumption in order to decrease its high dependency on imported fossil-based energy resources, and (2) RESs are critical for reaching climate change objectives (Bulut & Muratoglu, 2018, p. 240). These two reasons are briefly analysed below.

One of Turkey's priorities in the electricity market has been supporting RESs in electricity generation to reduce import dependency and improve security of supply (Dilli & Nyman, 2015, p. 117). High dependency on natural gas has historically constituted a security of supply problem for Turkey. EMRA's Natural Gas Report

2019 offers a statistic that may help clarify how heavily dependent Turkey is on imported natural gas. According to this report, 1.04% of Turkey's total natural gas supply was met from domestic production while the remaining 98.96% was met from imported resources (EMRA, 2020b, p. 21). In addition, while the share of natural gas in electricity generation generally decreased from 2018 to 2019, a considerable amount of electricity in Turkey is still generated from natural gas (EMRA, 2020a, p. 20). As might be expected, this remains a factor that increases the cost of electricity generation. Further, as rightly denoted by Bulut and Muratoglu, 'energy import dependency of Turkey is one of the essential reasons of high trade deficit and current account deficit' (Bulut & Muratoglu, 2018, p. 240).

A second reason for Turkey to support RESs is, unquestionably, to deal with the risks created by climate change. Climate change is a global threat that concerns every country. For this reason, countries around the world have long been working to take necessary measures to weather the threats posed by climate change at the global level, from the Kyoto Protocol to the Paris Agreement. The latter, which came into effect on 4 November 2016, is one of the most vital steps in this regard (United Nations Climate Change, n.d.-b). The Paris Agreement was built on the United Nations Framework Convention on Climate Change (UNFCCC) (United Nations Climate Change, n.d.-a). To date, 188 of the 197 parties to the UNFCCC have ratified the Paris Agreement (United Nations Climate Change, n.d.-b).[6]

Turkey's geographical position makes it vulnerable to the risks posed by climate change (UNDP Climate Change Adaptation, n.d.). On this point, the United Nations Development Programme (UNDP) Climate Change Adaptation notes:

> As part of the southern belt of Mediterranean Europe, the country [Turkey] is already facing an observed warming trend in temperatures and a decreasing trend in precipitation. This is having a major negative effect on water availability for food production and rural development, further exacerbating the social and regional disparities in a country characterised by a wide (and widening) gap between the eastern and southeastern provinces and the rest of the country. (UNDP Climate Change Adaptation, n.d.)

Likewise, Şen notes that Turkey will be one of the countries in the Mediterranean Basin most impacted by the threats stemming from climate change (Şen, 2013, p. 5).

4.3.1 Renewable Energy Laws in Turkey[7]

In the light of the factors discussed above, since 2005, Turkey has attempted to take necessary measures to reduce its dependence on imported energy and to tackle the risks posed by climate change. It is certain that RESs are of critical importance in this regard. The government has emphasised the crucial role of RESs in realising the goals of decreasing import dependence, diversifying the generation mix and meeting rising electricity demand in various legislation and policy papers, including EML No.

[6] It should be noted here that Turkey has not yet ratified the Paris Agreement.

[7] It is beyond the scope of this paper to examine the renewable energy incentive system in detail. Therefore, only a general assessment is provided here.

6446 and the 2009 Electricity Market and Security of Supply Strategy Paper (IEA, 2016a, p. 168). The definitive moment for RESs in Turkey began with Renewable Energy Law (REL) No. 5346, which has become the basic legal framework for RESs in Turkey. This Law was then comprehensively modified by the Law Amending the Utilisation of RESs in Electricity Generation (Law No. 6094), which took effect in 2011 (IEA, 2016a, p. 172). With Law No. 6094, a technology-specific feed-in tariff scheme (Renewable Energy Resources Support Mechanism [YEKDEM]) was established to provide 10 years of financial support for RESs commissioned before 31 December 2020[8] (IEA, 2016a, p. 172). In addition, local content support was provided limited to five years (IEA, 2016a, p. 172). These developments in the legal infrastructure had, as can be expected, a positive effect on investments in RES in Turkey. According to the IEA's report on Turkey, although the first legal framework for RESs was put in place in 2005, investment in RESs could only begin to increase with the introduction of YEKDEM, a technology-specific and long-term incentive mechanism, in 2011 with Law No. 6094 (IEA, 2016a, p. 170).

4.3.2 New Era for YEKDEM

The era of change in RES incentive mechanisms has already begun in many countries around the world. For instance, Dena, the German Energy Agency, notes with regard to this issue:

> The landscape of today's renewable energy (RE) market has changed dramatically over the last decade because of the rising demand for RE and steep decline of solar and wind cost. As the RE market has evolved, so have the financing mechanisms associated with it. (dena - German Energy Agency, 2019, p. 7)

Many jurisdictions are taking or preparing to take necessary steps to this end. However, this does not mean that RES investments will be left to unpredictable investment climates without adequate incentives. These technologies, as might be guessed, will continue to be supported by different and contemporary incentive models. For instance, China has decided to switch from the highly regulated feed-in tariff system to a new feed-in tariff mechanism wherein the price for each RES project is determined by auction (dena—German Energy Agency, 2019, p. 9). European Commission (EC) stated as early as 2013 that feed-in tariff mechanisms should be phased out and replaced with new types of incentive mechanisms that would force producers to adapt to market prices (European Commission, 2013, p. 15). In the light of other experiences around the world and considering its own conditions, Turkey has also had to update its incentive mechanism for RESs for the period after 2020. In this regard, it can be said that a new era has been started for YEKDEM with a law enacted at the end of 2020. The Law No. 7257 on amending the EML No 6446 and some other laws published in the Official Gazette (numbered 31322) on 2 December 2020 (Amendments to the Electricity Market Law and Some Other Laws No. 7257, 2020).

---

[8] This deadline was extended to 30 June 2021 by a presidential decree published on 18 September 2020 (Presidential Decree No. 2949, 2020).

Article 6 of the REL No. 5346 regulating YEKDEM was amended by Article 13 of the Law No. 7257. According to this change, incentive amounts for RESs, which will come into operation after 30 June 2021, will be determined in Turkish Lira. Also, it was decided that procedures and principles regarding YEKDEM and price updates will be determined by the President of the Republic of Turkey.[9] In addition to above, Article 6/B of the REL No. 5346 regulating local content support was amended by Article 15 of the Law 7257. According to this change, local content support determined in Turkish Lira will continue for RESs that will enter into operation after 30 June 2021. The amount and updates of local content support, the period to be applied and other procedures and principles will also be determined by the President of Republic of Turkey. How these critical changes will affect renewable energy investments in the upcoming period remains a vital research topic.[10] However, since it is out of the scope of this chapter, more discussions on this matter are not covered here.

4.4 *What Are CRMs and Why Might Liberalised Electricity Markets Need CRMs?*

CRMs are regulatory tools developed in liberalised electricity markets[11] to ensure generation adequacy, which can be defined as 't]he ability of resources including both supply and demand-side resources to meet total demand in the long term' (Şahin, 2018, p. 55). Generation adequacy is in actuality a sub-concept of reliability, which is an 'umbrella term that covers different dimensions from short to long term and different segments from transmission to generation' (Şahin, 2018, p. 48). CRMs are established to ensure generation adequacy as their general purpose. These markets have become increasingly popular among EU countries due to increasing concerns regarding generation adequacy. Considering that EC has made a total of 19 decisions since 2014 on whether CRMs established in various EU countries are in compliance with the EU State Aid Rules, one can form an idea of how quickly these markets have been established among member states (European Commission, n.d.). Due to several market imperfections—including the missing money problem, market power and entry barriers, the boom and bust cycle problem, lack of long-term contracts and inelastic demand structure—liberalised electricity markets cannot provide adequate incentives for market participants to ensure generation adequacy (Şahin, 2020, p. 95). In addition to these reasons, electricity markets have a political dimension, which

[9] In this regard, the Presidential Decree No. 3453 was published on 30 January 2021 regarding the incentive payments in Turkish Lira to be made for RESs to come into operation from 1 July 2021 to 31 December 2025 (Presidential Decree No. 3453, 2021).

[10] For a brief analysis concerning this issue, see (Kıl, 2021).

[11] In the literature on CRMs, the term 'energy-only market' is more common than the term 'liberalised electricity market'. The term 'liberalised electricity market' is used here rather than 'energy-only market' to out of a desire to maintain internal consistency within the paper.

means that it is very hard for politicians to remain indifferent to a sector such as the electricity market where almost everyone with voting rights is a consumer. On this matter, Rodilla and Batlle remark:

> [P]oliticians' risk aversion is by far larger than that of almost any power consumer. Regulated rates preclude the need for protection against high prices and even consumers initially exposed to spot market prices ignore reliability when making their decisions. There is a certain implicit assurance that leads consumers to believe that the regulator will never allow supply shortfalls or inordinately high prices that would jeopardise their interests. (Rodilla & Batlle, 2012, p. 183)

4.4.1 CRM in Turkey

For the abovementioned reasons, CRMs[12] have been established in many European countries in the last decade. Turkey also established a CRM in its electricity market with the Electricity Market Capacity Mechanism Regulation published in the Official Gazette dated 26 January 2018 and numbered 30,307 (repeated) (Electricity Market Capacity Mechanism Regulation, 2018). It should be clearly stated that the CRM in Turkey seems inadequate and structurally archaic, which can be justified with two basic principles.

Demand-Side Resources Are Not Allowed.

According to Article 6 of the Electricity Market Capacity Mechanism Regulation of Turkey, only supply-side resources (existing or new) can benefit from capacity payments for the subsequent calendar year. Put differently, only existing or new supply-side capacity resources can participate in the CRM in Turkey. However, the importance of demand-side resources is gradually increasing, especially in terms of providing security of supply. Moreover, it is critical that the demand side takes an active role in today's electricity markets, where the effects of climate change are increasingly felt. As can be seen in Fig. 1, demand-side resources can be involved in many areas of electricity markets, including CRMs and operating reserves.

It is obvious that CRMs can also play an important role in promoting demand-side resources. In fact, the following statement was made in an IEA report: 'Capacity markets have been more effective at developing demand response than energy markets' (IEA, 2016b, p. 160). In this sense, some examples in the USA indicate that demand-side resources can be reliable and cost competitive, as much so as generation capacity (Dimsdale et al., 2015, p. 1). According to the PJM Demand Response Monthly Activity Reports cited by Dimsdale et al., '[d]ata from the US ISOs suggests that the net availability of demand response is actually in the high 90% which represents far superior performance than that provided by generation' (Dimsdale et al.,

[12] There are many CRM designs, including capacity payments, capacity markets, capacity obligations, reliability options and strategic reserves. Since it is far beyond the scope of this paper to individually examine these types of CRMs, such an analysis will not be undertaken here. For a further discussion of the types, working principles, pros and cons of each of these CRMs, see (Şahin, 2018).

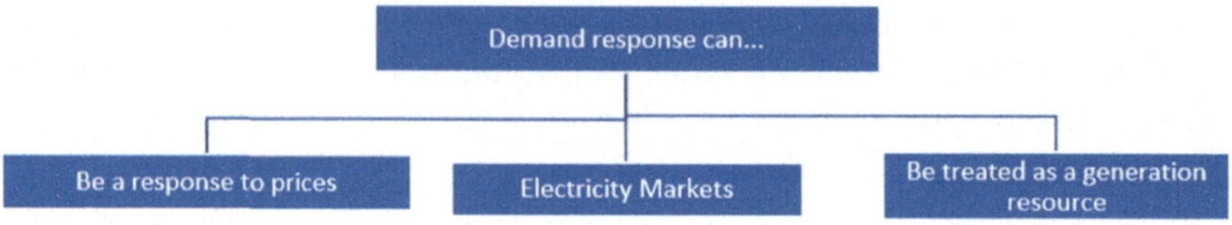

Reduced capacity requirements	Capacity	Participation as a resource in capacity markets
Time-of-use/dynamic pricing	Energy	Participation as a resource in energy markets
	Operating reserve	Reserve/ancillary services procurement open to demand

Fig. 1 Approaches to demand response. *Source* Adapted from IEA (2016b, p. 159)

2015, p. 5). Rationally, necessary legal measures have been taken to ensure that CRMs established in EU Member States take demand-side resources into account. For instance, according to Article 3.9.3 (Paragraph 226) of the Guidelines on State aid for environmental protection and energy 2014–2020 (the 'Guidelines'):

> The measure should be open and provide adequate incentives to both existing and future generators and to operators using substitutable technologies, such as demand-side response or storage solutions. The aid should therefore be delivered through a mechanism which allows for potentially different lead times, corresponding to the time needed to realise new investments by new generators using different technologies. The measure should also take into account to what extent interconnection capacity could remedy any possible problem of generation adequacy. (European Commission, 2014)

These relevant provisions in the Guidelines are vital, as they are used as criteria for determining whether CRMs in member states comply with EU State Aid Rules. For this reason, in the light of the aforementioned provision, it is a legal obligation for Member States to approach demand-side resources, including energy efficiency, under equal conditions with supply-side resources in their CRMs. Considering these examples, Turkey should reorganise its CRM so as to enable demand-side resources, including energy efficiency, to benefit from its capacity mechanism incentives.

Insufficient Lead Time Is Offered to Encourage New Investments.

Article 6 of the Capacity Mechanism Regulation of Turkey states that applications to benefit from capacity mechanisms are made for the next year. In these applications, the regulation does not recognise any difference between existing generators and new investors. The only difference between existing and new generators put forward by the regulation is that existing generators who want to benefit from the capacity mechanism in the following calendar year must fill out the candidate form and the generators whose provisional acceptance will be made in the next calendar year must fill out the candidate application form. However, it is obvious that adequate lead time should be offered in order to attract new generation investments. This issue is also raised by Article 3.9.3 (Paragraph 226) of the Guidelines mentioned above.

5 Conclusion

This chapter aimed to provide a general discussion of the regulation of wholesale electricity trading in Turkey. It began with a brief overview of why regulation is a necessary element of liberalised electricity markets. After this introduction, the historical background of reform efforts in the Turkish electricity market beginning in the mid-1980s was succinctly discussed. The chapter then discussed EML No. 4628, enacted in 2001, which was the turning point for the liberalisation of the electricity markets. In this context, the new institutional structures brought to the market by Law No. 4628 were examined. This review was followed by an examination of the New EML No. 6446, which came into force in 2013 and repealed Law No. 4628.

Following these discussions, it is clear that Turkey has achieved significant progress since the reform efforts that began nearly 40 years ago. A comprehensive legal and institutional framework has emerged, particularly as a result of the systematic liberalisation efforts in the last 20 years. The establishment of EPİAŞ for electricity trading was a critical success in this sense. In addition, it should be noted as an important success that marketplaces, including DAMs and intraday markets, began to function fully.

At the end of the article, there was a discussion of RES incentives and CRMs. Thanks to the generous and well-designed incentive mechanisms, Turkey has achieved significant gains in renewable energy investments, especially in the last decade. As discussed in this paper, RESs are strategically important for Turkey to reduce its dependence on imported energy and to deal with climate change. Therefore, it is essential that Turkey develop its post-2020 incentive mechanism for RESs. On the other hand, the CRM established in 2018 in Turkey, as noted in this article, is outdated and has a non-functional structure. First, this CRM must be redesigned to ensure that it is open to demand-side resources, including energy efficiency. Second, the CRM in Turkey does not take into account lead times of new power plants to encourage them to benefit from capacity mechanism. This is also a structural problem that should be solved. Undoubtedly, it is important to provide incentives so that existing power plants do not leave the market. It is also reasonable that the CRM should have such a purpose for the sake of generation adequacy. Indeed, fossil-based power plants, especially gas power plants, will be of strategic importance in dealing with the increased intermittency due to the flexibility they provide. Saygın made the following remarkable assessment in this regard:

> The share of renewable energy resources in Turkey's total electricity demand will increase, particularly, solar and wind, provided that planned investments are realised. Moreover, the share could even be higher if the demand for electricity continues to decline. This rapid transformation means that the power plants that operate on natural gas and imported coal may fail to have a business case in the marketplace. This puts the importance of having a flexible power system to manage the intermittent characteristics of wind and solar energy to the forefront, as this allows one to balance demand and supply. Thus far, Turkey has benefitted from the pliability that its current system offers, but, as the share of solar and wind power goes beyond the ten per cent threshold, new market design and flexibility technologies will be needed. (Saygın, 2020)

In the light of this seminal evaluation, it is clear that power plants that provide flexibility will increasingly be necessary.[13] It is also clear that not only existing fossil-based power plants, but also new power plants built with novel, more efficient technologies (and, of course, demand-side resources) can also provide flexibility, perhaps at a cheaper rate.

References

Acar, B., Selcuk, O., & Dastan, S. A. (2019). The merit order effect of wind and river type hydroelectricity generation on Turkish electricity prices. *Energy Policy, 132*(September 2018), 1298–1319.

Amendments to the Electricity Market Law and Some Other Laws No. 7257. (2020).

Atiyas, I., & Dutz, M. (2004). *Competition and regulatory reform in the Turkish electricity industry.* Retrieved August 18, 2020, from http://myweb.sabanciuniv.edu/izak/files/2008/10/atiyas-dutz-electricity-2004.pdf

Atiyas, I., Çetin, T., & Gülen, G. (2012). *Reforming Turkish energy markets: Political economy, regulation and competition in the search for energy policy.* Springer.

Batlle, C. (2013). Electricity generation and wholesale markets. In I. J. Perez-Arriaga (Ed.), *Regulation of the power sector* (pp. 341–396). Springer.

Bulut, U., & Muratoglu, G. (2018). Renewable energy in Turkey: Great potential, low but increasing utilization, and an empirical analysis on renewable energy-growth nexus. *Energy Policy, 123*(May), 240–250.

Cakarel, E., & House, J. (2004). *IPP investment in Turkey's electric power industry.*

Çakmak, Z. (2011). *Bağımsız İdari Otorite Olarak Enerji Piyasası Düzenleme Kurumu (in Turkish)* (1st ed.). Çakmak Yayınevi.

Cetin, T., & Oguz, F. (2007). The politics of regulation in the Turkish electricity market. *Energy Policy, 35*(3), 1761–1770.

Clair, E. (2020). *Looking at Covid-19 crisis from the EU electricity wholesale market.* Retrieved August 28, 2020, from https://fsr.eui.eu/looking-at-covid-19-crisis-from-the-eu-electricity-wholesale-market/

Coles, J. (2014). Season 2, Chapter 19. *United States of America: Netflix.* https://www.netflix.com/watch/70293584?trackId=200257859

Construction and operation of electrical energy generation facilities and energy sales in the build-operate model Law No. 4283. (1997).

DDK. (2010). *The State Supervisory Council (DDK) Report No. 2010/10.*

Dena—German Energy Agency. (2019). *How to use PPAs for cost-efficient extension of renewable energies—Experiences with Power Purchase Agreements from Europe and the U.S./Lessons learned for China.* https://www.dena.de/fileadmin/dena/Publikationen/PDFs/2019/dena-REPORT_How_to_use_PPAs_for_cost-efficient_extension_of_re.pdf

Dilli, B., & Nyman, K. (2015). *Turkey's Energy Transition: Milestones and Challenges.* Retrieved August 6, 2020, from http://documents.worldbank.org/curated/en/249831468189270397/Turkey-s-energy-transition-milestones-and-challenges

Dimsdale, T., Skillings, S., & Dufour, M. (2015). *Harnessing demand side resources in electricity markets: evidence from the United States.* Retrieved August 26, 2020, from https://www.jstor.org/stable/resrep17833?seq=1#metadata_info_tab_contents

Electricity Market Capacity Mechanism Regulation. (2018).

[13] For a further discussion of why conventional power plants are vital elements of energy transitions due to their flexible nature and why CRMs are critically important in this process, see (Şahin, 2020).

Electricity Market Law No. 4628, Pub. L. No. 4628. (2001).
Electricity Market Balancing and Settlement Regulation. (2009). *Turkey*.
Electricity Market Law No. 6446. (2013).
EMRA. (2020a). *Electricity Market Development Report 2019. Energy Market Regulatory Authority*. https://www.epdk.gov.tr/Detay/Icerik/3-0-24/elektrikyillik-sektor-raporu
EMRA. (2020b). *Turkish Natural Gas Market Report 2019. Energy Market Regulatory Authority*. Ankara. https://www.epdk.gov.tr/Detay/Icerik/3-0-94/dogal-gazyillik-sektor-raporu
EPİAŞ. (2016b). *Spot electricity market: intraday market: phases*. Retrieved August 22, 2020, from https://www.epias.com.tr/en/intra-day-market/phases/
EPİAŞ. (2016a). *Spot electricity market: intraday market: introduction*. Retrieved August 22, 2020, from https://www.epias.com.tr/en/intra-day-market/introduction/
EPİAŞ. (2016c). *Spot Elektrik Piyasası: Dengeleme Güç Piyasası: Genel Esaslar*. Retrieved August 22, 2020, from https://www.epias.com.tr/dengeleme-guc-piyasasi/genel-esaslar/
Erdogdu, E. (2007). Regulatory reform in Turkish energy industry: An analysis. *Energy Policy, 35*(2), 984–993.
EÜAŞ. (2020). *Elektrik Üretimi ve Ticareti Sektörü Raporu*. https://www.euas.gov.tr/tr-TR/sektor-raporu
European Commission. (n.d.). (2020). *Energy and environment: State aid to secure electricity supplies: individual state aid cases*. Retrieved August 25, 2020, from https://ec.europa.eu/competition/sectors/energy/state_aid_to_secure_electricity_supply_en.html
European Commission. (2013). *Communication from the Commission: Delivering the internal electricity market and making the most of public intervention,*. https://ec.europa.eu/energy/sites/ener/files/documents/com_2013_public_intervention_en.pdf
European Commission. (2014). *Guidelines on state aid for environmental protection and energy 2014–2020, C 200/1 Official Journal of the European Union §*. https://doi.org/10.1016/j.nucengdes.2011.01.052
Gürkaynak, R. (2020). *Covid-19 Sağlık Krizinin İktisadi Etkileri: Dünya ve Türkiye (in Turkish)*. https://www.youtube.com/watch?v=NBZI76m4NfA
Helman, U., Hobbs, B. F., & O'neill, R. P. (2008). The design of US wholesale energy and ancillary service auction markets: Theory and practice. In Fereidoon P. Sioshansi (Ed.), *Competitive electricity markets: Design, implementation, performance* (First Edit). UK: Elsevier.
Hunt, S. (2002). *Making competition work in electricity*. Wiley.
IEA. (2016a). *Energy policies of IEA countries: Turkey 2016 Review*.
IEA. (2016b). *Re-powering markets: Market design and regulation during the transition to low-carbon power systems*. Paris.
IEA. (2020). *Sustainable recovery: World energy outlook special report*.
Kıl, İ. F. (2021). Turkey: A brief analysis of the 2020 amendments of the Turkish energy acts. https://www.mondaq.com/turkey/renewables/1021288/a-brief-analysis-of-the-2020-amendments-of-the-turkish-energy-acts
Mäntysaari, P. (2015). *EU electricity trade law: The legal tools of electricity producers in the internal electricity market*. Springer.
OECD. (2002). *OECD reviews of regulatory reform regulatory reform in Turkey: regulatory reform in electricity, gas and road freight transport*. Retrieved August 18, 2020, from https://www.oecd.org/turkey/1840779.pdf
Özkývrak, Ö. (2005). Electricity restructuring in Turkey. *Energy Policy, 33*(10), 1339–1350.
Presidential Decree No. 2949. (2020).
Presidential Decree No. 3453. (2021).
Raineri, R. (2006). Chile: Where it all started. In F. P. Sioshansi & W. Pfaffenberger (Eds.), *Electricity market reform: An international perspective* (First Edit, pp. 77–108).
Rodilla, P., & Batlle, C. (2012). Security of electricity supply at the generation level: Problem analysis. *Energy Policy, 40*, 177–185.
Rodrik, D. (2011). *The globalization Paradox: Ghy global markets, states, and democracy can't coexist* (p. 2011). Oxford University Press.

Sahin, C. (2018). Consideration of network constraints in the Turkish day ahead electricity market. *International Journal of Electrical Power and Energy Systems, 102*(May), 245–253.
Şahin, T. (2018). *Capacity remuneration mechanisms in energy-only markets : in pursuit of creating a regulatory framework for the integration of capacity remuneration mechanisms in the european union electricity markets*. University of Dundee.
Şahin, T. (2020). Capacity remuneration mechanisms: regulatory tools for sustaining thermal power plants and the EU energy transition. In G. Wood & K. Baker (Eds.), *The Palgrave Handbook of Managing Fossil Fuels and Energy Transitions* (First Edit, pp. 85–105). Palgrave Macmillan.
Saygın, D. (2020). Turkey's renewables sector in light of COVID-19. Retrieved August 28, 2020, from https://www.atlanticcouncil.org/blogs/turkeysource/turkeys-renewables-sector-in-light-of-covid-19/
Şen, Ö. L. (2013). *A holistic view of climate change and its impacts in Turkey*. https://ipc.sabanciuniv.edu/Content/Images/CKeditorImages/20200327-01035961.pdf
Soysal, C., Pektaş, M., Şahin, S. Y., Tokgöz, E., & Erek, H. (2015). *Elektrik Toptan Satış ve Perakende Satış Sektör Araştırması (in Turkish)*.
Stoft, S. (2002). *Power system economics: Designing markets for electricity*. IEEE Press & WILEY-INTERSCIENCE A JOHN WILEY & SONS, INC.
Tunç, Z., Altunyuva Kehale, A., & Taşçı, T. (2019). *Electricity regulation in Turkey: Overview*. Retrieved August 20, 2020, from https://uk.practicallaw.thomsonreuters.com/0-523-5654?transitionType=Default&contextData=(sc.Default)&firstPage=true
UNDP Climate Change Adaptation. (n.d.). (2020). *Turkey*. Retrieved August 5, 2020, from https://www.adaptation-undp.org/explore/europe-and-central-asia/turkey
United Nations Climate Change. (n.d.-a). (2020). *Paris agreement: essential elements*. Retrieved August 6, 2020, from https://unfccc.int/process-and-meetings/the-paris-agreement/the-paris-agreement
United Nations Climate Change. (n.d.-b). (2020). *Paris agreement—status of ratification*. Retrieved November 12, 2020, from https://unfccc.int/process/the-paris-agreement/status-of-ratification

Taner Şahin is an assistant professor at Law School of Ondokuz Mayıs University. He holds a Business Administration degree from Marmara University, an LLM and a Ph.D. from the Centre for Energy, Petroleum and Mineral Law and Policy (CEPMLP), University of Dundee. His research interests focus on Energy Law and Policy. He published two book chapters on Capacity Remuneration Mechanisms and participated several international conferences to present his works. Since 2020, he has also been head of the Commercial Law Department at Law School of Ondokuz Mayıs University.

Regulation of Trade and Supply Contracts in the Turkish Electricity Market

Eylem Apaydin and Muzaffer Eroğlu

Abstract The early 1980s saw a drastic transformation in the economic model choice in Turkey. Prime Minister Özal declared that the mixed economic system would be replaced by a liberal economic system that was subject to the rules of free market (Duman in Sosyoloji Dergisi 23–24:105–123, 2011). The electricity market eventually benefited from this transformation. After some failed attempts in the 1980s and 1990s, the privatization of the state companies, the introduction of the Electricity Market Law (EML) in 2001, and the foundation of the Energy Market Regulatory Authority (EMRA) led to a liberal market for electricity in Turkey. The EML and EMRA promote and support the full liberalization of the electricity market. Apparently, the end users (the EML prefers the term "consumer") play important roles in expected fully liberated market. The contracts concluded by the end users have been privileged and regulated differently from the other contracts in the market. The EML regulates two different types of contracts for electricity supply to consumers in the electricity market: retail sale contracts and bilateral agreements. While retail sale contracts are concluded between the ineligible consumers and the supply company, the eligible consumers are entitled, but not obliged, to conclude a bilateral agreement with a supply company. The Electricity Market Consumer Services Regulation, issued by the EMRA, is intended to protect these consumers in the market. This chapter starts by examining the types of the contracts regulated by the EML and the Regulation in general. Secondly, the laws that are applicable to contracts for electricity supply and conflict resolutions are indicated, while noting that these are subject to private law and civil court jurisdiction. Thirdly, legal character of character and freedom of contract regarding these contracts are analyzed. Later, the content of contracts is investigated and the rights and obligations under the contracts for electricity supply to consumers are discussed. We then explain how the contracts for electricity supply are terminated is explained in accordance with the EML and

---

E. Apaydin (✉) · M. Eroğlu
Boğaziçi University, Faculty of Law, Istanbul, Turkey
e-mail: eyapaydin@yahoo.com

M. Eroğlu
e-mail: muzaffer.eroglu@boun.edu.tr

M. Eroğlu and M. Finger (eds.), *The Regulation of Turkish Network Industries*,
https://doi.org/10.1007/978-3-030-81720-6_4

the Regulation. We finish the chapter by examining competition in supply market, showing the increasing competition in the electricity market in Turkey.

Keywords Electricity trade · Supply contracts · Liberalization · Freedom of contract

1 Introduction

Over the 150-year history of the utilization of electricity, electricity markets have operated under different methods. The method of delivering electricity to end users also differed according to the variable economic factors over time in the world. At first, electricity was provided in accordance with the concession-based operating generation companies, which sold electricity to end users with private law contracts. However, electricity was commonly provided by the states to end users during most of the twentieth century, especially after World War One. However, it was still provided by private companies in some countries. In recent years, as a result of the collapse of socialist economic models and recent dominance of neo-liberal policies in market-based economies, the supply of electricity by private actors has become an established political and regulatory model. The important point in current political choices is the desire to create a competitive electricity generation and supply industry.

The utilization of electricity in Turkey within its hundred-plus-year history shows a similar characteristic to Western economies. Electricity generation and supply, which was first given as a concession to private-sector companies during the Ottoman Empire, was turned into public service provided by state institutions after the establishment of the modern Republic in 1923. Since the 1980s, with efforts to transition the Turkish economy to the liberal economic model, some steps have been taken to run the electricity markets in this model. The first initiations of privatization were generally related to the generation stage rather than opening supply contracts with end users to private companies.

Today, neo-liberal policies continue to determine the basic operating model in electricity markets. As the basis of these policies, while the role of the state in the electricity market is limited to regulation and supervision, it is essential that all other commercial activities are carried out by private-sector actors or by state-owned corporations. According to the political preferences established in the liberal economy, the energy companies owned by the state are first transformed into corporations and they operate under the same status as other privately held companies in the market. As a result, the state privatized many companies and ceased to be an active player in most portions of the market. The state, in neo-liberal terms, assumes the role of "steering" rather than "rowing" in the electricity market.

On the other hand, in accordance with the political preferences of recent years, it is preferred to leave the regulation and supervision to the independent administrative authorities, which are called the regulatory institutions, instead of to the control of the central government. These independent authorities from the central government are fulfilled with all the powers assigned by the law, especially the regulatory and supervisory powers. The main objectives of these authorities are to ensure the functioning of the market, which is planned to be fully liberalized in a competitive structure, and thus to provide public benefit through effective energy markets.

After some failed attempts in the 1980s and 1990s, Turkey successfully established the legal infrastructure for the transition to a liberal market by introducing the Electricity Market Law (EML) in 2001 and founding the Energy Market Regulatory Authority (EMRA). This law basically foresees a liberal and privatized market transition and adopts the opening of the market to competition, except for natural monopolies such as transmission and distribution. Since the introduction of the EML, a slow but important process has been completed and most of the electricity business is controlled by private-sector legal entities. As a result, it appears that the state's generation capacity is decreasing constantly and the state companies have almost no role in supplying electricity to end users. It is accepted that the main principle is that activities carried out by the state companies are also carried out under private law provisions with other companies; therefore, it can be said that the liberalization process of the market in Turkey is almost accomplished even though the privatizations are not yet fully completed.

The fact that the main purpose of the liberalization process is to create competitive markets at all levels leads the liberalization process for supplying electricity to end users. Therefore, the diversification on the demand side is also important, as is the diversification in generation and supply. In parallel with the liberalization of the supply market, the consumption side has also been opened to competition. At this stage, the choice is not fully provided on the consumption side, and only eligible consumers are free to choose their suppliers. However, whether they are eligible or ineligible, private companies provide electricity to end users with contracts that are subject to private law provisions.

Today, the main instrument in Turkish law that regulates the types of the private contracts in electricity market is the Electricity Market Act No. 6446 dated March 14, 2013. The secondary instrument on the matter is the Electricity Market Consumer Services Regulation issued by the EMRA ("the Regulation"). Its objective is to set forth the standards, procedures, and principles that will be applied for the end users that are connected to or that seek connection to the distribution and transmission system and the parties that provide service to these customers in accordance with the connection agreements, retail agreements, or bilateral agreements.

This chapter explains the contracts for electricity supply to end users. We should note at this stage that the EML and the Regulation define end users as consumers regardless of their legal personality such as real persons, companies, government bodies, or other institutions. Accordingly, in this chapter, the terms consumer and end user have the same meaning, unless otherwise specified.

2 Types of Contracts for Electricity Supply to Consumers in Electricity Markets

The EML regulates two different types of contracts for electricity supply to consumers in electricity markets: retail sale contracts and bilateral agreements. While retail sale contracts are concluded between the ineligible consumers and the supply company, the eligible consumers are entitled, but not obliged, to conclude a bilateral agreement with a supply company.

The Act defines retail sale as the sale of electricity to consumers (Art. 3/aa) and the consumer as a person purchasing electricity for his/her own purposes (Art. 3/mm). The Act defines retail sale, but it does not mention the retail sale contracts. Art. 4/1 (u) of the Regulation states that retail sale contracts cover the conditions and provisions relating to commercial activities for supply of electricity and/or capacity and obtainment of services between licensed supply company (production and supply companies) and customers in accordance with the provisions of Electricity Market Tariffs Regulation. Meanwhile, the Act states that bilateral agreements are commercial agreements between real and legal persons on trading electricity and/or capacity that are not subject to the approval of the Board, but to private law provisions.

On the other hand, eligible consumers are real or legal persons who are entitled to select their supplier as their yearly consumption amount is above the electrical power amount designated by the Board, or as they are directly connected to the transmission system, or as they are organized as an industrial zone legal entity (Art. 3/cc). Hence, a consumer who cannot be qualified as the eligible consumer since their consumption is below the limits determined by the Energy Market Regulatory Board (the Board) would be classified as an ineligible consumer. The Regulation defines an eligible consumer in parallel with the Act, as a real person or legal entity in a region that can purchase electricity and/or capacity only from distribution companies holding retail license or retail companies of that region (Art. 4/cc).

Art. 6 of the Regulation defines eligible consumers as those who meet the following conditions: (a) real or legal persons directly connected to the transmission system, (b) organized industrial zone legal entities, (c) consumers who have the right to use or acquire the right to use the place where electricity was consumed in excess of the eligible consumer limit in the previous calendar year or the current year. The Energy Market Regulatory Board (the Board) announced that the yearly consumption limit for the eligible consumer is determined as greater than 1400 kWh for 2020.[1] Hence, real or legal persons directly connected to the transmission system, organized industrial zone legal entities, or any real or legal person that consumes more than 1400 kWh of electricity in a year will qualify as an eligible consumer and be entitled to conclude an eligible bilateral agreement with a supplier. Being an eligible consumer has important advantages. For instance, in accordance with EML Art. 10/2, only eligible consumers may conclude bilateral agreement contract with a supply company without any area limitation. An eligible consumer may bargain

[1] https://www.epdk.gov.tr/Detay/Icerik/16/serbest-tuketici (accessed on October 15, 2020).

regarding the price, terms, and conditions of the electricity supply contracts with the supplier.

In accordance with the EML Art. 3/1 (j), the contract concluded between the eligible consumer and supply company is classified as a bilateral agreement and is only subjected to the private law rules and principles. They are exempted from the EMRA's approval. With this contract, the party can freely agree on the unit price of electrical energy and the rights and obligations arising from the contract (Özel et al., 2013, p. 2098).

Here, it is worth noting a concept regulated by the EML—the supply of last resort—which is defined in Art. 3/1 (ee) as the supply of electricity to eligible consumers who do not obtain electrical energy from a supplier other than the supplier license holder and authorized company as the last resort supplier. According to the Art. 8 of the Regulation, consumers with the following conditions are considered to be consumers in the scope of last resort supply: (a) a consumer who, despite having the qualification of an eligible consumer, does not procure electrical energy from a supplier other than the company that has the supply license, which is authorized as the supplier of the last source; or (b) a consumer who purchases electricity from the related supplier company due to the termination of the bilateral agreement of an eligible consumer who purchases electrical energy and/or capacity with a bilateral agreement. These consumers are originally eligible consumers, who are not currently in a bilateral agreement relationship with another supplier. Hence, these consumers would apply to the supply company that held the last resort license at the area. In that case, the responsible last resort supplier company is obliged to provide electrical energy and/or capacity to consumers within the scope of this article of last resort supply. Accordingly, each of 21 regions in Turkey has one determined last resort supply company, all of which are under the same ownership structure (but separate management) with the distribution company in the region.

According to Art. 7 of the Regulation, ineligible consumers are defined as: (a) consumers whose electricity consumption does not exceed the eligible consumer limit in the previous calendar year and in the current year, at the place of use for which they are entitled to use or acquired the right to use; (b) consumers who are eligible consumers and whose total electrical energy consumption for the previous calendar year is below the eligible consumer limit determined for the current calendar year; (c) consumers consuming electrical energy at the place of use with a connection agreement for the first time. Any real or legal person who consumes less than 1400 kWh (2020 limit) of electricity in a year will be classified as an ineligible consumer and would not be entitled to conclude a bilateral agreement with a supplier.

Ineligible consumers can only procure electrical energy and/or capacity by making a retail sales contract with the determined last resort supplier company in the distribution region. Art. 17/6 (d) of the EML authorizes the Board to regulate the retail sale tariffs. In respect of ineligible consumers, the tariffs shall include applicable prices, terms, and conditions without discrimination among equal parties. Hence, ineligible consumers do not hold bargaining power to discuss the terms of the retail sale contract regarding the supplier, prices, terms, and conditions. They all are determined by the tariff issued by the Board. It is apparent that there is no real freedom of contract regarding the retail sale contracts and ineligible consumers. The tariff is determined at national level, which makes its electricity cost the same for all ineligible consumers and all last resort consumers in Turkey. Even though national tariffs were supposed to be temporary (planned to end in 2015), the last date is repeatedly extended, most recently to 2025 by the latest amendments to EML.

The main difference between eligible and ineligible consumers is whether the consumer has the right to choose the supplier. While an eligible consumer has the right to choose the supplier, an ineligible consumer must purchase electricity from the last resort retail license holding company in the area (Yavuz, 2011). Accordingly, retail sale contracts with ineligible consumers, contracts with eligible consumers who still purchase electricity from last resort suppliers, and bilateral agreements between supply companies and eligible consumers are regulated by the EML and by The Electricity Market Consumer Services Regulation.

The Regulation divides eligible consumers into two categories: those with low consumption (annual electricity consumption below 100,000 kWh) and those with high consumption (those above that threshold). The reason behind this classification is to protect low-consumption consumers in their bilateral agreements with supply companies. Thus, the regulation introduces many protective measures regarding free contracts made between consumers with low consumption and supply companies. Accordingly, consumers with high consumption should be more cautious as the Regulation has fewer protective measures for them. On the other hand, a bilateral agreement between supply companies and eligible consumer with high consumption are more freely negotiated, as it is treated by regulation as a contract between parties with equal negotiating power. Thus, they might carry more creative measures.

This chapter mainly deals with the bilateral agreements made by eligible consumers with low consumption.

3 Applicable Laws to Contracts for Electricity Supply and Conflict Resolution

Recent years have seen an increase in conflicts among parties in bilateral supply agreements and retail sale agreements. This is due to changes in which laws are applicable law to these contracts, from administrative law to private law. Currently, the relationship between the parties, whether established by a bilateral agreement or retail sale contract or last resort supply, is a private law contractual relationship. Hence, Turkish private law governs these relationships and private law provisions will apply to all electricity supply contracts, regardless of their type. When a dispute arises between the parties, it concerns the law of obligations, commercial law, and consumer law in Turkish law and the dispute will be resolved in a civil court.

The general rules of the Turkish Code of Obligations apply to contracts for electricity supply as they apply to all private law relationships. For instance, the conclusion of a contract requires a mutual expression of intent by the parties (Art. 1).

When assessing the form and terms of a contract, the true and common intention of the parties must be ascertained without dwelling on any inexact expressions or designations they may have used, either in error or by way of disguising the true nature of the agreement (Art. 19). The contracts for electricity supply are subject to review of court under the provisions of general terms and conditions (Art. 20–25). A contract for electricity supply is void if its terms are impossible, unlawful, or immoral (Art. 27). Where there is a clear discrepancy between performance and consideration under a contract concluded as a result of one party's exploitation of the other's straitened circumstances, inexperience, or thoughtlessness, the injured party may declare within one year that he or she will not honor the contract and demand restitution of any performance already made (Art. 28). If one of the parties will is in error, or deceived or under duress, he or she may cancel the contract (Art. 30–39). Contracts for electricity supply can be made by agents (Art. 40) so on and so forth.

One of the parties to contracts for electricity supply is definitely a company, and the other side might be a consumer, a merchant or a company. The supplier is defined by the EML as a generation company providing electricity and/or capacity and the company holding a supply license (Art. 3/1 (hh). Hence, in accordance with the Turkish Commercial Code (the TCoC) Art. 16, it is apparent that the supplier is a capital company and thus a merchant. Art. 3 of the TCoC states that all transactions and acts that concern a commercial enterprise are commercial affairs with regard to the provisions regulated in this Code. Moreover, Art. 19/2 of the TCoC points out that contracts that are commercial affairs for only one of the parties are considered commercial affairs for the other, unless otherwise provided in the Code. Therefore, all the contracts for electricity supply are commercial affairs for both parties even if the other party is not a merchant at all. This article requires the application of the TCoC to all the contracts for electricity supply. The disputes arising between these merchants parties to contracts for electricity supply will be resolved at the competent Commercial Court of First Instance.

However, if the consumer party is a consumer in terms of the Law on the Protection of the Consumer, then the contracts for electricity supply are treated as consumer transactions. In the context of this Act, the consumer is a natural or legal person acting for non-commercial or non-professional purposes, and a consumer transaction is all kinds of contracts and legal transactions, including works, transportation, brokerage, insurance, mandate, banking, and similar contracts between consumers and real or legal persons acting for commercial or professional purposes. Based on these explanations, where a consumer acting for non-commercial or non-professional purposes concludes a contracts for electricity supply, whether it is a bilateral agreement or a retail sale contract or last resort supply, the contract is a consumer transaction and subject to the Law on the Protection of the Consumer. Disputes arising between these consumer parties and the supply company to contracts for electricity supply will be resolved at the competent Consumer Arbitration Committee or Consumer Courts depending on the monetary amount of conflict.

Whether the contract for electricity supply is a commercial affair or a consumer contract, if there is lacunae in the TCoC or Law on the Protection of the Consumer, respectively, the Turkish Code of Obligations will fill the gaps in all cases and resolve the matter between the parties.

On the other hand, each type of contract for electricity supply must comply with the mandatory content stipulated by the EML and the Regulation earlier. This does not have any ill effect on the private law nature of the contracts for electricity supply. The civil law courts will, of course, take the EML and the Regulation into account while trying a case.

4 Legal Character of the Contracts for Electricity Supply

As mentioned above, there are two kinds of contracts for the retail sale of electricity in Turkish law: bilateral agreements and retail sale contracts. The explanations below have been made considering these two different types of contracts. Except where the differences between these contracts are pointed out, the explanations are valid for both.

In accordance with the Turkish Code of Obligations Art. 209/1, "Any sale in which the object is not land, property or a right in rem entered in the land register is a chattel sale." On the other hand, Art. 762 of the Turkish Civil Code states that chattel ownership relates to movable physical objects and to forces of nature that may be the subject of legal rights and which do not form part of any immovable property. In Turkish Law, it is accepted that natural forces such as electricity and atomic energy are the subject of chattel ownership (Yavuz, 2011). Hence the contracts for retail sale for the electricity are chattel sales contracts in Turkish law.[2]

The contracts are established by concurrent declarations of wills of the parties to the contract in Turkish law. The first of two reciprocal and corresponding declarations of will is an offer. An offer is a declaration of will that is precise and complete as to the essential elements—not all the detailed terms—of the proposed contract (Markesinis et al., 1997). An offer should have specificity, definiteness (in some legal systems) and intention to be bound to be sufficient to conclude a contract with the confirmation of the offeree, who is the person to whom the offer has been addressed. The offer should include all the essential elements of the contract. These are, in general, goods, price, and consensus for the sale contract. On the other hand, the second expression to conclude a contract is an acceptance. Acceptance is a statement made by or other conduct of the offeree indicating assent to an offer. It is the final and unqualified expression of assent to the terms of the offer, which should be communicated to the offeror. Hence, the offer should include that the parties agree on the sale of electricity in turns of some certain amount of money; that is, price. This makes contracts for electricity supply to consumers onerous contracts.

[2] There are some other views on the legal character of the electricity supply contracts; such as sui generis contract view, mixed contract view, and framework contract view. For details, see Yavuz (2011).

The contract relationship between the ineligible consumers and the supply company is established by the consumers' acceptance of the company's public offer declaring consumers' demand for the service (Yavuz, 2011). In this context, there is no individually discussed content of the contract. These types of contracts, namely retail sale contracts, are called adhesion contracts.

A contract for electricity supply to consumers is concluded upon the mutual and congruent expression of the parties' will. The parties promise each other the performance that they undertake by the contract; hence, it is a obligatory transaction not an act of disposal. Meantime it is a *consensu contrahitur* contract, not a real contract (Özel et al., 2013, p. 2095). The formation of the contract does not require the transfer of possession of the sold electricity.

The parties to a contract for electricity supply to consumers are consumers and a supply company. The contract is a bilateral legal agreement that binds both parties. Both parties to a contract for electricity supply to consumers are under a principal debt, which has a reciprocal character. The performance of one of the parties constitutes the cause and quid pro quo of the performance of the other. The supply company is obliged to transfer the possession and ownership of electricity and, in return, the consumer has to pay the agreed price (Yavuz, 2011). The price is paid for the transfer of possession and ownership of the agreed amount of electricity, and the performances are interchanged. Hence, a contract for electricity supply to consumers is a synallagmatic contract (Özel et al., 2013, p. 2111; Yavuz, 2011).

In Turkish law, chattel sale contracts are contracts of instantaneous performance, not contracts of continuous performance (Eren, 2020; Zevkliler & Emre, 2017; Kılıçoğlu, 2019; Yavuz et al., 2016). The performances of the parties are conducted at a single time. The performances do not extend over the period. However, as an exception, contracts for electricity supply to consumers are deemed to be contracts of continuous performance (Özel et al., 2013, p. 2111; Yavuz, 2011). In electricity supply contracts, the performance of the supply company extends over the period. The seller in this contract must keep his/her performance ready over time continuously as the contract is applied. The performance should be carried out continuously within a certain time without interruption (Yavuz, 2011). This feature of contracts for electricity supply to consumers has important consequences regarding the termination of the contract and damages arising from the contract.

5 Freedom of Contract in Electricity Supply Contracts

Art. 35 of the Turkish Constitution regulates the right to property in Turkish law. It reads: "Everyone has the right to own and inherit property. These rights may be limited by law only in view of public interest. The exercise of the right to property shall not contravene public interest." Moreover, Art. 48 regulates freedom of work and contract, stating that "Everyone has the freedom to work and conclude contracts in the field of his/her choice. Establishment of private enterprises is free. The State shall take measures to ensure that private enterprises operate in accordance with national economic requirements and social objectives and in security and stability." Moreover, Art. 26 of the Turkish Code of Obligations states that "Parties can freely determine the content of a contract within the limits prescribed by law." It is clear that Turkey accepts liberal capitalism as an economic model. Hence, it is apparent that freedom of contract is one of the fundamental rights protected by the constitution in Turkish law. It is also an eminently practical principle and the inevitable counterpart of a free enterprise system (Kessler, 1943, p. 630).

The notion of freedom of contract is an internationally recognized fundamental contract law principle for both common and civil law systems. Owing to this principle, nobody can be forced to make a contract or choose a specific type of contract. Freedom of contract includes five contractual freedoms. It gives a party freedom to conclude a contract or not, freedom to choose the other party to the contract, freedom to choose the type and content of the contract, and freedom to terminate to contract (Ayrancı, 2013, p. 229). Freedom of formation is also an essential component of the freedom of contract in modern laws.

Freedom of contract enables the parties to freely enter into an agreement or bargain as equals. State regulation or intervention should be minimal. The freedom of contract is limited with the concept of good faith (Turkish Civil Code Art. 2), which requires that a party acting as a monopoly in a market conclude a contract with anyone who wishes to do so. Any person's avoidance of concluding a contract where the principle of good faith requires them to do so constitutes abuse of rights (Özel et al., 2013, p. 2113).

From this point of view, it is alleged that freedom of contract is quite limited in Turkish law regarding the contracts for electricity supply to consumers in the electricity market. Even there is an obligation to contract in an electricity market (Özel et al., 2013, p. 2095). The obligation to establish a contract refers to the obligation of the dominant person in the market to establish the desired contract if a person requests it (Ayrancı, 2013, p. 230). When a person who has the right to establish a contract owing to obligation to contract requests the establishment of the contract, the other party is obliged to do so.

Until the fully liberalization of the electricity market, ineligible consumers must purchase energy from authorized supply licensees through regulated tariffs. Hence, ineligible consumers do not have freedom of contract; they cannot choose the other party to the contract; they cannot bargain the price and other terms and conditions of the contract. The contracts with ineligible consumers and eligible consumer who have not chosen their supplier yet are subject to EMRA's retail tariffs. At the same time, legal entities holding a last resort supply license must conclude a retail sale contract with consumers who meet the conditions laid down by the EMRA. The function of the obligation to establish a contract in electricity supply contracts is to ensure the use of services under the same conditions in electricity markets, which are under strong economic or political influence (Ayrancı, 2013, p. 234).

The decisions of the Court of Appeal also recognize the obligation to contract in electricity supply contracts. In a decision,[3] the Court stated that the obligation to

[3] YHGK. T. 13.5.1977, E. 4–1976/480.

make a contract for services to provide electricity depends on whether the person who will benefit from it fulfills the required conditions and whether there are legal and technical possibilities in rendering the service (Yavuz, 2011).

6 Content of Contracts for Electricity Supply

As mentioned above, there are two kinds of electricity supply contracts in Turkish law: bilateral agreements and retail sale contracts. The discussion about thecontent of electricity supply contracts will be conducted in accordance with these headings. These contracts are regulated by the Regulation in detail. Parties to a bilateral agreement are eligible consumers and suppliers. On the other hand, parties to a retail sale contract are real or legal persons, who purchase electricity from the retail sales tariff or supply of last resort tariff and relevant assigned supply company.

Meanwhile, if an eligible consumer does not make a bilateral agreement with a supplier, the contract between the parties will be subjected to the last resort supply tariff. The EML 17/6 (f) states that this tariff would be established considering the retail tariffs in force and the market prices to encourage the consumers to supply electricity only from a company holding a supply license and authorized as the last resort supplier, despite being an eligible consumer, to ensure their engagement in the competitive market and to enable the last resort supplier to make a reasonable profit. Accordingly, the last resort supply company has to buy electricity that it sells under a last resort contract from a state-owned generation company. The net profit margin of this transaction is determined by EMRA and is set as 2.38% for 2021–2025.[4]The relationship between an eligible consumer and a supplier under the terms of the last resort supply tariff is almost the same as the one established between an ineligible consumer and a supplier by a retail sale contract. Both depend on a tariff and terms and conditions approved by the Board. Here we should also reiterate that the last resort supply companies are allowed to conduct bilateral agreements. For that reason, the last resort companies prefer eligible consumers to sign bilateral agreements when market conditions allow a better profit margin than the one set by the EMRA.

As for the formation of the contracts for electricity supply, the supplier must inform the consumer before the bilateral agreement is established and ensure that the consumer confirms that he/she has obtained the information in question. The fact that the information is not provided properly or at all is a rightful reason for termination for the consumer. The bilateral agreement should be signed with a handwritten signature or a secure electronic signature. A certified copy of the bilateral agreement and its annexes must be given to the eligible consumer on paper or via a permanent data storage method on the day of the agreement. For retail sale contracts, upon submission of a duly completed application of real or legal persons who want to buy electricity and/or capacity from the retail sales tariff or last resort supply tariff to the related

[4] Resmî Gazete 28 Kasım 2020 CUMARTESİ Sayı: 31,318.

supplier company with all the required information and documents mentioned in the Art. 21 of the Regulation, the retail sale contract is signed by the parties on the same day. The supply company then reports the application to the distribution company on the same day too. The contract is established and enters into force on the date the electricity supply begins.

6.1 Bilateral Agreements

Eligible consumers can make bilateral agreements with a separate supplier for each measurement point whose consumption exceeds the free consumer limit. The content of bilateral agreements is examined under three headings: mandatory terms, standard terms, and mutually discussed terms.

6.1.1 Mandatory Terms

Two main instruments in Turkish law regulate supply contracts: EML and the Regulation. It is apparent that there are many imperative rules in these regulatory instruments, which must be included in the bilateral agreements. The bilateral agreement and its annexes cannot contain provisions contrary to the relevant legislation. Provisions that are contrary to the relevant legislation in the bilateral agreement are invalid (Regulation, Art. 11/2).

According to Art. 10 of the Regulation, bilateral agreements can be made for a definite or indefinite period with eligible consumers with low consumption. However, the duration of fixed-term bilateral agreements cannot exceed three years. All of the documents and forms regarding bilateral agreements and their annexes should be written in an understandable language with at least 12-point font and in a clear, simple, and readable way. In accordance with the article, the offer for a bilateral agreement is made by the supplier and should be accepted by the consumer. The silence or inactivity of the consumer does not in itself amount to acceptance. The silence of the consumer or failure to expressly reject the proposal of the supplier regarding the establishment, renewal, and amendment of the bilateral agreement cannot be interpreted as acceptance of the proposal and the contrary provision cannot be included in the bilateral agreement. This is imperative.

In accordance with Art. 11 of the Regulation, it is mandatory that the bilateral agreement includes at least the following matters: supplier's name, title, address, MERSIS number, tax identification number, license number, telephone and fax numbers of consumer services centers, and website and e-mail addresses, consumer's first and last names, title, address, Turkish Republic identity or tax identification number, passport number or equivalent document number with international validity for non-Turkish citizens, telephone number or e-mail address of the consumer, if available, the date of establishment of the bilateral agreement and the starting date of the electrical energy and/or capacity sale, duration of the bilateral agreement, pricing based on electricity and/or capacity sales, transmission or distribution fee mediated in collection and funds, shares and taxes applied, security deposit/collateral purchase and application, the legal consequences of the default of the parties with the default interest foreseen to be applied in case of not paying the invoice on time, provisions regarding the collection or refund of the missing or excess amount in case of error in invoices and invoicing essential elements, the conditions regarding the exercise of the right of withdrawal within the scope of bilateral agreement and/or the application of the penalty clause, the duration and the withdrawal fee and/or the amount of the penalty clause, renewal or extension of the bilateral agreement, making changes in the provisions of the bilateral agreement, termination of the bilateral agreement, the rights and obligations of the supplier, consumers' rights and obligations, conditions regarding the termination of the bilateral agreement, resolution of complaints, and authorized resolution authority in case of bilateral agreement disputes.

In addition, provisions such as right of withdrawal, justified reasons for termination, and renewal of the agreement regulated in this regulation regarding the issues within the scope of bilateral agreements should be clearly included in the bilateral agreement.

6.1.2 Standard Terms

It is typical for bilateral agreements in electricity markets to be solely prepared by the supply companies, without the participation of the expected customers. It is observed that every supply company prepares, presents, and even imposes its own standard terms to the consumers. Indeed, in a simple explanation, a bilateral agreement is almost made of the company's standard terms and conditions.[5] These terms become the part of the bilateral agreement if the consumer agrees on them.

All of the bilateral agreements of the supply companies contain all the required terms and conditions laid out by Art. 10 and 11 of the Regulation. Hence, besides the companies' standard terms, all the mandatory terms in the legislation are formulated and incorporated into the bilateral agreement via standard terms and conditions.

TCO Art. 20 defines the standard terms as "general terms and conditions are the contractual provisions that are prepared in advance, unilaterally and submitted to the other party, in order to be used in many similar contracts in the future. The inclusion of these conditions in the text of the contract or its annex, its scope, type and form of writing is not important in qualification."

All of the standard terms of a company, apart from the mandatory terms (which are subject to administrative judiciary) are subject to the review under Articles 20–25 of the TCO. Art. 21 states that "the inclusion of general transaction conditions

[5] For examples, see https://trepas.com.tr/wp-content/uploads/2019/03/TREPAS-Dusuk-Tuketimli-ST-IA.pdf (accessed on November 22, 2020) https://www.enerjisa.com.tr/Media/GetFile?fileId=4436 (accessed on 22.11.2020) https://www.zorluenerji.com.tr/uploads/pdf/pdflist/oepsas-perakende-satis-sozlesmesi---ticari.PDF (accessed on November 22, 2020) https://enerji.gama.com.tr/wp-content/uploads/sites/2/2019/11/TEB_GATES_ENERJI_indirimli_Elektrik_Enerjisi_Satis_-Sozlesmesi.pdf (accessed on November 22, 2020).

contrary to the benefit of the other party within the scope of the contract depends on the fact that the organizer clearly informs the other party about the existence of these conditions and provides the opportunity to learn the content of these conditions and the other party accepts these conditions. Otherwise, the general operating conditions are deemed not written. General terms and conditions that are alien to the nature of the contract and the nature of the transaction are also deemed not written." When a company imposes a standard term that can be to the disadvantage of the consumer, it will be deemed as unwritten unless it is informed, read, and accepted by the consumer. In practice, the contract, including the standard terms, is presented to the consumers, and the consumer is expected to read them, understand them, and finally accept them. In many cases, this does not happen as expected. Usually, consumers sign whatever is presented to them without proper investigation. Hence, when a standard term is challenged in court, the courts decide in favor of the consumers.

Moreover, Art. 23 of TCO provides that if a provision in the general terms of business is not clear and understandable, or has more than one meaning, it is interpreted against the arranger and in favor of the other party. Hence the terms in a bilateral agreement are interpreted in favor of the consumers. Finally, in accordance with Art. 25 of the TCO, the arranger cannot impose a provision that is contrary to good faith and against the other party's advantage or which aggravates his/her situation in the general terms and conditions. For instance, terms empowering the arranger to impose a new provision or amend a provision of the contract unilaterally against the counterparty are deemed not written (Art. 24 of the TCO).

Art. 15 of the Regulation confirms the application of these rules in the contracts for electricity supply. It states that in bilateral agreements of definite duration, no changes can be made against the consumer in the agreement and its annexes during the agreement. Moreover, the bilateral agreement cannot include a provision authorizing the supplier to make unilateral changes against the consumer. These rules are imperative and apply to all contracts.

6.1.3 Mutually Discussed Terms

In theory, there is full freedom, but in reality, there is not much to discuss between the parties in a bilateral agreement owing to the mandatory and standard terms. When the samples of the bilateral agreements are examined, the consumer is expected to fill in personal details such as their name, ID number, and MERNIS address. The consumer should declare the place where the electricity will be consumed. The consumer must choose its status as a subscriber group from among household, commercial, industry and agricultural irrigation. Then the parties must decide on the duration of the contract, which can be made for an indefinite period or a fixed term of a maximum three years. Most importantly, the parties should decide on the unit price for the electricity. Even though they are free to choose any formula to determine the price, the parties usually agree on a certain percentage discount from the national tariff. Finally, the parties should discuss the security deposit. The consumer may provide a letter of credit, deposit cash, and deposit money in instalments. These are individually discussed terms and conditions of a typical bilateral agreement for electricity supply.

6.2 *Retail Sale Contracts*

In accordance with the legislation, an ineligible consumer may conclude a retail sale contract with the supplier. These contracts are subject to the tariff approved by the Board. Art. 17/4 of the EML states the terms and conditions of Board-approved tariffs covering all costs and service charges related to the relevant activity that shall bind all real and legal persons subject to these tariffs.

The Act provides that the retail tariffs include prices, terms, and conditions to be applied to consumers that do not qualify as eligible consumers without discrimination among the equals. Retail tariffs are comprised of charges covering all the costs and services within the scope of the execution of retail activities, such as active energy costs, invoicing and customer services costs, and retail service costs. These tariffs apply to all ineligible consumers and eligible consumers who choose to sign retail sale contracts.

All of the terms and conditions of the retail sale contracts are determined by the Regulation between Articles 21–31, including the formation and content of the contract, obligations of the parties, price, all other charges, deposits, and the termination of the contract. Taking all these rules into consideration, the Board prepares a standard contract that all of the parties have to sign. The parties cannot make any changes, amendments, additions, or omissions without the approval of the Board.

7 Obligations of Parties Under the Contracts for Electricity Supply

As mentioned above, contracts for electricity supply in Turkish electricity market are chattel sale contracts. Hence, the rights and obligations of the parties can be found in the Turkish Code of Obligations, Articles 207–236. Secondly, the rights and obligations of the parties are found in the EML and the Regulation.

7.1 *Primary Obligations of the Parties*

The primary obligation of the supply company under the contracts for electricity supply to consumers is to transfer possession and ownership of electricity as agreed by the contract. The supplier must provide uninterrupted electrical energy within the framework of the terms of the agreement, except for power outages.

The supplier must begin supplying the electricity on the date specified in the contract as at the time of performance. If there is no such a date in the contract, the supply should start immediately upon the onset of the contract (TCO Art. 90). The performance of the supplier shall continue until the contract is terminated. The performance shall be offered at the place agreed by the contract as the place of performance (TCO Art. 89).

The primary obligation of the consumer under the contracts for retail electricity supply is to pay the price for the electricity as agreed by the contract. While the price is unnegotiable for the ineligible consumers, as the price among other terms is determined by the tariffs announced by the Board binding both parties to the contract, the eligible consumers are able to negotiate the price with the supply company. Hence, the ineligible consumers and last resort supply consumers pay the price written in the tariffs and the eligible consumers pay the price agreed in the contract with the supplier.

7.2 Ancillary Obligations of the Parties

The ancillary obligations of the supplier are also regulated by Art. 38. For instance, upon the written request of the consumer, the supplier is obliged to submit the document regarding the payment of the invoices for the last 12 months on time. Moreover, regarding bilateral agreements, the retails sale contracts and supply of last resort, suppliers' websites must present the current tariff information and types offered to the consumer, and access of the consumer regarding bilateral agreements and/or retail sales agreements and related forms. They must also present the "Appeal or Complaint Application" service in an easily visible and accessible manner, the process of examination of complaints and resolution of disputes, and information on purchases from renewable energy resources included in the previous year's sales.

In accordance with Art. 40 of the Regulation, suppliers are obliged to establish communication channels through which the consumer's complaints are communicated, recorded, and followed by the consumer, and to inform the consumer by finalizing the requests submitted to them within 15 working days.

The other rights and obligations of consumers under bilateral agreements are regulated in Art. 39 of the Regulation. The consumer can request documents showing consumption amounts, consumption load curve and whether they use illegal and/or illegal electricity. Eligible consumers may purchase electrical energy within the scope of the supply of last resort from the supply company in charge in their area in case their bilateral agreements are terminated in any way or they request to do so. Lastly, eligible consumers are obliged to provide complete and accurate information about themselves within the scope of bilateral agreement. In case of any change in this information, the consumer is obliged to notify the supplier.

7.3 Payments

In accordance with the Regulation, the electricity meters are read by the distribution company once every calendar month for periods of at least 25 and at most 35 days. This reading is considered a monthly reading. A reading notification is left to the consumers by the distribution company. If the consumers demand from the distribution company, the reading notification is sent to the consumer electronically. The information included in the reading notification can also be edited in the invoice/payment notification. Reading notifications should contain first and last indices and reading dates for active and reactive consumption, and the amount of electrical energy consumed. Notice of reading is made in writing. If the consumer chooses, the notification of reading can also be sent to the consumer via the permanent data storage. In invoicing, the value found by multiplying the index difference between two consecutive periods by the multiplier factor is accepted as the consumer's electrical energy consumption. Except for cases such as using prepaid meters, the invoice period is determined based on the reading period within the scope of the retail sales agreement. The invoice period is determined by the parties in a bilateral agreement. Otherwise, the billing period is taken as the reading period.

Payment notification or electricity bills should include first and last indices for consumption and reading dates, the amount of electrical energy consumed, the daily energy consumption average, the current year's and the previous calendar year's consumption amounts of the consumption point as of the date of notification of payment, energy and capacity information for pricing consumption, electricity unit prices and other costs that are stipulated to be included in the invoices within the scope of the legislation regarding consumption cost, taxes, legal deductions and debts, payment deadline, payment methods and centers, debts or receivables for previous periods, and cutting-fastening cost. The payment notification or invoice issued by the supplier to eligible consumers with low consumption, non-eligible consumers, and consumers within the scope of last source resort shall be sent in writing at least 10 days before the due date. Invoice or payment notification can also be sent by e-mail if requested by consumers. The consumer must be informed about the invoice or payment using at least one of the other communication channels. Payment notification or invoices can be paid by credit card. The method of payment can be single payment or instalments. Expenses arising from the method of payment belong to the consumer. If the payment is made as a single payment, no cost is charged to the consumers. Suppliers include information on the costs of credit card payments on invoices. Detailed information on the payment notification is included on the website. Invoice payments are carried out through the company's cashier, service collection centers, telephone, mail, automatic payment, electronic fund transfer, or one of the banks.

Art. 35 of the Regulation concerns payment default of ineligible consumers and consumers within the scope of last resort supply. In the event that a consumer fails to make his/her payments for electrical energy consumption by the due date, the supplier company in charge shall notify the consumer in writing (second notification), including a payment period of at least five working days. This notification also includes the warning that if the payment obligation is not fulfilled within the specified period, the electrical energy will be cut. If the consumer fails to make the stipulated payments within the period specified in the second notification, the electricity of the place of use is cut off by the distribution company within five working days from the notification date, and a copy of the discontinuation notice is left to the place of use. If the consumer does not pay his/her accumulated debts within 30 days with delay increases or does not commit to pay within the schedule determined by the supplier, the supplier is obliged to inform the consumer, via at least one of the written or permanent data methods, that the debt will be deducted from the consumer's security deposit. If the payment is not made on time, and if the updated security deposit is equal to or higher than the total invoice amount including the delay interest applied since the last payment date, the updated security amount is deducted from the invoice amount. If the updated security deposit is lower than the total invoice amount, including the delay increase applied as of the last payment date, the updated security deposit is deducted from the total invoice amount. If the balance debt is not paid, the contract may be terminated and the debt that has become due will be collected by legal means.

As for the eligible consumers, the consequences of consumer's default of payment are determined in the contract agreed by the parties.

Finally, Art. 58 of the Regulation includes a provision regarding the jurisdiction of authority of the courts and enforcement offices where there is a dispute between the parties. In accordance with the article, the authority of the courts and enforcement offices cannot be determined in a way that will make it difficult for the consumer to exercise his right to seek rights and defense. Moreover, the courts and enforcement offices authorized by the Code of Civil Procedure and Code of Execution and Bankruptcy cannot be revoked in a bilateral agreement or a separate authorization agreement.

8 Termination of Contracts for Electricity Supply to Consumers

General reasons for the termination of a contract are valid for the contracts for retail electricity supply. For instance, TCO Art. 136 states that "an obligation is deemed extinguished where its performance is made impossible by circumstances not attributable to the obligor." If a supply company cannot offer its performance due to an incident that is not attributable to itself, such as force majeure (Özel et al., 2013, p. 2117), the obligation to supply electricity cease and the company will be relieved of its duty to perform and will not be held responsible for the damages.

The principle of freedom of contract enables the parties to terminate the contract upon mutual agreement. Henceforth, the contracts for retail electricity supply may end with the agreement of the consumers and suppliers. This is called a mutual rescission agreement (Eren, 2017). Then the contract is terminated *ex nunc*.

The contracts for electricity supply may end with the declarations of one of the parties to the contract where there is a rightful reason for the termination, which makes the continuation of the contract unbearable for the party. This is a legal consequence of the electricity supply contracts' legal character as contracts of continuous performance. The contract is terminated once the termination notice arrives. The justified reasons for the termination of electricity supply contracts are listed as examples in Art. 16 of the Regulation. A reason that is not listed there can be deemed as a justified reason for the termination.

The contract and Law may give the parties a right to avoid the contract. If one of the parties uses that right, the contract is terminated upon the arrival of avoidance notice.

The Regulation regulates the specific reasons for the termination of electricity supply contracts. Article 13 provides the eligible consumer with a right of withdrawal. Eligible consumers with low consumption have the right to withdraw from the agreement within 14 days from the date of establishment or renewal of the bilateral agreement without giving any justification and without paying the withdrawal fee by phone, e-mail, or written notice.

Article 14 provides that a definite-term bilateral agreement terminates automatically at the end of the period specified in the agreement. Moreover, provisions stating that the bilateral agreement will automatically renew or extend for the specified period cannot be imposed in a bilateral agreement with a fixed term. On the other hand, the parties can of course renew the bilateral contracts if they wish to do so.

As in all contract of continuous performances, the bilateral agreements can be terminated if there is a justified reason. Where there is a reason making the continuation of contractual relationship unbearable for one party, this party to the contract may terminate the contract with a unilateral declaration of the will. With the arrival of this notice, the contract is terminated for the future.

Article 16 states that the eligible consumer may terminate the indefinite-term bilateral agreement at any time without any justification and without paying a withdrawal fee and/or penalty.

However, to be able to terminate a definite-term contract, the parties have to indicate a justified reason. Article 16 of the Regulation provides the reasons for the termination of the contract with notice. An eligible consumer with low consumption may terminate the fixed-term bilateral agreement by giving advance notice before the end of the term without paying a withdrawal fee and/or penalty, based on one or more of the justifiable reasons mentioned in the article. These reasons are not limited, as other reason may enable the consumer to terminate the contract by a unilateral notice. For instance, the consumer may terminate the contract if the supplier refrains from providing the consumer with the information or notifications during the establishment, renewal, and amendment processes of the bilateral agreement; if the supplier does not pay any debt to the consumer on time; if the consumer moves to another place of use or evacuates the place of use without moving to another place of use; or if the supplier defaults.

On the other hand, the supplier may terminate the bilateral agreement without paying a withdrawal fee and/or penalty for a justifiable reason, which can include the consumer not paying the security deposit determined in the bilateral agreement; illegal electrical energy being consumed at the place of use; or where the bankruptcy of the consumer is finalized and a liquidator is appointed.

Art. 12 requires the supplier to inform the consumer before the bilateral agreement is established and ensure that the consumer confirms that he/she has obtained the information in question. The fact that the information is not provided properly or at all is a rightful termination reason for the consumer.

9 Competition in Supply Market

In Turkey, the number of consumers in the electricity market increased from 43,653,910 in 2018 to 44,958,924 in 2019 and latest number, as of August 2020, was 45,669,356.[6] This means that millions of supply contracts exist at any given moment. Even though electricity consumption increased significantly in the last decade, recent years saw some stagnation. Consumption dropped in 2019 and 2020 will not bring significant change due to COVID-19 related economic stagnation (Table 1).

The year 2019 saw an increase in eligible consumer consumption compared to 2018. In 2019, 93.9 TWh out of 229.6 TWh consumption was consumed by free consumers and 135.7 TWh by non-eligible consumers. The eligible consumer limit for 2019 was set at 1600 kWh, and the theoretical market openness rate on the demand side corresponding to this limit was calculated as 95.4%. While the actual market opening was 29.6% in 2018, this rate increased to 40.9% in 2019. At the end of 2019, the number of eligible consumers increased by 132% compared to the end of 2018 and reached 339,000. This number corresponds to 0.76% of the total number of consumers.

Figure 1 indicates the eligible consumer limits and theoretical market openness and actual openness in the last 10 years.

As seen, theoretical openness does not always trigger an increase in actual openness. This is mostly due to regulated tariffs. In periods when the Board sets tariffs at non-competitive levels and the electricity generation cost is high due to unforeseen reasons, actual openness in the market drops. For example, in 2018 there was a monetary crisis and the Turkish Lira dropped against foreign currencies, which increased electricity generation costs. This resulted in most of the bilateral contracts being terminated as they could not compete with regulated tariffs. However, when the market is stable we see an increase in eligible consumers. For example, when the monetary market settled in 2019, the number of eligible consumers increased significantly (Fig. 2).

Another table (Table 2) indicates that most of the eligible customers are industry and commercial customers. This is due to their consumption level and new tariff

[6] EMRA: Electricity Market Monthly Report August 2020.

Table. 1 Consumption in 2018 and 2019

Consumer type	Consumption amount of consumers connected to the distribution voltage level (MWh)		Consumption amount of consumers connected to transmission voltage level (MWh)		Total (MWh)	
	2018	2019	2018	2019	2018	2019
Non-eligible consumer	147,685,452.79	134,409,689.21	16,747,847.16	1,278,906.46	164,433,299.95	135,688,595.67
Eligible consumer	30,225,373.40	41,494,792.25	38,951,356.19	52,414,525.73	69,176,729.59	93,909,317.98
Total	177,910,826.19	175,904,481.46	55,699,203.35	53,693,432.19	233,610,029.54	229,597,913.65

Source EMRA: Electricity Market Report 2019

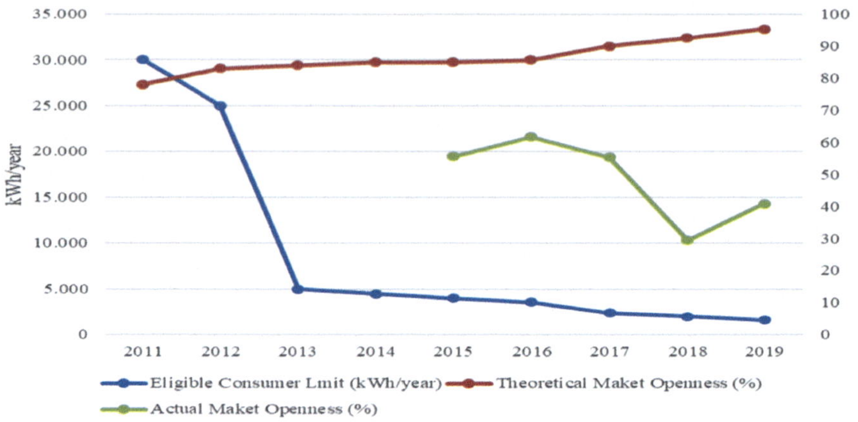

Fig. 1 Eligible consumer limit and market openness ratio by years (kWh-%). *Source* EMRA, Electricity Market Report 2019

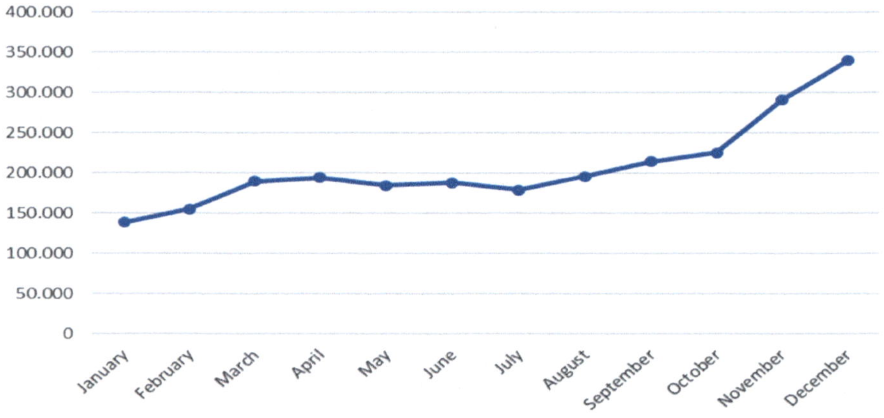

Fig. 2 Number of eligible consumers by month, 2019. *Source* EMRA: Electricity Market Report 2019

formula applied in recent years to yearly determined high consumers. There is a different tariff formula for the consumers that consume electricity higher than the consumption amount in the third paragraph of Article 5 of the Communiqué on the Regulation of the Last Resort Supply Tariff is applied as 50 million kWh/year for the residential consumer group and 7 million kWh/year for the other consumer groups in 2021. This consumption amount is determined by the EMRA and mostly reduced annually. Accordingly, EMRA sets two retail sale tariffs. One is a national fixed-price tariff, applied to all consumers consuming lower than 7 million kWh, and the second tariff is for high-consuming costumers. Based on this new model, the Board determines the second tariff based on the costs as a basis for the calculation of the last resort supply tariff and the calculation of the elements constituting the price.

Table. 2 Distribution of invoiced consumption by consumer type as eligible and non-eligible consumers, 2019 (MWh)

Consumer type	Consumption amount (MWh)					
	Non-eligible consumer	Share (%)	Eligible consumer	Share (%)	Total	Share (%)
Industry	22,006,655.22	16.22	72,456,043.56	77.16	94,462,698.78	41.14
Commercial	44,486,226.14	32.79	20,664,163.12	22.00	65,150,389.26	28.38
Household	56,219,053.60	41.43	170,721.62	0.18	56,389,775.22	24.56
Irrigation	7,949,887.45	5.86	603,479.99	0.64	8,553,367.43	3.72
Lighting	5,026,773.26	3.70	14,909.70	0.02	5,041,682.96	2.20
Grand Total	**135,688,595.67**	**100.00**	**93,909,317.98**	**100.00**	**229,597,913.65**	**100.00**

Source EMRA: Electricity Market Report 2019

As a result, if any consumer's annual consumption is above the limit determined by EMRA, this means that they will not be priced at the fixed-price national tariff, even if their tariff is an industry or a business as of the date determined.

For consumers that consume higher than the determined amount, the tariffs are set according to market conditions and in a punitive way, so almost all of them choose bilateral agreements in order to bargain better conditions with a supply company. Low-consuming consumers must choose between a fixed-price national tariff and the supply company's offered price. As seen from the latest numbers shown in Table 2, households prefer to stay with regulated tariffs due to two reasons: either they are ineligible or, even if they are eligible, they cannot find willing supply company to sign bilateral agreements or the market is not ready to convert household consumers. For consumers with better bargaining power, bilateral agreement might be more favorable than the national tariff.

As Table 2 shows, very few household consumers choose bilateral agreements, while the over whelming majority of industry prefers bilateral agreements.

10 Conclusion

There has been constant transformation of regulation of electricity supply contracts to end users since 2001, when the EML was first enacted. EMRA constantly lowered the limits of eligible consumers every year, which was set very low recently. As of 2020, more than 96% of consumed electricity can be subjected to free bilateral agreements. We expect the limit will be completely abandoned within few years' time when every consumer will be eligible.

From the first opening to market to free supply contracts, many private supply companies have aimed to sell electricity to eligible consumers. During recent years especially, an important number of industry and commercial consumers have procured their electricity from private supply companies through bilateral agreements. However, the willingness of end users to buy their electricity from supply companies rather than last resort suppliers mostly depends on the last resort tariff. The last resort tariff sometimes does not fully reflect the real cost of electricity in period in which most of the bilateral agreements are terminated and the end user is transformed to retail sale contracts. When the exchange rate and thus the electricity generation cost is stable, Turkey has seen an increase in bilateral supply contracts. In order to avoid this fluctuation, the law introduced a second tariff formula, which is applied to consumers with yearly consumption of more than 7 million kWh. The level is lowered every year by the EMRA. This tariff is determined according to reference price occurred in day ahead market. Thus, almost all of those high-consuming end users prefer to conduct bilateral agreements as they can find more competitive prices than that of the real-cost-based last resort tariff. We predict that every time this limit of consumption is lowered there will be an irrevocable transformation to bilateral contracts.

We predict even lower-consuming fixed tariff consumers will prefer bilateral agreements in near future. As the market is stable, fixed tariffs will be more expensive than those supply companies can offer.

Supply-side competition is increasing due to new entrants to the market. There are already many entrants to the supply side, not only by established electricity companies, but also by natural gas companies, mobile network companies, and internet and TV companies. This will increase competition for low-consuming end users as these companies are already in contact with them. In Turkey, concentration on the supply side is extremely low; even the leading company's market share was just 9.27% in 2019. In 2019, Harfindahl-Hirschman Index (HHI) values of supply license owners was only 413.16.[7]

Regulators are removing many hurdles for free contracts. For example, for household and lower-consuming commercial use, developments have occurred in recent years. As of 2020, contracts can be made through electronic means. This lowers contract negotiation expenses radically and opens ways to internet-based contracts and electricity marketing platforms. We believe that this will also transform many household and commercial consumers to bilateral contracts on the condition that EMRA tariffs are more costly than those offered by supply companies.

In conclusion, the laws and regulations are almost ready for fully functioning competitive electricity markets in Turkey. There will no ineligible consumers within a short time and tariffs are becoming more punitive for those not willing to convert to bilateral agreements. In order to see more creative contractual relationship, we believe that EMRA should leave regulation of private contracts and only regulate technical matters as the private law principle will be more suited to regulate the complex relationship between supply Company and consumers.

---

[7] The HHI values are evaluated as a competitive market indicator. If the HHI is under 1000, the market is competitive; medium intensity is an HHI between 1000 and 1800, and an oligopoly market exists if the HHI is above 1800.

References

Ayrancı, H. (2013). Sözleşme Kurma Zorunluluğu. *Ankara Üniversitesi Hukuk Fakültesi Dergisi, 52*(3), 229–252.
Duman, M. Z. (2011). Turgut Özal'ın Politikalarında Ekonomik Rasyonalizm. *Sosyoloji Dergisi, 23–24*, 105–123.
Eren, F. (2017). *Borçlar Hukuku Genel Hükümler*. Yetkin.
Eren, F. (2020). *Borçlar Hukuku Özel Hükümler*. Yetkin.
Kessler, F. (1943). The contracts of adhesion-some thoughts about freedom of contract role of compulsion in economic transactions. *Columbia Law Review, 43*, 629.
Kılıçoğlu, A. (2019). *Borçlar Hukuku Özel Hükümler*. Turhan.
Markesinis, L., & Danneman. (1997). *The German law of obligations Volume I the law of contract and restitution*. Clerandon Press.
Özel, Ç., Özcan Büyüktanir, B. G., & Özel, F. (2013). Elektrik Piyasalarında Elektrik Sağlama Amaçlı Sözleşmeler. *Journal of Yaşar University, 8*(Özel), 2075–2126. https://doi.org/10.19168/JYU.10900
Yavuz, C., Acar, F., & Özen, B. (2016). *Borçlar Hukuku Özel Hükümler*. Beta.
Yavuz, M. (2011). *Elektrik Tedarik Sözleşmeleri -Özellikle İkili Anlaşma*. Oniki Levha.
Zevkliler, A., & Emre, G. K. (2017). *Borçlar Hukuku Özel Hükümler*. Turhan.

Websites

https://enerji.gama.com.tr
https://trepas.com.tr
https://www.enerjisa.com.tr
https://www.epdk.gov.tr
https://www.resmigazete.gov.tr
https://www.zorluenerji.com.tr

Eylem Apaydin is an assistant professor at the Department of Civil Law, Faculty of Law, University of Kocaeli where he lectures on the Civil law, Comparative law and Intellectual Property Law matters. He graduated from University of Ankara, Faculty of Law and completed his legal practice training in Ankara Bar Association. He was granted his LLM degree from the University of Kent and his Ph.D. award from the University of Leicester in the UK. His academic research focuses on all areas of civil law, comparative law and intellectual property law. He is the author of "The Problematic Structure of Management of Co-Owned Properties in Turkish Law and Pursuance of Solutions" (2011) and "The Principle of Good Faith in Contracts" (2014) books. He has published a number of book chapters, articles and papers in the field of civil law, comparative law and intellectual property law both in English and Turkish. He acts as a legal expert for the Kocaeli Courts.

Muzaffer Eroğlu is an assistant professor at Boğaziçi University, Faculty of Law, he lectures on the Company Law, Energy Law and Competition Law. He holds an LLB from Ankara University, an LLM from the University of Kent and a Ph.D. from Queen Mary, University of London. He is a member of the Ankara Bar. His academic research focuses on company law, corporate governance, competition law and energy law, He published a book "Multinational Enterprises and Tort Liabilities". He has several publications in the field of company law, competition law and energy law both in English and Turkish. He also carried out researches as a visiting academic at the University of Oslo, Max Planck Institutes for Comparative and International Private Law in Hamburg and European Institute of Florence.

Regulation of the Natural Gas Grid

Bağdagül Kaya Caner

Abstract The Turkish experience regarding regulation of the natural gas grid dates back to 2001. The last 20 years can be defined as a progressive deepening and expansion in terms of the scope and attributes of the natural gas grid regulation. Although the main principles of the regulation concerning the natural gas grid, such as non-discriminatory third-party access (TPA), tariffs, transparency, and the role of the regulator, were embedded in the Natural Gas Market Law (NGML) No. 4646, the secondary legislation issued by the Energy Market Regulatory Authority (EMRA) made it possible to be in line with the evolving natural gas grid regulation that was taking place in the EU. This process started with the adoption of the entry-exit tariffs and promulgation of the Network Code and culminated with the introduction of the market-based natural gas balancing regime and the establishment of the Natural Gas Continuous Trade Platform. In line with the general principles in the Turkish Constitution, such as the right to property, freedom of contract, and supervision of the markets, the NGML and the secondary legislation issued by EMRA lay out the framework for the regulation of natural gas grid that would underpin deep-rooted doctrines such as essential facilities while enhancing the scope by paving way to new regulations that would cater to the needs of advanced and sophisticated new trade models such as future and forward contracts.

Keywords Natural gas grid · Regulations · Access to networks · Competition

1 Roots of Natural Gas Network Regulation in a Global Context

The energy sector is of great importance for national economies worldwide. However, the role of nation states in the energy sector has undergone major modifications in line with the radical changes that have taken place over the past four decades. In parallel with the tendency towards nationalization after the Second World War and

B. K. Caner (✉)
Head of Group at the Energy Market Regulatory Authority of Turkey (EMRA), Mustafa Kemal Mahallesi 2078, Sokak No. 4, Çankaya, 06510 Ankara, Turkey

M. Eroğlu and M. Finger (eds.), *The Regulation of Turkish Network Industries*,
https://doi.org/10.1007/978-3-030-81720-6_5

the oil crises in the 1970s, nation states' intervention and control in the energy sector increased. Starting in the 1980s, the decline of state intervention in these sectors, deregulation, and even privatizations were added the agenda (Roggenkamp et al., 2001). Deregulation is intended to allow markets to be more free and competitive by reducing government intervention in commercial life (Black, 1990).

In the 1990s, there was a general change in the perception regarding network sectors such as telecommunications, electricity, and natural gas. Parallel with this change, the idea that competition can be established in these markets has been strengthened by breaking away from traditional understandings. With this in mind, a trend emerged to create competition in electricity and natural gas markets, with the exception of sub-sectors that have a natural monopoly nature, such as transmission and distribution. The idea of a natural monopoly, due to the nature of technology and services in certain industries, is that the goods or services can be provided to the consumer with minimum cost and/or maximum net benefit only with a single company or a certain number of selected instruments. As for the activities with the nature of natural monopoly, it was projected to make regulations that help substitute a competitive environment by setting the rules of a fair game and ensuring that they act in accordance with these rules.

The purpose of liberalization can be described generally as making competition functional instead of regulation by enabling free confrontation of supply and demand in the market mechanism. In this context, the idea of *regulated competition* causes a reverse perception. However, in markets where natural monopolies exist, regulations provide substitution of competition as a result of competition not occurring naturally (Cameron, 2002). It can be argued that, in such markets, neither state control nor the so-called regulated competition can produce excellent results. However, it is necessary to choose between faulty competition and regulation. It is possible to say that the economic regulations applied with a soft approach can produce economic and productive results for the private sector and, at the same time, serve the public benefit in terms of economic and social policies (Phillips, 1993).

Generally, within the framework of the regulations concerning the network industries, it is necessary to prevent the network owner or operator from obtaining monopolistic rent over other players in the supply chain, by putting in place procedures and principles regarding the entry conditions and pricing of the mentioned rights (Cameron, 2002).

In the electricity and natural gas markets, market activities can be explained by sub-sectors such as imports, production, transmission, distribution, and wholesale and/or retail sales. In the case of the natural gas market, storage, and LNG terminal activities are also a part of the supply chain. It is possible to evaluate the market activities performed by importers, producers, wholesalers, and retailers within the scope of supply activity in general. In this context, it can be said that there is competition in terms of the mentioned activities. However, the transmission and distribution activities have the nature of natural monopoly. In case of LNG terminals and storage facilities, it is more difficult to generalize. In this respect, in countries that lack underground storage sites (given that such sites depend on geological formations), it can be assumed that this activity is a natural monopoly activity; otherwise, negotiated access may be employed in case of abundant capacity. In terms of LNG terminals, the approach differs according to energy policies and economic approaches of states, which are either regarded as merchant terminals, as is the case for the USA after the FERC decision of 2002 known as the Hackbery Decision (Vallee, 2003), or are to be regulated as in the EU's approach. In the American approach, however, they are regarded as merchant terminals; open access is required, so terminal owners must offer services on a first-come-first-served basis, and cannot discriminate against service requests to protect their own market activities.

The transmission and distribution networks of the electricity and natural gas markets require very high amounts of initial investment. Therefore, they have fixed costs that are high, so "duplication" of these networks does not seem economical, which makes it possible to talk about natural monopoly in these markets. In other words, in terms of production costs, natural monopoly is the case when it is cheaper to meet the demand in the market by a single firm rather than multiple firms. While the existence of scale economy is the condition of natural monopoly in the case of producing only one product, the existence of scope economy is essential for the formation of natural monopoly in cases where more than one product is produced in one facility.

Since electricity and natural gas are goods that are transported through networks, the possibility of the dominant undertaking holding the transmission and/or distribution network can possibly prevent its competitors from accessing the networks. Because the undertakings operating the transmission and/or distribution lines hold control of the essential facilities, such as the networks, they could avoid any other parties to access the network, which basically means entering and competing in the market. The essential facility means that, for the facility that must be utilized in order to perform an activity in the market, the source in question does not have any alternative or cannot be created economically and rationally. In accordance with the essential facility doctrine, an undertaking with a basic resource activity must sell this good or service to anyone on the basis of reasonable terms and prices. It can be said that the doctrine in question, which envisaged the idea that the dominant undertaking holding the essential facility, should make this element available to its competitors. This idea was first seen in the US competition law practices (Fine, 2002). It can be said that the essential facility doctrine first appeared in the US Supreme Court decision of 1912 on the use of railways. The doctrine in question has also been influential in EU competition law and has found application in many decisions of the EC Commission and European Court of Justice (ECJ). This doctrine is still the main pillar of regulated access to networks and facilities, which have natural monopoly attributes.

2 The EU Natural Gas Grid Regulation Then and Now: Where to Compete How to Compete?

The natural gas market structure has changed fundamentally with the liberalization process in the EU. Until the mid-1990s there were 15 different isolated, national natural gas markets. On the other hand, the number of market players in national natural gas markets in Member States had been fairly limited. Within this structure, incumbents (mostly public utilities) had been often granted supply concessions or monopoly transmission rights. On the other hand, gas production companies had the predisposition to make long-term contracts in which gas prices were adjusted to mainly crude oil prices. Long-term gas import/supply contracts were concluded by incumbents that were part of a well-established supply chain, extending from gas producers to end users. Downstream markets, covering transportation, distribution, and storage, had been mostly national in scope and dominated by former monopolists. The latter had been usually vertically integrated and would also control the pipeline network (Albers, 2005). This situation changed gradually with the liberalization of natural gas markets in the EU, aiming to achieve a competitive liquid and transparent natural gas internal market where energy suppliers compete with each other in supplying natural gas to customers throughout EU with non-discriminatory third-party access to natural gas transmission networks. Accordingly, the first legislative tool, Directive No. 98/30/EC Concerning Common Rules for Internal Market in Natural Gas, created new market structure. The first natural gas directive introduced, among other things, promotion of supply side competition, opening natural gas networks to third-party access, empowering customers to choose their suppliers.

In the subsequent years, taking into account the experience accumulated in the new market structure in order to overcome its deficiencies of the old vertically integrated tradition and enhance competition, the second Directive No. 2003/55/EC Concerning Common Rules for Internal Market in Natural Gas, was introduced by repealing the former Directive No. 98/30/EC. The second directive contained relatively more ambitious provisions, such as regarding market opening and unbundling. In this framework, legal unbundling became the minimum requirement for the companies that are active in both transmission and wholesale or supply actives. Additionally, all non-household customers were declared to be eligible to choose their natural gas supplier by July 2004. Household customers were entitled to be eligible to choose their natural gas suppliers at least by July 2007, thus paving way to enhance the wholesale markets within all EU Member States. In the coming years, Regulation (EC) No. 1775/2005 on Conditions for Access to the Natural Gas Transmission Networks came into force, in order to ensure non-discriminatory access to transmission networks, thus making it possible for all of the suppliers/wholesale companies that utilize the transmission networks within EU to be subject to pre-determined, transparent, and non-discriminatory standards. Although, this regulation contributed a great deal towards creating a common perception on access to transmission networks, it did leave the issue of creating and implementing the natural gas network codes to the national authorities, which did not provide EU-wide access that would underpin the single natural gas market. The network codes can be called the constitution of network utilization; determining all the commercial and technical aspects of third-party access to transmission networks; that is, capacity allocation, balancing, and rules on tariffs.

Then came the third directive and a revised Regulation as part of the so-called third energy package. In this framework, Directive No. 2009/73/EC foresaw a greater separation of transmission interests from generation, production, and supply, albeit without making the ownership unbundling obligatory to Member States, and instead placing it as a third option alongside independent transmission operator (ITO) and independent system operator (ISO). It is believed that without effective separation of transmission networks from the activities of generation, production, and supply, there is a risk of discrimination, not only in the operation of the network but also in the incentives for vertically integrated undertakings to invest in their natural gas networks. On the other hand, under the revised Regulation on Access to Transmission Networks, European-wide network codes, which set common arrangements for cross-border, gas flows across Europe are foreseen. Ensuring non-discriminatory and transparent network access is indispensable in the well-functioning European-wide wholesale markets, since natural gas is a commodity that is dependent on infrastructure; therefore, being able to utilize the network capacity is as important as having natural gas itself.

Although the key legislative measures are directives and regulations, applying competition rules to these markets have been indispensable in achieving internal natural gas market. In this regard, competition policy has a complementary role. The main instruments used have been Articles 81, 82, 86, 87, and 88 of the EC Treaty[1] and the Merger Regulation. Energy liberalization requires comprehensive use of all these instruments in order to bring about and ensure competition in natural gas markets, especially in avoiding abuse of the dominant position in the natural gas market and practices that restrict competition in the market. The decisions of the European Commission and the Court of Justice of the European Union on EU competition rules have been crucial in achieving internal natural gas market and in creating functioning natural gas wholesale markets. (Cabau et al., 2005).

On the other hand, it can be argued that, since the directives are mostly chosen as the key legal instruments, and owing to the principle of 'subsidiarity', differences in implementations of the directives by Member States are observed. In this respect, the approach of the European Commission is crucial in terms of its using its powers concerning monitoring the Member States' implementation of the directives and initiate infringement procedures against the Member States if their national legislation is not in compliance with the directives. This is especially important concerning the natural gas market rules; these are often in the form of directives, which necessitate transposition by Member States according to their national legislation.

---

[1] The EC Treaty has been amended by the Lisbon Treaty as the Treaty on The Functioning of The European Union Articles 101–109.

3 Turkish Natural Gas Network Legislation: A Comparative Analysis in the Constitutional Context

The main pillar of natural gas network regulation is the Natural Gas Market Law (NGML) dated 2001. The NGML came into force at a time when the EU had its first so-called energy package. In this regard, it is possible to say that, at that time, the NGML contained more ambitious provisions on network regulation than the first gas directive of 1998. In this respect, it is possible to say that the NGML had the model of regulated third-party access rule in place in 2001, whereas it was optional under the first gas directive. Indeed, the first gas directive gave two options to Member States between regulated third-party access and negotiated third-party access, which was abolished with the second gas directive, which made regulated third-party access the only option for transmission network access.

Moreover, the NGML also put in place legal and ownership unbundling provisions between transmission and supply activities in 2001, whereas the first gas directive had merely foreseen account unbundling for vertically integrated activities. This was improved with the second gas directive by introducing ownership unbundling and further developed by the third gas directive of 2009, which introduced ownership unbundling, independent transmission operator, and independent system operator options. Finally, in the EU's case, the compulsory establishment of the regulatory authorities that are to be the guardians of ensuring non-discriminatory third-party access rules, determining the network utilization tariffs, and protecting consumers' rights, was promulgated with the second Directive of 2003. In the Turkish case, the establishment of the independent regulatory authority was foreseen in 2001. Until the Lisbon Treaty, the founding document of the European integration, which is the EC Treaty, did not have a specific provision on internal natural gas market. The Lisbon Treaty has a chapter on energy for the first time and it encompasses the internal gas and electricity markets. Until this provision at the EU level, starting in the 1990s, the internal gas and electricity markets were dealt with under the umbrella of a single market encompassing free movement of goods and services.

In the context of natural gas network regulation, three provisions of the Turkish Constitution are of the utmost importance: freedom of contract, right to property, and surveillance of the markets. The Turkish constitution stipulates that everyone has the right to own and inherit property.[2] On the other hand, it also states that the right to property may be limited by law only in view of public interest; moreover, it concludes that the exercise of the right to property shall not contravene public interest. The right of property in Turkish law is an exclusive *right in rem*, which gives the owner the rights of utilization (*usus*), exploitation (*fructus*), and disposition (*abusus*) over the property. In this respect, the provisions of the NGML foreseeing non-discriminatory third-party access to transmission networks and regulated tariffs of network connection and utilization interact with the essence of the property rights when natural monopolies are in question. In other words, intervention over the property rights manifests public

[2] Article 35 of the Turkish Constitution of 1982.

interest rationale when there is a natural monopoly in question under Turkish law. Similar to the approach of NGML regarding the transmission, concerning the LNG terminals (including FSRUs) and storage facilities, the NGML foresees regulated access by putting in place the obligation of third-party access that will be applied equally and impartially as long as the system allows.

Freedom of contract is guaranteed under the Turkish Constitution[3] and is a well-established principle of Turkish law. The Turkish Code of Obligations stipulates that the parties are entitled to freely determine the content of the contract in line with the boundaries foreseen by the law. The Code of Obligations also stipulates that contracts that are contrary to mandatory provisions of the law, moral values, and public order are null and void.

In order to create a level playing field, avoid any potential abuse of the network owner over the network users, and provide continuity of supply and attain safe and secure network operation, the NGML contains several ex-ante provisions on determining the content and parties of the contracts that are defined within the law. In this framework, the NGML clearly defines two agreements that shall be concluded in order to render transmission network services, which are transportation and delivery agreements. According to the NGML, the transportation contract is the contract to be concluded between the system users and transmission companies for the purpose of transporting the natural gas. The delivery agreement is also directly defined in the NGML as the agreement executed between the system users or those acting on behalf of them and transmission companies, or between the storage company and transmission companies, or in between the transmission companies for the hand-over of natural gas. Furthermore, the NGML lays down the parties and determines the content of these contracts; the transmission companies shall enter into a transportation contract with the import companies, wholesale companies, production, and export companies. The transmission companies further enter into a delivery contract with the production companies, eligible consumers, storage companies, and other transmission companies. The NGML also states that procedures and principles determined by the authority shall be taken into account for the contracts entered into or to be entered into and no provision may be included that would prevent or hinder the functioning of the natural gas system that encompasses the entire natural gas supply chain.

Furthermore, the Turkish Constitution also lays down provisions on supervision of the markets. Within this framework, the State shall take measures to ensure and promote the sound and orderly functioning of the markets for money, credit, capital, goods, and services; and shall prevent the formation of monopolies and cartels in the markets, emerged in practice or by agreement.[4] Apart from Law No. 4054 on Protection of Competition, in terms of natural gas markets, with a view to ensuring and promoting the sound and orderly functioning of the market and prevent the formation of monopolies and cartels in the markets, the NGML contains sector-specific provisions.

[3] Article 58 of the Turkish Constitution of 1982.

[4] Article 167 of the Turkish Constitution.

In this regard, the NGML put in place several ex-ante precautions in order to avoid companies having dominant positions in the supply side of the market. In line with this rationale, there are static precautions concerning market share of the import and wholesale companies to 20 percent of previous years' national natural gas consumption. However, it is possible to argue that the well-established principle of competition law—that is, the single economic unit—was not utilized by the lawmaker in this approach, which would have been a more rigid approach for an ex-ante provision. In addition to these limitations, the lawmaker also addressed the status quo of the incumbent, BOTAŞ, by the temporary articles by introducing contract and volume release provisions and limiting supply agreements to be concluded by BOTAŞ, until its market share was decreased to the same 20% as the newcomers. Additionally, the reconstruction of BOTAŞ to supply, transmission, and storage companies and ownership unbundling were also within the scope of this Temporary Article 2 of the NGML.

4 New Trading Models: Access to Network as the Key

Liberalization reconstructed the "supply chain" through independent competitive offerings of each of the relevant segments, which is free to operate independently of one another (Jensen, 2003). In this regard, the natural gas wholesale market evolved on a global scale. That is to say, along with long-term natural gas sale and purchase contracts, relatively short-term sale and purchase contracts, spot transactions, and future/forward contracts transactions started to be utilized by the trading parties.

Within this framework, the commercial practices in the natural gas wholesale markets led to harmonization of the provisions used in the energy trade transactions and network utilization. Most of these harmonization attempts resulted in the creation of master agreements in LNG trade, standard OTC contracts, and organized exchange transactions. These agreements can be deemed a compilation of the trading practices. Thus, they reflect the technical, commercial, and financial aspects relating to natural gas wholesale activities in different mediums, such as LNG trade or wholesale OTC trade, or exchange traded transactions. On the other hand, the success of these wholesale developments is dependent on transparent and non-discriminatory rules on access to networks and natural gas facilities such as coherent network codes, LNG, and storage codes.

Natural gas wholesale markets can be defined as "*markets where participants such as producers, regulated and unregulated utilities, traders buy and sell natural gasfor immediate or near future physical delivery, or for delivery at distance future. The latter may not involve financial transactions that will set prices, premiums, and so on*" (Office of the People's Council, 2010). The type and duration of the contract depends on the standpoint and commercial needs of the market participants. Currently, when the market participants in the European wholesale level are examined, it can be said that energy utilities, natural gas producers, importers, proprietary traders, banks, and funds are active players at the wholesale level (Heather, 2010).

When trading channels in the current EU wholesale natural gas structure is examined, three trading channels are observed: negotiated contracts, over the counter trades (OTC), and exchanges. The latter two are new trade channels that evolved out of liberalization. Accordingly, the main characteristics of these trading channels are defined as follows (Heather, 2010).

1. Negotiated contracts can be defined as direct, non-regulated, non-standardized bilateral contracts; they are also called "*old-world contacts*". Under these contracts, all of the terms and conditions, including the quantity, quality length of contract, and price of natural gas, are negotiated. These negotiated contracts tend to cover medium or long-term periods. On the other hand, it can be said that negotiated contracts have relatively high counterparty risks.
2. OTC is also non-regulated, but the use of standardized bilateral contracts is prevalent. OTC trade can either be physical or paper-based. That is to say, while physical OTC trade always ends up with actual natural gas delivery, paper-based OTC trade usually end up with financial settlement. Physical OTC transactions can be either for spot or forward purposes, whereas paper-based OTC transactions in natural gas markets are used for swaps or options.
3. The third trading channel in natural gas wholesale markets are the exchanges. It can be said that exchange transactions are regulated by financial regulatory authorities and standardized. They can be described as paper-based transactions, although physical or financial settlement may be possible. It can be said that exchange-based transactions are made by buying or selling futures or options.

The so-called OTC trade is performed directly between two parties—the buyer and the seller of natural gas—whereas in the exchange-based transactions the exchange acts as an intermediary between the buyer and the seller of natural gas. In the exchange-based trade, both sides of the energy trade enter into agreement with the exchange instead of entering into agreements with one another. It can be said that this situation decreases the risks related to the counterparty. On the other hand, it also makes the trade relatively less disputable and fast (Edwards, 2009).

In this respect, it is worthwhile to say that "*old-world contracts*" were designed and utilized in an un-liberalized world where isolated and national natural gas markets existed. Thus, these contracts gave the opportunity to foreclose the markets via holding all the capacity of the natural gas transmission networks. Indeed, after the liberalization started, the so-called grand-father rights related to these contracts were regarded as questionable in terms of competition and liberalization.

In this framework, it is crucial to put in place non-discriminatory rules concerning network access. Firstly clear, non-discriminatory, preferably market-based capacity reservation and allocation rules are the key to having a well-functioning wholesale market containing bilateral contracts, OTC, or exchange-based transactions. In this regard, putting aside the issue of existing long-term contracts that grant capacity rights, which involves analysis for every national legal system concerning how far the constitution protects the sanctity of the contracts, it is possible to outline several capacity reservation models, such as first-come-first-served, pro-rata, or auction systems. In the EU system, until the adoption of the third energy package, most of the EU countries had different choices on capacity allocation regime, which hampered cross-border trade and the functioning of an EU-wide integrated natural gas market (ACER, 2012). After the third package, with the gradual introduction of four network codes (Network Codes on Capacity Allocation, Balancing, Inter-Operability, and Tariffs), the auction mechanism was set as the sole mechanism for the inter-connection points, bundled entry-exit rules, market-based balancing regimes, and rules, and transparency. The same rules on allocation of the gas and regulated tariff structures based on entry-exit system created coherence and helped achieve an EU-wide wholesale market.

5 Access to Natural Gas Network Rules: EMRA as the Locomotive

EMRA was established in 2001 and equipped with vast regulatory powers compared to its EU equivalents, going far beyond ensuring third-party access, licensing, tariff setting, solving disputes, and protecting customers. Indeed, EMRA encompasses powers to regulate all the segments of natural gas supply chain independently. In the 19 years since its establishment, it has become a catalyst for ensuring competition and deepening regulation. NGML did not face major amendments apart from those of 2008, which freed spot LNG imports, and 2018, which gave EMRA the power to regulate spot pipeline imports. However, with powers vested by the NGML, EMRA have taken steps to deepen the market and ensure TPA by way of secondary legislation independently regarding issues such as by laws, principles and procedures, and numerous Board decisions. Besides, the NGML vested EMRA with powers to determine tariffs to natural monopoly segments such as transmission and connection. EMRA also has the authority to determine tariffs and/or principles of formation where there is a lack of or no competition.

The market structure before the NGML was in place was a monopolistic one similar to most countries of continental Europe, driven by state-owned companies; the exception was the United Kingdom, which started energy market liberalization in the late 1980s. It is possible to say that Boru Hatları ile Petrol Taşıma A.Ş. (BOTAŞ) enjoyed being a state-owned monopoly on import, transmission, and sale of natural gas under Decree Law No. 397 dated 1990. The Decree Law in question also laid down the rules on legal entities entitled to perform natural gas distribution. In this regard, the BOTAŞ and relevant municipalities may be granted the authorization to perform distribution activities. As a result, the natural gas market before liberalization was dominated by BOTAŞ across the supply chain, besides the five distribution companies.

The monopolistic natural gas market structure was abolished with the enactment of the NGML in 2001; introducing free market entry through licensing of EMRA, regulated TPA and administrative powers granted by EMRA including independent regulation and surveillance of the market. However, it is necessary to point out that the NGML paved the way to new entrants to the market while following a gradual approach to decreasing the market share of the ex-incumbent. Likewise, a calendar was foreseen in the NGML Temporary Article 2, concerning the unbundling of the ex-incumbent, BOTAŞ.

In line with the EU Directives in force, the NGML established a legal framework for all market activities such as transmission, distribution, import, sales, LNG terminals, and storage of gas, and set out the bases for, inter alia, licensing tariff distribution, and customer affairs by related by laws issued by EMRA. In 2004, the network code of the Transmission System Operator (TSO), BOTAŞ was approved by an EMRA Board Decision. The Network Code, similar to the Irish one in place at that time, established rules including a capacity allocation mechanism based on pro-rata principle giving a maximum one year capacity right, congestion managements rules, rules on allocation, billing, quality, and dispute resolution mechanisms. Additionally, in 2004, EMRA determined the transmission tariffs based on entry/exit system; that is, determining the capacity charge for each transmission entry and exit point, plus the commodity charge for each cubic meter for which transmission service was rendered by the transmission company. As a result of these steps, initial steps TPA to the gas infrastructure has been granted in a non-discriminatory and transparent manner for the system users that were licensed after the liberalization.

When the liberalization of the Turkish natural gas market commenced, all of the natural gas consumed in the country, apart from a relatively small amount produced domestically, was supplied via the long-term import contracts, which were signed before the enactment of the NGML. In line with the temporary articles of the NGML, in order to decrease the ex-incumbent BOTAŞ's market share the contract release tender was held in 2005. As a result of this tender, natural gas sale and purchase agreements corresponding to 4 bcm were transferred. These private importers started importing in late 2007, from Russia via the Malkoclar entry point; in accordance with the Network Code, the gas they imported was transported via the transmission network. Following the contract release tender, wholesale prices were liberalized to be negotiated by parties, at the beginning of 2008, by an EMRA Board Decision, in line with the relevant article of the NGML that dictates that the wholesale tariffs shall be set freely between the parties if the EMRA Boards deems there is necessary competition in the market.

Another important step towards a more liberalized market was taken in 2008 by the NGML amendments made in July 2008 that liberalized spot and long-term LNG import conditions by lifting the limitations in the law including the source countries. Following this amendment, in May 2009, EMRA issued the "By law on Establishing Basic Usage Procedures and Principles of Liquefied Natural Gas Storage Facilities", and in 2010 approved two separate "Basic Usage Procedures and Principles" LNG Codes that regulate non-discriminatory TPA to the two active LNG terminals.

Similar to the UK model, the Turkish National Balancing Point was introduced by EMRA, in the transmission network that would serve as virtual title transfer point in 2011. Thus, trading between the supply companies for balancing and mere trading purposes in the transmission network paved the way for the over the counter natural gas sales.

Furthermore, in 2012, EMRA published a Board Decision declaring that residential customers with annual consumption above 75,000 Sm^3 are eligible customers, as are all non-household customers. In 2014, EMRA issued the Model Transport Agreements and Model Delivery Agreements, which determine the technical and commercial rules that shall be included in contracts that will be enacted between suppliers and distribution companies, basically serving as a distribution network code. These developments, along with the deregulation of wholesale tariffs and entry of new wholesales firms, benefited the market opening and liberalization considerably.

After the expiry of BOTAŞ's contracts with Russia, worth 6 bcm/year in 2012, the opportunity was given to private companies to enact new contracts with this supplier for the same amounts in line with the NGML. As a result, the number of private firms holding long-term import licenses to eight and share of private importers increased to 10 bcm/year. Another development in 2013 that further contributed to liberalization and source diversification in Turkey was the first import license given for supplying gas from a country with which BOTAŞ had never had contracts before.

In March 2016, the EMRA Board issued a decision on the licensing regime of FSRUs, introducing the FSRUs to the Turkish gas market. Basic Usage Procedures and Principles of the first private FSRU was approved by the EMRA Board and published in November 2016. These developments led to growing interest in FSRUs, hence Turkey's second FSRU was licensed and its Basic Usage Procedures and Principles was published in 2017. The Board rendered another important decision in March 2016, enabling transportation of natural gas in LNG and CNG areas where neither transmission nor distribution networks exist. Along with the project, which aimed to widen the distribution network to all districts that have enough population to support the necessary investments, these significant changes not only allowed the development of distribution system operators (DSOs) to regions where network extension is not feasible but also provided additional flexibility to the market players.

Concerning the LNG terminals and the storage facilities, the general rule in the NGML is negotiation concerning the tariffs. On the other hand, Law No. 4628 gives power to the EMRA Board to determine the tariffs and to make decisions on price formation in cases where there is no or insufficient competition. In this framework, taking into account the storage capacities until December 2017, the EMRA Board determined the tariffs for both underground storage and LNG facilities. By that time, two land LNG terminals with yearly capacity of approximately 14.2 bcm (one of which is privately owned), one FSRU (privately owned) with yearly capacity of approximately 6 bcm, and one underground storage, were serving the natural gas market. When the second FSRU came into the market with 6 bcm yearly send-out capacity, the EMRA Board determined that the need to determine due to insufficient LNG capacity is no longer in question; thus, by 2018, LNG terminal access tariffs were being negotiated by the parties. However, third-party access to LNG terminals and storage facilities was regulated by the EMRA in line with the NGML and the relevant by laws concerning access and utilization of these capacities and individual LNG Terminal and Storage Codes approved by EMRA.

Another major milestone in Turkish natural gas market occurred in 2018 with the establishment of the organized natural gas market with a market-based balancing regime by EMRA. This was in line with the EU's Balancing Network Code, before all the countries in its region and neighboring countries with natural gas market structure in place, in order to underpin the idea of making Turkey a natural gas trading center stretching from the Mediterranean to the Balkans, the Black Sea Region, and beyond.

The By law on Organized Natural Gas Wholesale Market was adopted by EMRA in 2017, establishing the legal basis for the foundation of the Turkish Continuous Trade Platform (CTP), which is similar to the On-the-Day Commodity Market (OCM) of the UK. The market operations on CTP started on September 1, 2018. STP, operated by Enerji Piyasaları İşletme Anonim Şirketi (EPİAS), allowed the market players to trade natural gas anonymously in an organized liberal market, as well as let the transmission system operator balance the system as the "Residual Balancer". Market design of EMRA based on the continuous trading principles allows import, wholesale, and export companies to perform day-ahead, intra-day, and after-the-day operations within a single trading window. Since these contracts concern physically delivered gas and the fact that a new balancing regime was created, EMRA adopted major changes in the Network Code of BOTAŞ. Under this new market structure, natural gas transmission system users can make trade nominations in the CTP and buy or sell gas in order to be balanced daily. If the system users are not balanced, the transmission operator will enter CTP in order to buy or sell gas to achieve system balance. The supply of balancing gas by TSO from the CTP as a residual balancer allows the pricing of the balancing gas under market conditions in an objective and transparent manner. The Daily Reference Price (GRF), which is formed as a result of the spot trading activities on CTP, including residual balancing operations, is an indicator of the Turkish Reference Natural Gas Price. The GRF has been published daily by the market operator since September 1, 2018. By the amendment of the Market Operation Code in January 2020, Balance of Week, Weekends, and Working Days Next Week contracts were introduced to the organized market. The new products were made available as of June 2020, creating the basis of Weekly Reference Price, providing another benchmark for gas prices in the Turkish gas market. Furthermore, the legislation on physically settled derivatives markets is planned to be published in late 2020 by EMRA.

The foundation of the Organized Natural Gas Wholesale Market is an important step towards establishing Turkey as an international gas trade center, allowing the trade of gas from different sources. The Organized Natural Gas Wholesale Market is the first organized natural gas market that generates market-based price signals in the region. As of the end of 2019, the total transaction volume in the organized market was 2.84 billion TRY and the amount of the traded natural gas was 1.87 bcm. Around 30,000 thousand orders were given by the 44 active traders and more than 10,000 matchings were cleared in the first 16 months of the market (EMRA 2020).

In line with this prerogative, in 2018 the NGML was amended to allow spot import via the pipeline entry points, vesting the power to regulate the spot import and capacity regime to EMRA. In September 2019, EMRA issued the "The Principles and Procedures on Determination of Spot Pipeline Import Methods and Quantities". This regulation introduced spot import capacity auctions for yearly, quarterly, and monthly capacity products to the Turkish Natural Gas Market in January 2020. The new capacity mechanism, which is in line with main ideas such as the auction and use-it-or-lose-it and transparency requirements of the EU's Capacity Allocation Network Code, is expected to encourage new suppliers and result in greater source and price diversity.

Along with the flexibility provided by the organized spot gas market, the increase in cross-border trade, including the spot imports, will be beneficial for Turkey's aim of becoming a regional natural gas trade center. Furthermore, the natural gas discovery of 320 bcm in the Black Sea at the Sakarya field in August 2020, will also have a great impact on Turkey's aim to be a natural gas trading center.

6 Conclusions

The last 20 years have seen a major transformation in Turkish natural gas market, moving from a monopolistic market structure to a competitive one. Although the primary law creating this market had not undergone as many changes in the last 20 years as the EU, the natural gas market secondary legislation adopted by EMRA has deepened and evolved the market in line with the developments in the EU *aquis*. The powers provided to EMRA under the NGML, in terms of independent regulation, has given EMRA the flexibility to shape the market rapidly.

Within this framework, in addition to the rules in the NGML concerning non-discriminatory compulsory third-party access to transmission networks, LNG terminals, and storage facilities, by way of secondary legislation EMRA made it possible to create a legal framework in which natural gas title transfers are made within the transmission virtual points, either at the OTC or exchange-based. Furthermore, EMRA created a market-based balancing regime where the TSO acts as a residual balancer of the system and buys or sells gas in the CTP similar to the Balancing Network Code regime of the EU. Additionally, the transparency rules envisaged both in the Network Code and the organized market legislation laid down by EMRA makes it possible for the market players to have a level playing field in the transmission activities.

Furthermore, the regime introduced by EMRA for regulating spot pipeline imports is reminiscent of the rolling auctions regime of the EU's Capacity Allocation Network Code, which enabled yearly, quarterly, and monthly capacity allocation regime at the import entry points, making new entry to the market possible.

The regulations issued by EMRA concerning third-party access to LNG terminals and storage facilities not only provided a non-discriminatory capacity utilization and capacity trade regime within these facilities but also made it possible for new investments such as new FSRU and storage facilities to be built and operated under this structure.

The vision of the Temporary Article 2 of the NGML on unbundling of BOTAŞ has not been fulfilled until now. However, the vertically integrated structure of BOTAŞ has not caused any hindrance on the utilization of the transmission network by the third parties and the establishment of the new organized market by EMRA. This new market has so far enabled around 40 market participants to use the transmission network and trade natural gas within this structure every year. Moreover, EMRA has designed the organized market and the new balancing regime in a way that clearly separated the transmission role and the supplier role of BOTAŞ. Thus, transmission licensee BOTAŞ, prescribed to act as a residual balancer, must buy or sell gas at the CTP in which the gas is traded anonymously on a pay-as-bid basis. The introduction of future contracts to the organized market will surely deepen the market. The introduction of the market maker to the organized market is foreseen in order to boost liquidity in the organized natural gas market, with the amendment that passed from the Turkish Grand National Assembly in November 2020.

However, the supply side competition is still fairly limited since all of the newcomers holding an import license are importing gas from the same country and their market share is around 20 percent. However, the new spot pipeline gas structures on the basis of capacity auctions may yield fruitful results in the future by supplying gas on a short-term basis, hopefully with a market-based competitive structure.

Although the TPA rules are in place, the LNG terminals have not been utilized by third parties other than BOTAŞ, except in 2009–2011. Taking into account, the increase in the number of the spot LNG licensees granted by EMRA and the effects of the shale gas revolution in the global LNG markets, it would be to the benefit of the market in terms of competition if new and competitive gas supply was spurred into the market by newcomers.

Along with all the regulatory environment provided, the natural gas discovery of 320 bcm in the Black Sea at the Sakarya field in August 2020 will also have a great impact on Turkey's aim to be a natural gas trading center, making Turkey the only country in the region to be both a natural gas source country and have a well-established natural gas market legislation and tradition of regulation.

References

Albers, M. (2005). Energy liberalisation and EC competition law. *Fordham 28th Annual Conference of Antitrust Law and Policy*. https://ec.europa.eu/competition/speeches/text/sp2001_028_en.pdf

ACER Market Monitoring Report. (2012). https://acer.europa.eu/Official_documents/Acts_of_the_Agency/Publication/ACER%20Market%20Monitoring%20Report%202012.pdf

Black, H. C. (1990). Black's law dictionary (6th ed.). West Publishing Co.

Cabau, M., Hancher, L., Jones, C., Landes, V., & Van der Woude, M., (2005). *EU energy Law—EU competition law and energy markets, Claeys & Cateels* (Vol. 2).

Cameron, P. D. (2002). *Competition in energy markets: Law and regulation in the European Union*. Oxford University Press.

Edwards, D. (2009). *Energy trading and investing: Trading, risk management and structuring deals in the energy market*. McGraw-Hill Professional.

EMRA Natural Gas Market Sector Report. (2020). https://www.epdk.gov.tr/Detay/Icerik/3-0-94/dogal-gazyillik-sektor-raporu

Fine, F. (2002). A logical application of essential facilities doctrine. *Practicing Law Institute "Intellectual Property Antitrust Conference"*.

Heather, P. (2010). The Evolution and functioning of the traded gas market in Britain. *Oxford Institute of Energy Studies*, NG 44. https://www.oxfordenergy.org/wpcms/wp-content/uploads/2010/11/NG44-TheEvolutionandFunctioningOfTheTradedGasMarketInBritain-PatrickHeather-2010.pdf

Jensen, J. (2003). The LNG Revolution. *Energy Journal of the International Association for Energy Economics, 24*(2), 1–45. https://svn.eeni.tbm.tudelft.nl/Education/spm3x20/spm3x20_th_II_dictaat_en_leeswijzer/spm3x20_th_II_doc_7_economie_Jensen_2003.pdf
Phillips, C. F. (1993). *Regulation of public utilities—Theory and practice.* Public Utilities Reports Inc.
Office of the People's Council. (2010). http://www.opc-dc.gov/market-monitoring/natural-gas-market/the-wholesale-natural-gas-market
Roggenkamp, M., Ronne, A., Redgwell, C., & Del Guayyo, I. (2001). *Energy law in Europe-National.* Oxford University Press.
Vallee, J. E. (2003). FERC hackberry decision will spur more U.S. LNG terminal development. *Oil & Gas Journal, 101*, 64–68.

Bağdagül Kaya Caner is the Head of Natural Gas Network Operation Regulations Group at the Energy Market Regulatory Authority of Turkish Republic (EMRA). She has over 17 years of experience in natural gas and energy markets. She is currently the Chair of the ICER (International Confederation of Energy Regulators) Virtual Working Group on Gas and Other Fuels and Vice-Cahir of MEDREG (Mediterranean Energy Regulators) Institutional Working Group. She is a lawyer by training. She holds a Ph.D. degree in Private International Law (Thesis on International Arbitration in Disputes arising from Energy Investments) from Ankara University where she also holds, LLB., and MA on the European Union and International Economic Relations (Thesis on Competition in the EU Electricity and Natural Gas Markets and Its Reflections to Turkey). She also holds a Master of Laws degree from the Queen Mary, University of London (Thesis on Renewable Energy and International Investment Protection Regime). She has postgraduate certificates from the Florence School of Regulation-European University Institute and Institute of Public Utilities- Michigan State University. She has published several articles on both national and international publications and she has written book chapters, on issues such as cross-border natural pipeline projects, investment arbitration, competition in energy markets and carbon trading, climate change. She gives lectures in universities and NGOs on issues related to energy law, EU law, and international investment law.

Regulatory Aspects of Natural Gas Distribution in Turkey

Mehmet Kürkçü and Mehmet Suat Kayikçi

Abstract Turkey added natural gas to its primary energy mix for the first time in 1976 by utilizing locally produced gas in industry, but made the actual leap in 1987, owing to the gas procurement contract between BOTAŞ, Turkey's state-owned gas company, and Soyuzgazexport of USSR, which was signed in 1986. However, until the introduction of Natural Gas Market Law in 2001, the local natural gas distribution remained limited to a handful of big cities, where gas distribution was carried out via companies owned and operated by local municipalities and BOTAŞ. Before the law, natural gas was supplied in seven distribution zones in six cities, namely in İstanbul (two distribution zones), Ankara, İzmit, Adapazarı, Eskisehir and Bursa. The law was introduced to ensure liberalization of the natural gas market to create a financially strong, stable, and transparent natural gas market, and to ensure an independent regulation and supervision in this market in order to make natural gas available to consumers with high quality, continuity, economy, and environmental sensitivity. After the introduction of the law, Energy Market Regulatory Authority, the energy regulator of Turkey which was founded one year earlier to regulate the electricity market, took on responsibility in restructuring and liberalization of natural gas market, including local distribution among other market segments. The law defines the distribution as "*Transfer of natural gasfor purposes of delivery to clients through local gas pipeline network and its retailing*" and requires the activity to be carried out by licensees selected through a tender process for regions to be determined by Energy Market Regulatory Authority and upon privatization for the existing distribution regions. This chapter aims to outline the regulatory aspects of natural gas distribution in Turkey.

The views expressed in this article are those of the author and do not necessarily reflect the official policy or position of the institution he is affiliated with.

M. Kürkçü (✉)
Energy Market Regulatory Authority of Turkey, Mustafa Kemal Mahallesi 2078, Sokak No:4 Çankaya, 06510 Ankara, Turkey
e-mail: mkurkcu@epdk.gov.tr

M. S. Kayikçi
Partner At LBF Law & Consultancy, Buyukdere Caddesi No: 108/1 Kat:7, Şişli, 34394, 34347 İstanbul, Turkey

M. Eroğlu and M. Finger (eds.), *The Regulation of Turkish Network Industries*,
https://doi.org/10.1007/978-3-030-81720-6_6

Keywords Natural gas distribution · Regulation · Access to network · Gas retail · Licence

1 Introduction

The natural gas market of Turkey has undergone a reform process since the adoption of Natural Gas Market Law in 2001 (NGML) to lay the foundations for the reform and restructuring process. The objective of the reform is to establish a financially sound, stable, and transparent natural gas market, based on competitive rules with independent regulation to achieve effective, continuous, environment-friendly, and economic natural gas delivery, involving the progressive withdrawal of the state and introduction of competition. The harmonization of the domestic legal framework with that of European Union's (EU) is also aimed through the reform process (Erdogdu, 2009: 5).

To this end, the NGML sets forth rules regarding restructuring and liberalization of the market, transparent market entry conditions as well as non-discriminatory Third Party Access (TPA) to the gas infrastructure and as well, licensing obligations, licensees' rights and obligations, the extension of Energy Market Regulatory Agency's (EMRA) duties and powers, tariffs, audit of the market and sanctions to be imposed. Import, transmission, distribution, storage, marketing, trade and export of natural gas have been determined as market activities and are regulated by EMRA.

The law defines the distribution as "*Transfer of natural gas for purposes of delivery to clients through local gas pipeline network and its retailing*" and requires the activity to be carried out by licensees selected through a tender process for regions to be determined by EMRA and upon privatization for the existing distribution regions. The licensees are not entitled only to operate the distribution network but also to supply natural gas to captive consumers (household customers) exclusively based on tariffs determined by EMRA and eligible consumers with bilateral sale contracts at prices freely determined between parties.

Turkish natural gas distribution market has been divided into 72 regions so far and the licenses have been awarded for all regions. Apart from İstanbul Gaz Dağıtım A.Ş. (İGDAŞ, a municipality owned company) all licensees are private enterprises.

In this study, it is aimed to analyze the Turkish Natural Gas Distribution Market from a regulatory perspective. In this context, a brief history of the market development and regulatory works will be submitted first. Then comes the tender process and licensing obligations. Subsequently, the main rights and obligations of licensees will be dealt with. Tariffs will also be thoroughly evaluated. Final part of the study will be the supervision of the market and sanctions that the distribution licensees may face in case of acts against law.

2 Development of Distribution Network and Legal Framework

Despite its close location to many actual or potential gas producing countries, e.g., Russia, Iran, Azerbaijan, and Turkmenistan, Turkey's indigenous natural gas resources have been very scarce and no significant gas reserves have been explored so far[1] leaving aside the reserve claimed to be discovered in the Black Sea in August 2020. Such that domestic production can meet only 1–3% of the country's natural gas consumption. Due to the lack of local resources, natural gas industry in Turkey has developed relatively late compared to the European countries and the country is highly dependent on imports to meet its demand.

The use of natural gas in Turkey started in 1976 at the Pınarhisar cement factory by using the gas discovered in Hamitabat and Kumrular natural gas field in 1970 by Turkish Petroleum Corporation (TPAO). However, meaningful consumption began in 1987 with the importing of natural gas from Russia in an effort to diversify electricity generation sources and prevent air pollution in big cities, specifically Ankara, due to coal-dependent heating. Following the import agreement with Soyuzgazexport of USSR, the use of natural gas, has gradually increased in the fields of electricity production, housing, industry, and various other uses. Natural gas use for residential purposes began first in Ankara in 1988, and continued with Istanbul and Bursa in 1992, and then Eskisehir and Izmit in 1996. As Turkey's natural gas consumption increased rapidly, several new long-term agreements were signed with countries such as Russia, Iran, Turkmenistan, Azerbaijan, Nigeria, and Algeria. Besides new gas pipelines and LNG terminals were constructed in order to carry and deliver the gas subject to these long-term agreements. (Ertürk, 2009: 5–6).

The first important projects were the natural gas pipelines between Russia and Turkey (Malkoçlar, West Stream), the pipeline from Malkoçlar (on the Bulgarian border) to Ankara (1988), the Hamitabat CCGT Power Plant (1989), and natural gas delivery to Istanbul and Bursa (1992), Blue Stream Pipeline between Russia and Turkey (Durusu, Samsun). The Marmara Ereglisi LNG Terminal project also began in the same period as another important investment in natural gas infrastructure and the plant was commissioned in 1994 (The World Bank, 2015: 127).

The Turkish natural gas industry was state owned and vertically integrated through the 1980s and the 1990s. In 1987, the duties and responsibilities of Boru Hatları ile Tasima A.S. (BOTAS), which was founded to transport Iraq's crude oil in 1974, were expanded to include natural gas transportation and trade activities. It was granted monopoly rights on natural gas import, distribution, sales, and pricing in 1990 with Statutory Decree numbered 397 and dated 2 January 1990. As a state-owned company, it controlled transmission and distribution as well (Cetin & Oguz, 2007). Before the NGML the major distribution companies were EGO (renamed as BASKENTGAZ) in Ankara, IGDAS in Istanbul, IZGAZ in Izmit, AGDAS in

[1] Turkey holds only 0.13 trillion cubic feet (Tcf) of proven gas reserves as of 2017, ranking *84th* in the world and accounting for about 0.002% of the world's total natural gas reserves of 6,923 Tcf. https://www.worldometers.info/gas/turkey-natural-gas/.

Adapazari, BURSAGAZ in Bursa and ESGAZ in Eskisehir. Distribution companies operated in Bursa, Eskisehir and Adapazari were owned by BOTAŞ; whereas in Istanbul, Ankara and Izmit provinces the distribution companies were municipality owned entities.

Fiscal crises, inadequate investment, poor quality of service, negative effects of rent seeking, and external pressures provided an impetus for reform in the last decades. As a part of energy market restructuring, the legal structure of the natural gas industry, which was state owned and vertically integrated through the 1980s and the 1990s, was reformed in 2001 with NGML (Law no: 4646). (Erdogdu, 2009: 5–6, PWC, 2014).

NGML aims to ensure liberalization of the natural gas market to create a financially strong, stable, and transparent natural gas market, and to ensure an independent regulation and supervision in this market in order to make natural gas available to consumers with high quality, continuity, economy, and environmental sensitivity. To this end, the NGML sets forth rules regarding restructuring and liberalization of the market, transparent market entry conditions as well as non-discriminatory TPA to the gas infrastructure and as well, licensing obligations, licensees' rights and obligations, the extension of EMRA's duties and powers, tariffs, audit of the market, and sanctions to be imposed. Import, transmission, distribution, storage, marketing, trade, and export of natural gas have been determined as market activities and are regulated by EMRA.

As for distribution segment, the law describes;

- The natural gas distribution as "*Transport and retail sale of natural gasthrough the local gas pipeline network, to be delivered to customers*".
- The Distribution system operator as "*a legal person authorized to carry out the activities of natural gasdistribution and transportation through the local gas pipeline network in a specified city*",
- The Distribution network as "*natural gasdistribution facilities and pipelines operated by a distribution company in its designated area*".

Since all legal entities that would engage in natural gas market activities are obliged to obtain the necessary licenses, operating the distribution network is also subject to the license requirement. The principles and procedures related to the licenses are laid down in the Natural Gas Market License Regulation published in the Official Gazette on 7 September 2002. In compliance with NGML, the License Regulation requires the distribution activity to be carried out by licensees selected through a tender process for regions to be determined by EMRA. Whereas the main rights and obligations of the distribution licensees are laid down in the License Regulation, the details of tender process are regulated in the Natural Gas Distribution and Consumer Services Regulation dated 3 November 2002.

The distribution companies are not entitled only to operate the distribution network but also to supply natural gas to captive consumers (household customers) exclusively based on tariffs determined by EMRA and eligible consumers with bilateral sale contracts at prices freely determined between parties. The tariffs to be applied by distribution licensees are calculated and confirmed according to the principles and procedures specified in the Natural Gas Market Tariffs Regulation published in the Official Gazette dated 13 October 2016.

In addition to the abovementioned main legislative pieces, following regulations also include several rules pertaining to the distribution activity and licensees from different aspects:

- Regulation on Notifications in the Energy Market dated 27 May 2014
- Regulation on Natural Gas Market Facilities dated 26 October 2002
- Regulation on Audit and Preliminary Investigation in the Natural Gas Market dated 31 October 2008
- Natural Gas Market Interior Installment Regulation dated 18 September 2002
- Natural Gas Market Transmission Network Regulation dated 26 October 2002
- Several EMRA Board Decisions.

3 Licensing and Tender Process

License is a permission certificate given to legal entities by the Energy Market Regulatory Board (EMRA Board) to engage in market activities respectively for each market activity in accordance with the Law. Accordingly, there are 7 types of natural gas market licenses in Turkey which are import license, transmission license, storage license, wholesale license, distributor license, CNG license, and exportation license.

Local natural gas distribution service is taken on by the winning company for the license period to be determined by EMRA, taking into account the development level of the region, consumption capacity, and the number of users, including the ownership of the local natural gas distribution network.

EMRA is responsible for organizing tenders for new natural gas distribution licenses in the zones declared. Prequalification for tendering is based on the financial strength and experience of the potential licensees. Evaluation of the tenders are based on the unit service and depreciation charge (USDC) for supplying one kWh of natural gas to consumers (¢/kwh).

The main characteristics of these tenders can be summarized as follows:

- Distribution licenses are granted through a tender process for 30 years. Prequalification to participate in the tenders is based on financial strength and experience of the company.
- Bidding is based on the "unit service and depreciation charge" (expressed as an US cent) (USDC) for supplying one KWh of natural gas to consumers for the first 8 years. The firm with lowest USDC bid wins the tender and is qualified to set up all natural gas distribution infrastructure and supply gas to all consumers in its predefined distribution zone.
- The distribution companies may charge consumers a one-time "connection fee" when they are connected to network for the first time. The upper limit of this charge to be implemented in the tender period is determined by EMRA and specified in conditions of tenders.

- USDC and connection fee are the primary sources of cash flow for the winning company for the first 8 years.[2]
- The firm is required to start investment within six months and provide gas to consumers within eighteen months and cover all consumers in five years. (Erdogdu 2009, p. 14).

The gas distribution tenders for all 81 provinces of Turkey were completed by 2017. (EMRA, 2017; Rzayeva, 2018).

Main rights and obligations of distribution licensees are stipulated in Article 28 of in the Natural Gas Market License Regulation as follows:

(i) Distribution companies are responsible for the planning, projecting, construction, expansion, and operation of the distribution network in accordance with the procedures and principles specified in the legislation and the standards stipulated.
(ii) Distribution companies perform services related to the sale and delivery of natural gas to these consumers by enabling subscribers and eligible consumers within their area of responsibility to access the distribution network in accordance with the procedures and principles in the relevant *regulations* and communiqués published by EMRA.
(iii) Distribution companies are obliged to connect the consumers to the system if requested by the consumers within their area of responsibility.
(iv) The legal entity that receives a distribution license may sell the distribution network under its ownership to another legal entity, with the approval of the Board, in accordance with the principles and procedures specified in the Distribution and Consumer Services Regulation.
(v) Distribution regions may be combined under a single license, or existing distribution regions may be divided into more than one license area if requested by the distribution licensee legal entities and the Board approves technically and economically appropriate.
(vi) Distribution companies may provide natural gas transportation and ancillary services to eligible consumers and the suppliers they choose, upon their request.
(vii) Distribution licensees shall make natural gassupply programming and system balancing in order to fulfill the demands of users who buy natural gas from the regional distribution network and to ensure the reliability of the system, and ensure the natural gas supply security of those who buy natural gas from it.
(viii) Distribution companies shall establish a dispatch control center for the distribution networks under their responsibility excluding cities where consumption capacity is determined by the Board to be insufficient.
(ix) Distribution companies shall issue certificates to real or legal persons in accordance with the relevant legislation within the scope of the authority to

[2] So far, only one exception was made to this rule. The tender period for one distribution company was set as five years, instead of eight.

be obtained from the relevant Authority regarding the internal installation and service lines, and audit the real or legal persons they have issued certificates.

(x) Distribution companies shall submit to EMRA the information and documents proving that they supply the natural gas*offered to the subscribers from the most economical source.*

(xi) Unless otherwise agreed by the Board, a licensee may purchase at most fifty percent of the gas that it will distribute within one year from a legal entity.

(xii) Distribution companies cannot mix any substance other than the odorizing agent to be mixed with the natural gas delivered to the distribution network for safety reasons.

(xiii) Distribution companies must prove to EMRA that they operate in a safe and efficient manner.

(xiv) Distribution companies shall provide the necessary information to change the supplier to their subscribers who have the right to become an eligible consumer according to the eligible consumer limits to be determined by the Board each year.

(xv) EMRA may direct, supervise and monitor the distribution activities of the property owner distribution company and, when necessary, may purchase these services from real and legal persons at the expense of the distribution company.

(xvi) Distribution companies can have licenses in only two cities nationwide. However, this number can be increased by the decision of the Board, taking into account issues such as the development status of the cities, consumption capacity, and a number of users.

In addition to the abovementioned ones, other rights and obligations of the licensees selected through tender are specified separately in their licenses, taking into account the tender dossier and offers.

Although any company can have distribution license only in two regions nationwide, NGML allows the Board to increase this number, taking into account issues such as the development status of the cities, consumption capacity, and a number of users. The Board has extended this limit to 20 regions with Board Decision dated 27 December 2007 and numbered 1436/5 such that 30 distribution regions are currently operated by only three groups of companies.

Natural gas distribution companies are authorized to carry out natural gas distribution activities within the distribution region specified in their licenses, and are obliged to carry out distribution activities in all municipal areas of the cities corresponding to the distribution region. The defined distribution region of a distribution company can be redefined or expanded without being tendered by the Board in a way that does not exceed the provincial borders, taking into account technical and economic requirements. This provision has been put into practice many times, as a result of governmental policies that favor promotion of natural gas for residential and industrial use, resulting in expansion of natural gas distribution regions to include smaller towns.[3] According to NGML, if distribution companies, operating within the same provincial boundary for the relevant city, request expansion at the same time, the Board gives priority to the distribution company with a higher number of subscribers in the entire distribution region. The Board may divide a city into more than one distribution region with defined boundaries according to the population density and can tender each region separately. In the event that distribution companies operating within the boundary of a province do not request to expand the distribution area for the relevant city, an additional distribution license tender may be made for the relevant city, if deemed appropriate by the Board.

Distribution regions may be combined under a single license, or existing distribution regions may be split into more than one license area if requested by the distribution license holder. The Board approves such requests in case the merger or split of the regions are technically and economically viable. Network integrity and regional proximity are taken into account in the evaluation of applications regarding merger requests of distribution companies in terms of operational efficiency. The NGML requires that the procedures and principles regarding such application are regulated by a regulation to be enforced by the Board.

Natural gas distribution companies can carry out distribution activities in organized industrial zones by making network and connection line investments for organized industrial zones with the demand and consent of organized industrial zones. The procedures and principles regarding the conditions under which these activities can be carried out are determined by the Board by taking the opinion of the Ministry of Energy and Natural Resources and the Ministry of Industry and Technology. Investments and operating expenses to be made by distribution companies in accordance with the determined procedures and principles are taken into account in their tariffs.

4 Liability of Service and Access to Network

Distribution companies are obliged to connect consumers within the distribution region to the network if requested. However, the obligation to make the connection depends on the capacity of the system, and on condition that

- The consumer performs the procedures foreseen in the Natural Gas Distribution and Consumer Services Regulation and,
- The connection is possible technically and economically.

In case of disputes between the distribution company and the consumer on this issue, the Board is authorized to decide whether the connection is technically and economically feasible. The subject consumer, whose connection request is rejected by the distribution company, may apply to EMRA about the rejection. After receipt

[3] Turkey's gas supply reached 500 settlement areas as of the end of 2019, where total number of subscribers reached nearly 16.5 million (Gazbir, 2020).

of the defense of the distribution company on the issue, the Board decides if the rejection of the connection is based on a valid reason or not. The distribution company is required to comply with the Board's decision on the case.

The Natural Gas Distribution and Consumer Services Regulation governs rights and liabilities of distribution companies pertaining to connections and corresponding charges. In this context, the Regulation lays down the connection liabilities of distribution companies and,

- Reemphasizes the connection liability of distribution companies, in case the request of the consumers is compatible with the network capacity and the connection is technically and economically viable,
- Stipulates that the connection agreement between the company and the consumer governs the transfer of possession of the connection line, service line, and the meter to the company,[4]
- Points out that the connection fee is collected once during the signing of the connection agreement, and this charge cannot be refunded after the construction of the service line has started. If the connection fee has been taken for the same address before, this charge cannot be applied again. However, the cost of additional costs that may arise as a result of the expansion of the building or facility is covered by the owner of the property.

According to the Regulation, the design and construction of the internal installations are outsourced by the owner of the consumer facility[5] to the certificate holders. The subscription agreement to be made during the first subscription is signed after the submission of the necessary documents to the distribution company following the approval of the internal installation project and the accrual or collection of the security deposit determined within the framework of the legislation on tariffs. Preparation and approval of projects, their construction, conformity control, and commissioning of the internal installations are explained in the following sections.

In order to guarantee their claims, distribution companies are allowed to receive a security deposit from the subscribers using mechanical meters only once, during the signing of the subscription agreement. However, no security fee is collected from subscribers using prepaid meters. The security deposit is collected from the persons who make a contract to use natural gas. The methodology and unit charges for calculation of security deposits are determined by EMRA. Distribution companies are obliged to calculate security deposits for new subscriptions based on these guidelines.

[4] "Service line" refers to the line connecting the customer facilities to the distribution network. The cost of the service line is covered by the customers under the connection fee but remains an asset of the distribution company. "Connection line", on the other hand, is again the line connection the customer facility to the distribution network, but is built specifically for a single consumer. This regulation also points out that connection lines can be branched to service other consumers only if the consent of the original consumer is received and the capacity of the original consumer is not impaired.

[5] In implementation of this provision, owner refers to the owner of an independent section, the owner of a building or a facility, or their authorized representatives. An independent section stands for the initial 200 m^2 of the consumption facility and each consecutive 100 m^2 section.

4.1 Consumer Complaints

Consumers are required to submit their complaints regarding the network issues to the distribution companies firstly. Distribution companies, on the other hand, are required to respond to the complaints, taking into account the customers' requests no later than in fifteen days as of the date of application, and to take necessary actions. Records of consumer complaints and corresponding transactions are kept by the distribution company and forwarded to EMRA upon request. According to 2019 figures, over one million complaints were delivered to gas distribution companies.

The distribution companies are also obliged to establish a call center to provide uninterrupted service to consumers around the clock. Distribution companies are allowed to outsource this call service provided to consumers, remaining liable for all related conditions imposed by the legislation. Distribution companies also establish a "Complaint Application System" on their websites, through which consumers can submit their complaints, provide easy access and follow-up of applications.

4.2 Technical Requirements

On the technical side, natural gas distribution companies are obliged to establish a dispatch and control center to manage gas flows through their distribution networks. However, this condition is not required in cities where the consumption capacity is determined by EMRA to be insufficient.

Technical rules pertaining to internal installations of natural gas users are covered by the Regulation on Natural Gas Market Facilities. The purpose of this implementing regulation is determination of the procedures and principles regarding the establishment and control of all kinds of natural gas devices and related installations to be placed in the buildings or land for the consumption of natural gas. The regulation stipulates the compliance with any of the TS, EN, ISO, IEC standards referred regarding the design, construction, installation, control, commissioning, and operation of the interior installations. In the event that none of these standards are available, other standards accepted by Turkish Standards Institution can be referred.

Projects and construction of interior installation are done by certified companies at the expense of the consumers. The companies are certified by either distribution companies or other companies accredited by EMRA, according to the terms and conditions set out in the corresponding legislation, Regulation on Natural Gas Market Certificates, which covers the procedures regarding certification of real or legal persons carrying out the interior installations and service lines and construction and service activities in the natural gas market. The execution of the activities within the scope of the certificate, cancelation, termination, renewal, amendment of the certificates, and the rights and obligations of certificate holders are governed by this regulation.

The project approval, construction, conformity control, and commissioning of the internal installation made by the customer are under the responsibility of the distribution company. The distribution company may fulfill its control obligation through its own technical staff or through certified auditing companies working on its behalf, provided that the responsibility remains with it. Distribution companies may have their own technical personnel and certified companies working on their behalf to control the internal installations built for consumers, including residences, businesses, and industry, to use natural gas. If it is determined that the internal installation does not comply with the internal installation regulation to be published, the company may refuse to supply gas to the prospective customer or stop supply for existing customers. In such a case, the consumer has to renovate installation to comply with the legislation and apply for certification again. Distribution companies are not responsible for any damages and losses that may arise due to unauthorized modifications to the internal installations, improper and abusive use, incorrect and damaged equipment, installations not covered by the internal installation project, and lack of maintenance of the installation.

5 Price Regulation and Tariffs

In the context of its objectives, NGML aims liberalization of natural gas market in Turkey via vertical unbundling of supply and network services, where supply is opened to competition and network services are subject to regulation. In this scheme, retail price is composed of wholesale natural gas price, distribution charge, and taxes, wholesale price being the aggregate of gas (commodity) and midstream (mainly transmission and storage) service costs. While wholesale natural gas price applied to distribution companies and eligible consumers are freely determined between buyers and sellers, the distribution charge is subject to regulation. Therefore, this section outlines the price structure in Turkish natural gas market and the schemes in which the distribution charge is determined throughout the whole license period of a distribution company.

5.1 Pricing During Tender Period

As a result of the tender made by EMRA, the retail sale price to be applied to the subscribers of the legal entities that received the local natural gas distribution license for the period specified in the tender specifications is set in via the tender. The unit natural gas purchase price cannot exceed the upper limit of the retail sale price, which is the sum of gas price, system usage costs, and financial liabilities such as taxes. The system usage costs to be applied by these legal entities during the period specified in the tender specifications are the unit service and depreciation cost (USDC) they have offered in the tender. The USDC to be applied to eligible consumers by these legal entities cannot exceed the upper limit of the USDC they have offered in the tender.

The connection fee to be applied by distribution companies to their customers cannot exceed the connection fee they have offered in the tender, to be applied for the period specified in the tender specifications.

5.2 Legal Framework of Tariff Making

The law defines the fundamental principles of price regulation and sets the types of tariffs to be enforced by EMRA. Among these, distribution companies are obliged to implement their "retail price tariff", which is composed of unit gas price (commodity price), the distribution charge, and taxes. In carrying out the distribution service, the distribution companies are responsible for procuring gas from the most economical source and ensure financial viability.

Retail prices and tariff principles, which will be composed of the unit gas purchase price, unit service fee, depreciation charges, and other factors of the distribution company, are determined by the Authority. Consumers cannot be charged under any name other than the specified retail price. Retail tariffs may be recalculated in case distribution companies apply to the Authority, taking into account inflation and other issues. In determining these prices, the Authority takes into account the service cost, a reasonable rate of return that will allow investment, and the current natural gas purchase prices in the market and similar situations. The terms and conditions of the tariffs approved by the Board bind all real and legal persons subject to these tariffs.

In line with the principles specified in this article, the Authority prepares a tariff regulation for all tariff types. Within the framework of this regulation, tariff proposals are prepared by the relevant legal entities and submitted to the Authority. The Authority determines the tariffs based on the financial data and tariff recommendations of the relevant legal entities and market data. Relevant legal entities apply the tariffs approved by the Board. Tariff principles and limits can be readjusted by the Agency, taking inflation and other issues into consideration.

5.3 Price Components and Regulated Charges

Eleventh article of the NGML stipulates that EMRA is in charge of setting retail prices and corresponding methodologies, where retail price is composed of "unit gas (commodity) price, unit service charge, depreciation charges, and other factors" by its legal definition. As a result of this legal provision, retail natural gas price applied by distribution companies consists of wholesale gas price and the distribution charge, which is calculated by taking into account all capital and operational expenditures

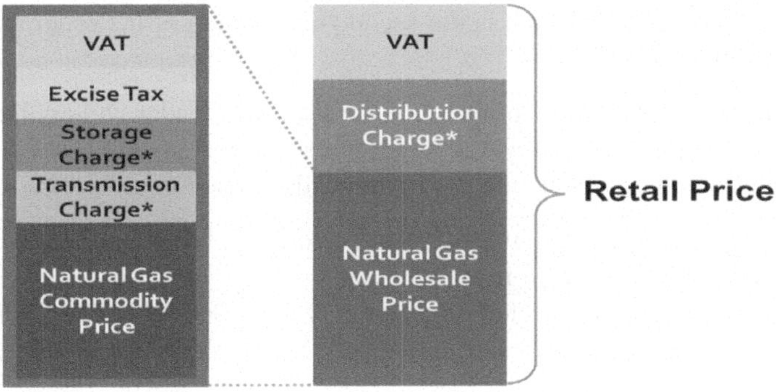

Fig. 1 Components of retail natural gas price[7]

[7] The charges shown with (*) are determined by EMRA, storage including underground gas storage only. Figure taken from (Kürkçü, 2018).

undertaken for distribution services.[6] The details of natural gas price components are illustrated in Fig. 1.

According to eleventh article of NGML, EMRA sets conditions for establishment of the first component of retail price, the "natural gas wholesale price". The law defines a "wholesale tariff", but requires that EMRA is responsible for setting the elements of this price and related conditions to be applied in the market. Indeed, the NGML does not let EMRA set wholesale tariff, but points out that the wholesale prices are freely determined among parties, according to the conditions set by EMRA. In corresponding Board Decision (No. 3577), it is stipulated that wholesale prices are set as a single price covering all expenses (gas price, transmission charges, storage charge, taxes, etc.). The distribution companies are obliged to pass wholesale prices to consumers directly, without and profit or loss over it. According to the provisions of NGML, distribution companies are obliged to prove that they buy natural gas at the lowest price possible on the market and meanwhile operate efficiently. The Boards Decision No. 3577 requires that,

- Natural gas wholesale prices are freely determined between the parties. In determining the prices, the principles of transparency and non-discrimination between equal parties are taken into account,
- In case suppliers make sales to distribution companies, the suppliers determine the final wholesale prices as a single price, including the costs of the transmission and dispatch control to be applied for each distribution company in the relevant month, the storage fee, if any, and the costs that may arise from commitments such as minimum purchase-maximum withdrawal. Suppliers notify distribution companies on the first day of each month.

[6] This is a result of the fact that, unlike many EU countries, unbundling rules did not apply to natural gas distribution in Turkey (Ardiyok & Turan, 2015).

The distribution charge which is the second component of the retail natural gas price, on the other hand, is

- Offered by the distribution company in the licensing tender and applied for the tender period, which is eight years for all but a few exceptional cases,
- Set by EMRA to be implemented after the tender period ends.[8]

The distribution charges set by EMRA are calculated based on the methodology described in the following section.

The final component of the retail price, namely the Value Added Tax, is regulated under the fiscal legislation, therefore lies totally beside the scope of the regulatory authority.

5.4 *Principles and Methodology of Tariff Setting*

Based on Article 11 of NGML, EMRA is authorized to set the retail price of natural gas and gas distribution companies are disallowed to charge any fee except the retail price regulated by EMRA. When setting the retail price, EMRA takes into account cost of service, reasonable rate of return to sustain investments and wholesale gas prices observed in the market. Determination of the regulated component of the retail price necessitates calculation of the revenue requirement for a specified period of service and forecasting demand for the same period, according to the general principles and methodology explained below.

5.4.1 General Principles

In the regulation of tariffs, the aim is providing natural gas to customers in a reliable, sufficient, high quality, continuous and low-cost manner, while ensuring financial sustainability and reasonable profitability that will allow investments. In doing so, supporting the development of competition, reflecting the benefit obtained from economic efficiency and increased competition to customers, avoiding cross subsidies, principles of fundamentality, non-discrimination among equal parties, and transparency are taken as basis as well. Following additional principles are strictly applied in tariff making:

- The tariffs do not include any items that are not directly related to the market activities of legal entities.
- Administrative fines applied to legal entities subject to tariff regulation are not included as a cost element in the tariffs to be prepared.

[8] There is one exception to this scheme. The distribution charge of one distribution company, BASKENTGAZ, which was founded before the NGML, is fixed by the temporary provisions of the NGML to be implemented for eight years after the privatization of the company.

- Legal entities whose tariffs are subject to regulation cannot apply any fees other than those determined within the scope of the relevant legislation.

The retail price to be applied by the distribution company consists of unit natural gas purchase price, the distribution charge, and financial liabilities such as taxes. This price constitutes the upper limit of the retail price to be applied by the distribution company. The distribution company cannot demand any price from the customers under any name other than the retail price, except the prices specified in the relevant legislation.

In making tariffs, both capital and operational expenses are addressed, while a reasonable rate of return to meet the capital expenditures and to sustain the investments in accordance with the provisions of the relevant legislation is foreseen. In this context, the distribution charge is also set to the extent that the efficiency targets estimated for each distribution company are reached. The distribution charges for different consumer groups are determined taking into account the burden and costs brought by the customers to the system and/or the consumption levels determined by the Board.

The distribution charge to be applied by distribution companies for eligible consumers who buy natural gas from a supplier cannot exceed the price determined by the Board. In other words, when implementing distribution charges, distributor companies cannot discriminate between eligible and captive consumers.

The Board may rank the distribution charge to be implemented by a company, taking into account consumption levels and/or customer groups. Consumption levels can differ among different distribution companies, depending on the characteristics of the distribution zones.

In some cases, distribution companies are allowed to supply gas to consumers in the immediate neighborhood, but outside their distribution zones. In such cases, the distribution charge applied by the distribution company for the customers whose connection is made outside the distribution zone cannot exceed the charge determined by the Board for that zone.

The distribution charge is billed,

- Directly to consumers who buy natural gas from the distribution company,
- To the suppliers, in case eligible consumers buy natural gas from a supplier through the distribution network, instead of the distribution company.

5.4.2 Tariff Calculation Methodology

As illustrated in Fig. 2, the tariff setting process is based on calculation of the revenue requirement for a specified period of service, forecasting demand for the same period, and calculation of distribution charge in Turkish Liras per standard cubic meter.

The revenue requirement consists of two main components, namely,

- Capital expenditures (CAPEX), which accounts for the opportunity cost of capital used to finance investments to carry out the distribution service and the depreciation of investments,

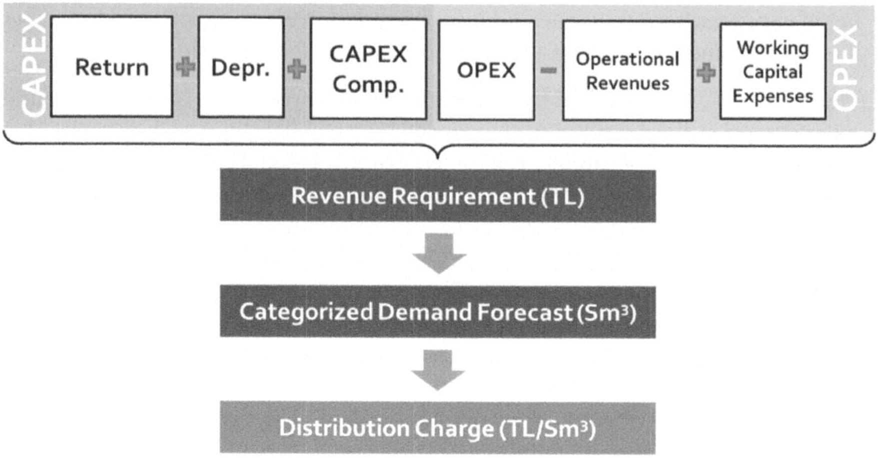

Fig. 2 Summary of tariff setting process[9]

[9] Figure taken from (Kürkçü, 2018).

- Operational expenditures (OPEX), which account for costs associated with regular execution of distribution service, such as overhead costs, cost of outsourced services (for instance metering, billing, information technology costs, rentals, consultancy, etc.), energy costs, some taxes, and charges, etc.

So far, tariffs have been calculated for two five-year periods (2012–2016 and 2017–2021). The length of the first period differed among distribution companies depending on the date their tender period ended. Tender period of few companies ended in the second tariff period. As most up-to-date facts, following parameters were taken into account for the second period of tariff calculations:

- The deprecation period was set as 22 years. Considering that the amortization period of natural gas distribution equipment in fiscal legislation is 30 years, and the first 8 years of the license duration is the tender period, the Board decided to set the depreciation period as 22 years.
- The reasonable rate of return was calculated as the weighted average cost of capital (WACC) of the distribution activity. WACC was calculated by capital asset pricing model and pre-tax WACC appeared to be 12.074% for the second tariff period.

In calculation of CAPEX component, companies' investments for past years were analyzed and benchmarked against statistically determined unit costs, when assessing the monetary value of their investments. The purpose of this benchmark is to compare companies' unit costs for different investment items (such as pipes of different diameters, valves, meters, etc.), disqualify excessive cost items in investments, and encourage efficiency in investments. Furthermore, when assessing the value of investments, companies' revenues (such as revenues from connection fees or scrap sales) were deducted from their costs to calculate their net investments. Companies' future investments were also taken into account to allow for investments for expanding gas networks.

Calculation of OPEX, on the other hand, aimed to predict the operational expenditures for the upcoming years based on the operational costs that took place in last four years. To this end, statistical and econometric models were used to relate operational expenditures to physical quantities (such as network size, volume of gas delivered, number of customers served, etc.). Companies' past costs were statistically benchmarked with data envelope analysis to generate an efficiency factor (so called X-factor in the literature), to force companies to operate more efficiently in the future. When modeling operational expenditures, companies' operational revenues were also taken into account to calculate their net costs. These included, but were not limited to revenues from meter activation/deactivation and fees charged for servicing internal installations. Considering the time lag between the operational expenses born and the receipt of distribution charges, financial cost of operational expenditures, which corresponds to one-month equivalent of WACC, is also added to the OPEX component.

Next step in tariff calculation is the demand forecast, which is done for every distribution region with two techniques:

- Total demand for consumers whose consumption is upto 10 million standard cubic meters was forecasted by using time series analysis, which is an econometric tool to forecast future value of a series of data based on past data.
- Demand for consumers whose consumption exceeds 10 million standard cubic meters was done individually, as such consumers are big industrial facilities that are few in numbers and whose demand is dependent on specific conditions, needing extra considerations.

Finally, both the revenue requirement and forecasted demand are segregated into consumption levels, to calculate distribution charges for each consumption level. When allocating revenue requirements to consumption levels, the burden of capital and operational expenditures placed by each consumption level on the company is estimated and corresponding allocation factors are generated. When segregating demand, demand for consumers whose consumption is upto 10 million standard cubic meters is split among consumption levels based on past data.

After entry into force, distribution charges are indexed with consumer price index on monthly basis. Except indexation, tariffs can be revised on the occasion of,

- Amendment of governing legislation in a way to require revision of tariffs,
- Amendment of license terms that potentially affect tariffs,
- Occurrence or end of cases of force majeure,
- Any significant change in tariff parameters.

6 Audit and Sanctions

The acts considered against law and the Administrative Fines and Sanctions to be applied are specified in Article 9 of the NGML. The forbidden acts are listed as provision of incorrect information, breach of the legislation, failure in performing license conditions, provision of misleading document and information, prohibited affiliate relations, performing activities outside of the license conditions, failure about informing the Board about changed conditions.

The fine to be applied to the market players that commit abovementioned acts ranges between 370.000TL and 740.000TL. EMRA is empowered to de-regulate and increase the amount of the fines annually if necessary. In case of repetition of the infringement within two subsequent years, the Board may double the fines. Besides monetary fines, the EMRA Board has the authority to impose other administrative sanctions like suspension of the activity or termination of the license.

The EMRA Board can start investigations against licensees either based on a complaint or on of its own accord. The audit of market takes place according to the NGML and the Regulation on Audit and Preliminary Investigation in the Natural Gas Market and includes:

- Examination of the compliance of all activities, practices, transactions, accounts, and financial statements of real or legal persons operating in the natural gas market with the provisions of the relevant legislation and generally accepted accounting principles,
- Determination of possible errors, deficiencies, irregularities and abuses, and
- Implementation of the sanctions stipulated in the NGML.

The investigation process is carried by EMRA personnel authorized on behalf of the Board and concluded with a Board decision whether to impose any sanctions or not. The audit process is expected to ensure that natural gas is made available to consumers within the framework of quality, continuous, cheap, and competitive principles in a manner that does not harm the environment, and create a financially strong, stable, and transparent natural gas market.

The decisions of EMRA regarding imposing sanctions, like other administrative acts, are subject to judicial control. The related market player can challenge any sanction decision of EMRA before the competent court within 60 days from notification. The competent court to see the NGML related cases against EMRA is Ankara Administrative Court. The decisions of Ankara Administrative Court are appealable.

7 Conclusion

Since incorporation of natural gas into its primary energy mix in 1987, Turkey has evolved into a remarkably established natural gas market with a total annual volume of about 45 billion cubic meters. After the introduction of NGML local gas distribution networks dramatically expanded to provide gas to all 81 provinces in the country. Today, gas distribution companies in Turkey have nearly 16 million subscribers, according to 2019 figures. Same figures point out that only household consumers consume more than 14 billion cubic meters of gas annually.

This picture was the outcome of ambitious policies to establish a financially strong, stable, and transparent natural gas market, and to ensure an independent regulation and supervision in this market in order to make high quality, continuous, cheap, and environmentally benign natural gas available to consumers. The legal framework of this policy was drawn by the NGML and EMRA became the architect of this sound natural gas market by developing and implementing the secondary legislation summarized above.

References

Ardiyok, Ş., & Turan, T. (2015). *Turkey: Distribution in Turkish Natural Gas Market*. Balcioglu Selçuk Ardiyok Keki Attorney Partnership. https://www.mondaq.com/turkey/oil-gas-electricity/444522/distribution-in-turkish-natural-gas-market

Cetin, T., & Oguz, F. (2007). The reform in the Turkish natural gas market: A critical evaluation. *Energy Policy, 35*, 3856–3867.

EMRA (2017), Energy Market Regulatory Authority 2017 Activity Report

Erdogdu, E. (2009). *A review of Turkish natural gas distribution market* (MPRA Paper 19088). University Library of Munich.

Ertürk, M. (2009). *Efficiency analysis of Turkish natural gas distribution companies by using DEA method* [Unpublished Master Thesis]. Middle East Technical University.

Kürkçü, M. (2018). *New incentive-based tariff methodology for natural gas distribution in Turkey* [PowerPoint slides]. ERRA Tariff & Pricing Committee.

GAZBİR-Natural Gas Distributors Association of Turkey (2020). 2019 Natural Gas Distribution Sector Report. https://www.gazbir.org.tr/2019-NATURAL-GAS-DISTRIBUTION-SECTOR-REPORT/

PWC (2014, February). *Liberalising natural gas in Turkey*.

Rzayeva, G. (2018, January). Gas supply changes in Turkey. *The Oxford Institute for Energy Studies*. https://www.oxfordenergy.org/wpcms/wp-content/uploads/2018/01/Gas-Supply-Changes-in-Turkey-Insight-24.pdf

The World Bank (2015). *Turkey's energy transition milestones and challenges* (Report No: ACS14951). http://documents1.worldbank.org/curated/en/249831468189270397/pdf/ACS14951-REVISED-Box393232B-PUBLIC-EnergyVeryFinalEN.pdf

Mehmet Kürkçü is working as the Group Head of Foreign Relations in Strategy Development Department of Energy Market Regulatory Authority of Turkey. His main responsibilities include managing relations of EMRA with international organizations and regional associations, establishing and managing bilateral relations with partner regulators, and monitoring developments in international energy markets. He previously served as an energy expert on determination of natural gas distribution and storage tariffs, monitoring natural gas and petroleum prices, and developing secondary legislation on natural gas and petroleum markets. Mehmet holds Bachelor of Science and Master of Science degrees in Chemical Engineering from Middle East Technical University. Later he conferred the degree of Master of Science in Energy Science, Technology and Policy from Carnegie Mellon University in the USA.

Mehmet Suat Kayikçi LLM Partner at LBF Partners Law & Consultancy: Mehmet Suat Kayikci graduated from Ankara University Law Faculty in 2001. He completed his LLM degree at Queen Mary, University of London in 2011. After working as an associate environmental expert at the Ministry of Environment for two years, Mr. Kayikci began to work as an energy specialist/lawyer at the Energy Market Regulatory Authority of Turkey in September 2004. In 2014, he left the Energy Market Regulatory Authority and joined LBF Partners as a founding partner. Kayikci occasionally teaches energy law in leading certification programs held by several universities and institutions. He extensively writes on environmental and energy law issues.

Regulation of Natural Gas Contracts

Burak Kepkep

Abstract Each market has its own specific nature driving market players and commercial relationships. To understand the market dynamics and rules governing natural gas contracts, the fundamentals of the Turkish natural gas market must be ascertained. Based on this foundation, it is possible to discuss the functioning and rules governing supply contracts, with a specific focus on the main characteristics surrounding import and wholesale contracts. The natural gas market in Turkey started its liberalization in 2001 with the enactment of the Natural Gas Market Law No. 4646. The principles, targets, and methodology put in place by the Law were incompatible with the actual market conditions. A quick review of the 20 years since the entry into effect of the Law and comparison with Turkish electricity market, which started its privatization in the same period, shows the failure to establish a liberal natural gas market in Turkey. The explanations contained in this study will, regrettably, reveal the prevention of legal evolution of natural gas supply agreements due to Turkey's natural gas dependency on foreign sources and a failure to establish a liberal market due to BOTAŞ's quasi-monopolistic position in the market.

Keywords Natural gas · BOTAŞ · EMRA · Take-or-pay · Back-to-back recourse · Antitrust laws · Grid operation principles · Settlement and balancing · Imbalance · General transaction terms · Organized natural gas wholesale market · Limitation of liability

1 Introduction

Coal-based energy production resources have been used since the Industrial Revolution, especially in industrial production and heating. Consequently, environmental pollution has increased over the years. Studies on the negative effects of global warming have been advanced and solutions have been proposed due to the negative impact that various methods of obtaining energy have had on the environment.

B. Kepkep (✉)
Polat Enerji Sanayi ve Ticaret AŞ, 1321-70 Roehampton Avenue, Toronto, ON M4P 1R2, Canada
e-mail: burak.kepkep@polatenerji.com

M. Eroğlu and M. Finger (eds.), *The Regulation of Turkish Network Industries*,
https://doi.org/10.1007/978-3-030-81720-6_7

In addition, governments, non-governmental organizations, and private institutions have cooperated to reach energy from renewables such as solar, wind, water, and biological sources.

Among fossil fuels, which includes coal, natural gas remained at the top of the list of 'clean energy' sources. Currently, residential consumers in Turkey use natural gas in their day-to-day activities, with all of the country's 81 cities having access to natural gas. In addition to direct individual use, another use of natural gas is converting it into electricity by processing it through combined cycle power plants. However, Turkey imports almost all of its natural gas from abroad in either liquid or gas state, which hinders its capacity to provide sustainable domestic policies. Considering the important place natural gas occupies in today's conditions, it is necessary to describe the legal aspect of natural gas trading from the moment of entering Turkey's borders up until its consumption.

Examining different markets in a uniform way will lead to legally inconclusive results. In the field of energy, commercial relations must be evaluated in line with each market's own conditions. Since oil, liquified petroleum gas (LPG), electricity, and natural gas each constitutes separate markets, the legal procedures and administrative decisions governing these markets must be examined in line with the underlying respective economic market conditions.

In this study, general explanations about the functioning of the Turkish natural gas market will be explained in order to understand the economic aspect of natural gas trading, followed by an analysis of natural gas supply agreements executed in the Turkish domestic market.

2 Functioning of the Turkish Natural Gas Market

2.1 Turkish Natural Gas Market at a Glance

Turkey sits at the crossroads of Europe, the Middle East, and the Caspian Sea and is an energy transit nation that links Caspian and Central Asian suppliers with European customers. The legal framework governing natural gas market was prepared by considering these factors as well as the country's specific needs and the physical conditions of this commodity.

2.1.1 Natural Gas in Turkey

Natural gas, which had very limited use in Turkey in early 1980s, is now used for heating, electricity generation, and industrial production. A large percentage of natural gas imported to Turkey is consumed in electricity generation through combined cycle power plants or CCGTs because of low establishment costs of these facilities.

The Natural Gas Market Law No. 4646[1] (**Law**) governs natural gas market activities in Turkey. Under the Law, natural gas is defined as *all natural hydrocarbons in gaseous state, generated or can be generated from the ground and other states of gas which has been liquefied and pressurized or physically processed by various methods (except for liquefied petroleum gas, LPG) in order to offer to market*. When evaluated chemically, natural gas belongs to the hydrocarbon family and fossil fuels. As part of hydrocarbons, production (exploration and processing) of natural gas is governed by Turkish Petroleum Law No. 6491[2] and is excluded from the scope of the Law.

In its gaseous form, natural gas enters the domestic market (or national grid) via pipelines from one of the entry points. Depending on the identity of its owner, natural gas can be exported, sold to third parties in domestic market, or consumed directly.

Liquefied natural gas (**LNG**) is natural gas transported by sea or land by means of specially designed tankers in cases where the transmission of natural gas through pipelines is not possible. LNG demand is increasing day-by-day due to limited transportation capacity of pipelines, and the share of LNG in the global natural gas market is expected to increase in the future.

Transactions involving compressed natural gas (**CNG**) include operations related to the purchase and compression of natural gas from wellhead, transmission, and sales. CNG licensed companies operating in this field are responsible for the operation of CNG filling, transportation, and discharge in accordance with the applicable legislation. In this way, as with LNG, natural gas can be transported to other places in different physical forms.

This study only focuses on natural gas supply agreements involving natural gas in its gaseous state; thus, transported exclusively through pipelines. Supply or sale agreements involving LNG and CNG are excluded from the scope of this study.

2.1.2 Natural Gas Supply Sources and Foreign Dependency of Turkey

When the percentage of energy consumption of different sources in Turkey is examined, natural gas and coal occupies first places with approximately a 30% share each. While natural gas consumption was almost negligible in early 1980s, its share of total energy consumption started to increase in 2000 with the organization of natural gas distribution tenders. Natural gas reserves and production volume in Turkey is insufficient to meet these consumption levels.

Almost 99% of the natural gas consumed in Turkey is imported. In 2019, Russia was Turkey's main natural gas supplier followed by Azerbaijan[3] and Iran. Algeria and Nigeria remain Turkey's principal LNG suppliers.

---

[1] Published in the Turkish Official Gazette dated 2 May 2001 numbered 24390.

[2] Published in the Turkish Official Gazette dated 11 June 2013 numbered 28674.

[3] In 2020, with the increased operations under the Turkish Anatolian Gas Pipeline (TANAP) project, Azerbaijan is expected to become Turkey's leading natural gas supplier. According to the mid-year report issued by the Natural Gas Companies Association of Turkey (*Gazbir*), as of 30 June 2020,

In order to meet Turkey's natural gas needs, natural gas is acquired from producer countries bordering Turkey through pipelines and from remote producers in liquefied form. Despite attempts to prevent dependency on the same supplier and, thus, to distribute risk, this situation has not changed Turkey's foreign dependency. Foreign dependency also hinders Turkey's decision-making capacity when adopting or implementing sustainable natural gas market policies.

2.1.3 Natural Gas Storage

The natural gas transportation capacity of pipelines is limited. Considering that almost the full volume of natural gas consumed in Turkey is imported from abroad,[4] natural gas entering from entry points and delivered to customers cannot always meet demand. This situation is easily observed during peak periods (namely, winter), when natural gas consumption in residences, businesses, and offices increases significantly.

Natural gas must be stored in order to compensate daily and seasonal changes and to ensure balanced operation of the national grid. This storage can be either underground or above ground. As of 2020, there are five operational storage terminals in Turkey, four of which are used for LNG storage.[5] By considering Turkey's foreign dependency in procuring natural, the number of storage facilities is insufficient to satisfy domestic market's needs and to create security of energy supply.

2.1.4 Natural Gas Supply Chain

The amount of natural gas produced in Turkey (473.87 million Sm^3 in 2019[6]) is disproportionate to actual consumption; hence the current natural gas high import volume.

When natural gas is imported, it is transmitted to Turkey via two different ways through import contracts. If there is a geographical connection between the producer country and Turkey, the natural gas is imported via pipelines. In the absence of direct geographical or natural gas pipeline connection, the natural gas produced is transported via the producer's national pipeline and converted to a liquified state, constituting LNG. LNG is transported to LNG terminals within the borders of Turkey by specially produced tankers (with LNG import or spot LNG contracts), where it is transformed to a gaseous state. Finally, LNG transformed to gas is injected into the national grid and transported to any location.

Azerbaijan is the leading supplier with a 24% market share, followed by Russia with 21% (Gazbir, 2020, page 2).

[4] In 2019, total natural gas consumption in Turkey was 45,285.50 million Sm^3. Total imports were 45,211.47 million Sm^3 (Energy Market Regulatory Authority of Turkey (EMRA) (2020), page iii).

[5] EMRA (2020), page 8.

[6] EMRA (2020), page ii.

2.2 *Legal Framework*

The competitive structure of natural gas market had not been designed in Turkey until 2001, the enactment of the Law. Today, natural gas market activities are carried out in accordance with the provisions of the Law and secondary legislation. Current natural gas market legislation has a regulatory structure that requires strict control of the state.

2.2.1 Role of EMRA

The organization charged with regulating and supervising the natural gas market is the Energy Market Regulatory Authority of Turkey (**EMRA**), which can directly affect the market through its decisions. EMRA is a public legal entity with administrative and financial autonomy.

EMRA regulates the market by preparing secondary decisions and issuing its board decisions with the aim of creating a financially strong, stable, and transparent energy market that can operate in accordance with the provisions of private law in a competitive environment. EMRA aims to ensure an independent regulation and supervision in this market, in order to make it available to consumers in a sufficient, high-quality, continuous, low-cost, and environmentally compatible manner. As of 2020, it is highly debatable whether EMRA reached these ambitious targets concerning natural gas market.

2.2.2 Instruments Assisting Supply Agreements

The signature of a supply agreement between a supplier (seller) and its customer (buyer) is not sufficient to realize the sale relationship between the parties. In other words, the supply agreement alone is not enough for the supplier to deliver natural gas.

A customer that wants to consume natural gas must establish a natural gas connection to its place of consumption. A *connection agreement* is executed to establish a natural gas connection for the relevant customer between the grid or relevant distribution network and the eligible consumer's place of work or a gas pressure regulating and metering station (**RMS**). Suppliers may take delivery of natural gas at a virtual transfer point (**VTP**) or at a relevant RMS and resell it to their customers at these points or any other contractually agreed location.

Standardized transport agreements (**STAs**) are standard-form (adhesion) contracts issued by BOTAŞ, the Petroleum Pipeline Company of Turkey, in accordance with the provisions of Transmission Grid Operation Principles of BOTAŞ (**GOP**) and are complementary to supply agreements. Shippers (such as sellers or suppliers) are obliged to sign STAs, which are classified as general transaction terms (**GTT**) in accordance with the Turkish Code of Obligations No. 6098[7] (**TCO**), the counterpart having no possibility to introduce any changes. Through STAs, natural gas is transported over the grid until the delivery point and duly delivered to the customer. If the delivery point is in a license area of a company holding a natural gas distribution license, in addition to STAs signed with BOTAŞ, a system usage agreement (**SUA**) must be signed with the relevant distribution company. In accordance with the SUA, the distribution company grants the supplier the right to transport natural gas over its own distribution network in return for consideration.

In addition to supply contracts and STAs, capacity transfer agreements and end-of-day trading operations are conducted as an indirect result of the supply relationship between the parties.

2.3 BOTAŞ's Dominant Position in the Turkish Natural gas Market

Players operating in a particular market must remove the barriers to entry in that market and comply with antitrust laws. The system that the Law aims to put in place for Turkish natural gas market was in this direction and, on its face, legislation seemed to provide all the right mechanics to implement the necessary measures for the functioning of a proper market. Still, principles enacted in 2001 have not been implemented as of 2020. As result, barriers to entry into the Turkish natural gas market still exists.

2.3.1 Vertically Integrated Structure of BOTAŞ

BOTAŞ was incorporated in 1974 and is currently engaged in the import, export, transmission, transportation, and wholesale of natural gas. The vertically integrated structure of BOTAŞ could not be changed despite the mandatory provisions of the Law.

As determined by the Turkish Competition Authority[8] nearly a decade ago, BOTAŞ has a monopoly in the import, export, and wholesaling of natural gas. Private sector activities continue only in the field of wholesaling of natural gas, whereas other areas are not open to competition. BOTAŞ's trade receivables arising from the market are treated as public receivables and, according to Article 12-g of the Law, they are collected in accordance with the provisions of Law No. 6183 on Collection Procedure for Public Receivables.

Therefore, BOTAŞ must be restructured in a way that separates import, transmission, and wholesaling activities from each other. To achieve this goal, BOTAŞ

[7] Published in the Turkish Official Gazette dated 4 February 2011 numbered 27836.

[8] Turkish Competition Authority, July (2012), Natural Gas Sector Report, Ankara, page 119–145.

must cease to be a public economic entity and its public law prerogatives must be removed.

2.3.2 BOTAŞ and Natural gas Prices

As result of Turkey's foreign dependency in natural gas, almost all Turkey's natural gas consumption is imported. In 2019, BOTAŞ's total market share of total natural gas imports were 96.42%.[9] The monopolistic role of BOTAŞ in the import of natural gas causes the transmission network operator to preserve its role as a supplier in the Turkish domestic market and affects natural gas prices substantially.[10]

This price-making capacity reveals itself in the relationship between BOTAŞ and distribution companies, which are obliged to provide natural gas to their residential subscribers; that is, the final consumers. Since BOTAŞ has also committed to supply natural gas to natural gas distribution companies that cannot find suppliers in the natural gas market due to a shortage of natural gas supply, BOTAŞ sells natural gas to distributors at the price it determines.

2.3.3 Contract Releases

In accordance with the provisions of Temporary Article 2 of the Law, contract release means the transfer of natural gas import contracts executed by BOTAŞ with all their rights and obligations to private players in the market.

Transfer of contract is the transfer of all rights and liabilities included in a contract and requires the written consent of the counterparty. In the past, attempts have been made to organize these contract release tenders. As part of the eligibility requirement, EMRA requires a "Seller's Consent" document to be submitted in the application file to guarantee the completion of the transfer. As one might imagine, this document, which must be issued by the counterpart of the contract (such as Gazprom), is practically impossible to obtain for most of the domestic market players due to a lack of commercial relationship or necessary financial capability. As a result, some of the tenders have failed to close and BOTAŞ's monopolistic role in the market has been preserved.[11]

In this context, we also find it useful to mention an alternative proposal referred to in the sector report issued by the Turkish Competition Authority. The Competition Authority recommended[12] completely abandoning contract releases and proposed that BOTAŞ arrange volume releases instead. Accordingly, BOTAŞ will continue

[9] Other natural gas importers are *Ege Gaz AŞ*, *Akfel Gaz Sanayi ve Ticaret AŞ*, *Batı Hattı Doğalgaz Ticaret AŞ*, *Bosphorus Gaz Corporation AŞ*, *Kibar Enerji AŞ*, and *Shell Enerji AŞ* (EMRA, 2020, page 12).

[10] See Nalbant et al. (2020), page 5 and elsewhere.

[11] The Economic Policy Research Foundation of Turkey (TEPAV) (2009), page 65–68.

[12] Turkish Competition Authority (2012), page 22.

to remain a party of the import contract and will transfer the volumes to wholesale companies licensed in Turkey. Although this seems to create a temporary solution, it will not eliminate the presence or decision-making capacity of BOTAŞ in the market.

2.3.4 BOTAŞ's Status Under the Law

The current status of BOTAŞ does not correspond with the provisions of Temporary Article 2 of the Law. In order to create a competitive market environment, the deadline for reducing BOTAŞ's market share to 20% and ending its vertically integrated structure was 2009. Although provisions of Temporary Article 2 of the Law related to BOTAŞ's market share have not been amended since 2001, the natural gas market is currently far from reaching this unbundling goal.[13]

Market players that started to operate in the Turkish natural gas market based on the clear provision of the Law has incurred losses from this conflicting situation. In addition, BOTAŞ's current situation creates an entry barrier in terms of antitrust laws.[14] Failure to comply with these provisions also damages the financial status of BOTAŞ and may lead to BOTAŞ's financial insolvency in the short term. By acknowledging this fact, the Executive transferred the ownership of BOTAŞ to Turkey Wealth Fund in 2017.

2.3.5 Is There a Natural gas Market in Turkey?

For a market to be formed, it is necessary to consider the existence of competition between existing market players, bargaining power of customers and suppliers in the market, barriers to entry of new players into market, and provision of spare goods and services. In view of all these factors, the foundations of a competitive market can be laid in a healthy way. Although the geostrategic position of Turkey in terms of access to natural gas resources (for example, its proximity to Russia, Iran, and Qatar, the largest producers in the world) stands out as an advantage, while its weaknesses are low production, the existence of weak infrastructure facilities, and an obsolete legal framework.

The enactment of the Law in 2001 was the first step in the liberalization of the natural gas market in Turkey. With the first natural gas distribution tenders—Kayseri and Konya—held in 2003, which are the first natural gas distribution tenders, the first major step was taken towards creating a liberal natural gas market. Despite positive developments of 2000s, the natural gas market did not reach the targets defined by the Law. The principal reason for this is BOTAŞ's vertically integrated structure, which prevented new entrances to the market.

As revealed by the Competition Authority, if the development of natural gas markets are considered in four stages as introduction, growth, development, and

---

[13] Eroğlu (2010), page 142.

[14] Energy Community Secretariat (2015), page 27.

maturity,[15] the Law skips the stages of birth and growth and jumps directly to development. The Law defines a competitive market but does not prescribe a forward-looking strategy that includes transition periods. This situation is like constructing a building with an unstable foundation. Therefore, the Law, in its current form, cannot reach a liberal and competitive market and must be amended or reissued accordingly.[16]

3 Natural Gas Supply Agreements

3.1 Legal Nature

The supply of natural gas can be defined as the delivery of natural gas from a seller to a buyer as a commodity. In accordance with the applicable natural gas market legislation and the conditions of the licenses in the market, delivery of natural gas can occur in various ways: from natural gas distribution companies to final consumers; from BOTAŞ to natural gas distribution companies or wholesalers, or from producers located abroad to importers in Turkey.

Prior to the promulgation of the Law, the right to carry out natural gas wholesale activities in Turkey belonged exclusively to BOTAŞ. In accordance with the Law, wholesale means the sale of natural gas to distribution companies and eligible consumers.[17] Provisions of the Law and the Natural Gas Market Licensing Regulation[18] allow wholesale companies to engage in wholesale of natural gas with other wholesalers, importers, eligible consumers, and distribution companies. Our explanations in this study related to supply agreements should be understood to include sales between companies entitled to engage in wholesale activities (for example, among wholesalers and/or importers or between wholesalers and distribution companies) excluding natural gas sales to residential consumers. Furthermore, import contracts executed with the producers located abroad and wholesale supply agreements executed in the domestic market are based on quasi-identical contractual and commercial principles.

Natural gas supply is a form of movable sale and is subject to general provisions regarding the sale of movable goods under the TCO in addition to the provisions of the natural gas market legislation. Therefore, complementary legal rules provided under the TCO shall be applicable to supply agreements to the extent that certain issues are not regulated by the parties.[19]

Under a natural gas supply agreement, a supplier's obligation is not related to the delivery of a determined *object*, but a determined *type*, which is natural gas

---

[15] Turkish Competition Authority (2012), page 15–21.

[16] Rzayeva (2014), page 65–68.

[17] EMRA set eligible consumer limit in natural gas market for (2020) as persons consuming at least 75,000 m^3 of natural gas.

[18] Published in the Turkish Official Gazette dated 7 September 2002 numbered 24869.

[19] Ayrancı (2010), page 139 et al.

possessing predetermined technical specifications.[20] The simple fact that natural gas in a grid is owned by a supplier is enough to fulfill the supplier's obligation towards its customer. For instance, if a supplier purchasing natural gas at the Malkoçlar Entry Point[21] sells this to a customer in İzmit, legal transfer of ownership, as well as physical transmission of the gas, must be realized by executing an STA with BOTAŞ. Even so, the natural gas that the seller delivers to its customer is not actually the natural gas at the Malkoçlar Entry point, but the gas circulating in the grid.

Supply agreements further establish a continuous performance obligation between the parties. The obligation of the supplier to sell natural gas and the obligation of the customer to pay the contract price continues throughout the contract period.

Natural gas supply agreements can be qualified as synallagmatic contracts where each party to the contract is bound to provide something to the other party. Consequently, if the contract is terminated before its term, the continuous performance of the contract results in a termination *ex nunc*, and for this reason, rescission (or termination *ex tunc*) is not possible.

3.2 Parties

There are two parties to a natural gas supply agreement: a seller, authorized by EMRA to conduct natural gas wholesale activities, and a buyer (customer).

The seller must hold necessary license in order to sell natural gas to its customer. Under the applicable natural gas market legislation, the following entities can engage in wholesale of natural gas: BOTAŞ, wholesale license holders, import license holders, export license holders, producers, and distribution license holders.

If both parties to a natural gas agreement is incorporated in Turkey, then the agreement must be executed in Turkish language in accordance with the provisions of Law No. 805 on Mandatory Use of the Turkish Language for Corporations.[22] Natural gas import agreements usually include a supplier incorporated outside Turkey allowing the execution of the relevant agreement in a language other than Turkish. For wholesale supply agreements executed in the domestic market, the parties do not have such option and must execute the agreement in Turkish.

[20] For a similar analysis of electricity supply agreements, see Yavuz (2011), page 93 et al.

[21] Malkoçlar is one of the nine natural gas entry points to Turkey's natural gas grid. The other entry points are *Durusu*, *Gürbulak*, *Türkgözü*, *TANAP Seyitgazi*, *Marmara Ereğlisi* LNG terminal, *Egegaz* LNG terminal, *Dörtyol* floating storage and regasification unit (FSRU) terminal.

[22] Published in the Turkish Official Gazette dated 22 April 1926 numbered 353.

3.3 Term

Natural gas market legislation does not limit the term of supply agreements, so the parties are theoretically free to determine the term. Under current market conditions, however, the term of natural gas supply agreements are usually determined as one gas year, which corresponds to one calendar year. In exceptional cases, natural gas supply contracts are concluded for shorter or longer periods. The primary reasons for limiting the contract period to one gas year are as follows:

- One year contract term limitation of importers with their own suppliers due to supply-demand uncertainty in international and domestic markets
- Uncertainty of natural gas price in international and domestic markets preventing long-term commitments
- Impossibility for suppliers to collect more than one calendar year of demand from the market
- Annual change in eligible consumer limits.

In some cases, supply agreements are concluded for longer periods, especially between wholesalers who have signed contracts with importers. In very exceptional cases and considering the commercial relationship and mutual trust between the parties, or if the companies are members of the same group of companies, the term of the contract can extend to 10 years. The term of the contracts executed with eligible consumers or industrial customers, on the other hand, does not exceed one year.

These market realities also shape the contract structure. If the seller is confident about its own capacity to allocate natural gas to that particular buyer and its financial capacity, then, as a commercial strategy, the contract includes automatic renewal clauses with the aim of not losing the relevant customer.

3.4 Major Provisions

Natural gas market legislation does not provide mandatory content to natural gas supply agreements. Although this situation is compatible with the development and maturity periods of a market, it is not beneficial for Turkey where a competitive strategy is not determined. Due to the absence of regulation, seller and buyer will be able to include provisions they wish within the framework of the principle of freedom of contract. This situation creates an obstacle to have solid foundations in Turkish natural gas market.

3.4.1 The Principle of Freedom of Contract

The principle of freedom of contract is directly included in Article 4-(4)-(e)-(4) of the Law. Therefore, the parties to a supply contract are free to choose the contracting party and to determine the form of the contract.

The parties to a natural gas supply agreement is mostly private legal persons in the sense of civil law. Pursuant to Article 55 of the Turkish Civil Code No. 4721, public legal entities are subjects of public law and are individuals and property groups subject to public law rules. Contracts to be executed between private law legal entities will be subject to the rules of private law, and the implementation of the principle of freedom of contract in these contracts, even in the absence of provisions, arises from the identity of the contracting parties. The supply agreements between a private legal entity (as supplier) and a public legal entity (as customer) will be subject to the provisions of private law. Therefore, public law will not be applicable to natural gas supply agreements.

In accordance with the clear provision of the Law, the parties to the supply agreement are free to determine the unit price of the natural gas, payment terms, amount, and form of the collateral to be issued by the buyer in favor of the supplier. In the past, attempts have been made to oblige the parties to execute their supply agreements in writing, mainly to collect Turkish stamp duty (approximately 1% of the contract price) over these agreements and to disclose the commercial terms to EMRA. However, these attempts have never seen the light of day and, under the current legislation, the supply agreements are not required to be executed in writing.

In accordance with the broad contractual freedom recognized by the Law, the parties are free to determine commercial terms to the sale relationship. However, the principle of freedom of contract in the Law is not unlimited.

In general, principles regarding the validity conditions of the contract provided under the TCO are applicable to these contracts. For example, provisions of a supply agreement should not be contrary to the invalidity conditions referred to in Article 27 of the TCO. Likewise, provisions of the supply agreements must comply with ancillary agreements accompanying supply agreements (such as STA, connection agreement, SUA) and natural gas market legislation, especially EMRA Decisions, provisions of GOP, and antitrust laws.

3.4.2 Delivery of Natural Gas

Transfer of ownership of movable property such as natural gas occurs with transfer of possession. Correspondingly, delivery of natural gas from a supplier to its customer is required to complete the sale agreement and to fulfill supplier's obligation to its customer.

Delivery Per Se

Operation of a natural gas grid is like the functioning of a swimming pool. Water entering a pool should be discharged immediately to prevent overflow. To keep the national gas grid's pressure balanced, natural gas entering the grid at one of the entry points must leave at the same rate by consumption or through exports.

In most cases, natural gas is not physically delivered from supplier to its customer. There may be two different delivery possibilities to transfer ownership of natural gas on grid: VTP or RMS stations.

A VTP is a virtual point on the grid. These points also referred to as national balancing entry or exit points under GOP, are places that do not physically exist in the grid and are intended to allow suppliers to trade with each other in order to correct their imbalance. The parties to a supply agreement may agree on the delivery of natural gas on this VTP. In this case, natural gas circulating in the grid and owned by supplier is delivered at a point at the network determined by the parties in the contract. In order to perfect this transfer, the contracting parties must make mutual entries to the electronic bulletin tables (**EBT**) held by BOTAŞ, the transmission system operator. This capacity reservation is carried out in accordance with the provisions of GOP. BOTAŞ, which is responsible for the balancing of the transmission network, can follow the natural gas supply and consumption or sub-supply movements of the parties through the EBT system, and can identify the party responsible for this in case of imbalance.

RMS are established to reduce gas pressure and to measure the gas volume during the transportation of natural gas from one point to another. They are classified as RMS-A, RMS-B, and RMS-C according to their pressure ratios. The parties to the supply agreement can agree that the natural gas in the transmission system can be transferred at these RMS. In this case, natural gas in the grid occurs at the RMS point and the measurement will be made at this point. Finally, within the scope of supply agreements, it is possible to deliver natural gas at the customer's natural gas meter, as in subscription agreements.

These delivery processes, over VTP and RMS, are conducted in an abstract manner (that is, parties to the supply agreement do not have physical access to natural gas) and are frequently used in practice by importers and wholesalers.

Once the natural gas reaches the delivery point, natural gas becomes the property of the customer. The transfer of rights and liabilities in supply contracts within the meaning of Article 208 of the TCO takes place at these delivery points.

Poor Performance in Supply Contracts

In civil law, poor performance is generally defined as a sub-classification of positive violation of a contract and includes defective performance leading to the breach of ancillary obligations and protection duties in a contract.[23]

---

[23] For more detailed analysis on poor performance in civil law, refer to Aral (2011), page 106–112.

When supply agreements are examined from a poor performance standpoint, it is almost impossible, in practice, for a supplier to have defective performance. In a natural gas supply agreement, the main obligation of the supplier is to sell natural gas, and the main obligation of the customer is to pay the contract price. In addition to these reciprocal obligations, the parties also have ancillary liabilities such as provision of guarantee and confidentiality. If one reverts to the original contractual obligations of the parties, the customer's payment obligation cannot be affected by defective performance.

From a supplier's perspective, considering that natural gas is consumed from grid and gas quality at the entrance points to the grid is measured by BOTAŞ, it does not seem possible to encounter a situation where natural gas does not correspond to the technical specifications. The actual process of delivery of natural gas and how the balance of the system is ensured is of great importance. Considering this process, the supplier delivers natural gas to its customer by utilizing the transmission grid owned by BOTAŞ and the distribution network owned by the natural gas distribution company, which is the continuity and a part of national grid. Although natural gas to be injected into the grid must meet certain technical criteria in accordance with the provisions of GOP, the inspection of whether this natural gas meets these technical criteria is done at BOTAŞ metering stations located at the entry points where the natural gas enters Turkey. Therefore, it is technically impossible to have in the grid natural gas that does not meet technical requirements provided under GOP.

Failure to fulfill ancillary obligations (such as provision of payment security) committed by the parties under the contract, other than defective performance, is always possible, and if poor performance occurs in this way, the parties will be able to use all their contractual and statutory rights against each other.

3.4.3 Measurement of Sale

Pursuant to the Natural Gas Market Tariffs Regulation,[24] energy for the measurement of natural gas is shown based on (kWh) and the upper calorific value of 9155 kcal/m^3 based on volume (m^3). Natural gas consumption is normally measured using meters. The consumption value is calculated by multiplying the consumption volume calculated via meters by the unit price in the contracts and invoiced accordingly.

If parties agree to transfer the ownership of natural gas at RMS or VTP, the measurement of the gas sold is determined in accordance with the provisions of GOP and in line with the capacity reservations of the supplier and the customer on BOTAŞ EBT. Detailed technical criteria for the operation of this system are provided under GOP.

The measurement process to be carried out in both cases is important for determining whether the supplier has fulfilled its procurement obligation in accordance

[24] Published in the Turkish Official Gazette dated 13 October 2016 numbered 29856.

with the terms of the contract and how much the buyer has fulfilled its purchase commitment in the contract.

After determining the contractual commitments related to measurement, BOTAŞ determines whether the transmission system is balanced.

3.4.4 Determination of the Contract Price

In accordance with the Law and the Tariffs Regulation, natural gas wholesale tariffs can be freely determined by the parties. Despite this clear provision of the Law, EMRA has determined the upper limit for BOTAŞ's wholesale tariffs and applicable criteria in the calculation of tariff until the end of 2007.

Since 2008, EMRA has not intervened in the determination of wholesale prices. It is not possible to discuss a single tariff in the wholesale of natural gas. Connection tariffs, transmission, and shipment tariffs, storage tariffs, wholesale tariffs, and retail tariffs may also be involved in a supply relationship. The connection tariffs and the tariffs regarding transmission and shipment control are determined in accordance with the provisions of the Tariffs Regulation. Considering the vertically integrated structure of BOTAŞ in the natural gas market, transmission and shipment costs are determined by EMRA board on BOTAŞ recommendation.

Under natural gas supply agreements executed by and between commercial private law companies, the price of imported natural gas sales are generally fixed to the price of Brent crude oil. For instance, the price of natural gas supplied by Gazprom is determined in this way. The fact that the contract price is indexed to Brent crude oil constitutes a substantial risk for the parties and should be evaluated in detail. In addition, these price mechanisms are generally determined in hard currency, which is exempted under the provisions of Decree No. 32 on the Protection of the Value of Turkish Currency[25] (that is, natural gas sales being movable sale). This imposes a mutual diligence obligation on the contracting parties in order to determine who will bear the currency risk in payments. Finally, in the contracts concluded between the importer and its customer (that is, the first supplier of the imported natural gas) to the market, profit sharing price determination mechanisms can also be negotiated.

BOTAŞ determines the natural gas sales price according to the import contracts to which it is a party. However, these prices are, of course, directly affected by price increases adopted by the Executive.

Most of natural gas supply agreements executed in the domestic market include price mechanisms that are indexed to natural gas sales prices determined by BOTAŞ. Under the back-to-back recourse principle, it is possible to encounter situations where suppliers reflect the price mechanisms of import contracts to their customers in the domestic market. However, this situation is seen in exceptional cases involving high sale volumes and, therefore, the main supplier carries too much price risk.

[25] In accordance with the provisions of Decree No. 32, contract price and other payment obligations arising from certain agreements between Turkish residents cannot be denominated in, or indexed to, foreign currency except limited exempted circumstances (such as a movable sale).

Finally, it is also worth mentioning spot natural gas sale markets established in accordance with the provisions of the Regulation on Organized Natural Gas Wholesale Market.[26] Wholesale, import, and export license holders have been authorized to register with the organized wholesale market operated by EPİAŞ, *the Energy Exchange Istanbul*, since September 2018 to engage in natural gas wholesale in the spot market for the balancing and settlement of natural gas sales and where the natural gas price is determined daily.

3.4.5 Force Majeure Under Supply Agreements

Force majeure can be defined as an unforeseeable circumstance that is completely outside the control of the parties preventing a party from fulfilling a contract. Force majeure circumstances included in natural gas supply agreements materialize provisions of Article 112 of the TCO.

A force majeure circumstance that is frequently encountered in the natural gas market is Hard Day, defined under GOP as the day when the system balance is disrupted due to the increase and/or decrease in the natural gas consumption amounts of the shippers and/or the cessation of natural gas intake at any entry points. BOTAŞ notifies all market suppliers when Hard Day application is activated in line with GOP. In such cases, suppliers are obliged to stop natural gas supply; likewise, the customer should stop its consumption. Hard Day manifests itself in peak periods, generally from December to February, when the supply-demand balance in Turkey is imbalanced. Due to the imbalance in the system during peak periods, suppliers stop natural gas supply to industrial customers and eligible consumers, but distribution companies continue to supply natural gas to residential consumers.

Political crises that may be experienced with or between natural gas supplier countries can also manifest themselves as force majeure events. It is undisputed that natural gas can be used by producer countries as a foreign policy tool. The slow progress of various natural gas pipeline projects to which Turkey is a party affected by regional and global conflicts. This situation is reflected in supply agreements executed in domestic market.

Any event that may directly or indirectly affect natural gas transmission system or its electronic infrastructure is considered force majeure and brings forward the importance of infrastructure and cyber security to create a sustainable energy market.

Circumstances accepted as force majeure under STAs or system usage contracts to be signed with the transporter (BOTAŞ or the company holding natural gas distribution license) are also considered as force majeure.

[26] Published in the Turkish Official Gazette dated 31 March 2017 numbered 30024.

3.4.6 GOP and Supply Agreements

BOTAŞ has issued GOP in order to operate the transmission network in an economical, efficient, and safe manner. GOP provides a set of rules that contain procedures and principles for the physical balancing of the natural gas system. It is obligatory to comply with the provisions of GOP in every direct or indirect operation related to natural gas transmission network.

GOP includes provisions for dealing with technical and operational issues related to the transmission network. In summary, GOP regulates *introduction to the system, capacity allocation, shipment control, and system balancing, transport amount notification and schedules, determination of measurement and transport quantity, conditions of use, characteristics of the transmission network, and daily operation and maintenance requirements.*

As previously noted, natural gas sold by contracting parties is entered mutually on the EBT system, which was established to monitor market movements in the supply process of natural gas and operated by BOTAŞ. This process includes how and until when the parties can make their capacity reservations and, if there is idle capacity in the system, how this capacity can be used.

Capacity Transfer Contracts

Capacity transfer in GOP refers to when the shippers making capacity reservations on the transmission network do not use their reserved capacities and transfer them to other shippers. Suppliers often do not know the exact volume of natural gas they will register to the system in the next gas year until the last notification date of the capacity reservation because they have not executed all supply arrangements with their customers and/or they do not know exactly the amount of natural gas they will sell and market.

Capacity reservation can be considered as an annual lease agreement to use grid. Shippers may transfer the leased but unused capacity to other shippers under other capacity transfer agreements in accordance with GOP. This situation adds flexibility to the functioning of the market. Otherwise, the parties will face commercial losses and monetary sanctions.

End-of-Day Trading

End-of-day trading fulfills the function of eliminating the imbalance that occurs in the system between the suppliers, thus preventing one of the parties from bearing the consequences due to imbalance.

In some cases, suppliers may be short of natural gas supply since their buyers consume more natural gas than the maximum volume allowed in their contract. If the volume of natural gas that a supplier registers with EBT is below the volume of natural gas consumed, it can be compensated within the scope of end-of-day trading. In other words, after the consumption is realized, the supplier can determine its own imbalance to the system and purchase missing natural gas volume from other suppliers that provide more natural gas input to the system. In such cases, end-of-day trading agreements may be executed with other suppliers through framework agreements and these operations may be conducted at the organized natural gas wholesale market for the settlement and balancing. In this way, the supplier can legally be compensated for the imbalance in the system and, naturally, the consequences related to this imbalance.

Complementary Nature of STAs

It is not possible for natural gas to be transported from one point to another on the transmission system spontaneously without any legal basis. STAs are executed to complement the supply agreements and to perfect the passing of title. If STAs are not concluded, it will not be possible to determine whether the supplier has delivered the natural gas.

Provisions of GOP and any amendment are automatically binding on transmission system users. Even if there is no reference to the provisions of GOP in a natural gas supply agreement, these provisions remain binding on the parties. In other words, although GOP provisions are part of the secondary legislation in the natural gas market, there should be no hesitation regarding their mandatory nature and their direct application to supply agreements and transactions related to grid.

3.4.7 Take-or-Pay Obligation

The take-or-pay obligation is a *sine qua non* provision included in natural gas supply agreements.[27] This obligation, also known as *take or, if not take, pay*, is included in all supply agreements.

Rationale of the Obligation

Natural gas supply agreements are executed for an initial term of one calendar year. All contracts in a supply chain, from production of natural gas to its consumption, are linked with each other commercially. All parties from production to consumption are somehow interdependent in the process. Any commercial default that may occur in the sale process will affect all contracting parties in the supply chain.

When considered commercially, the contractual commitments of a seller under the supply agreement executed with its initial supplier are closely linked to this seller's obligations to its customers. Compliance of a seller's customer with its contractual commitments is critical for that seller to avoid any imbalance costs in the transmission system and to fulfill its own contractual commitments to its initial supplier. For this

[27] Rzayeva (2018), page 9.

reason, take-or-pay provisions have been introduced to supply agreements to ensure the customer's fulfillment of natural gas consumption commitments towards their suppliers.

Minimum and Maximum Purchase Commitments

Each natural gas supply agreement includes minimum and maximum purchase commitments of the customer to materialize its take-or-pay commitment.

Parties to a natural gas supply agreement cannot exactly determine natural gas volume to be consumed within a calendar year (that is, to the Sm^3). Accordingly, the volume of natural gas that can be consumed by customer is determined with a margin by including minimum and maximum consumption values. The customer commits to consume minimum volume of natural gas, which is determined at a rate over the annual contract amount. Considering the practices in the natural gas market, this rate is generally determined as 80% of the annual contract volume.

Maximum consumption volume allows a customer to consume natural gas at a certain rate above the annual contract volume determined by the parties and to meet the potential needs of the customer in the contract period. This rate had been 20% above the annual contract volume until the end of 2011, in accordance with natural gas market practice. However, due to the natural gas shortage experienced in Turkish market in 2011, these rates were revised in a way to allow consumption up to 10% above the annual contract volume. The purpose of having maximum volume in a supply agreement is to determine maximum exposure of the seller under the respective agreement. In this way, the supplier will know the maximum volume of natural gas it may be obliged to provide to its customer, enter into its contracts with its own (initial) supplier taking these amounts into account, and ensure that the transmission system (and therefore itself and its customer) does not fall into imbalance under GOP, and does not pay respective penalties.

Consequences

Under natural gas supply agreements, the supplier is obliged to sell natural gas and the customer is obliged to pay the contract price. In addition to the obligation to pay the contract price, the customer commits to consume the minimum natural gas volume included in the contract. In case the minimum purchase commitment is not fulfilled by the customer for reasons other than those defined in the contract, the cost of the missing consumption volume will be invoiced to the customer at the end of the contract period. In other words, in a supply agreement where the minimum purchase commitment volume is determined as 80% of the total contract volume, the difference between the total of the customer consumption below this rate (for example, 70%) and 80% (=10%) will be qualified as missing consumption. The customer will not have fulfilled his contractual commitment and the contract price corresponding to this missing consumption volume will be invoiced by the supplier to its customer.

The take-or-pay mechanism functions as a contractual penalty in the sense of Article 179 of the TCO. A customer who fails to fulfill its minimum consumption commitment is obliged to pay the difference to its supplier as a contractual penalty. While determining the payment amount, it is irrelevant to determine whether the supplier incurred a loss or not. The underlying commercial reason is simple: the supplier concludes its contracts with its own suppliers in accordance with the demands and contractual obligations of its customers. If any of the supplier's customers fail to fulfill the minimum consumption obligation, it will have to resell the amount of natural gas in the market or lose it. Considering the contracted natural gas consumption volumes and the monetary values they represent, the penalty conditions for take-or-pay are indispensable conditions of supply agreements. In accordance with the provisions of Article 22 of the Turkish Commercial Code No 6102[28] (**TCC**), since the parties to a supply agreement are legal entities (that is, merchants), the reduction of the penalty amounts cannot be requested from the court.

3.4.8 Make-Up or Carry-Forward Clauses

Within the scope of take-or-pay obligation, suppliers may give their customers the right to compensate their missing consumption volumes in the following year(s) instead of invoicing missing consumption as a contractual penalty. Contractual provisions that allow customers to compensate their missing consumption are also called *make-up* or *carry-forward* clauses.

Under these contractual arrangements, the consumption volume below the minimum purchase commitment is compensated in the following gas year(s). The special conditions of this may vary depending on the commercial relationship between the parties. For example, a certain proportion of the contract price corresponding to the missing consumption volume can be paid at the end of the first gas year, whereas the remaining portion can be paid at the time of compensation; that is, at the time of the consumption of the missing volume.

3.4.9 Back-to-Back Recourse Principle

From production to final consumption, a chain of contractual relationships exists involving different market players. Any issue, especially related to supply or price, that may arise from the contract between a supplier and its own (initial) supplier, may successively trigger a dispute between the supplier and its customer. In serial procurement contracts, each supplier endeavors to transfer the risks arising from the previous supply agreement to the contract with its own customer. Consequently, the price mechanism of natural gas (other than the profit rate), take-or-pay provisions, minimum and maximum purchase commitments, conditions on default, provisions

[28] Published in the Turkish Official Gazette dated 14 February 2011 numbered 27846.

on invoicing and payment, and force majeure events are exactly reflected in the successive agreement.

Under the back-to-back recourse principle, a risk arising from the producer may affect all contracts in the procurement process and can be reflected exactly to the final consumer, leading to a domino effect.

Natural gas is a commodity that is not abundant or accessible to all. It is often not possible for a supplier to procure natural gas from elsewhere and fulfill its obligations towards its customer if it cannot receive natural gas from its own (initial) supplier. For instance, in terms of invoicing and payment, if the supplier receives late payment from its customer, it will be in default to its own (initial) supplier. Correspondingly, while designing supply agreements, the back-to-back recourse principle is implemented systematically by considering the conditions of the initial supply contracts. Otherwise, the supplier (and indirectly its customers) may incur irrecoverable damages.

In cases where a customer consumes more than the maximum consumption volume, its supplier makes every effort to meet the customer's demand and to ensure the continuity of natural gas supply. These efforts are generally realized through contracts to be made in the end-of-day trading. Under normal market conditions, if a customer makes consumption more than maximum consumption volume referred to in the contract, this will cause an imbalance on the grid. GOP attributes various consequences to this imbalance. If a supplier cannot procure natural gas in the end-of-day trading with preferred conditions, BOTAŞ will charge the imbalance penalty as its capacity of system operator. Thus, if the customer consumes more than the maximum consumption volume, additional costs will be reflected by BOTAŞ to the supplier, and back-to-back from the supplier to the customer.

The back-to-back recourse principle that prevails in the natural gas market allows the supplier to reflect these additional costs as is. Since the additional costs that may arise are diverse, in case the customer exceeds the maximum consumption volume, the supplier should have the contractual right to have recourse to its customer to compensate its losses.

If the back-to-back recourse principle is not implemented, the supplier will be required to bear these additional costs or penalties that occurred due to its customer's breach of contract, leading to its financial ruin.

Limitation of Liability

Provisions limiting the liability of the supplier are often included in supply contracts. With these agreements, the supplier aims to limit its liability arising from provision of natural gas for direct or indirect damages arising from gross negligence or negligence.

Nevertheless, the provisions of Article 115 of the TCO include clear rules on the limitation of liability. Article 115 explicitly stipulates that the agreements regarding the limitation of liability of a debtor will be void due to the damages that may arise from gross negligence. Therefore, any agreement removing supplier's liability for gross negligence will be automatically void.

In accordance with the last paragraph of Article 115 of the TCO, if a service, profession, or art requiring expertise can only be performed with the permission given by the law or competent authorities, then the limitation of liability for negligence will be void. Since the supplier can only engage in natural gas wholesale activities based on a license issued by EMRA, its legal liability for negligence cannot be contractually limited. Therefore, contractual provisions limiting the supplier's liability for negligence shall also be null and void to the extent they are directly related to the exercise of the licensed activity.

3.5 *General Transaction Terms (GTTs)*

One of the novelties brought to Turkish law by the TCO is the regulation of GTTs among the validity conditions sought for contracts.

3.5.1 Definition and Scope

GTTs have been adopted by the legislator considering that the contracting parties do not always have equal bargaining power and one party can impose its contractual terms on the other due to its position in the market. Pursuant to Article 20 of the TCO, the contract provisions are considered as GTTs if they are (i) drafted for the purpose to be used in similar contracts, (ii) previously prepared individually, and (iii) submitted to the other party.

Provisions of the TCO are designed to apply to all contracts subject to private law, including commercial agreements executed between merchants. Even if both parties are merchants in accordance with the provisions of the TCC, the provisions of the TCO regarding GTTs can apply.[29]

3.5.2 Legal Consequence

The TCO considers that the conditions that are contrary to the benefit of the party with low bargaining power are not clearly and specifically notified, or the conditions that give the party the right to unilaterally change the contract are 'deemed unwritten'.

Natural gas supply agreements are executed by merchants and do not involve consumers. Therefore, in the event of a possible dispute arising from a supply agreement, the parties may request the conditions issued by the supplier to be reviewed in accordance with GTTs.

Companies that have the right to wholesale natural gas sometimes have more than 100 customers. It is literally not possible to sign different supply agreement

[29] Oğuzman et al. (2018), Borçlar Hukuku Genel Hükümler Cilt I, Istanbul, page 172.

forms for each customer. Therefore, it is inevitable that these contracts will be subject to GTT review.

The first situation that needs to be proved when reviewing GTTs is proof of whether the customer is clearly and privately informed of the terms of the supply agreement. The first way is to prove that the agreement was negotiated with the other party during the preparation phase of the supply agreement. Electronic correspondence of the parties in this direction and the black-lined texts showing proposed changes by the parties can be accepted as the commencement of written proof.

Another method is to prove that the other party has examined the contract. However, Article 21 of the TCO brings a presumption according to which a clause indicating that contractual provisions are accepted through negotiations will not exclude GTT classification of these provisions. Another acceptable method may be to enter a record on each page or each provision of the contract with a handwritten note indicating that the respective provision has been negotiated.

Another circumstance in which GTTs are deemed unwritten is the unfamiliarity with the contract type. The question that should be asked here is whether the provisions regarding imbalance costs (or penalties) that may arise from GOP are relevant to a natural gas supply agreement. Given the complex structure of the natural gas market, this question will need to be answered positively. Despite the explicit provisions of Article 22 of the TCC, even if a supply agreement is executed between merchants, a penalty condition or default interest rate may be deemed invalid in accordance with Article 1530 of the TCC.[30] This situation is in line with the reasoning of the provisions of the TCO regarding GTTs. However, due to statutory penalties that may be imposed on the supplier for its customer's default on fulfilling the minimum purchase commitment within the scope of the take-or-pay obligation or customer's consumption in excess of the maximum consumption volume, we feel that the imbalance fees should not be sanctioned within the scope of GTTs. These contractual provisions stem from the nature of natural gas trade, and enforcement of GTTs against these provisions may result in commercial ruin of market players.[31]

4 Conclusion

Evolution of natural gas supply agreements are driven by economic, legal, and geopolitical developments.

Strong fundamentals and rules of law are required to establish a market. The natural gas market in Turkey is in its infancy and will require time to compete with European markets. To further liberalize the Turkish natural gas market, the domination of BOTAŞ must be reduced. In addition to contract releases, it is necessary

[30] In accordance with Article 1530-(6) of the TCC, provisions stating that the debtor is not required to pay interest for late payment, or is required to pay a very small amount of interest that can be considered to be grossly unfair, are void.

[31] For a similar solution adopted in relation to loan agreements, refer to Kuntalp (2012), page 93–102.

for BOTAŞ to organize capacity release tenders. In order to reduce BOTAŞ's market share, a solid legal foundation based on country and existing market dynamics must be established with the contribution of sector professionals, economists, scholars, and NGOs.

It was reported recently that Turkey had discovered the Sakarya Gas Field in the Black Sea region, which is estimated to have a natural gas reserve of 320 billion cubic meters (bcm). Putting aside the technical issues around extracting this gas, Turkey needs a strong legal foundation and a market reform to progressively establish a liberal natural gas market. Failing to implement this foundation, new discoveries will not be enough to move Turkish natural gas market forwards.

Natural gas distribution companies have recently announced their plan to blend hydrogen (at a rate of 6%) in natural gas distribution network as a means of increasing the output of renewable energy systems. Nevertheless, any introduction of a hydrogen blend or any other matter for this purpose would require extensive studies and monitoring. Considering these upcoming challenges, a clear legal framework governing natural gas supply must be adopted to allow natural gas market players to make foreseeable commercial projections for the future.

As we have seen, the importing and wholesaling of natural gas are governed by similar—but not identical—legal frameworks. Various legislations and legal mechanisms impacting natural gas supply agreements become a challenge for market players. Correspondingly, legislators or EMRA should intervene to create a more transparent contractual framework for natural gas wholesalers. This intervention could be by means of adopting new legislation or amending the existing one, or via issuing or obtaining circulars or opinions from relevant third parties. For instance, as a first step, EMRA may issue a ruling stating that if provisions of a natural gas supply agreement conforms to applicable natural gas market legislation, supply agreements will not be subject to GTT review. Alternatively, EMRA may adopt a specific or derogative legal framework governing supply agreements. As a second step, however, the natural gas trade must evolve to straightforward trading activity with the use of standardized forms issued or online systems managed by regulatory authorities. This will help achieve transparency in the natural gas sector and compliance of market players to domestic and international norms.

At the same time, while market relaxation is expected with liberalization, global concerns and security of supply in natural gas production negatively affect the development of natural gas supply contracts. Therefore, as long as the current market conditions are not restructured, principles of take-or-pay or back-to-back recourse will continue to govern supply agreements. This should incentivize global policymakers to find common ground with their counterparts for the realization of largescale energy projects, including building cross-border natural gas pipelines. For instance, recent commissioning of the Turkish Anatolian Pipeline Project (*TANAP*) is a good move forward to reach supply security. At the domestic level, building natural gas storage facilities with involvement of experienced contractors will also help increase supply security. These developments will also help reduce Turkey's foreign dependency in natural gas supply.

Finally, liberalization of the natural gas market and unbundling of BOTAŞ's monopoly must bring the optimization of the transportation and transmission costs of natural gas to be delivered and market consolidation. We also expect that, with the entry into effect of the changes proposed herein, parties to supply agreements will not be required to enter into separate agreements for the transportation and transmission of the natural gas and bear these costs separately. Market players must make their strategic investment decisions with these upcoming changes in mind.

References

Aral, F. (2011). *Türk Borçlar Hukukunda Kötü İfa*. Ankara.

Ayrancı, H. (2010). *Enerji Sözleşmeleri*. Ankara.

Energy Community Secretariat (2015, October 1). Energy Governance in Turkey. *Report on Compliance with the Energy Community Acquis*, Brussels. https://www.energy-community.org/dam/jcr:361337d2-d6c0-424b-b161-56b70ad92d0e/EnC_%20Turkey_compliance_2015.pdf

Energy Market Regulatory Authority of Turkey (EMRA). (2020). *Natural Gas 2019 Sectoral Report*. Ankara. https://www.epdk.gov.tr/Detay/DownloadDocument?id=Sbf8i4dRZHw=

Eroğlu, M. (2010). *Unbundling in the Energy Sector*. Ankara. https://papers.ssrn.com/sol3/Delivery.cfm/SSRN_ID2736858_code948771.pdf?abstractid=2736858&mirid=1

Gazbir, (2020, January–June, August). *Natural gas sector report*. Ankara. https://www.gazbir.org.tr/uploads/page/GAZBIR-2020-1-yariyil-rapor.pdf

Kuntalp, E. (2012). Bankalar ve Genel İşlem Koşulları, Türk Hukukunda Genel İşlem Şartları Sempozyumu 8 Nisan 2011. *Banka Ve Ticaret Hukuku Araştırma Enstitüsü, Ankara, 2012*, 79–102.

Nalbant, H., Kayalica, M. Ö., Kayakutlu, G., et al. (2020). An analysis of the natural Natural gas pricing in natural Natural gas hubs: An evaluation for Turkey. *Energy Systems*. https://doi.org/10.1007/s12667-020-00395-8

Oğuzman, M., Kemal Öz, & Turgut, M. (2018). *Borçlar Hukuku Genel Hükümler Cilt I*. İstanbul.

Rzayeva, G. (2018, January). *Gas supply changes in Turkey*. Oxford Institute for Energy Studies. https://www.oxfordenergy.org/wpcms/wp-content/uploads/2018/01/Gas-Supply-Changes-in-Turkey-Insight-24.pdf

Rzayeva, G. (2014, February). *Natural gas in the Turkish domestic energy market policies and challenges*. Oxford Institute for Energy Studies. https://www.oxfordenergy.org/wpcms/wp-content/uploads/2014/02/NG-82.pdf

The Economic Policy Research Foundation of Turkey (TEPAV) (2009). *Türkiye'de Doğalgaz Sektörünün Yeniden Yapılandırılması*. Ankara. https://www.tepav.org.tr/upload/files/1295967350-8.Turkiyede_Dogalgaz_Sektorunun_Yeniden_Yapilandirilmasi.pdf

Turkish Competition Authority (2012, July). *Natural gas sector report*. Ankara. https://www.rekabet.gov.tr/File/?path=ROOT%2f1%2fDocuments%2fG%c3%bcncel%2fsektorarastirma.pdf

Yavuz, M. (2011). *Elektrik Piyasası Kanunu'nun Öngördüğü Hukuki Rejim ile Elektrik Tedarik Sözleşmeleri – Özellikle İkili Anlaşma*. İstanbul.

Burak Kepkep is an attorney-at-law and holds an LLB from Université Paris 1 Panthéon Sorbonne, an LLM from the City University of Hong Kong and a Ph.D. from Marmara University. He is a member of the Istanbul Bar. He published a book "Natural Gas Supply Agreements". He has several publications in corporate law, insurance law and real estate law, both in English and Turkish. He provides legal consultancy to many companies regarding corporate law, energy law and real estate law matters.

Regulation of Domestic Water in the Framework of Provincial Greater City Municipality Water and Sewerage Administrations

Nilgün Görer Tamer

Abstract The development and expansion of water infrastructure have been a fundamental requirement of urbanization since the mid-nineteenth century. As urban areas have grown, the protection of water resources, the collection, and treatment of wastewater and rainwater from urban areas have become as significant as potable water supply. Urban water as a public service provided by the municipalities in Turkey. Water management in Turkey moved away from the public service approach under regulatory changes enacted after 1980 as an extension of worldwide neoliberal policies. The prerequisites of obtaining credit other than from public resources, and specifically through international organizations, and technical assistance, have led to structural changes in water management at the central and local levels. This article discusses the structural change in water management as part of the integration of the national economy with global capital between 1980 and 2000 and within the framework of local economies becoming integrated with global capital after 2000. In 1981, the first arrangement was made for the commercialization of water services with the Water and Canal Administration, the model prerequisite of the World Bank infrastructure loan for the metropolitan city of Istanbul. In 1986, water and sewerage administrations (Su ve Kanal İdareleri-SKİs-) were founded under all greater city municipalities as organizations with an independent budget and public legal entity. After 2000, with the adoption of the EU Water Framework Directive, new institutional arrangements were made, watershed management and provincial water management were established, and the commodification of water and the environment continued under environmental legislation. Today, SKİs are strong actors in water management that provide rural and urban water services in Turkey's 30 provinces. However, after the rescaling made by overlapping the metropolitan boundaries with the provincial boundaries, the operational efficiency of SKIs and the affordability of water service by user was opened up to discussion.

Keywords Domestic water · City municipality · Regulation · Sewerage

---

N. G. Tamer (✉)
Department of City and Regional Planning, Faculty of Architecture, Gazi University, Ankara, Turkey
e-mail: gorer@gazi.edu.tr

M. Eroğlu and M. Finger (eds.), *The Regulation of Turkish Network Industries*,
https://doi.org/10.1007/978-3-030-81720-6_8

1 Introduction

Leonardo Da Vinci described water as "the driving force of nature". As early as the sixteenth century, he demonstrated the importance of water for nature and society and the mutual interaction among these phenomena (Pfister et al., 2009). In the twentieth century, the idea that the environment is an area that should be managed and that all human activities and services should be "sustainable" has come to prominence. However, the international agreements and regulations that have been adopted and the technological developments that have taken place have been inadequate for achieving the sustainability targets. Furthermore, global environmental problems have resulted in the emergence of another global issue—climate change—as the most significant problem of the twenty-first century. The system that is most affected by climate change is water. Regional- and local-level water policies and management models of nations play a role in the protection of the planet's water resources.

The first time that water was openly described as a commodity that should be priced according to market conditions was in the concluding declaration of the Dublin International Conference on Water and the Environment. The declaration also proposed a water management model in which capital would be the determinant stakeholder. Today, the provision of basic urban infrastructure tops the list of fields that have been rapidly abandoned by the social state in line with liberal policies. With the privatization of infrastructure services, in particular, the social aspect of service provision has been replaced by the goal of profitability and citizens have come to be seen as consumers. The downsides to urban service provision for profit become especially visible in the context of drinking water, and sewerage services in the form of "water poverty" or "water deprivation". This human problem stems from the implementation of a new policy that sees water as an *economic commodity* and makes it easier to implement demand-oriented policies. This means that natural resources that constitute global public goods are being converted into private goods, and natural resources are being increasingly commercialized (Prasad & Ramachandraiah, 1999). This policy has not only led to water being regarded as an economic commodity to be priced at its true cost, but has prevented it from being seen, firstly, as a common good and service and, secondly, as a human necessity. Privatization policies in the drinking water and sewerage sectors have been based on this way of thinking.

During the worldwide integration of national economies with global economies after 1980, "economic restructuring" and its discourse of "free movement of capital" have targeted urban areas. The fundamental element that pointed capital in this direction and turned urban areas into zones of attraction was urban infrastructure. During this period, it became compulsory to subject urban administrations to new legal and administrative regulations. The distribution of services between central and local administrations that occurred at around the same time ensured that service provision became local, but also commercialized and privatized. Although the environment-oriented and economically focused policies that shaped the sector often appeared to contradict one another, they were, in the end, collectively responsible for the commercialization and privatization process that took services out of the public sphere. The principles of "user pays" and "polluter pays" that were promoted through environmental protection policies contributed to a new water resources management policy shaped by international actors that saw water as an "economic good".

Within the general framework provided above, I assess water management in Turkey as a channel of integration with global capital and as an extension of the restructuring policies that ensured the transition to a neoliberal economy after 1980. This study, which covers the period from 1980 to 2020, approaches the subject from a historical perspective, with a special focus on the metropolitan municipality water and sewerage administrations (SKİs) that were at the heart of the arrangements that commercialized urban water services. The period from 1980 to 2000 is seen as the one in which the national economy was integrated with global capital, while the period since 2000 is seen as one of integration of local economies with global capital through water management. In the first decade of the 2000s, the structural change in water management was influenced by the implementation of financing models that paved the way for the free market provision of resources and local services; since 2010, it has been influenced by the inclusion in the process of European Union adjustment laws and institutions and by the contemporaneous restructuring of the central and local administrations.

In Turkey, domestic water services are provided as a local public service overseen by the municipalities. With the adoption of the Water and Sewerage Administration model, established for the first time for the metropolitan city of Istanbul in 1981, the first regulatory steps were taken towards the commercialization of water services. In 1986, amendments to Law 2560 on the Establishment and Duties of the Istanbul Directorate General of Water and Sewerage Administration established "water and sewerage administrations as an organization with an independent budget and as a public legal entity" under all the metropolitan municipalities. Under what is now known as the SKİ model, urban water management organizations of this kind are currently operating in Turkey's 30 provincial greater city municipalities.

After 2000, institutional arrangements were made at the central level, introducing watershed management and provincial water management as steps towards the implementation of the European Union Water Framework Directive. Amendments were made to environmental legislation that were important in terms of commodifying water and the environment. As a result of the re-regulation of local administration in Turkey under Law 6360 of 2014, the number of municipalities with greater city status was increased and the areas served by the SKİs were expanded to cover the entire province surrounding each city. The justification given for extending municipal boundaries to include the entire province was a desire to benefit from "economies of scale". The excessive growth of urban areas in terms of population and area has caused them to exceed the optimal size and has resulted in more losses than gains (Keleş, 2015: 253). The increase in the areas served by the provincial greater city water and sewerage administrations after the 2014 local elections effectively gave them responsibility for providing rural water services. Following the new legislation, which included adding rural areas into the scope of their responsibilities, the water and sewerage administrations became important actors in the field of water management.

After 2014, the affordability of water services for both rural and urban dwellers became a controversial issue. The extension of the areas served by the administrations was reflected in an increase in user charges. Increasing public dissatisfaction with higher water bills led political parties and individual candidates of all political viewpoints to prioritize social policy proposals for water bills in their campaigns during the 2019 local elections. The Covid-19 pandemic has again focused attention on the issue of a public service approach to the management of water services due to the increased visibility of the role of hygiene in public health. In the shadow of the "user pays" principle, access to water has come to be provided through a social approach to municipal management.

The remainder of this study is divided into two sections. Section 2 assesses the process of commercialization and commodification of water services in its historical process with reference to the SKİ model. In Sect. 3, the SKİ model is discussed in the context of the rescaling of local administrations and water management.

2 A Review of Urban Water Management in Turkey After 1980 from a Historical Perspective

Graham and Marvin (1999), who studied infrastructure policies from a historical perspective, have shown that there is a connection between the urban infrastructure policy preferences of countries and the structural transformations of the world economy. In their study, the authors named the pre-1980 period, marked by the isolated development of national economies, as the "nationalization" period, and the period after 1980, during which national economies focused on policies to integrate with the global economy, as the "global-localization" period (Graham & Marvin, 1999: 98). Turkey has experienced the same periods and the same connection.

The municipalities have significant responsibilities as the regulators and operators of services that are essential for healthy urban life. In Turkey, waterworks were placed under municipal responsibility for the first time under Law 831 on Water, dated 1926. Through Law 1580 on Municipalities of 1930 and Law 1593 on Public Health of 1930, the authority and responsibility for the construction of drinking water and sewerage systems, the collection of refuse, and the protection of water resources in urban areas for the protection of public health was entrusted to municipalities. Water services at the municipality level were generally run by departments within the municipalities, while specific water management organizations were established under the municipalities of Ankara, İstanbul, and İzmir in 1947.

A financially and technically sound infrastructure bank was established in 1945 to carry out the necessary infrastructure investments to enable the municipalities to provide these services. This infrastructure bank was İller Bankası (the "Bank of Provinces"), a central organization that would serve provincial special administrations and villages as well as municipalities. İller Bankası was made responsible for assisting municipalities with the financing, development, and construction of infrastructure projects, including projects for drinking water and sewerage. With the establishment of the Municipalities Fund two years after its foundation, İller Bankası acquired a solid source of finance.

The Municipalities Fund, which was the main provider of resources for the investments of municipalities, was scaled down from 1993 onwards in line with the wishes of the IMF and was finally disbanded at the end of 2001. As a result, İller Bankası withdrew from this field of investment. The 2002 World Bank report *Municipal Sector Assessment: Turkey* contained a discussion of the restructuring of İller Bankası. The report argued that the current responsibilities and activities of İller Bankası were disrupting the municipal financing sector and impeding the bank from acting as an efficient development organization. The Governing Regulations on the Law on İller Bankası that entered into force in 2004 emphasized the need for local administrations to *decide on and operate their investments themselves* and enabled them to borrow directly from foreign financial organizations. The transformation of İller Bankası into a development and investment bank was carried out under the "Municipal Services Project" signed with the World Bank on February 8th, 2006 (Çınar, 2006a, b: 74).

As a result of further preparations, which continued up to 2011, the Directorate General of İller Bankası was re-established as İller Bankası Inc., or İLBANK A.Ş., under Law 6107. With the reorganization of İller Bankası, its founding purpose of providing aid and technical consultancy to municipalities in water, sewerage, and infrastructure work was abolished, together with the system of subsidized public credit including financial support through the Municipalities Fund.

In summary, under the public-private partnership model imposed by the World Bank after 2000, the Directorate General of İller Bankası featured only as an intermediary organization with close ties to the infrastructure sector and expertise in financing, tasked with extending credit to municipalities and monitoring the credit extended. After 2010, İLBANK A.Ş. only participated in water management as a gateway for credit from international markets. It was also given a seat on the central water management board, the central watershed management board, and the provincial water management coordination boards.

After 2000, a significant amount of the funds used for the establishment of infrastructure and facilities for the provision of water and sewerage services were obtained in the form of external credit from international funds and financial institutions. As the contribution of public resources declined, the operation of the service was opened up to private investors under attractive conditions. This ensured that water and sewerage services, along with infrastructure facilities, were transferred to the private sector as concessions. After 2010, in support of this liberalization process, legislative changes were made that extended the areas the local administrations served and changed their scale and administrative structures.

The first of these legislative changes was a restructuring of local administrations that altered municipal boundaries and increased their areas of service in order to provide a sufficiently large population and land area over which the water and sewerage system and wastewater treatment plants could be operated efficiently.

The second change envisaged that water management in Turkey should be carried out at the watershed level. This led to the establishment of institutions for managing water in each of 25 basins or watersheds, in line with the geographical distribution of water assets. These institutions did not have a strict hierarchical relationship with the central administration, but were responsible for it.

2.1 The Integration of the National Economy with Global Capital Through Water Management (1980–2000)

The influence of the post-1980 "global-localization" period on infrastructure services in Turkey began with the commercialization of services. The legislation envisaged that services would be priced in order to recoup costs and to earn profits, thereby adapting to liberal market conditions; that they would be managed through autonomous administrations that were not part of the general public administration; and that financial resources for infrastructure investment would be obtained exclusively from the private sector. These arrangements created the structure that international organizations required as a prerequisite for extending credit.

The space left behind by İller Bankası for the provision of public credit for urban infrastructure was filled by international financing organizations. This led to an increasing localization of water and sewerage investments, the transfer of services to the private sector, and privatization. From the beginning of the 1990s, the effects of liberal policies on water were felt in more and more cities. Privatization in the guise of localization, and external borrowing in the guise of alternative financing became the norm (TMMOB, 2009).

The SKİ Model

The organizational structure of water provision and sewerage systems was affected by the restructuring of liberal economic policies after 1980. Under the credit agreements for the Istanbul Urban Development and Water Provision Project of 1972 and the Istanbul Water Provision and Wastewater Projects of 1987, the World Bank recommended the establishment of an autonomous operations model that would ensure the recouping of costs and adopt demand management. The proposed model was implemented for Istanbul in 1981 through Law 2560, which represented the first step towards the establishment of the Water and Sewerage Administrations (SKİs).

The World Bank argued that if municipalities were to provide water procurement and sewerage services directly, their performance would be low and they would make a loss. According to the World Bank, municipalities perform poorly in the provision of services for reasons such as higher physical and commercial losses due to inadequate maintenance, inefficient management of water services by public officials who are ineffective and have limited financial management capacities, having too many employees, and pricing policies that do not reflect the true costs of the services due to political considerations (World Bank, 1992, 1994). The World Bank further proposed the adoption of arrangements that would enable the private sector to undertake environmental protection measures in the supply of water and sewerage, such as the collection and treatment of wastewater. As a result, the SKİs came to be established through Law 2560 on the Establishment and Duties of the Istanbul Directorate General of Water and Sewerage Administration as an institutional model that would provide water and sewerage services under market conditions. The article of the said law concerning the determination of charges foresaw, notably, "the determination of different rates for the costs of the sale of water, the discharge of wastewater in areas where sewerage facilities exist and the emptying of septic tanks, with the charges to reflect the management and operating costs as well as such renewal, improvement and extension cost the amortization of which requires them to be directly recorded as costs (not capitalized), and with the addition of a profit margin of no less than 10 percent to the bill."[1] That article established the *minimum* profit margin, leaving the maximum to market conditions. Therefore, this arrangement clearly signaled the commercialization of water and sewerage facilities and the rolling back of the public service approach. Furthermore, the regulation paved the way for the itemization of the costs of the discharge of wastewater and the disposal of solid waste based on the amount of water used (Görer, 2000).

The Spread of the SKİ Model

Law 2560 on the Establishment and Duties of the Istanbul Directorate General of Water and Sewerage Administration of 1986 made it possible to implement the Istanbul SKİ (İSKİ) model in other metropolitan municipalities. Another piece of legislation that would serve as a means for the dissemination of the model, and would ensure that water and sewerage services in large cities would be provided as a commercial, rather than a public, service, was Law 3030 of 1984 concerning metropolitan municipalities. Under this law, based on Article 127 of the Constitution of 1982, the model of municipal administration that had applied in İstanbul, Ankara, and İzmir before 1984 was also introduced for Adana, Konya, Bursa, Kayseri, and Gaziantep, making a total of eight cities. In 1993, Antalya, Diyarbakır, Erzurum, Eskişehir, İzmit, Mersin, and Samsun were also declared metropolitan municipalities, followed by Adapazarı in 2000, bringing the total number to 16. Meanwhile, Law 3305 of 1986 added an additional article (Additional Article 4) to the Law on the Establishment of İSKİ, stating that this law would also apply to other metropolitan municipalities. In this way, the process of first establishing metropolitan municipality administrations and then SKİs attached to these administrations got underway. The increase in the number of metropolitan municipalities implied a simultaneous rolling out of the SKİ model in water management.

With the establishment of metropolitan municipalities during this period, and in accordance with the political economy of the times, water services came to be financed by international organizations rather than from the public purse, the provision of subsidized services was replaced with a full cost recovery approach, and the private sector was included in service provision.

[1] Article 23-Amended: 5/6/1986 – 3305/2 art.

The World Bank, Multinational Corporations, and the Different Models Applied to Urban Water Management

The "user pays" and "polluter pays" approaches to wastewater collection and disposal and the collection and storage of solid waste, derived from environmental discourse, were added to the World Bank agenda under the heading of environmental infrastructure.[2] In Turkey, waste disposal usually falls under the responsibility of a municipal unit dealing with environmental cleanliness. In 1993, an additional amendment to the Law on the Environmental Cleanliness Tax and Municipal Incomes allowed municipalities to price wastewater services on the condition that the price should not exceed the price of the drinking water consumed. It is worth noting that solid waste services were presented here under the same umbrella as water and wastewater services. Given that solid waste has recently become the area of urban services offering the highest returns, as well as the need for these services to be provided equally to everyone for the sake of public health, this decision to unite urban services of water supply, sewerage, and garbage disposal and transfer them to the private sector was of a very strategic nature. The World Bank credits obtained by the Bursa and Antalya metropolitan water and sewerage administrations may also be cited as examples of this.

Bursa Metropolitan Municipality reached a loan agreement with the World Bank for the financing of the Bursa Water and Environmental Health Project in 1993. The World Bank project loan covered two inter-related projects. One of the projects to be carried out under the agreement was a solid waste project to be operated by Bursa Metropolitan Municipality. The other was a project for the provision of drinking water and sewerage that was to be operated by the Bursa Water and Sewerage Administration (BUSKİ).

The loan agreement listed the aims of the Bursa Water and Environmental Health Project as follows: to improve environmental conditions and reduce environmental health risks in the Bursa metropolitan area; to improve the institutional structure for the management of municipal water distribution; to meet the demand for drinking water, sewerage, flood protection, and solid waste systems, including the demand from the poor living in the outskirts of Bursa; to increase water use efficiency by reducing the amount of uncollected water bills, and to implement optimal cost recovery policies.

Another initiative for the privatization of drinking water and sewerage services was the "twinning"[3] agreement included in the loan agreement. The twinning firm was to guide BUSKİ in the preparation of its program for reducing water losses and in the commercial pricing of water.

Another attempt at water services privatization was made in conjunction with a World Bank Environmental Health loan, which differed from the Bursa scheme,

---

[2] In 1993, the Bursa Water and Environmental Health Project (Official Gazette dated July 4, 1993 issue 21627) was carried out with a loan agreement with the World Bank. Bursa Solid Waste Loan 3566-TU and Bursa Water and Sanitation Project Loan 3565-TU.

[3] Twinning is a technical assistance contract between the water administration and the consortium formed by international consultants and an international operator favorably viewed in its field.

and concerned the Antalya Metropolitan Municipality Directorate General for Water and Wastewater Management (ASAT). In 1995, as part of the Antalya Water and Environmental Health Project, the city's water operations were leased to a French multinational company as a prerequisite for a credit line secured by the World Bank. In keeping with the loan agreement concluded with the World Bank in 1995, the powers of ASAT with respect to water were transferred to a company called Antalya Altyapı Yönetim ve Danışmanlık Hizmetleri Sanayi ve Ticaret A.Ş. (ALDAŞ), which was to be responsible for the city's drinking water and sewerage services on behalf of ASAT. ASAT had been established to provide services in the fields of water and sewerage and the collection and disposal of garbage and all kinds of solid waste within the area served by the Antalya Metropolitan Municipality under a 12-year contract (1996–2008). ALDAŞ, in turn, transferred its water operations to Antalya Su İşletmeleri A.Ş. (ANTSU) under a 10-year contract (1996–2006) in order to carry them out with private sector involvement. Thus, ANTSU became the organization responsible for the operation, maintenance, and management of the city's water and wastewater infrastructure.

ASAT differs from the other SKİs in its name and its staffing structure. In contrast to the SKİ model, ASAT was established with a smaller staff suitable for running a management office that would deal mainly with administrative matters. ASAT was assigned just 19 staff positions.

The World Bank set three basic conditions for its lending to the Antalya Water and Environmental Health Project. Upon the establishment of ASAT, with its new institutional structure, the water and wastewater operations would have to be transferred to the private sector in accordance with the French *affermage* (leasing) model. Prices would have to be adjusted, and ASAT would have to be audited by independent auditors. In the privatization of water services in Antalya, the World Bank arranged an international tender for the operation of services before any physical investments had started.

As a result of the tender held to involve the private sector in water and wastewater operations under the Loan Agreement signed between ASAT and the World Bank, the 10-year contract was awarded to a partnership between Suez-Lyonnaise des Eaux of France and Enka of Turkey. In February 1997, ANTSU A.Ş. began providing water and wastewater management services in Antalya.

However, water prices soon increased to a level that users could not afford (Görer, 2003; Çınar, 2006a, b; Güler, 1999). In response, the Antalya Metropolitan Municipality terminated the contract between the French company and ASAT, which was due to last until 2007 on June 1, 2002, citing Article 324 of the Turkish Criminal Code. The water company went to international arbitration and demanded damages. The Board of Arbitration ruled on the arbitration case between ASAT and ANTSU A.Ş. in 2014. A court case was then filed with the Antalya Regional Court of Appeal for the annulment of the arbitration ruling. The case was rejected by the Regional Court of Appeals in its ruling 2018/1 E-2019/1 K. ASAT filed an appeal with the Court of Cassation. Following this appeal, the case was finalized by ruling 2019/943 Esas-2019/2348 of the 15th Legal Chamber of the Court of Cassation, issued on May 16, 2019, and the award of damages was upheld (ASAT, 2019 Activity Report: 173).

Apart from the SKİ model, the idea of setting up unions (associations) for urban infrastructure came onto the agenda in 1989 under the Mediterranean and Aegean Tourism Infrastructure and Coastal Management (ATAK) project. This project resulted in the establishment of GATAB, the Marmaris-Armutalan-İçmeler Municipalities Union, and the Çeşme-Alaçatı Infrastructure Union for the purpose of investing in water, sewerage, treatment, and solid waste, and operating these services (TODAİE YYAEM, 1999: 202–203). The Çeşme and Alaçatı Environmental Protection Infrastructure Facilities Construction and Operations Union (ÇALBİR) was an agreement that made water services conditional on a transfer of operating rights. In addition to the credit agreement signed between ÇALBİR and the World Bank in 1998, a Guarantee Agreement was signed between the Republic of Turkey and the World Bank and the Republic of Turkey guaranteed to perform ÇALBİR'S obligations as the main debtor if ÇALBİR became unable to perform its obligations. Like similar loan agreements, the credit agreement with the World Bank aimed to privatize service provision and deliver services under market conditions (Güler, 1999: 6). To this end, ÇALBİR signed a 10-year contract for the operation and maintenance of the water supply and wastewater system with a consortium consisting of the French company Veolia Eau- Compagnie Generale des Eaux (CGE) and TEKSER İnşaat of Turkey. As a result, Alaçatı-Çeşme Su İşletmeleri San. Tic. A.Ş. (ALÇESU) was established in February 2003. For a period of 10 years, ALÇESU undertook to produce and distribute water, maintain and repair the systems, and manage customer relations within the boundaries of the municipalities of Çeşme and Alaçatı. With a private company taking over the operation of the service, unit operating costs for water doubled (TMMOB, 2009: 62). Coinciding with the end of the 10-year period, the water and sewerage services of all municipalities within the boundaries of İzmir Metropolitan Municipality came under the duties and responsibilities of the İzmir Water and Sewerage Administration and the concession agreement lapsed.

The commercialization of water, which began with the establishment of the Istanbul Water and Sewerage Administration in the 1980s, and the process of turning it into a field of investment for multinational companies through privatization, continued into the 2000s. The methods used to privatize services varied, as illustrated by the examples of the Antalya Water and Sewerage Administration and ÇALBİR, dating from the 1990s.

During this period, credit lines obtained through intergovernmental bilateral agreements, without the involvement of the World Bank, are also observed to have changed the structure of infrastructure service provision. One example of this is the Project for Improving the Quality of Local Administration Services. This project got underway on December 11th, 1998 under a bilateral technical assistance agreement between Turkey and Germany, with credit provided by GTZ. In the credit agreements it signed with municipalities, GTZ (KfW) insisted that water and sewerage administration with an autonomous budget should be formed under the municipality in the case of municipalities that did not have metropolitan municipality status. The stated aim of this arrangement was to ensure the transparency of revenues and expenditures and conduct separate bookkeeping for the investments that would need to be repaid.

As can be seen from the above, credit conditions played a key part in accelerating changes to the way in which water and sewerage services were provided, especially in those municipalities with a population of between 100,000 and 300,000 that offered profitable economies of scale.

2.2 The Integration of Local Economies with Global Capital Through Water Management During the European Union Adjustment Process After 2000

New legal arrangements continued to be made for local administrations after 2000. Law 5216 on Greater City Municipalities, ratified in 2004, extended the boundaries of existing greater city municipalities and the areas they serve. The process of increasing the scale of municipalities in terms of population and geographical reach continued with Law 5393 on Municipalities. Under this law, the minimum number of inhabitants required for a settlement to have its own municipality was raised from 2,000 to 5,000, and municipalities with fewer than 2000 inhabitants were closed down (Çınar et al., 2013).

Under Law 5216, the areas to be served by the SKİs were extended in parallel with the extension of the boundaries of the 16 greater city municipalities. The boundaries of the İstanbul and Kocaeli metropolitan municipalities were extended up to the provincial borders, while the greater city municipality boundaries in the other affected provinces were expanded to encompass circular zones, the radius of which depended on their populations. The areas to be served by the Ankara and İzmir municipalities, each with a population of over 2 million, were extended to a radius of 50 km; those of Adana and Bursa, which had populations of between 1 and 2 million, were extended to a radius of 30 km; and those of municipalities with populations of up to 1 million were extended to a radius of 20 km.

The Law on Municipalities created significant opportunities for transferring the duties of municipalities to provide public services to third parties. The transfer of public services to third parties generally takes place through a service purchase agreement within the framework of the public tenders law or through the transfer of public services as concessions. The arrangements made in the Law on Municipalities were appropriate to either option. Paragraph A of Article 14 of the Law on "the duties and responsibilities of Municipalities" lists virtually all the duties of municipalities that constitute public services, and states that the municipalities will "carry out these services themselves or have them carried out". Among the services listed are urban infrastructure such as water and sewerage, the environment and environmental health, and cleaning and solid waste services.

In the case of water, the adopted approach focuses on the preservation and efficient and effective use of resources, while also viewing water as more of an economic commodity than a public good.[4] In this way, environmental values have been used to justify the supply of water, which is an indispensable resource and a fundamental requirement of all ecosystems, as an economic commodity subject to market conditions, and to provide capital with a new opportunity for profit (Keleş et al., 2012: 219).

After 2000, the use of grants regulated by agreements with the EU and the World Bank became as prevalent as the use of the public-private partnership mechanism as a financing model for the development and improvement of water and sewerage infrastructure projects. Turkey has mainly secured grants under the EU's Pre-Accession Financial Assistance Program for Turkey. The water and sewerage infrastructure facilities of many settlements have been financed through such grants.

Loans extended under the World Bank's Country Assistance Strategy (CAS), which operate in parallel with EU funding, also played an important role in the transformation of the institutional structures through which municipalities provide water and sewerage services, as well as in the restructuring of İller Bankası in this period. Loans extended under CAS were used to finance the sewerage networks and treatment plants needed to conform to EU wastewater treatment criteria, as well as for the overall expansion of the water and sewerage infrastructure. This form of support was also used for the construction of sewerage and wastewater treatment facilities in line with the Millennium Development Goals (Görer-Tamer, 2007: 77).[5]

The Watershed Management Model in the EU Harmonization Process

The principles of integrated water resources management and watershed level management form the last link in the chain that began with the implementation of the principles accepted at the Dublin International Conference on Water and the Environment in 1992. In world water forums and all international meetings after 2000, it was argued at the administrative and political level that water resources should be managed by local administrations at the watershed level and with the participation of multiple actors.

In Turkey, the geographical, watershed level model for managing water was formulated as part of the EU harmonization process.[6] The process of adjustment to the EU acquis simultaneously served to further the adoption of the legislative and regulatory texts needed to open up this field of service to the free market. The EU requires candidate Member States to bring their water policies into line with the Water Framework Directive. The proposals for water policy recommended in

[4] Under the General Agreement on Trade in Services (GATS) of 1994, water services were no longer classed as public services, and water and water-related services were treated as a market commodity that could be bought and sold on the global marketplace like other goods and services.

[5] With the end of the UN Millennium Development Goals period in 2015, they were replaced by the UN Sustainable Development Goals (SDGs) under Agenda 2030. SDG 6 is on "Ensur[ing] availability and sustainable management of water and sanitation for all" and emphasizes the rising importance of water and sanitation conditions on the global political agenda.

[6] This approach based on the watershed level governance model was realized under the 7th Environmental Action Plan (2012–2020) dealing with EU acquis chapters Regional Policy and Coordination of Structural Instruments (Chap. 22) and the Environment (Chap. 27).

the Directive—such as the development of water resources, more effective use of existing resources, demand management, and alleviation of environmental impact—have been regarded, along with integrated watershed management, as arrangements that will remove the public monopoly over the management of water resources and promote the participation of capital under the coordination of the EU (Görer-Tamer, 2015).

The proposed water management system has central and local components and is fully oriented towards managing water resources at the watershed level. The watershed management structures established for 25 watersheds have fundamentally played the role of assigning water and soil resources to the needs of the marketplace. What the World Bank proposed for privatization policy in the management of water resources was a watershed management approach that had top-down control and regulation and bottom-up decision making governance structures with an optimum level of relationship and distribution of tasks between the center and local parts, that enabled the sale (assignment) of water resources within the same watershed with which they formed a hydrological whole (Kartal, 1999: 115).

This structure was put in place in 2015 with the amended Directive on the Formation, Duties, Working Procedures and Bases of Watershed Management Committees, and as its local pillars, the *Province Water Management Coordination Boards* were established alongside *Watershed Management Committees.* Furthermore, the *Central Watershed Management Board* was established as the body responsible for the coordination of watersheds. The Watershed Management Boards are presided over by the coordinating provincial governor and consist of the governors or deputy governors of the other provinces concerned, SKİ general directors, provincial mayors, the representative of the Directorate General of Water Management, the representative of the Directorate General of State Hydraulic Works (DSİ), the DSİ regional director responsible for the coordinating province,[7] a representative of the Ministry of Foreign Affairs in the case of cross-border watersheds and representatives of universities, organized industrial zones, and civil society organizations. The duties of these committees founded on multi-party participation are to prepare watershed protection action and management plans and flooding and drought management plans, and conduct monitoring and evaluating activities at the watershed level. The *Province Water Management Coordination Board* is tasked with preparing watershed protection action and management plans and flooding and drought management plans, and with monitoring and evaluation, to be done at the provincial level (Görer-Tamer, 2015).

[7] The Directorate General of State Hydraulic Works, an important national investor organization in the supply of drinking and irrigation water, was established under Law 6200 of 1953. With the transition to a presidential government system in Turkey in 2018, the central administration responsible for water was restructured. Statuary Decree 703 repealed the Statuary Decree 645 on the Organization and Duties of the Ministry of Forestry and Hydraulic Works dated June 29, 2011. Under Presidential Decree 1 on the Organisation of the Presidential Structure dated July 10, 2018, the Ministry of Agriculture and Forestry was established. The general directorates of Water Management, which has watershed-based water management tasks, and the State Hydraulic Works were placed under this ministry as part of the restructuring.

Another step taken towards implementing the Water Framework Directive was preparation of the Watershed Protection Action Plans. An outcome of this process as 2012 Regulation on the Protection of Watersheds and the Preparation of Management Plans. As per the regulation, Watershed Protection Action Plans (HKEP) were prepared for 25 watersheds. These plans took the form of a feasibility study for the wastewater treatment plants to be established in the watersheds. Article 6 of the Regulation stated that the necessary planning would be carried out with European Union countries for cross-border watersheds, thus using the watershed to overcome national borders. The Regulation also used the expressions "user pays" and "polluter pays", thereby juxtaposing the basic free market language with the environmental protection approach in water services.

3 "Rescaling" in Urban Water Management: The Provincial SKI Model

A common feature of the new laws that governed local administration units after 2010 was that they created changes of scale in terms of land areas and populations served and the impact areas of services. The amendment regarding metropolitan municipalities that led to rescaling in urban water management was introduced under Law 6360, adopted in 2012. Many cities that did not have metropolitan characteristics were turned into metropolitan municipalities. Under the new legislation, metropolitan municipalities were established in the provinces of Aydın, Balıkesir, Denizli, Hatay, Malatya, Manisa, Kahramanmaraş, Mardin, Muğla, Ordu, Tekirdağ, Trabzon, Şanlıurfa, and Van, which had populations over 750,000, with the same name as the municipality and with the same boundaries as the administrative boundaries of the corresponding provinces.

The increase in the number of greater city municipalities and the extension of their boundaries to cover the whole of the corresponding provinces led to an increase in the number of municipalities implementing the SKİ model. In the greater city municipalities, which administer entire provinces, rural and urban water management were unified. However, due to the size of the areas for which they were now responsible, and the increase in the populations that they were obliged to serve, the provision of water services to rural settlements affected the level of efficiency of the operations.[8] The water units of district municipalities that did not previously

[8] The Mersin Water and Sewerage Administration (MESKİ) can be given as an example of the rescaling of service areas. Mersin Municipality became a greater city municipality in 1993 and, in the same year, MESKİ was established by a decision of the Council of Ministers. In 2004, under Law 5216 on Greater City Municipalities, 19 municipalities and 47 villages were included within MESKİ's service area. As a result of this change, MESKİ's responsibilities extended to 296 settlement units in total, including neighborhoods and villages, spread across an area of 650 km^2. On March 30th, 2014, MESKİ's service area was extended to cover the whole province. Four central districts, as well as nine outlying district municipalities, additionally 509 villages and 41 towns that had lost their legal entity status and become neighborhoods, included in the MESKİ service area.

fall within the boundaries of a greater city municipality were appended to the water management institutions as branch directorates.

With the new regulations, the Province Special Administrations, town (*belde*) municipalities and villages lost their status as legal entities. Similarly, in terms of water services, the water and sewerage services of the Province Special Administrations, which had lost their legal status, were transferred to the Water and Sewerage Administrations of the metropolitan municipalities. With the SKİ model, the pricing and provision of water services under market conditions, and the possibility of having them provided by third parties, was extended to service hinterlands to province border, including rural areas. This has made the Water and Sewerage Administrations significant actors in the context of watershed management.

3.1 *Duties, Authorities, Responsibilities, and Sources of Income of the Provincial SKİs*

Under Law 5216 of 2004, the duties and responsibilities of greater city municipalities regarding water management were defined as:

> to ensure the protection of the environment, agricultural areas, and watersheds in keeping with the principle of sustainable development; to run water and sewerage services; to establish, have established, and operated the necessary dams and other facilities; to rehabilitate watercourses; to market spring water or water obtained through treatment; to meet the needs for drinking, clean and industrial water using all forms of underground and surface resources; to establish the facilities needed to ensure the supply of water from the source to the customer, or to have such facilities established or take over facilities already established, and to operate these facilities and maintain them or have them maintained; to disperse used water and rainwater from settlements without damage, and to establish all forms of facilities for the reuse of such water or have them established.

The authorities and privileges of the metropolitan municipalities include providing drinking, clean, and industrial water, dispersing wastewater and rainwater, establishing and operating the facilities necessary for these purposes, or having them established or operated by others, and processing spring water or having it processed. Greater city municipalities may either provide these services with their own revenues or established corporations for the duties and services entrusted to them. The use of these authorities has led to the establishment of subsidiary companies under the SKİs. The most common service purchased from such companies as meter reading and billing. The municipality may transfer these services under concession agreements for a maximum duration of 49 years, subject to the opinion of the Council of State and a decision of the Ministry of the Interior.

Following this change, the service area of MESKİ's increased from 650 to 15,512 km^2, a multiple of almost 24. The number of water utility subscribers doubled from 348,494 (in 2013) to 724,602 (in 2014) (MESKİ, 2020). Despite the large increase in the service area, the increase in the number of subscribers was limited.

Within the context of the duties given to the greater city municipalities, the duties and authorities of the provincial SKİs are: to provide the required drinking, clean, and industrial water from underground and surface sources, ensure its healthy transport to the required locations in line with the needs of society and maintain the necessary lines and facilities; to collect wastewater, ensure its dispersal from settlements, and store it where necessary; to take the necessary physical measures for the protection of all surface and underground water resources from pollution; to manage the material assets of the organization, and maintain the continuity of legal financial operations, and to manage expropriation processes and transactions to meet its needs in fulfilling its duties.

The tasks that the SKİs are obliged to undertake are managed by three departments acting in cooperation. The Water and Sewerage Administration Department is responsible for operating all water resources from which drinking and clean water are procured, ensuring the delivery of water to the public, analyzing data collected from drinking water monitoring points and reporting it to the wastewater treatment unit, keeping statistics of the amounts of water delivered to settlements, reducing losses and leakages, and taking precautions according to the warnings of disaster units. The Treatment Facilities Department is responsible for establishing treatment plants, laboratories for the control of wastewater systems, and wastewater lift stations, and for ensuring the safety of the stations. The Environmental Protection and Control Department has the duties of protecting drinking water watersheds, preventing polluting activities, and seeking the expropriation of existing structures.

In summary, the provincial greater city municipalities and the SKİs have become the solely responsible organizations in all fields of water services, including rural water distribution.

Sources of Income

Pricing serves two fundamental purposes in the provision of water and sewerage services. Firstly, it aims to manage demand (through graduated scale pricing) in order to reduce waste and to orient users towards rational water use. Secondly, it aims to provide resources for the long-term maintenance of the network over which the service is delivered and for new investments. In the early 1980s, two more factors were introduced when determining water prices. The first was the calculation of environmental protection costs and their addition to water bills under the principle of the "polluter pays". The second involved charging the user the *true cost* of the service under the "user pays" principle, which legitimizes the commercialization of public services.

In provincial greater city municipalities, all kinds of costs are included in water bills under various headings. Regular additional taxes and cost elements are included on the bill under the heading "contributions to expenses", alongside the costs of drinking water, infrastructure and sewerage work, etc., while the cost of fixing burst pipes is included under the heading "repair costs".

The internal revenues of the provincial SKİs consist of the receivables they charge according to the price tariff for the sale of water and the dispersal of used water. Water is priced not as a basic need, but as an economic commodity. The clearest indicator of this is that a minimum profit margin of 10% is added to water procurement service costs to determine the unit (m^3) price of water.

Water bills price wastewater in addition to drinking water. The amount of wastewater to be charged for is determined on the assumption that half of the water consumed will return to the system as wastewater. Water bills also include costs for the collection and disposal of solid waste. SKİs determine their prices for these in line with the Regulation on the Procedures and Principles to be Followed in Determining Price Tariffs for Wastewater Infrastructure and Domestic Solid Waste Disposal Facilities of 2010.

Provincial SKİs transfer 1% of the price of the net volumes of water and wastewater amounts on sales of water, the disposal of used water in areas with sewerage facilities, and the emptying of septic tanks to the general (national state) budget as an environmental contribution. Under Article 43 4/d of the Regulation on the Monitoring and Collection of Environment Revenues and the Use of the Payments Derived from their Collection, which is based on Article 18 of Law 2872 on the Environment, the environmental contribution is defined as an "amount to be collected to prevent environmental pollution, to improve the environment and to support investments related to the environment". This contribution corresponds to 1% of all amounts collected for the net cubic meters of water sold and used water disposed of stripped of all taxes and fines including environmental cleanliness tax and value-added tax (Alıcı, 2017: 110). SKİs collect the Environmental Cleanliness Tax from all buildings used as residences within the boundaries of metropolitan municipalities as part of the water bill. Of the amounts of this tax that they collect, 80% is transferred to the district municipalities that are responsible for waste collection and the remaining 20% goes to the greater city municipalities to be used in the establishment and operation of waste disposal facilities.

The established procedure among municipal councils in Turkey is for subscriber water meters to be read at intervals of 30–40 days and for consumption to be billed accordingly. By using increasing block tariffs, municipalities aim to differentiate pricing for different groups of subscribers and to ensure water savings. The use of the "prepayment water metering system" has become widespread, especially in the SKİs. Under this system, only those who pay in advance can access the service—a commercial practice that makes access to services dependent on the ability to pay. This practice also hides the number of consumers who are unable to pay and are excluded from the system and face "water deprivation". The acceptance of "user pays" system in Turkey in the 1980s and the commercialization of water services have affected low-income households (Görer, 2003).

3.2 *Operational Problems of Provincial SKİs*

Provincial SKİs have undertaken services such as the provision of water services and the collection and treatment of wastewater, which are still an important problems for rural settlements. While there are advantages to the unified provision of these services, there are operational problems with respect to planning, implementation, monitoring, auditing, financing, losses of water in the network and unaccounted water problems, and increased water prices.

Management of Water Losses and Leaks

Of the 54 billion m^3 of water used in Turkey, 13% (approximately 7 billion m^3) is used for drinking-clean water (domestic) purposes. Regarding potable water, it has been found that water losses and leaks in water management units amount to around 40%, resulting in annual losses of approximately TRY 7.5 billion (SUEN, 2019). Regulations have been introduced to protect water resources and to prevent economic loss to businesses.

Steps to prevent water losses and leaks in Turkey have been added to the agenda in the context of compliance with the European Union's Water Framework Directive. The Regulation on the Control of Water Losses in Drinking Water Provision and Distribution Systems was issued in 2014 to minimize losses and leaks and bring them under complete control. The regulation governs the procedures and principles for the effective use of drinking and clean water and for the control of water losses in the provision and distribution of drinking and clean water to ensure the preservation of water resources and increase efficiency.

According to the regulation, the greater city municipality Water and Sewerage Administrations are obliged to reduce their losses to a maximum of 30% by 2019 and 25% by 2023, while other municipalities are obliged to lower losses to a maximum of 30% by 2023 and 25% by 2028. The aim is to reduce the levels of urban water losses and leakages to that of the developed countries. One of the SKİs that has taken significant steps in this direction is Ankara SKİ (ASKİ), which reduced its water losses and leakages from 42% in 2014 to 37% in 2019. The percentage of physical losses was 35% and the percentage of administrative losses was 2.4%.

Service Areas of Larger than Optimal Size

The optimum size of an area to be served by a water administration depends on the criteria of physical area and population density. For example, the area served by the Antalya water and sewerage administration ASAT increased from 2033 to 20,909 km^2 with Law 6360 and the population within this area from 1.1 to 2.2 million. ASAT's service area grew tenfold and the population it served doubled. The density of the population in the area served fell from 541 to 105 persons/km^2. In ASAT's former service area, 541 people benefited from each km^2 of services or investment; in 2016, 113 people benefited. Following the rescaling, the issues of optimum area and population have emerged as factors that have tended to increase the operational costs of the network.

According to the findings of the Turkish Water Institute (SUEN) (2019), 92% of the budgetary expenses of the SKİs are covered by their budget revenues. However, this proportion varies widely, from 60 to 179%. An examination of the incomes and expenditures of the Provincial Water and Sewerage Administrations shows that until the boundaries of the areas they service were extended to coincide with the administrative boundaries of the corresponding provinces, the revenues of most of them exceeded their expenses, and the resulting budget surpluses were transferred to their greater city municipalities. Following the extension of their boundaries, however, costs have increased, which has affected the budget balances. Deficits in the budgets of the SKİs have resulted from tariff exemptions and the prioritization of investments in providing services to rural areas as legally required. The legal arrangements in question have resulted in major losses for SKİs with large rural areas to cover (Çapar & Demir, 2017). Low population density, distance to water resources, and geographical conditions make the provision of water and sewerage services to rural areas costly, causing operational costs to rise.

In addition to the above, provincial greater city municipalities and the units attached to them are obliged by law to devote at least 10% of their investment budgets for a period of 10 years to infrastructure services for these settlements that have been newly included within the boundaries of the municipalities. There are also limitations on the pricing of water services. "Contributions to expenses" are not charged on services provided to rural areas, and water is priced at not more than 25% of the lowest tariff applicable in central areas for settlements that used to be villages and are now known as neighborhoods and at not more than 50% for settlements that used to be known as towns and are now known as neighborhoods. This has affected the water income of the provincial greater city municipality SKİs. Besides these lower water prices, the proportion of the population that is provided with sewerage services to the total population is lower in rural areas and outdated drinking water facilities need to be replaced. For reasons such as these, it will take time for the SKİs to balance their budgets (Alıcı, 2017; TBB, 2020). Since budget deficits have increased due to investments in rural areas, and deficits are expected to grow due to legal limitations, the internal revenues of the SKİs will inevitably tend to fall as a proportion of total revenues.

In summary, as a result of the arrangements that have led to rescaling, the areas served have expanded, population density has fallen, and the cost of investment per user has increased. Following rescaling and the related increased investment costs, the uninterrupted provision of water services and the access of users to affordable and healthy water through the network are becoming operational issues for SKİs. In the provision of water services, rescaling has not resulted in the expected benefits of economies of scale.

4 Conclusion

After 2000, there was a deviation from the principle of localization, which had started with structural adjustment policies in the second half of the 1980s and continued with the EU integration process, and which envisaged the provision of local common needs at the level closest to the people and the strengthening of local authorities to this end. In the literature, this deviation has been evaluated as allowing for the free movement of trans-national capital, on the pretext of providing credit and technical support to urban infrastructure investments for the strengthening of the local administration within the framework of the integrity of government in unitary states (Geray, 2001: 28). It has been argued that the arrangements did not help to strengthen local administrations, but instead appeared to lead to "local centralisation" (Duru, 2013: 33). The boundary changes for local administrations introduced after 2005 tend to corroborate these views. The closure of municipalities with populations under 5000 and the extension of the boundaries of the areas served by the greater city municipalities to meet rural and urban needs throughout the corresponding provinces have resulted in the establishment of new market-based economies of scale in the geographical areas affected. In terms of water services, the water, and sewerage services of the Provincial Special Administrations, which lost their status as legal entities, were transferred to the Water and Sewerage Administrations of the provincial greater city municipalities. This paved the way for 30 of these Water and Sewerage Administrations to provide services to rural areas under market conditions or to have such services provided by third parties.

In 2004, under Law 5216 on Greater City Municipalities, the areas served by the İstanbul and Kocaeli metropolitan municipalities were overlapped with the boundaries of the corresponding provinces. In 2014, with the newly established metropolitan municipalities, this practice was implemented for a total of 30 greater city municipalities. A detailed examination of the areas to be served in terms of their geographical extent and rural populations shows that the degree of restructuring varied from one provincial greater city municipality SKİ to another. The SKİs of all the 14 greater city municipalities that existed before the regulation of 2014 were given the task of serving rural areas. This extension caused problems for some provincial SKİs, such as Antalya, with hinterlands where significant agricultural activity takes place. In the 14 new greater city municipalities serving the whole of their provinces established under Law 6360, SKİs were established for the first time. These provinces are not metropolitan in character and have large rural populations. Besides organizational issues, these SKİs began operating in provinces with rural areas that are quite large compared with the urban areas. This situation has raised questions about the optimum scale of services under the SKİ model and about the original intention behind it.

Turkey is currently undergoing a process in which the rights of access to water and its use are being reorganized around watersheds in both rural and urban areas. The Provincial SKİs, which are institutional powers within watershed water governance, have from the beginning adopted a commercial model based on the guiding principle of profitability and the "user pays" approach. As of 2019, the population being served within this new structure, modeled on the example of the provincial SKİs, has surpassed 64 million. As much as 77% of Turkey's population uses domestic water as a commercialized public service. According to a report on affordability by the Turkish Water Institute (SUEN), water and wastewater bills account for an average of 1.03 and 3.02% of the household income of consumers in the middle- and low-income groups, respectively. Affordability data for low-income consumers show that in seven of the 27 SKİs, the share of water and wastewater expenditures as a proportion of total household income is over 4%. Yet, according to various studies, water bills should not amount to more than 2% of household expenses if they are to be affordable. This should be kept at 1.25% for low-income households (Smets, 1999).

The adoption of demand-oriented policies for the provision of drinking water and sewerage services has caused the problem of access to healthy water to deteriorate from the realm of scarcity to that of deprivation. The problem is no longer one of being able to find sufficient amounts of water to meet one's needs, but one of not being able to access the water, or being deprived of it, due to high water bills. Ensuring universal access to healthy water is essential and requires the service to be priced at a level affordable to all sections of society, without making any profit. This inevitably calls for subsidies. Regulation is necessary to ensure that the service is provided without any distinction between the rich and the poor. In such cases, subsidies for the service are generally met out of government funds (Neutze, 1997: 212–214).

Keleş (2017) has emphasized that the implementation of the free market model negatively affects the right to water. He recalled that the United Nations defined the provision of access to clean water and sanitation as a human right on July 24, 2010. He has therefore called for the repeal of Article 23 of Law 2560 on SKİs, which obliges local administrations to set water tariffs with a minimum profit margin of 10%, and Law 4736 on the Tariffs for the Goods and Services Produced by Public Bodies and Organizations, which "criminalizes" discounted water tariffs.

Access to secure and clean potable water and hygienic conditions is necessary for the full enjoyment of life and human rights. Therefore, water services should be provided as a public service. The increase in water bills that followed rescaling has become a significant issue for mayoral candidates in local elections. Following the 2019 local elections, the additional maintenance costs included in water bills under names such as "branch road maintenance" were removed from water bills in Ankara, resulting in an average reduction of 25%. During the COVID-19 pandemic, the importance of access to water as a public service has come to be better understood once more. Several emergency measures have been taken during the pandemic to ensure the public's uninterrupted access to water. These have included suspending water cuts due to unpaid bills, reinstalling removed meters, providing mobile access for users on pre-paid cards, making bills payable in installments, and waiving the delay interest charged on bills paid late. ASKİ, the water, and sewerage administration of Ankara Greater City Municipality has charged water users receiving welfare benefits just TRY 10 for the first 10m^3, thereby protecting the right to water with a social approach to municipal affairs. Moreover, the water bills of 72,000 households in Ankara have been paid by Ankara residents following a call from Mayor of Ankara Greater City Municipality for social solidarity in view of the pandemic. Similarly, İstanbul Greater City Municipality campaign "Bill on the Hook" (Askıda Fatura) has helped 22,000 families in İstanbul to have their outstanding bills paid.

Water meters that had been removed because of the failure of the household to pay its bills were re-installed and, as it was not possible to implement the progressive tariff during the COVID-19 pandemic, VAT was limited to 8% for all levels of consumption (ASKİ, 2020). Some municipalities have allowed citizens to pay the bills of other households anonymously as a form of social support for those financially affected by the pandemic, while others have met the water bills of low-income citizens out of municipal assistance funds. With the pandemic, the reading of meters on location was interrupted to reduce exposure to the virus, and instead, average consumption levels were used to calculate bills (SUEN, 2020).

The pandemic conditions have made water deprivation visible. Despite the promotion of privatization and public-private partnerships by international financial organizations and national governments, it seems inevitable that the re-municipalization of water as a social policy will become a political option. Direct experience of high water costs, and of how the rise in the prices of water under privatized water management has made it unaffordable for users, has revitalized the view that water services can once again be provided in a spirit of public service as a human right.

References

Alıcı, V. O. (2017). Su ve Kanalizasyon İdarelerinin Bütçe Gerçekleşmeleri Üzerinden İl Mülki Sınırı Kapsamında Hizmet Sunumunun Değerlendirilmesi:MESKİ Örneği. *Uluslararası Sosyal Araştırmalar Dergisi, 10*(52), 904–914.

ASAT (2019). Faaliyet Raporu. https://www.asat.gov.tr/images/editor/FaaliyetR/2019%20YILI%20ASAT%20FAALIYET%20RAPORU%20290520_.pdf

ASKİ Genel Müdürlüğü (2020, Haziran). *Bir Damla Faaliyet Bülteni.* http://yayinlar.aski.gov.tr/pdf/13%20temmuz%20eb%C3%BClten.pdf

Çapar, S., & Demir, R. (2017). Türkiye'de Büyükşehir Yapılanması ve 6360 Sayılı Kanuna Uygulayıcıların Bakışı. *Türk İdare Dergisi, 484*, 39–90.

Çınar, T., Duru, B., Çiner, C. U., & Zengin, O. (2013). *Belediyelerin Sınırları.* TODAİE.

Çınar, T. (2006a). Türkiye'de İçmesuyu ve Kanalizasyon Hizmetleri: Yönetim ve Finansman. In T. Çınar, H. K. Özdinç (Eds.), *Su Yönetimi Küresel Politika ve Uygulamalara Eleştiri* (pp. 227–253). Yayed.

Çınar, T. (2006b). Su Yönetimi ve Finansmanında Strateji, Model ve Aktörler. In Tayfun Çınar ve Hülya K. Özdinç (Ed.), *Su Yönetimi: Küresel Politika ve uygulamalara Eleştiri* içinde (pp. 43–91). Memleket.

Duru, B. (2013). Büyükşehir Düzenlemesi Ne Anlama Geliyor? *Güneydoğu Anadolu Bölgesi Belediyeler Birliği Dergisi, 8*(12), 33–37. http://www.gabb.gov.tr/doc/gabb_bultenler/2013_06_birlik_dergi. pdferişim tarihi: May 06, 2014.

Geray, C. (2001). Küreselleşme. *Yerelleşme Ve Yerel Yönetimler. Kent Kooperatifçiliği, 133*(135), 27–29.

Görer-Tamer, N. (2015). Neoliberal Politikalar, Çevre ve Kent Bağlamında Türkiye'de Su Yönetimi. In A. Çoban (Ed.), *Yerel Yönetim Kent ve Ekoloji Can Hamamcı'ya Armağan* içinde (pp.193–227). İmge Yayınevi.

Görer-Tamer, N.(2007). Türkiye'nin gündemindeki su sorunlari. *Planlama*, 67- - 81.

Görer, N. (2003). Commercialization and privatization of urban water and sewerage services in Turkey: Poverty reduction view. In O. Ünver, R. Gupta, & A. Kibaroğlu (Eds.), *Water development and poverty reduction* (pp. 175–195). Kluwer Academic Publishers.

Görer, N. (2000). *Kentsel Altyapı Politikalarının Değişimi: İçmesuyu ve Kanalizasyon Sektörü Örneği.* (Unpublished Ph.D. dissertation). Ankara Üniversitesi/Sosyal Bilimler Enstitüsü, Ankara.

Graham, S., & Marvin, S. (1999). Cities, regions and privatised utilities. *Progress in Planning, 51*(2), 91–161.

Güler, B. A. (1999). Yerel-Küresel: Dış Kredilerin Yürütme Gücü. *Yerel Gündem, 3*, 6–23.

Kartal, F. (1999). Su Yönetimi: Son Dönemdeki Politika Arayışları. *Çağdaş Yerel Yönetimler, 8*(4), 101–108.

Keleş, R. (2017). *Su Kaynakları ve Su Hakkı, Sağlıklı Kentler Birliği Kahramanmaraş Konferansı Bildiriler Kitabı*, 39–41.
Keleş, R. (2015). *Kentleşme Politikası* (14th ed.), İmge Yayınevi, Ankara.
Keleş, R., Hamamcı, C., & Çoban, A. (2012). *Çevre Politikası* (7th ed.), İmge Yayınevi, Ankara.
MESKİ (2020), 2020–2024 Stratejik Planı. https://www.meski.gov.tr/files_upload/belgemenu/2020202024.pdf
Neutze, M. (1997). *Funding urban services*, Allen-Unwin, Australia.
Pfister, L., Savenije, H. H. G., & Fenicia, F. (2009). *Leonardo Da Vinci's water theory: On the origin and fate of water*. International Association of Hydrological Sciences, IAHS Special Publications 9. ISBN 978-1-901502-34-3
Prasad, S., & Ramachandraiah, C. (1999). New economic policy: Implications for water, *Economic and Political Weekly*, May 22, 1251–1256.
Smets, H. (1999). Implementing the right to drinking water in OECD countries. *OECD Seminar on the Social and Environment Interface*, TA 1251:22/11/99–12/01/00.
SUEN (2020). COVID-19 Salgınının SU Hizmetlerine Etkileri. https://drive.google.com/file/d/186NNZd65Dw2G2To_t5-uvFFWtZAJ0him/view
SUEN (2019). *Su ve Kanalizasyon İdareleri Mukayeseli Değerlendirme (Benchmarking) Veritabanı ve Yazılımının Geliştirilmesi.* Değerlendirme Sonuçlarının Analizi ve Raporu http://www.suen.gov.tr
Türkiye Belediyeler Birliği (TBB) (2020). *Şehirlerde Su ve Atıksu Hizmetleri Yönetimi*. In A. Esen, O. V. Alıcı (Eds.), Ankara https://www.tbb.gov.tr/online/kitaplar/Su_ve_Atiksu_Hizmetlerinin_Yonetimi/
TMMOB (2009, March). *Küresel Su Politikaları ve Türkiye* (p. 35). Su Raporu, Ankara.
TODAİE (1999). Yerel Yönetimler Araştırma ve Eğitim Merkezi. *Su Hizmetleri Yönetimi: Antalya İncelemesi*, TODAİE, Ankara.
World Bank (1994, November 5). Managing urban water supply and sanitation operation and maintenance. *Lessons & Practices*, Operations Evaluation Department, online. http://www.worldbank.org/html/oed/lpoushtm
World Bank (1992). *Development and the Environment*, World Development Report, Washington D.C.

Nilgün Görer Tamer is a Professor at Gazi University, Faculty of Architecture, Department of City and Regional Planning. She received her Ph.D. from Ankara University Political Science and Public Administration, Urban and Environmental Sciences, and her BA and MA from Middle East Technical University, Department of City and Regional Planning. She lectures applied and theoretical courses in urban infrastructure, urban water planning, spatial research, and urban design. She has publications on spatial growth and climate change, water poverty, water management, and public health.

Airport Regulations in Turkey: An Efficiency Review

Burak Keskin

Abstract Network industries can be defined as in which a firm or its product consists of many interconnected nodes, a node is a unit of the firm or its product, and the links between nodes define the character of the trade in the industry. Telecommunications, transport, and energy are examples of network industries. Network industries play important roles in the modern economy because the infrastructure and services they provide form the basis of the functioning of modern economies. To improve the functioning and efficiency of these sectors, local and national governments issue some regulations on these sectors. The main objective of these regulations is to ensure that the activities are carried out in the most efficient and effective way possible. The rapid increase in trade due to globalization has made the aviation sector one of the most important elements of the global economy. Therefore, in this study, regulations issued on aviation sector in Turkey are mentioned on the basis of slots, charges, passenger rights, ground handling, and passenger services. Later, an efficiency review analysis is built on studies carried out during the last fifteen years on airports located in Turkey. Then, statistics related to methods, input, and output parameters are provided.

Keywords Airports · Regulation · Efficiency · Privatization

1 Introduction

Network industries represent a significant part of the global economy. Network industries can be defined as those in which a firm or its product consists of many interconnected nodes, a node is a unit of the firm or its product, and the links between nodes define the character of the trade in the industry. Examples of network industries include telecommunications, computer software, hardware, data services, transport, and energy. Network industries play an important role in the modern economy because the infrastructure and services they provide form the basis of the functioning of modern economies. To improve the functioning and become more efficient

B. Keskin (✉)
Faculty of Economics and Administrative Sciences, Department of Business Administration, Cankiri Karatekin University, Cankiri, Turkey
e-mail: burakkeskin@karatekin.edu.tr

M. Eroğlu and M. Finger (eds.), *The Regulation of Turkish Network Industries*,
https://doi.org/10.1007/978-3-030-81720-6_9

in these sectors, local and national governments have issued regulations regarding these sectors. The rapid increase in the amount of trade in the world as a result of globalization has made the aviation sector one of the most important elements of the world economy.

By connecting sectors to each other by providing inter-sectoral connections in the realization of a sustainable global economy, the aviation sector is one of the most fundamental sectors in the development of the global economy.

Considering that all countries share a common goal of developing industry and trade and reducing poverty, the key role of the transport sector and especially the aviation sector, which enables them to connect for this purpose, has emerged.

In addition to the positive impact of the aviation sector on economic development, it also has the potential of a sector that is more important than its counterparts thanks to it being more environmentally friendly than other transportation methods. The rapid transport of people and products between distant geographies and the development of production-service-oriented investments without boundaries add dynamism to the region by ensuring the development of the regions where the airports are located in terms of export and import. It also enables the expansion of international trade and faster and easier travel of people and the development of tourism (Aviation Sector Report, 2020).

The aviation sector, which enables the swift transportation of passengers and high value-added products, is constantly increasing its economic value and is playing a dominant role in the development of both countries and all sectors.

A comfortable transportation opportunity is provided to international entrepreneurs and investors with airway transportation, which is of key importance in making investments across country borders without discrimination between countries and regions. It can be expressed as an important benefit that the shortness of air transportation times contributes to the increase of communication between different cities, regions, and cultures, and to the enrichment of culture.

Due to the strategic importance of the aviation sector, countries that incorporate advanced and new technologies gain great advantages in commercial, cultural, and military fields and become important powers (Aviation Sector Report, 2020).

Governments and international sectoral organizations established to regulate wide variety of aviation activities and to protect the rights of passengers and businesses issue regulations such as charges, ground handling services, slot allocation, passenger services, and passenger rights. The main purpose of these regulations is to ensure that the activities at the airports are carried out in the most efficient and effective way possible.

2 A General Overview of Civil Aviation in Turkey

The Turkish civil aviation industry has developed significantly since 1983 liberalization. With the advent of the Republic of Turkey, more attention was paid to the aviation industry. The first step in considering civil aviation in the Republic was the establishment of the Turkish Aeronautical Association on February 16, 1925 (Gerede, 2010; Saldıraner, 1992).

One of the most important developments in the history of Turkish aviation was the establishment of the State Airlines Administration, which forms the basis of the General Directorate of State Airports Authority, on 20 May 1933. Its mission was to provide both air transport and airport operations. Thus began air transportation between major cities in Turkey.

It was observed that the design and manufacturing activities developed rapidly between 1930 and 1950. Thanks to this rapid development, the Turkish aviation industry became Europe's third-largest aviation industry in the mid-1940s.

The development of air transport in Turkey in the 1950s—"air transportation" and "airport operations" has required the separation of functions. Duties were separated within the scope of Law No. 6623 of 21 May 1955, when air transportation was entrusted to Türk Hava Yolları A. Ş. Thus, Turkish Airlines was restructured as a company managed and operated by private law. In 1958 the state-owned Aircraft Service Corporation (USAŞ) was established under the management of Turkish Airlines to provide ground handling and catering services to airlines. The developments in the Turkish Civil Aviation Industry slowed down in the 1950s and there was no significant development until 1983 (Saldıraner, 1992).

Before 1983, with a few exceptions, the civil aviation sector in Turkey were only allowed to operate public companies. Turkish Airlines was the country's only airline and dominated the domestic market. All airports were state-owned but operated by public companies. The market was controlled by the state-owned ground handling service company, one of the country's two ground handling companies (Gerede, 2010).

Until the mid-1980s, with some exceptions, air transportation activities continued under the Turkish Airlines (THY) monopoly and the state implemented restrictive policies on civil aviation activities. Changes in economic policies in Turkey that occurred in 1983 affected the civil aviation industry. Later, "Civil Aviation Law" No. 2920 entered into force and the liberalization process in Turkish civil aviation began. This law grants the private sector the right to air transport and management for commercial purposes at the national and international levels (Battal et al., 2006).

The Turkish aviation sector was negatively affected by the 1990 Persian Gulf crisis, the economic crisis in the countries of the Far East in 1998, the 2001 economic crisis in Turkey, and terrorist attacks of September 11, 2001, in the United States. In connection with these, the Turkish aviation industry has faced some major challenges, such as the bankruptcy of companies and increase unemployment rate in the aviation sector (Yılmaz, 2020).

Civil aviation activities have shown rapid growth in recent years in Turkey. From 2010 to 2019, flight traffic including overflight increased by 1.68 times, the number of passengers carried increased by 2.02 times, and the amount of cargo carried increased by 2.81 times.

The busiest airports in terms of domestic passenger traffic in 2019 are as follow: Istanbul Sabiha Gökçen (22%), Istanbul (13%), Ankara Esenboğa (11%), Izmir Adnan Menderes (9%), Antalya (7%), Istanbul Atatürk (4%), and Adana (4%).

In the same year, the leading airports in terms of their proportion of total international passenger traffic were Istanbul (36%), Antalya (26%), Istanbul Sabiha Gökçen (13%), Istanbul Atatürk (11%), Muğla Dalaman (3%), Izmir Adnan Menderes (3%), Esenboğa (2%), and Milas Bodrum (2%) According to the passenger traffic results of 2019, Turkey was ranked fourth in Europe and tenth in the world (Aviation Sector Report, 2020).

Since 2005, passenger, flight, and freight traffic has increased significantly in Turkey. The total number of passengers has increased by 507% to 209 million, aircraft traffic has increased by 284% to 2,030,291 passengers, and the total amount of cargo has increased by 256% to 3,436,423 tonnes in the last 17 years. Additionally, the number of aircraft has increased by 237%, seat capacity by 276%, cargo capacity by 659%, and the total number of flight points flown to has reached 384 in domestic and international (SHGM, 2020).

3 Privatization in Turkish Civil Aviation

The word "privatization" first appeared in the ninth edition of Webster's New Collegiate Dictionary in 1983 and was defined as "making it private, transferring control and ownership in industrial or commercial life from the public sector to the private sector." The first use of the word was in the form of "rapivatization" in Peter F. Drucker's work "The Age of Discontinuity", published in 1969; Robert W. Poole abbreviated this term as "privatization" in 1976 and used it in his work entitled "Reason Foundation" (Ketenci, 2014).

Although there is no official definition for privatization, the effect of government control is emphasized in all classifications made and this is considered an impressive factor in the choice of method. In some cases, ownership is transferred to the private sector, but considerable public sector influence can still be felt, especially in the context of operating. This kind of privatization is seen during the transfer of facilities that are not well financed with public-sector resources and require renewal of the private sector, especially in developing countries.

Alternatively, governments can use the partial privatization method; for example, if they do not want to completely give up their rights over airports (Graham, 2011).

Both theoretical and applied studies have sought to answer the question of whether privatization activities will bring about an improvement in efficiency. Theoretically, although the general opinion is that the productivity of privatized enterprises will be higher, most applied studies have not supported this theory (Villalonga, 2000).

It is difficult to respond to the needs of the aviation sector for the public, which has such a high growth rate, without private sector support. For example, it is difficult for the public sector to find a solution along with many issues such as the capacity problem brought about by rapid growth and the need to build new terminals as a result, wide runways where wide-body aircraft with high passenger carrying capacity can easily land and take off, and the need for technological innovation. At this point, the implementation of privatization policies at airports are becoming inevitable.

Airport management is a comprehensive and highly costly commercial activity. Although the main purpose is to carry out the aircraft-passenger-freight operations in accordance with the stipulated international rules, institutions and companies have their own different objectives. Services determined by international rules are intertwined with activities aimed directly at profit. The most important resource in meeting all these costs is seen as private sector capital.

Various models are used in the privatization activities of airports. The models usually applied in the airport privatization program in Turkey are Build-Operate-Transfer (BOT) model and Lease-Operate-Transfer (KID) projects. Privatization process of airports started with law enacted numbered 3996 in 1994 in Turkey. DHMI General Directorate, which operates the airports on behalf of the public, started privatization activities by privatizing the Antalya Airport International Terminal in 1996 for a period of 9 years and 45 days. The main point to note here is that DHMI privatizes only the terminals, not the entire airports. The purpose of this is to meet the needs of the terminals such as maintenance repair and technological renewal without using public resources. Airport privatization examples in Turkey are given on Tables 1 and 2.

After the construction of the Antalya Airport International Terminal was completed and put into service in 1998, private sector companies started to show great interest in airport privatization. After Antalya, privatization activities were carried out in Istanbul Atatürk, Ankara Esenboğa, Izmir Adnan Menderes, Muğla Dalaman, Milas-Bodrum, Zafer, Alanya-Gazipaşa, Zonguldak-Çaycuma and Istanbul Sabiha Gökçen Airport. Istanbul New Airport is also built and operated by private sector.

4 Slot Regulation and Slot Allocation System

The term "slot" refers to planning the scheduled arrival and departure times for an airport at a specific date and a specific time. In other words, it is the definition of the right to take off and land from an airport in terms of time and date. Slots can never be considered as property in the civil aviation sector and are lost when an airline cannot use them efficiently. This efficiency rate is the use of at least 80% of existing slots in one season. The analysis of this ratio is made by the authority of the current country under the procedures and rules from ICAO. If this ratio falls below 80% for a specific airline, the slot becomes void and can be bought by another airline through auction or by purchasing according to the brand image and size of the airline.

Slots are distributed in two ways under the primary and secondary allocation systems. In the primary allocation system, a regulatory agency or airport authority distributes slots to airlines. Some airline companies can also use slots that are lost in the inherited system from airlines that do not continue their activities. This right is called grandfather rights. Airlines that use slots obtained through inheritance must use at least 80% of these slots; otherwise, other airlines may request for this slot received by inheritance. The determination of the airlines that receive this right is made by the regulatory agency or airport authority (Inan, 2020).

Table 1 Transferring of operational rights projects

Project name	Project cost	Period	End of period
Atatürk Airport passenger terminals	$2,543,000,000 + VAT	15.5 year	03.01.2021
Antalya Airport passenger terminals	€2,010,000,000 + VAT	Stage I: 17 years, 3 months, 17 days Stage II: 15 years, 3 months, 8 days	31.12.2024
Zonguldak Çaycuma Airport	1.06% of endorsement for renting fee + $32,291 facility use charge	25 years	20.08.2032
Alanya-Gazipaşa Airport	65% of net profit for renting fee + $50,000 facility use charge	25 year	13.07.2034
Adnan Menderes Airport passenger terminals	€610,000,000 + VAT	Domestic: 20 years, 11 months, 29 days International: 17 years, 11 months, 21 days	31.12.20232
Aydın Çıldır Airport	7% of net profit for renting fee + €20,000 facility use charge	20 years	20.07.2032
Milas-Bodrum Airport passenger terminals	717.000.000 € + VAT	Domestic: 21 year 5 month 16 days International: 20 year 2 month 3 days	31.12.2035
Dalaman Airport passenger terminals	705.000.000 € + VAT	Domestic: 26 Year 4 Month 30 Days International: 25 Year 8 Month 3 Days	31.12.2040
Total	**8.041.225.675 $**		

Source DHMI (2020)

The time period needed by the carrier to take off or land in order to fly is called a slot, while allocation of this time period within the framework of certain order and rules is called slot allocation. In technical terms, the slot allocation "row/time" refers to the permission taken from the competent authority in order to benefit from infrastructure services at predetermined time periods, especially at airports with heavy air traffic.

Through the allocation of the time period, the carrier can use its right to flight based on national or international legislation. The time allocated is important for the carrier to use its commercial activities efficiently and effectively. Therefore, the allocation of "row/time" has an economic meaning for carriers.

Row/time allocation first started in the United States. In 1968, when air traffic density reached its highest level, flight quotas were introduced for busy airports and time periods were determined for these airports. However, there is no coordination

Table 2 Built-operate-transfer projects

Project name	Investment/renting amount	Period	End of period
Antalya Airport I. International terminal	$75,902,000	9 years, 45 days	13.09.2007
Antalya Airport II. International terminal	$85,386,000	3 years, 5 months, 26 days	22.09.2009
Atatürk Airport international terminal	$397,793,500	4 years, 10 months, 15 days	02.07.2015
Dalaman Airport international terminal	$91,997,688	8 years, 2 months, 17 days	28.04.2015
Adnan Menderes Airport international terminal	$181,941,685	7 years, 4 months, 26 days	10.01.2015
Esenboğa Airport domestic and international terminal	$247,200,350	15 years, 8 months	24.05.2023
Milas-Bodrum Airport international terminal	$116,122,330	3 years, 9 months	22.10.2015
Zafer Airport	$65,500,000	29 years, 11 months	21.03.2044
Çukurova Airport	$464,764,505	9 years, 10 months, 10 days	Under construction
İstanbul Airport	€10,247,000,000 Investment Amount + €22,152,000,000 renting fee	25 years	Under construction (only first stage is opened)
Total	**$1,718,640,372**		

Source DHMI (2020)

or preliminary process for the allocation of these time periods. After the Airline Liberalization Act in 1978, competition increased and the demand for busy airports started to be unmet. Consequently, the system of allocation of row/time was limited to 9000 covering 22 airports starting in 1981 (Cengiz, 2010).

In the fierce competitive environment in the civil aviation sector, airlines apply various strategies to survive. One such strategy is the cooperation formed within the framework of bilateral agreements. Thanks to these collaborations (code share, etc.), airlines contribute to each other. Apart from bilateral agreements, the most important issue for airlines is the slot concept. Nowadays, due to the rapid growth of the aviation industry and the filling of airport capacities, it has become increasingly difficult to obtain good slots, and therefore more important for airlines. In addition, the "use it or lose it" system shows that the purchased slot can be lost over time, which encourages airlines to apply strategies correctly in the intensive competition conditions (Inan, 2020).

Slot coordination in Turkey began in 1992 and was appointed as the coordinator of this process by the Ministry of Transportation Turkish Airlines. Private airlines competing with THY faced slot problems due to increased traffic at Atatürk Airport after the deregulation in 2003. Some private airlines claimed that THY, as the coordinator, blocked the entrance to the Istanbul Atatürk Airport-Ankara market. Later, the Ministry of Transport gave the slot coordination authority to an independent commission established under the chair of the DGCA in February 2006. The commission is composed of the State Airports Administration, Turkish Airlines and Turkey Slot Coordination Unit Private Aviation Enterprises Association representatives, terminal operators, airlines, and ground handling services and their representatives. Establishment of the slot coordination center is an important development in terms of removing barriers to entry to the domestic market and the monopolistic position of THY at Atatürk Airport (Gerede, 2010).

In 1990s, after the collapse of the USSR, intense air transportation began in Turkey, especially at Istanbul Atatürk Airport. Turkish civil aviation was unprepared for this situation and was insufficient. Since it is not possible to increase the capacity of airports in the short term, various solution proposals were developed. For example, the terminal problem was sought to be solved with the temporary charter terminal built at Istanbul-Atatürk Airport. In addition, Tekirdağ-Çorlu, Bursa-Yenisehir and Izmit-Cengiz Topel Airports were opened to civilian traffic. Significant investments were made in these airports and some of the traffic was tried to be shifted to these airports, but this application was also not successful.

In those days, because the aircraft and passenger traffic towards Istanbul-Atatürk Airport was concentrated on certain days of the week and certain hours of these days, the airport could not be used efficiently and there was unnecessary congestion. For this reason, in order to ensure the safety of life and flight at the highest level, it is necessary to ensure the optimum use of the airport by spreading the heavy air traffic in airports as much as possible, every hour of the day and every day of the week. In other words, the "slot" application was launched in order to ensure that the airport facilities could be used at the highest level in the departure and departure time zones that can be allocated to the aircraft.

The slot allocation was initially applied only at Istanbul-Ataturk Airport, and later started to be used at Antalya, Ankara-Esenboğa, Izmir Adnan Menderes, Muğla-Dalaman, Kayseri-Erkilet and Bodrum-Milas Airports.

In 1992, the Slot Coordination Unit was established by the Directorate General of Civil Aviation (SHGM) in Turkey. Due to the fact that the State Airports Authority (DHMI) of that day did not accept this task, the coordinator position was assigned to Turkish Airlines (THY), taking into account the coordinators in other countries where aviation was developed. However, because of the developments in the EU Legislation and the increase in aircraft and passenger traffic resulting from the liberalization initiated in domestic lines in 2003, it has become necessary to reorganize the capacities of airports and to make slot coordination more independent. In this context, DGCA Slot Coordination Center at the Istanbul-Atatürk Airport International Terminal was activated in February 2006 (Uslu, 2015).

The slots are allocated by the General Directorate of States Airport Authority (DHMI) Slot Coordination Center in Turkey. Slot allocation is carried out in order to use the available capacity in the most effective and efficient way at airports where it is not possible to meet all flight requests at the same time to all domestic and foreign airlines that operate scheduled and non-scheduled domestic and international flights. The airports with allocated slots are Istanbul Atatürk Airport, Antalya Airport, Esenboğa Airport, Izmir Adnan Menderes Airport, Milas-Bodrum Airport, Dalaman Airport, and Istanbul Sabiha Gökçen Airport.

Slot applications cover all types of flights of all domestic and foreign airlines operating on scheduled and non-scheduled domestic and international flights to carry passengers, cargo, and mail. Up to three hours of slot time is given to passenger aircraft and up to five hours for cargo aircraft, depending on the aircraft type. The priority criteria in evaluating the slot demand are as follows:

(a) Vested rights
(b) Scheduled flights
(c) Non-scheduled flights
(d) Demand from airlines that are organizing flights to the airport for the first time or are included in the category of first time flights according to the current number of flights.

Slot applications cover all types of flights of all domestic and international airline companies that travel for the purpose of carrying passengers, cargo, and mail on scheduled, non-scheduled domestic, and international flights. Slot times are given for all kinds of flights up to three hours for passenger aircraft and up to five hours for cargo aircraft, depending on the aircraft type. If the maximum take off weight (MTOW) of the aircraft is 100,001 kg, a slot time of up to seven hours is given upon request. Slot time can not exceed seven hours, including overtime.

In slot applications, it is essential that the requests made in accordance with the bilateral air transport agreements are met within the requested time zones to the extent permitted by the aerodrome capacities or within the closest time zones, if not possible, within the same day.

The slot capacity of each airport is determined by the participation of air traffic, airport management, terminal, ground services, customs, and security organizations in order to allocate slots to as many aircraft as possible in unit time and to maintain efficiency and quality in services.

While determining the capacity, restrictive criteria such as runway capacity, the number of aircraft parking places, the number of gates, passport control, customs, security, ticketing, and baggage procedures are taken into account.

If it is determined that the airport requires planning in its current situation, the Slot Technical Committee decides on the implementation of coordination or tariff regulation and makes the necessary announcement to IATA and airline operators within the framework of a schedule that will allow all relevant parties to make the necessary arrangements (DHMI, 2020).

5 Ground Handling Services Regulation

The indispensable part of airline transportation, which has been among the fastest growing sectors in recent years, is ground handling services offered to passenger and cargo aircraft that land and depart from thousands of airports around the world.

In aviation, "ground handling service" refers to all services provided to the aircraft in the terminal area from the time the aircraft's landing wheels touch the runway to the departure of the aircraft at an airport. Ground handling services at airports comprise all services at an airport from landing to take off, including passenger traffic, catering, and technical services within the terminal for passenger aircraft preparing for flight or landing, and palletizing operations in the warehouse for cargo aircraft (Eski & Tasus, 2018).

Ground and catering services are carried out under coded SHY-22 regulation at Turkish airports. The purpose of this regulation is to regulate the procedures and principles to be applied to ensure that airport ground handling services are carried out at the international level and to determine the principles of authorization and supervision for these services and the charges for ground handling services. This regulation covers public institutions and organizations that operate airports and/or terminals, real and legal persons, organizations that obtain working licenses, air carriers, and other organizations operating within the scope of this Regulation by meeting the conditions specified in this Regulation (DHMI, 2020).

Types of airport ground handling services are grouped as follows:

(a) Representation
(b) Passenger services
(c) Load control and communications
(d) Ramp
(e) Cargo
(f) Aircraft line maintenance
(g) Flight operations
(h) Transport
(i) Catering services
(j) Supervision and administration
(k) Aircraft private security service and audit.

The details of the airport service types mentioned above are determined by the aviation instructions to be issued by the General Directorate.

Havas is the first ground handling company in Turkey; it was established in 1933 by the government to serve airlines. Sixty percent of Havas's ground handling shares were sold to the Park Holding Group in 1995, and the catering services were privatized in 1987 under the name USAS. Turkey's first private ground handling company, Celebi, was established at Ankara Esenboğa Airport in 1958. TGS ground services, established by the partnership between Turkish Airlines and Havas has been in the ground handling business since 1 January 2010.

Ground handling companies that obtain a group working license must give the final bank letter of guarantee to DHMI for an indefinite amount of USD 1,000,000, the content and form of which will be determined by DHMI for liabilities that may arise from the service contract. Ground handling companies that receive Working License Group B or C are required to provide DHMI with the same type of bank letter of guarantee for USD100,000.

6 Tariff Regulation

The tariffs will be applied at the airports operated by DHMI are determined by the DHMI general directorate. Tariffs to be applied to airports operated by public-private partnership (PPP) enter into force with the approval of the General Directorate of DHMI.

While applying the international landing tariff, airports are divided into two groups:

Group 1: İstanbul Atatürk, Ankara Esenboğa, Antalya, İzmir Adnan Menderes, Muğla Dalaman, Muğla Milas Bodrum, Adana, Trabzon, Erzurum, Gaziantep, Kayseri and Hatay Airports.

Group 2: Bursa Yenişehir, Denizli Çardak, Diyarbakır, Elazığ, Erzincan, Isparta, Kahramanmaraş, Kapadokya, Kars, Konya, Malatya, Ordu Giresun, Samsun Çarşamba, Sinop, Van ile geçici hudut kapısı olan Adıyaman, Ağrı, Amasya Merzifon, Balıkesir Koca Seyit, Batman, Bingöl, Çanakkale, Iğdır, Kastamonu, Kocaeli, Mardin, Muş, Siirt, Sivas, Şanlıurfa GAP, Şırnak, Tekirdağ Çorlu, Tokat, Uşak, and other airports that will be opened for international flights soon.

(a) International landing charges apply as follow:

- Between 0 and 500 landings: 7.14 Euro/tonne for Group 1; 4.44 Euro/tonne for Group 2; 9.28 Euro/tonne for Istanbul; 5.55 Euro/tonne for Zafer; and 4.44 Euro/tonne for Alanya Gazipaşa, Zonguldak Çaycuma, and Aydın Çıldır Airports.
- Between 501 and 1000 landings: 7.14 Euro/tonne for Group 1; 4.44 Euro/tonne for Group 2; 8.89 Euro/tonne for Istanbul; 5.30 Euro/tonne for Zafer; and 4.24 Euro/tonne for Alanya Gazipaşa, Zonguldak Çaycuma, and Aydın Çıldır Airports.
- Between 1001 and 2000 landings: 7.14 Euro/tonne for Group 1; 4.44 Euro/tonne for Group 2; 8.35 Euro/tonne for Istanbul; 5.04 Euro/tonne for Zafer; and 4.03 Euro/tonne for Alanya Gazipaşa, Zonguldak Çaycuma, and Aydın Çıldır Airports.
- Between 2001 and 4000 landings: 7.14 Euro/tonne for Group 1; 4.44 Euro/tonne for Group 2; 7.95 Euro/tonne for Istanbul; 4.79 Euro/tonne for Zafer; 3.83 Euro/tonne for Alanya Gazipaşa, Zonguldak Çaycuma, and Aydın Çıldır Airports.

- 4001 and more landings: 7.14 Euro/tonne for Group 1; 4.44 Euro/tonne for Group 2; 7.56 Euro/tonne for Istanbul; 4.54 Euro/tonne for Zafer; and 3.63 Euro/tonne for Alanya Gazipaşa, Zonguldak Çaycuma, and Aydın Çıldır Airports.

Passenger service charges include the usage fees of the airport facilities and services provided to departing passengers within the Terminals.

(b) International passenger service charges apply as follow:

- From January 1–April 30 and November 1–December 31st, 3.75 Euro; May 1–October 31, 7.50 Euro for Adana, Erzurum, Hatay, Gaziantep, Kayseri, and Trabzon Airports; January 1–April 30 and November 1–December 31, 3.00 Euro; May 1–October 31, 6.00 Euro for other airports; 20 Euro for Istanbul Airport, 15 Euro for Esenboga, Antalya, Adnan Menderes, Milas Bodrum and Mugla Dalaman Airports, 10 Euro for Zafer Airport, 12 Euro for Alanya Gazipasa Airport and 5 Euro for Zongulda Caycuma Airport.

(c) Domestic passenger service charges apply as follows:

- 8.75 TL for Adana, Diyarbakır, Erzurum, Gaziantep, Kayseri, Samsun-Çarşamba, Trabzon and Van Airports; 5.83 TL for Adıyaman, Ağrı, Balıkesir Koca Seyit, Batman, Bursa Yenişehir, Denizli Çardak, Elazığ, Erzincan, Hatay, Iğdır, Kahramanmaraş, Kapadokya, Kars, Konya, Malatya, Mardin, Muş, Ordu Giresun, Sivas, Şanlıurfa GAP, Şırnak Airports; 2.92 TL for other airports. 3 Euro for Istanbul Airport, 3 Euro for Esenboga, Antalya, Adnan Menderes, Milas Bodrum and Mugla Dalaman Airports, 2 Euro for Zafer Aiport, 10 TL for Alanya Gazipasa Airport and 2 TL for Zongulda Caycuma Airport.

(d) Ground handling mandatory service charges are given in the Table 3 for passenger aircraft.

7 Passenger Rights Regulation

Under the Civil Aviation Regulation in Turkey, the purpose of regulation is to determine and regulate the rights of passengers traveling by airline and the situations in which these rights are valid, and their minimum rights in cases where passengers are not admitted to the plane, flights are canceled, or delayed.

The air transport operator performing the flight makes an announcement asking for volunteers to waive their reservation in return for the benefits to be agreed upon between the passenger and the air transport operator performing the flight in case of denial to the aircraft in a flight. Volunteers are compensated for the following issues and given the right to choose between the following options:

Table 3 Ground handling mandatory service charges

Seat capacity	Passenger service	Load control and communication	Ramp	Cargo and mail
1–50	€7 (€9.10 for Istanbul)	€2 (€2.60 for Istanbul)	€8 (€10.40 for Istanbul)	€3 (€3.90 for Istanbul)
51–100	€17 (22.10 for Istanbul)	€2 (2.60 for Istanbul)	€23 (29.90 for Istanbul)	€10 (13 for Istanbul)
101–150	€39 (€50.70 for Istanbul)	€4 (€5.20 for Istanbul)	€47 (€61.10 for Istanbul)	€20 (€26 for Istanbul)
151–200	49 (€63.70 for Istanbul)	4 (€5.20 for Istanbul)	60 (€78 for Istanbul)	27 (€35.10 for Istanbul)
201–250	€67 (€87 for Istanbul)	€7 (€9.10 for Istanbul)	€77 (€100.10 for Istanbul)	€33 (€42.90 for Istanbul)
251–300	€77 (€100.10 for Istanbul)	€7 (€9.10 for Istanbul)	€94 (€122.20 for Istanbul)	€39 (€42.90 for Istanbul)
301–350	€87 (113.10 for Istanbul)	€7 (9.10 for Istanbul)	€103 (133.90 for Istanbul)	€50 (65 for Istanbul)
351+	€99 (€128.70 for Istanbul)	€8 (€10.40 for Istanbul)	€123 (€159.90 for Istanbul)	€53 (€68.90 for Istanbul)

(a) Refund of the entire ticket fee at the price purchased and a return flight free of charge to ensure that the passenger returns to the first starting point of their journey as soon as possible;
(b) Under similar transportation conditions, ensuring the route change to the final destination at the earliest opportunity;
(c) Changing the route to the final destination, depending on the availability of seats, at a later date deemed appropriate by the passenger, under similar transport conditions;
(d) In the event that passengers offer a flight to an alternative airport, the air transport operator performing the flight must cover the cost of transferring the passenger from that alternative airport to the airport where the reservation was made or to another nearby destination accepted by the passenger.

If there are not enough volunteers to allow the remaining passengers to be admitted to the flight, the air transport operator performing the flight may refuse to accept the passengers against their will. In the event that the passengers are refused admission to the flight against their will, the air transport operator performing the flight must assist the passengers under following issues. Compensation is given to the passengers equivalent to 100 Euros in Turkish Lira for domestic flights. On international flights, the compensation is:

(a) Turkish Lira equivalent of 250 Euros for all flights of 1500 km or less
(b) Turkish Lira equivalent of 400 Euros for flights between 1500 and 3500 km
(c) For flights longer than 3500 km, compensation of 600 Euros is given in Turkish Lira.

If passengers are offered a route change to their final destination with an alternative flight that does not exceed the planned arrival time of the booked flight within the following limits, the air transport operator performing the flight may reduce the compensation by 50%.

(a) Two hours for flights of 1500 km (including 1500 km) or less
(b) Three hours for flights between 1500 and 3500 km (including 3500 km)
(c) Four hours for flights longer than 3500 km.

In case of cancelation of a flight, the air transport operator performing the flight is obliged to assist the passengers within the conditions stated above.

If a delay is expected within the following limits from the scheduled departure time of a flight, the air transport operator must assist the passengers (SHGM, 2020).

(a) Shorter than 1500 km (including 1500 km) and two or more hours for domestic flights
(b) Three hours or more for flights between 1500 and 3500 km (including 3500 km)
(c) Four hours or more for flights longer than 3500 km.

8 A Review on Efficiency Analysis Studies Carried Out on Turkish Airports

Airports, as vital national resources, play an important role in the transportation of people and products in the regional, national, and international arena and trade. Many airports are responsible to the governments of other countries as well as to their own regional and central government. Because airports are systems that are in constant communication with each other and cannot be considered independently from each other, it is important for countries to operate such important resources efficiently (Keskin & Koksal, 2019).

The most important aim of public- or private sector enterprises that have the right to operate the airports is to operate this important resource in the most efficient way possible. Regulations, instructions, etc. are all based on the aim of making efficient use of available resources. Moreover, it is necessary to measure how these resources, which are outlined on paper with regulations or instructions, are used effectively in operational terms. These measurement processes are usually carried out with multi-criteria decision-making methods and/or data envelopment analysis.

This section reviews the methods used to measure the efficiency of Turkish Airports, the variables, and the studies carried out over the last 15 years about airports.

Firstly, the studies were obtained by using keywords such as "performance analysis of Turkish airports", "efficiency analysis of Turkish airports", "performance and efficiency analysis of Turkish airports", and "efficiency analysis of Turkish airports with DEA". Later, each of the studies reached was examined separately, and Table 4 was prepared regarding the methods, variables, and results obtained.

The most used method is Data Envelopment Analysis (DEA), as seen in Table 4. The DEA methodology is based on an interesting application of linear programming. DEA was originally developed for performance measurement. The principles of DEA date back to Farrel (1957) and the recent series of discussions on this topic started with Charnes et al. (1978). This technique aims to measure how efficiently a DMU uses the resources available to generate a set of outputs (Charnes et al., 1978).

Number of employees (32%), terminal area (28%), number of runways (21%), and operational expenses (19%) are the most used input variables. And the most used output variables are number of passengers (33%), aircraft movements (31%), total cargo or freight (23%), and operational revenues (13%) according to Table 3.

9 Conclusion

Airports are important national resources for countries. Because of this importance, all countries in the world publish their own regulations and instructions. Besides that, they should care about international aviation rules as well. Generally, the purpose of these regulations and instructions are to operate airports efficiently and not waste potential time, effort, labor, and money.

After the first actual airport privatization, where a government authority transferred to the private sector by the British Government in 1987, the interest of the private sector has increased gradually over the years to airports. Due to limited public resources, private sector interest has been attractive for governments. Because building new terminals, runways and many facilities at airports are quite expensive investments, they must be operated efficiently as possible. Theoretically, although the general opinion is that the productivity of privatized enterprises will be higher, this theory has not been supported in most of the applied studies.

Many researchers have investigated airport efficiencies over the years. While some of them focus on whether airports are operated by the public or private sector, others have focused on airports' size, tourism, distance to the city, etc. Nevertheless, there is no clear conclusion regarding this discussion. Besides these factors, regulations and instructions published by the governments are also important factors for airport efficiencies. Regulations in airports are issued in accordance with countries' internal legal and economic situations. Generally, the objectives of regulations are as follows:

- Reduce the State's costs in performing its economic regulatory functions
- Increase consumers' benefits and choices
- Improve air connectivity
- Create more competitive business opportunities in the marketplace.

Table 4 Efficiency analysis studies in Turkish airports, 2005–2020

Author	Method	Inputs	Outputs	Results
Kıyıldı and Karaşahin (2006)	DEA	Number of check-in counters, number of X-rays, passenger usage area, parking capacity, runway and apron size, aircraft capacity, taxiway airport capacity	Number of aircraft	An efficiency analysis was employed between 1996 and 2002 and just a few airports were found to be efficient
Peker and Baki (2009)	DEA	Parking capacity, number of runways, airport size, employees	Number of passengers, total cargo	Efficiency of the airports located in regions with high tourism, industry and work flow is higher
Koçak (2011)	DEA	Operational expenses, staff number, annual flight traffic, number of passengers	Number of passengers/area, total flight traffic/runway, cargo traffic, revenues	As a result, İstanbul Atatürk, Antalya, Denizli Çardak, Sinop, Kayseri, Malatya and Van Ferit Melen Airports were found to be efficient

(continued)

Table 4 (continued)

Author	Method	Inputs	Outputs	Results
Ar (2012)	Total factor productivity	Staff number, aircraft capacity per runway/apron, passenger capacity per area	Aircraft traffic per runway, passenger traffic per area, freight traffic	According to the results, TFP increased by 11.8% between 2007 and 2011 as a result of technological developments
Ömürbek et al. (2013)	DEA	Expenses, service area, passenger and parking capacity, vehicle parking, apron capacity, aircraft capacity, number of rescue devices, staff number	Flight traffic, commercial flight traffic, cargo traffic, passenger traffic, sales revenues	The study concluded that airports should be evaluated by dividing them into three groups—large, medium and small airports—considering the flight traffic and passenger traffic
Öztürk et al. (2014)	DEA and MOORA	Number of employees, number of runways	Number of passengers, number of aircraft movements, total freight	Atatürk, Antalya and Diyarbakır airports were found to be the most efficient airports
Avcı and Aktaş (2015)	DEA and total factor productivity	Staff number, terminal area	Number of passengers, freight traffic, total aircraft traffic	When the summer and winter periods are evaluated in terms of efficiency, it is concluded that the airports generally show higher performance in summer

(continued)

Table 4 (continued)

Author	Method	Inputs	Outputs	Results
Ülkü (2015)	DEA	Staff costs, other operating costs, total runway area	Number of passengers, air traffic movements, commercial revenues	The results indicate higher average efficiency levels at Spanish airports, but private involvement enhances efficiency at Turkish airports
Yazgan and Karkacıer (2015)	DEA and Tobit Model	Number of employees, operating expenses, terminal area, number of runways, number of aprons	Operating income, number of passengers, freight and aircraft traffic	As a result of the Tobit model, operating costs, terminal area, and number of runways have a negative effect on efficiency in all years when they are significant; Freight traffic and number of aprons have a positive effect on efficiency
Bolat et al. (2016)	DEA and artificial neural networks	Number of check-in counters, number of baggage conveyors, number of gates, number of runways, terminal size, employees, parking capacity	Number of passengers, load amount, commercial flight traffic	The results of DEA show that 19 airports in 41 airports work efficiently in Turkey. After DEA analysis, an artificial neural network model was built that helps to predict the efficiency of existing and alternative airports by utilizing same data
Bolat et al. (2016)	Data Envelopment Analysis (DEA) and Malmquist productivity index	Number of runways, dimension of runway units, passenger terminal area	Annual number of flights, annual passengers, annual cargo	Measured operational efficiency of 21 Turkish airports between 2009 and 2014. The findings showed that the efficiency and productivity of most Turkish airports increased during the research period

(continued)

Table 4 (continued)

Author	Method	Inputs	Outputs	Results
Gümüştekin et al. (2016)	DEA and Tobit regression model	Number of employees, total expenses, number of check-in counters, number of aprons	Number of passengers, total revenues, aircraft traffic, baggage traffic	This paper measures and compares the efficiency of 25 Turkish airports for the period 2010–2014. Despite a decline in the 2011–2012 period, efficiency scores of Turkish airports have increased again since 2013
Koç and Çalıpınar (2017)	FarePrimont, DEA, Malmquist TFP	Number of employees, operational expenditures	Number of passengers and total freight	FarePrimont analysis was applied for the first time in Turkish airports with the results compared with the Malmquist productivity index. According to the results, the Malmquist TFP index and the FarePrimont TFP index had the largest increases in total factor productivity between 2011 and 2014 in Isparta Süleyman Demirel Airport, and the greatest decrease is in Tekirdag Corlu and Çanakkale Airports
Çınaroğlu and Avcı (2017)	DEA and MOORA	Number of employees, terminal area, aircraft capacity	Number of passengers, freight traffic	Atatürk and Adana airports were found to be efficient for domestic flights, Atatürk and Antalya airports were found to be efficient for international flights. This situation may be caused by tourism effect

(continued)

Table 4 (continued)

Author	Method	Inputs	Outputs	Results
Keskin and Köksal (2019)	AHP/DEA-AR	Number of employees, number of gates, runway area, terminal area, operational expenditure	Number of passengers, amount of cargo, aircraft movements, total revenue	Results show that none of the public airports are efficient. To cope with this situation, it may be a useful policy to apply a regional airport system
Ateş et al. (2018)	DEA, Malmquist TFP	Runways, terminal area	Passenger traffic, load transmission, aircraft traffic	According to the results, total factor productivity increased by an average of 3.6% in the relevant period
Asker et al. (2018)	TOPSIS and DEA	Number of runways, terminal area, number of check-in counters	Aircraft movements, number of passengers	An efficiency analysis was conducted on 15 airports and the results were compared using the TOPSIS and DEA methods
Ulutaş (2018)	DEA	Amount of emissions (undesirable output)	Terminal area, aircraft capacity, total amount of air traffic, distance to the city	The data consist of 22 airports that have a green airport certificate. The mean efficiency score was 0.38, which is quite low

(continued)

Table 4 (continued)

Author	Method	Inputs	Outputs	Results
Yalçın (2018)	Stochastic Frontier analysis	Number of commercial flights, number of employees, number of check-in counters	Number of passengers	Cobb–Douglas production function was used out of production functions as it is statistically significant. The function results show that the number of commercial airliners and staff number parameters were statistically significant while the check-in counter number parameter was not statistically significant in any model analyzed
Şahin (2019)	DEA and Malmquist TFP	Number of employees, operating expenses, terminal area, runways, number of aprons	Passenger traffic, operating income, aircraft traffic, freight traffic	According to the findings of study, airports reached their best mean efficiency score in 2015
Köleoğlu ve Demirel (2019)	DEA	Employees, employee costs, number of flights, terminal capacity	Tourist numbers, tourism revenue	As result, Atatürk, Dalaman, Bodrum, Isparta, Nevşehir, and Denizli airports were found to be efficient
Uludağ (2020)	EATWOS-DEA	Total costs, number of runways, number of staff, total passengers, cargo, and general aviation terminal area	Number of passengers, cargo traffic	The second-group airports are more productive than the first-group airports. This is due to the low theoretical capacity of the second-group airports, the lower inputs that they use, and the higher demand for air transport

Source Author

Thereby contributing to the sustainable economic development and to the expansion of trade and tourism (ICAO, 2020).

References

Ar, I. (2012). Türkiye'deki Havalimanlarının Etkinliklerindeki Değişimin İncelenmesi: 2007–2011 Dönemi İçin Malmquıst-Tfv Endeksi Uygulaması. *Atatürk Üniversitesi İktisadi Ve İdari Bilimler Dergisi, 26*(3–4), 143–160.

Asker, V., Kiracı, K., & Yaşar, M. (2018). Entropi Temelli TOPSİS ve Veri Zarflama Analizi Yöntemleriyle Etkinlik Ölçümü: Türkiye'deki Büyük Havalimanları Üzerine Bir Uygulama.

Ateş, S. S., Ulufer, S., & Altınbaş, Ş. (2018). Effıcıency evaluation of Turkısh aırports wıth DEA and Malmquıst-TFP index approach. *İstanbul Aydın Üniversitesi Dergisi, 10*(3), 83–94.

Avcı, T., & Aktaş, M. (2015). Türkiye'de Faaliyet Gösteren Havalimanlarının Performanslarının Değerlendirilmesi. *Journal of Alanya Faculty of Business/Alanya Isletme Fakültesi Dergisi, 7*(3).

Aviation Sector Report. (2020). *DHMI 2019 aviation sector report 2019*. Retrieved August 17, 2020, from https://www.dhmi.gov.tr/Sayfalar/HavaYoluSektorRaporlari.aspx

Battal, Ü., Yılmaz, H., & Ateş, S. S. (2006). Türkiye'de İç Hatlarda Serbestleşme ve Geleceği. In *Conference paper presented on Kayseri VI Havacılık Sempozyumu*.

Bolat, B., Temur, G. T., & Gürler, H. (2016). Türkiye'deki Havalimanlarının Etkinlik Tahmini: Veri Zarflama Analizi ve Yapay Sınır Ağlarının Birlikte Kullanımı. *Ege academic review, 16*.

Cengiz, A. (2010). *Avrupa Birliği ve Türkiye'de Slot (Sıra/Zaman) Tahsisi Uygulaması ve Hukuki Değerlendirme*.

Charnes, A., Cooper, W. W., & Rhodes, E. (1978). Measuring the efficiency of decision making units. *European Journal of Operational Research, 2*(6), 429–444.

Çınaroğlu, E., & Avcı, T. (2017). Türkiye'de Faaliyette Bulunan Büyük Havalimanlarının İç Ve Dış Hat Performanslarının Karşılaştırılması. *BMIJ, 5*(4), 55–75.

DHMI, (2020). Devlet Hava Meydanları İşletmesi (General Directorate of State Airport Authority). Retrieved August 25, 2020, from www.dhmi.gov.tr

Eski, S., & Tasus, H. S. (2018). *Havaalanlarında Sunulan Yer Hizmetlerinin Avrupa Ekonomisine Etkisi: Türkiye, Almanya ve İngiltere Uygulamaları*.

Farrell, M. J. (1957). The measurement of productive efficiency. *Journal of the Royal Statistical Society: Series A (General), 120*(3), 253–281.

Gerede, E. (2010). The evolution of Turkish air transport industry: Significant developments and the impacts of 1983 liberalization. *Journal of Management & Economics, 17*(2).

Graham, A. (2011). The objectives and outcomes of airport privatisation. *Research in Transportation Business & Management, 1*(1), 3–14.

Gümüştekin, S., Senel, T., & Dunder, E. (2016). The Evaluating the efficiency performance of Turkish airports (2010–2014). *International Journal of Sciences: Basic and Applied Research, 30*(4), 460–473.

ICAO (2020). International Civil Aviation Organisation. Retrieved August 28, 2020, from https://www.icao.int/Pages/default.aspx

Inan, T. T. (2020). Slot Kavramının Havalimanları ve Havayolları Açısından İncelenmesi. *International Journal of Management and Social Research, 7*(13), 46–55.

Keskin, B., & Köksal, C. D. (2019). A hybrid AHP/DEA-AR model for measuring and comparing the efficiency of airports. *International Journal of Productivity and Performance Management*.

Ketenci, S. T. (2014). *Özelleştirmelerin Finans Piyasaları Üzerine Etkisi: Türkiye Örneği*. Yüksek Lisans Tezi, Selçuk Üniversitesi.

Kıyıldı, R., & Karaşahin, M. (2006). Türkiye'deki hava alanlarının veri zarflama analizi ile altyapı performansının değerlendirilmesi. *Süleyman Demirel Üniversitesi Fen Bilimleri Enstitüsü Dergisi, 10*(3), 391–397.
Koç, A. G. E., & Çalıpınar, H. (2017). Fareprımont Ve Malmquıst Verimlilik Endeksleri İle Türk Havalimanlarının Etkinliklerinin Karşılaştırılması. *International Journal of Academic Value Studies, 3*(7), 77–87.
Koçak, H. (2011). Efficiency examination of Turkish airports with DEA approach. *International Business Research, 4*(2), 204.
Koleoglu, N., & Demirel, E. (2019). Türkiye'nin Önemli Turizm Kentlerindeki Havalimanlarının Etkinliklerinin Veri Zarflama Analizi Yöntemiyle Ölçülmesi. *Seyahat Ve Otel İşletmeciliği Dergisi, 16*(3), 352–365.
Ömürbek, N., Demirgübüz, M. Ö., & Tunca, M. Z. (2013). Hizmet Sektöründe Performans Ölçümünde Veri Zarflama Analizinin Kullanımı: Havalimanları Üzerine Bir Uygulama. *Süleyman Demirel Üniversitesi Vizyoner Dergisi, 4*(9), 21–43.
Örkcü, H. H., Balıkçı, C., Dogan, M. I., & Genç, A. (2016). An evaluation of the operational efficiency of Turkish airports using data envelopment analysis and the Malmquist productivity index: 2009–2014 case. *Transport Policy, 48*, 92–104.
Öztürk, E., Durak, H. A., & Bal, H. (2014). Evluation of Turkish airport efficiencies using data envelopment analysis and Moora method. *Data Envelopment Analysis and Performance Measurement, 189.*
Peker, G. & Baki, B. (2009). Veri Zarflama Analizi İle Türkiye Havalimanlarında Bir Etkinlik Ölçümü Uygulaması. *Journal of the Cukurova University Institute of Social Sciences, 18*(2).
Saldıraner, Y. (1992). *Sivil Havacılık Faaliyetleri ve Türk Sivil Havacılık Otoritesi İçin Organizasyon Yapısı Önerisi.* Eskişehir, Anadolu Üniversitesi SHMYO Yayınları.
SHGM, (2020). Sivil Havacılık Genel Müdürlüğü (Directorate General of Civil Aviation) 2019 Annual Report. Retrieved August 15, 2020, from http://web.shgm.gov.tr/tr/kurumsal/4006-faaliyet-raporlarimiz
Şahin, İE. (2019). Türkiye'deki Havalimanlarının Veri Zarflama Analizi ve Malmquist Toplam Faktör Verimliliği Endeksleri İle Finansal Etkinliklerinin Analizi. *Selçuk Üniversitesi Sosyal Bilimler Enstitüsü Dergisi, 42*, 33–47.
Uludağ, A. S. (2020). Measuring the productivity of selected airports in Turkey. *Transportation Research Part E: Logistics and Transportation Review, 141*, 102020.
Ulutas, B. H. (2018). Efficiency analysis of airports in Turkey considering environmental effects. *The Central European Review of Economics and Management, 2*(1), 65–78.
Uslu, S. (2015). Türkiye'de Havaalanlarındaki Slot Sorunları ve Çözüm Önerileri. *Akademik Bakış Uluslararası Hakemli Sosyal Bilimler Dergisi, 50*, 219–230.
Ülkü, T. (2015). A comparative efficiency analysis of Spanish and Turkish airports. *Journal of Air Transport Management, 46*, 56–68.
Villalonga, B. (2000). Privatization and efficiency: Differentiating ownership effects from political, organizational and dynamic effects. *Journal of Economic Behavior & Organization, 42*(1), 43–74.
Yalçın, E. (2018). Stokastik Sınır Analizi İle Havalimanlarının Etkinliklerinin Ölçülmesi: Türkiye Örneği. *Mustafa Kemal Üniversitesi Sosyal Bilimler Enstitüsü Dergisi, 15*(42).
Yazgan, A. E., & Karkacıer, O. (2015). Veri Zarflama Analizi İle Etkinlik Ölçümleri Ve Havalimanı İşletmeciliği Sektöründe Bir Uygulama. *Journal of Alanya Faculty of Business/Alanya Isletme Fakültesi Dergisi, 7*(2).
Yılmaz, F. (2020). *Türkiye'de sivil havacılık sektörünün tarihsel gelişimi ve 2003–2018 yılları arasında sektörün değerlendirilmesi.*

Burak Keskin is an Assistant Professor at Cankiri Karatekin University since 2017. He was a Research Assistant at Cankiri Karatekin University between the years 2010–2017. He lectures on Mathematics, Statistics, Research Methods and Performance Evaluation, and Benchmarking. He holds master degree and Ph.D. in Business Administration at Akdeniz University. His academic research focuses on efficiency, productivity, and multi-criteria decision-making methods. He published book chapters on “Industry 4.0 and Big Data” and “Meta-Analysis”. He has a Post-Doctoral scholarship (2219) provided by The Scientific and Technological Research Council of Turkey (TUBITAK) and he continues his research at Worcester Polytechnic Institute (WPI), Massachusetts.

Air Traffic Management (ATM) Regulation in Turkey: What is the Next Step?

Engin Zeki

Abstract This chapter focuses on the historical development of air traffic management (ATM) regulation in Turkey, by looking into the milestones and relating the current ATM regulation in Turkey with the case of the European ATM. The milestones are presented with a focus on the historical progress of liberalization in the Turkish ATM. Specifically, Turkish ATM evolved from having a public air traffic governmental structure towards a competitive liberalized air traffic environment. By presenting the two inter-related regulatory environments, the chapter specifically explains Turkey's Local Single Sky Implementation (LSSIP), focusing on the latest developments and technologies in the roadmap. Concretely, Turkey establishes coordinated development of air traffic with the support of the key partners and supranational organizations in the region. Currently, the global ATM environment is changing due to the push from emerging technologies. These technologies have the ability to disrupt the current provision of services by enabling digitalization and virtualization with the support of open service-oriented architectures and dynamic resource allocation. Following closely the developments in the European ATM, Turkish ATM serves as an important air traffic hub in the region. Clearly, the next step for the Turkish ATM is to align for the latest developments in the region, adopting new technologies and adapting regulations accordingly in a fast-paced technology-driven ATM landscape.

Keywords Air traffic · Aviation · Regulation · Competition · Liberalisation

1 Introduction

Turkish air traffic management (ATM) has evolved progressively since the start of civil aviation in Turkey. Institutional, technical, and economic changes paved the way towards the liberalization of air traffic components, such as airlines, airports, industry, and ground services. Through this evolutionary process, a number of milestones have paved the way towards liberalization. Initially, This chapter focuses

E. Zeki (✉)
Ecole Polytechnique Fédérale de Lausanne, Lausanne, Switzerland
e-mail: engin.zeki@epfl.ch

M. Eroğlu and M. Finger (eds.), *The Regulation of Turkish Network Industries*,
https://doi.org/10.1007/978-3-030-81720-6_10

on these milestones by providing a brief history of ATM regulation in Turkey. Following these milestones, I discuss how the changes in ATM regulation affected the airlines, airports, industry, and ground services in Turkey. I then present the two inter-related regulatory ATM environments of the EU and Turkey, focusing on how Turkey closely follows the SESAR implementation in the EU. The chapter concludes with a description of the emerging technologies in global ATM environment, and how these developments could be reflected inside the Turkish ATM.

The chapter focuses on the Turkish ATM regulation with a past, current, and future outlook. In order to provide this outlook, I rely on the inter-relation between the EU and Turkish ATM regulations, and how the two complement each other. The latest developments in the European ATM are discussed, focusing on the emerging technologies and their impact on European aviation. Furthermore, the importance of these developments for the Turkish ATM are pointed out with suggestions for how future regulation could deal with rapid technological developments.

Turkey is a key player in terms of global air traffic, mostly as a geo-strategic air traffic hub connecting Europe to Asia and the Middle East. For this reason, Turkey collaborates with partners in the region to develop and coordinate air traffic and related services. This puts Turkey at the forefront of the European air traffic, being a Member State of EUROCONTROL (the European Organization for the Safety of Air Navigation). By inter-relating Turkish ATM regulation with that of the EU, this chapter aims to explain how the developments in Single European Sky (SES) are followed by the Turkish ATM in the form of Local Single Sky Implementation (LSSIP).

Globally, in the ATM landscape, there are emerging technologies, such as virtualization, cloud-based services supported by open service-oriented architectures, and dynamic resource allocation. These technologies have a disruptive nature, possibly altering the current structure of the actors and their objectives. In conventional civil air traffic, air traffic service and control are provided locally by the national air navigation service providers (ANSPs). New technologies have the potential to alter the local provision of services by enabling location-independent and remote provision of services, which, in turn, will affect the global air traffic service provision. The regulatory environment would need to align with the technological landscape in order to adapt and accommodate these technologies in the system.

The chapter concludes with recommendations for future regulation in the Turkish ATM, addressing the rapid rise of digital and virtual technologies globally in the ATM landscape. Considering their effect on the actors in the European ATM, it is expected that there would be a similar impact on the Turkish ATM and the relevant actors involved. Turkish ATM closely follows the implementation of the SES and new technologies to support the future of ATMs. Among these, the latest implementation of voice over internet protocol (VoIP), update of systems and infrastructure to support system-wide information management (SWIM), and free-route show that Turkish ATM is ready to support the future of ATM.

2 Structure of ATM and Aviation Regulation in Turkey

Turkey's national ATM environment consists of the geographical properties, including airspace structure, classification, organization, service provision, infrastructure, and systems supporting the ATM landscape, national stakeholders, and the regulatory aviation framework (EUROCONTROL, 2019). In order to support the national aviation environment and international cooperation, Turkey has become a Member State of the following supranational organizations: ICAO in 1945, NATO in 1952, ECAC in 1955, and EUROCONTROL in 1989 respectively.

In the current ATM regulatory structure, Turkey operates in two distinct flight information regions (FIRs)—Ankara FIR and Istanbul FIR—without any upper or lower airspace separation. These two FIRs encompass 27 terminal maneuvering areas (TMAs) covering the Turkish airports. The main national stakeholders include the Ministry of Transport and Infrastructure, which is responsible for the general civil aviation in Turkey focusing on economic and environmental aspects. The Directorate General of Civil Aviation (DGCA) is responsible for rule making, safety oversight, and airspace. Turkey has a single ANSP, Devlet Hava Meydanlari Isletmesi (DHMI), which is responsible for the provision of air navigation and air traffic control for the Turkish airspace.

In line with the ICAO airspace classification implementation, a reorganization of the airspace structure is planned. Currently, the area control center (ACC) in Ankara controls a total of 38 sectors that comprise the en-route operations throughout the Turkish airspace. Yaman et al. (2015) examined the structural features and capacity of Turkish airspace. In order to provide the analysis, Yaman et al. (2015) analyzed the digitized data of the Turkish airspace, including the sector boundaries, routes, and waypoints. Accordingly, Yaman et al. (2015) offered a restructure of the airspace, focusing on the balancing of the controller workload among the sectors. This is mainly due to the fact that, over the Turkish airspace, controller workload is not evenly distributed considering the type and number of aircraft. Additionally, specifically for the Istanbul ACC, where the traffic density is the highest, there is an increase in controller workload and traffic complexity due to frequent transition from upper-lower sectors to intermediate sectors.

2.1 Service Provision and Air Traffic Control (ATC) Systems

The sole responsible state-owned entity for civilian airspace service provision is the Turkish ANSP, DHMI. DHMI provides air traffic control services, including en route, approach, and aerodromes. In addition to the ATC provision, the ANSP provides the aeronautical information services (AIS) and communication, navigation, and surveillance (CNS) required within the Turkish airspace. Housing the ATCOs inside DHMI, the ANSP also provides the necessary training for the ATCOs for the seamless provision of ATC services. Furthermore, DHMI is responsible for the management of the airports functioning inside the Turkish airspace. In support of the ATC services provided by DHMI, the Turkish State Meteorological Service provides the necessary meteorological services. In order to provide ATC services, DHMI relies on flight data processing systems (FDPS) and surveillance data processing systems (SDPS) that are manufactured by SELEX and deployed in 2015 for Ankara ACC.

Besides the main Turkish ANSP (DHMI), the airports are also responsible for the provision of the services encompassing approach (TMA), airfield, tower, and ground control of the aircraft for both general air traffic (GAT) and operational air traffic (OAT). The Turkish State Meteorological Service provides the necessary meteorological information for the corresponding airports. For both OAT and GAT, the Turkish Air Force performs search and rescue (SAR) operations.

2.2 *National Stakeholders and Regulation*

The current regulatory ATM landscape in Turkey includes a number of stakeholders that play a specific role for the aviation in Turkey. The main responsible state body for aviation in Turkey is the Ministry of Transport and Infrastructure of the Republic of Turkey, which represents Turkey internationally in the EUROCONTROL Commission (EUROCONTROL, 2019). The regulatory aviation authority is the Directorate General of Civil Aviation (DGCA) in Turkey, which represents Turkey in the Provisional Council of EUROCONTROL. As described above, DHMI acts as the national service provider for aviation as a fully state-owned entity. Additionally, the military is represented by the Turkish Military Authority and the Turkish Air Force. Additionally, specifically for the airports, the General Directorate of State Airports Authority (GDSAA) is the responsible regulatory entity (Çetin & Benk, 2011). Considering the international aviation regulation, Turkey works closely with supranational organizations such as ICAO and EUROCONTROL.

The civil ATM regulators, the Ministry of Transport and Infrastructure of the Republic of Turkey, the DGCA, the Ministry of Transport Environment and Urbanism of the Republic of Turkey, and the Investigation and Assessment Commission are responsible for the following regulatory roles (EUROCONTROL, 2019):

- Rule making—Creation and management of regulation considering social, technological, and economic aspects of civil aviation, while ensuring public interest and national security (Çetin & Benk, 2011) (DGCA): The Law No: Presidential Decree No. 4 and 5431 (Legal Basis)
- Safety oversight (DGCA): Law No. 5431 Presidential Decree No. 4 (Legal Basis)
- Enforcement actions in case of non-compliance with safety regulatory requirements (DGCA): Presidential Decree No: 4 and 5431 (Legal Basis)
- Airspace (DGCA): Presidential Decree No. 4 and 5431 (Legal Basis)
- Economic (Ministry of Transport and Infrastructure)
- Environment (DGCA and Ministry of Environment and Urbanism)
- Security (DGCA): The Law No. 5431 Presidential Decree No. 4 (Legal Basis)

- Accident investigation (Investigation and Assessment Commission)

Being the main regulatory body, the DGCA functions within the directives of the Civil Aviation Law and supporting technical documents (Çetin & Benk, 2011). Besides its regulatory roles, the DGCA is responsible for the following licensing, certification, and oversight roles (EUROCONTROL, 2019):

- Licensing of aircrew, engineers, ATCOs, and air traffic safety electronics personnel (ATSEP): Regulating terms, conditions, and duration of licenses of persons who perform activities in Turkish airspace
- Certificate of Airworthiness (CoA) of aircraft flying in Turkish airspace
- Oversight and certification of airports, ensuring technical qualities and operating conditions
- Approval of airborne electronic equipment and air traffic communication services
- Approval and management of all ATM rules and regulations in correspondence with the Ministry, service provider DHMI, military, and other relevant actors.

Specifically, for the airports, the GDSAA is the responsible body for the regulation and inspection of the airspace and borders of airports in Turkey (Çetin & Benk, 2011). Founded in 1933, the GDSAA transformed into an autonomous state economic enterprise affiliated to the Ministry of Transport, having responsibilities for both ensuring the civil international rules and regulations in relation to airport entrepreneurship and navigation services. Furthermore, the GDSAA is the sole responsible institution for the provision of air navigation services in Turkish airspace, while the DGCA is mainly responsible for regulation and rule making (Çetin & Benk, 2011).

An important actor in Turkish ATM is the military, which has established civil-military cooperation in order to manage the reserved military airspace. For this reason, a civil-military coordination group was formed between the service provider, DHMI, and the military. A similar concept to flexible use of airspace (FUA) is implemented by the military and the service provider, DHMI. A number of airports are reserved for the use of military, while some of the airports are used by both the civil and military air traffic. An ATM modernization project, named SMART, targets the implementation of FUA to increase Turkish airspace considering the future needs of the actors, including the service provider, DHMI, and the Military.

3 History of ATM Regulation in Turkey

Looking through the history of ATM Regulation in Turkey, a number of milestones that paved the way for liberalization of the air traffic components, including airlines, airports, and ground services. This section focuses on these milestones by reviewing the previous literature and presenting the path towards liberalization of ATM in Turkey.

3.1 Beginning of Civil Aviation in Turkey (1926–1980)

Civil aviation in Turkey started with the foundation of the Turkish Aeroplane Society in 1926 (Gerede, 2010). This marked the first official organization for the air traffic, industry, and training of staff. Within the activities of the society, Turkish air operations gained traction, moving beyond the military operations by enabling civil air traffic. This was followed by the foundation of the State Airlines Administration in 1933, which paved the way for the creation of Turkish Airlines and the DGSAA (Gerede, 2010). The main aim was to start civilian air traffic between the major cities in Turkey.

Between 1930 and 1950, there was tremendous growth in the number of aircraft, and air traffic operations in Turkey (Saldiraner, 1992). By the mid-1940s, Turkey had become one of the first three aviation industries in Europe. In 1955, Turkey joined the European Civil Aviation Conference (ECAC), which aimed to harmonize civil aviation policies and practices between Member States. Within the rise of global operations and the complexity of air traffic, air transport (airlines) and airport services were separated. Turkish Airlines had responsibility for the air transport, and a new entity called the Directorate General of State Airports took the responsibility for the air traffic control, communications, and air-ground services in support of Turkish Airlines. For the provision of the air-ground services, a public company called Ucak Service Anonim Sirketi (USAS) was formed in connection to support the operations of the Turkish Airlines.

3.2 Moving Towards Liberalization (1980–1990)

Until the 1980s, the civil aviation in Turkey was mostly run by public companies. Given that all airports are state-owned, the only airline operating was Turkish Airlines. All of the services that supported air traffic, including air traffic control, air-ground services, and catering, were provided by single state-owned companies. In 1983, the first attempt to liberalize the air traffic market appeared in parallel with the reforms as in other network industries (Gerede, 2010). With the rise of air traffic, especially for touristic travel, private companies were allowed to enter into air traffic market with the establishment of liberalization law for air traffic services. This allowed new private airlines to be formed in competition with Turkish Airlines. This was a critical milestone for Turkish aviation, with the deregulation of air traffic market and enabling entry and competition for service provision.

The air-ground services were liberalized following an irregular pattern of half-sharing the ownership of the provision with state-owned companies (Gerede, 2010). This created discussions about how to enable full liberalization. Following that, CELEBI was established to provide air-ground services for the first time, being a private company in the area. The state-owned companies already had a substantial market share, with the two state-owned companies, USAS and HAVAS, providing the air-ground services (Saldiraner, 1992). Following privatization in other areas, USAS became a candidate for privatization within the scope of an enlarged privatization campaign in other industries. Consequently, USAS was privatized following the privatization of the catering services for air traffic, setting a milestone as the first privatization in the air traffic area.

In the early 1990s, privatization of Turkish Airlines (THY) entered the picture. Back in 1984, Turkish Airlines was regarded as a state economic enterprise, which paved the way towards its privatization. Consequently, in 1990, initially around 2 percent of Turkish Airlines was privatized, and further privatization of the airline was on the way (Saldiraner, 1992). Similar to the developments of airlines in the Europe, such as the privatization of the Air France, this paved the way towards further liberalization of airlines in Turkey. Consequently, privatization of the ground services gained traction, and the two state ground handling companies, USAS and HAVAS, entered into the scope of privatization, in order to compete with the only private ground service provider, CELEBI (Gerede, 2010). The privatization of the ground services continued progressively, with HAVAS becoming fully privatized by the end of 1998.

3.3 *Privatization of Airport and Services (1990–2000)*

In the 1990s there were no definitive regulations on the entry of private companies into the air traffic market, namely capital, staff, experience requirements (Gerede, 2010). The Ministry of Transport established higher barriers of entry in order to stabilize and make the entrant private companies stronger in terms of their capital and fleets. Following the revision of the related amendments in regulation, leasing, seating, and ownership of aircraft by the private companies were specified. In light of the deregulation of the domestic air market in 1983, except for price tariffs, Istanbul Airlines entered the market in 1987. Following that, Onur Air also entered the market in competition with the two existing airlines at the time.

However, in 1996, private airlines were only allowed to operate for flights that were not scheduled; scheduled traffic was only reserved for the Turkish Airlines to operate. There were a number of conditions for private airlines to operate that would later become a topic for discussion between the stakeholders to find the optimal solution for both THY and the other private airlines. Additionally, there emerged a need to increase terminal capacities of airports, while ensuring that the airports could accommodate forecasts for the following years. For this reason, a new management procedure, including the deployment, operating, and transfer of the airport services emerged.

Following the success of this procedure, the same principles were applied to the airports for the major air traffic hubs in Turkey. Moreover, the liberalization of airlines and ground services gained momentum. Overall, the 1980s and 1990s also saw important developments in global aviation with increasing coordination and collaboration. Turkey, as one of the main hubs in the network, signed bilateral international agreements with the United States of America, and the European States, specifically Belgium and Germany. In 1989, Turkey became a Member State of EUROCONTROL and European Joint Aviation Authorities (JAA). This extended the responsibilities of Turkish civil aviation and the flag carrier THY, as well as carrying THY inside the competitive EU market.

3.4 Enabling Competition in the Air Traffic Market (2000–2020)

An important milestone for civil aviation in Turkey is the full liberalization of the domestic flight prices in 2001. Before that, airlines were not able to determine the prices for domestic flights freely. In 2003, deregulation enabled private airlines to enter into the market, competing for domestic and international flights. The private airlines were able to compete freely with the flag carrier THY and had reduced taxes, and ticket prices that were beneficial for both the users and the airlines. The number of airlines increased rapidly due to the demand for domestic and international flights overall. The competition brought ticket prices down, and passengers started to travel more frequently on domestic lines (Gemici & Alpkan, 2015). Thus, liberalization of the domestic flight market is regarded as one of the most important milestones for the Turkish ATM.

Now operating in a competitive market, privatization for THY became inevitable, and in 2004, a quarter of its share started to be traded on the Istanbul Stock Exchange. All airports were state-owned and operated under the regulations of State Airports Authority. By 2004, the newly constructed Sabiha Gokcen airport was unable to attract an estimated portion of air traffic in Istanbul from the main airport Ataturk Airport. Due to the rise of air traffic, congestion, and slot problems at Ataturk Airport, THY migrated some of its domestic lines to Sabiha Gokcen Airport. Following the operations in the new airport, Pegasus Airlines entered the market as a low-cost airline, mainly operating from the Sabiha Gokcen Airport with low-cost airport services. This also created a milestone for the Turkish ATM with the first low-cost airline operating in a liberalized domestic market.

Slot coordination and assignment are among the major obstacles for the capacity and performance of the major airports in Turkey. Especially, the airports in Istanbul, Ataturk and Sabiha Gokcen Airport could not accommodate for the increased air traffic. In order to solve this problem, in 2006 Ministry of Transport established an independent commission with representatives from State Airports Authority, THY Slot Coordination Unit, Turkish Private Aviation Enterprises Association, terminal operators, airlines, and ground services. Regarding the importance of the slots, the formation of a slot coordination center is also an important milestone for the liberalization of air traffic services in Turkey.

Privatization of the airports continued between 2000 and 2010, with both internal and international stakeholders showing interest in obtaining shares for the major airports. Specifically, airports with foreign tourism traffic, such as Antalya, were privatized in connection with international partners. Moreover, the privatization step in 2005 for Ataturk Airport solved the capacity and financial problems and this paved the way for further privatization of the services in Sabiha Gokcen airport and other airports around Turkey. Furthermore, private companies started to compete for the provision of the services, which increased the cost efficiency, quality, and the delivery of the services overall.

Turkish ATM regulation introduced competition in the airlines and ticket prices. Increasing competition in the airline market created specific competition between full-service network carriers and low-cost carriers in the Turkish airline market (Acar & Karabulak, 2015). Acar et al. (2015) highlighted this competition by presenting different strategies. Specifically, these strategies include airlines in Turkey adapting towards cost leadership and differentiation in business strategies, particularly focusing on low-cost. Apparently, Turkish airlines is considered as a full-service network carrier, while Pegasus Airlines creates competition by being a low-cost carrier. The main competitive force between the two airlines is the establishment of cost leadership, combined with effective cost control system and forming strategic alliances.

In order to increase the air traffic capacity of Istanbul as the main air traffic hub connecting Europe and Asia, Turkey built Istanbul Grand Airport (IGA). As Saldiraner (2015) noted, Turkey's two existing airports, Ataturk Airport and Sabiha Gokcen Airport had limitations on their capacity and runways, coupled with the increasing air traffic in Istanbul. Specifically, Ataturk Airport previously had capacity problems and, according to the studies carried out by DHMI, an additional runway could not be built for the airport. A relatively small airport, Sabiha Gokcen, operates on a single runway, terminal, and cargo. An additional runway is planned for this airport, but it could not accommodate the air traffic in Istanbul.

The need for the new airport in Istanbul arose due to the insufficient capacity and runways of the two existing airports. The previous air traffic estimation provided that, in the future, the capacity of the two airports would not be able to accommodate the European traffic. For these reasons, the new airport offered extended capacity for aircraft, passengers, and cargo overall. Compared to its European counterparts, such as London Heathrow, Amsterdam Schiphol, and Frankfurt, the conceptual idea of an airport is moving towards creating city and cargo hub airports. Thus, the new Istanbul airport brings opportunities in the areas of available slots for new flights, new regional planning, very large aircraft flights, and improved transit flight and passenger services (Fig. 1).

4 ATM Regulation in the EU and Turkey

In the foundation **of** the European Union (EU), air transport was not directly addressed, and harmonization of air traffic was not an initial aim in the European Communities (EC) Treaty. No Europe-wide aviation policy existed until 1984. In the

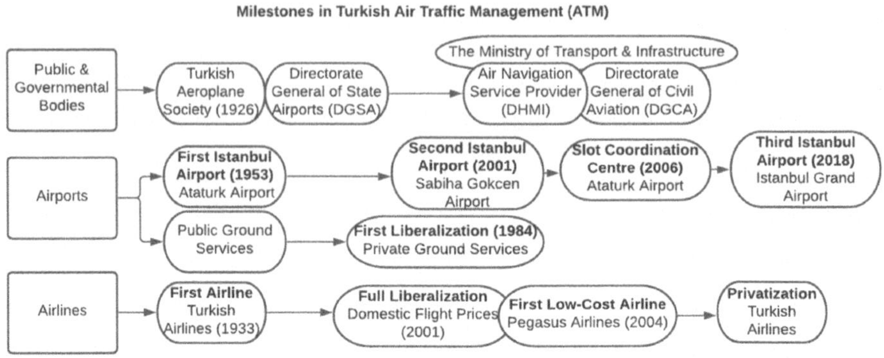

Fig. 1 Milestones in Turkish Air Traffic Management (ATM)

scope of the foundation of the European Economic Community (EEC), civil aviation was directly addressed by forming a common air transport policy (Servantie, 2015). This policy aimed to improve the efficiency, quality, and cost-effectiveness of the air transport sector. This marked the first step for the creation of the civil air traffic market and cross-border operations between the Member States of the EU (Servantie, 2015).

That year, 1984, marked an important milestone for European aviation by creation of the single EU aviation market (Servantie, 2015). Economic and regulatory changes between the EU Member States were required to implement such an EU-wide aviation policy. As a result, the actors in the European ATM aimed to increase the number of routes and businesses with better quality and lower prices. Bilateral agreements for the international routes were established between the EU States and the other countries. However, the routes and airlines were limited and there was limited competition compared to the internal local market of the Member States.

The single aviation market of the EU was successively formed by the implementation of EU-wide aviation policy, including three packages (Servantie, 2015). The areas included in the packages were mostly related to market formation, access, air traffic fares, and licensing of airlines. Previously, cross-border operations within the European States had limitations due to the number of airlines operating, pricing, and regulation of the air traffic in the EU scale overall. The introduction of competition in the Pan-European scale opened new routes between the EU states, making it easier for the passengers to prefer air traffic rather than other transport methods, such as rail or road.

Following the consolidated views of the EU Member States to form an EU-wide aviation policy and the progress of the first two packages between 1987 and 1990, the third Aviation package (1992) aimed to implement Pan-European scale liberalization with the creation of a single aviation market, and the economic services to support it (Servantie, 2015). This led to the formation of the EU's 30-year policy on creating a single, competitive, and liberalized market in the region. Consequently, between the 1990s and the 2000s, the EU revised its aviation policy towards maintaining a common aviation roadmap for the whole of Europe by designing the grand project Single European Sky.

Starting from 2002, the EU advanced its aviation policy, establishing a Pan-European approach for its Member States and connecting with the rest of the world via bilateral agreements. Specifically, the EU followed three pillar approach to developing aviation with its international partners (Servantie, 2015):

- Establishment of competition of international flights that were regulated for flights outside the scope of the EU
- Enlargement of the Common Aviation Area, with the inclusion of the southern, south-eastern, and eastern neighboring countries
- Bilateral agreements and negotiations for the integration of EU air traffic with international partners.

Following these steps, the EU aimed to enlarge its vision with external partners, mainly in the scope of the Open Skies initiative, which in 2002 created the path towards the formation of an external EU aviation policy (Servantie, 2015). For international air services agreements, the EU would act with the Member States to implement a common roadmap. This initial roadmap followed the alignment of the previous bilateral agreements of the EU with external partners, creation of the Common Aviation Area with neighboring countries to the EU, and new agreements with key partners.

In order to provide a common roadmap for Pan-European civil aviation, the European Commission (EC) initiated the Single European Sky (SES), as a grand European air traffic project, which started to take shape from 1999. The SES covers institutional, technological, operational, and management areas by providing a consolidated approach to research and development of the air traffic in Europe. In order to provide a common roadmap, SES sets high-level target objectives for the key performance areas. These high-level target objectives of the SES include three-fold increase in capacity, improvement of safety by a factor of 10, reduction of the cost of the provision of ATM services to airspace users by 50 percent, and a 10 percent reduction in environmental effects.

During its progress, the SES also introduced additional pillars, namely SESII and SESII+, which aimed to introduce a gate-to-gate approach and functional airspace blocks (FABs) in order to defragment neighboring European ANSPs. A common roadmap, called the European Master Plan, took form guided by the Single European Sky ATM Research (SESAR). In order to support the institutional pillar, new actors emerged inside the SESAR framework. Consequently, a SESAR-Joint Undertaking (SESAR-JU) was created by the joint coordination between the European Commission (EC) and EUROCONTROL. The main aim was to create synergies between public and private actors towards adopting new technologies and solutions provided under the SESAR framework.

4.1 The EU–Turkey Aviation Market

A key milestone for the removal of nationality restrictions between the EU and Turkey was established in the context of a signed horizontal agreement in March 2010. Within this agreement, and as far as the bilateral agreements between the EU Member States and Turkey allow, any EU airline became able to fly between Turkey and any EU Member State (Christidis, 2016). Moreover, the agreement brought forth the adaptation of the aviation law according to the increased competition with regard to the EU law.

Following this, the air traffic and passenger volumes between the EU Member States and Turkey increased, as reflected in the increase of competition for the airports. The routes between the EU Member States increased, starting with the most frequent destinations, such as Germany, the Netherlands, and a number of Western EU Member States (Christidis, 2016). Additionally, in Turkey there exists a concentrated airport competition, where airports in Istanbul, as the geographic hub, and Antalya holds the largest flight and passenger volume, followed by a number of touristic locations.

The footsteps of 2003 deregulation movement in the airline industry created competition in all fields of the aviation industry (Çetin & Benk, 2011). Removing the barriers for entry, the legal structure of Turkish aviation become compliant with the EU aviation regulations. This resulted in restructuring of pricing and tariffs policy, increased demand for air transport, creation of new routes, and the abolition of monopolies. Furthermore, deregulation in the ATM environment brought forth a significant increase in air traffic, a decline in ticket prices, and opportunities for new jobs, which was reflected in an increase in GDP (Çetin & Benk, 2011). In the case of unutilized airports, deregulation increased the effective demand since 2003.

Even though 2003 deregulation movement brought competition, the flag carrier, THY, still had the majority of the market share, with a decline of the percentage of passengers through 2009 (Çetin & Benk, 2011). However, the seat capacity of THY increased, as it did with the nine new competing airlines that increased their seat capacity due to local increase in Turkish civilian air traffic. The significant increase in number of passengers is the direct benefit of the deregulation movement in 2003. This, in combination with the increase in competition, proved that the deregulation steps are successful, bringing positive impacts to Turkish aviation.

4.2 Local Single Sky Implementation (LSSIP)—Turkey

The progress of SES has been closely followed by Turkey in the form of local Single European Sky Implementation (LSSIP), in close cooperation with EURO-CONTROL since 2013. The Ministry of Transport and Infrastructure represents Turkey in the EUROCONTROL Commission. The Turkish aviation regulatory authority, the DGCA represents Turkey in the EUROCONTROL Provisional

Council. These implementations are planned by the Turkish air navigation service provider (ANSP), namely Devlet Hava Meydanlari Isletmeleri (DHMI), being the responsible state-owned entity for service provision.

For international cooperation, Turkey became a member of International Civil Aviation Organization (ICAO) in 1945. In European aviation, Turkey has also been a member of the European Civil Aviation Conference (ECAC) since 1955, and a member of EUROCONTROL since 1989. Additionally, a relatively new actor, the European Aviation Safety Agency (EASA), included Turkey as a Pan-European partner (PANEP), in which EASA works with the non-EASA European countries cooperating for the implementation of the EU's safety rules (EASA, 2020).

In line with the progress of SESAR, Turkey progressively implemented a number of SESAR projects and concepts progressively. In 2019, Turkey adapted its infrastructure and systems for the new communication system of European ATM, New Pan-European Network Service (NewPENS). Turkey also deployed a new system for information exchange en route in support of arrival management (AMAN), in line with the new SESAR concepts. Currently, the Turkish ANSP, DHMI is working on the implementation of Voice over Internet Protocol (VoIP) for en route and enhanced tactical flow management.

For future implementation, the Turkish ANSP, DHMI, aims to support the Free-Route Airspace by implementing extended flight plan and direct routing. For the new communication systems, a European-wide network service NewPENS is expected to support these new concepts. According to the LSSIP plan, Turkey expects to extend VoIP implementation to the airports by 2023, and also connect to the System-Wide Information Exchange (SWIM) platform supported by standardized aeronautical data and information.

LSSIP provides a detailed description of the implementation of new projects and implementations in line with the SESAR program. LSSIP reports that Turkey establishes coordination and cooperation with all its neighboring states in order to optimize ground networks, data exchange, airspace design, and management to increase regional capacity, safety, and quality. In 2012, Turkey established a Memorandum of Cooperation (MoC) between the DHMI and the Bulgarian Air Traffic Services Authority (BULATSA). The cooperation is on the common understanding and adoption of international requirements relevant to the ATM domain and establishing future cooperation within an effective route network, common operational and technical projects. A similar cooperation agreement between the DHMI and the Ukrainian State Air Traffic Service Enterprise (UkSATSE) was signed, aiming for regional cooperation on ATM, communication, navigation, surveillance, economic, and financial support.

Turkey, being a hub in the region, also implemented a number of enhancements in the air traffic flow with the neighboring countries. Specifically, in order to implement airspace changes, Turkey cooperated with Bulgaria, Georgia, Iran, and Iraq. Ultimately, these changes optimize the routes for the air traffic between the Middle East, the Gulf Area, Asia, and Far East. Additionally, for the new Istanbul Airport, new routes have been established between Turkey and neighboring countries in close coordination with EUROCONTROL for the airspace changes (Düzgün & Tanyaş,

2014) (Saldıraner, 2014). This close link with EUROCONTROL makes Turkey an important actor for the European air traffic, given Istanbul's important geographical location as a hub connecting Europe and Asia (EUROCONTROL, 2015).

5 Implementation of New Technologies and Their Reflection in the Turkish ATM Regulation

The current landscape of global ATM is changing due to the rise of new disruptive technologies. Specifically, digitalization and virtualization enable totally new ways of providing services, shifting the paradigm of ATM from legacy systems towards open service-oriented architectures. Ultimately, these technologies pave the way towards sharing of infrastructure, hardware, and software between the stakeholders supporting dynamic resource allocation. Mainly, the fast-paced developments in information and communication technologies (ICT) affect all the network industries, as in the case of air traffic. Given the slow adoption rate of new technologies in the air traffic domain, the reflection of their effect in regulation also lasts longer.

From a global perspective, these new technologies are supported by cloud-based services and open service-oriented architectures, and flight-centric operations are supported by free-route and business trajectories. These new technologies have a disruptive nature on the current provision of ATM and ANS services globally. Both point towards the defragmentation of infrastructure, hardware, and software. Virtualization offers virtual defragmentation by creating a shared and standardized virtual layer on top of the conventional physical layer of air traffic. Flight-centric operations offer vertical and horizontal defragmentation of sectorized structures of air traffic in parallel with free-route. Furthermore, these drastic changes could alter the business models of the actors that are involved in the ATM domain.

Digitalization and virtualization are expected to bring a number of benefits for European traffic management. Since the beginning of the Single European Sky, defragmentation of the European ATM has been one of the main targets of the initiative. With virtualization, these could be established in a virtual layer with standardized services with shared infrastructure, hardware, and software between the service providers. Furthermore, virtualized provision of services opens up new frontiers for ATC provision. Location-independent, remote provision of services from new entrants could be possible with the new virtualized configuration. Even large ANSPs could provide a number of services for small to medium-sized ANSPs.

Turkey closely follows the SESAR implementation by having a local roadmap to support the concepts in the European ATM. Modernization of the communication systems, connecting to SWIM infrastructure and services, points towards the preparation of new technologies in the European ATM. Virtualization is expected to be one of the key technologies for altering current regulation and moving towards competition in ANS and ATM services. In this respect, in the future, ANSPs could provide services for each other, or dynamically allocate resources for the dynamic capacity allocation. New entrants may be interested in providing services as third party providers, as long as they can satisfy quality, safety, and security of air traffic operations. Liberalization of the ANS and ATC services could be possible under a virtualized configuration.

The future may bring substantial changes in terms of how we currently perceive air traffic. From the beginning of the civil air traffic, local ANSPs were solely responsible for the provision of the national airspace. However, new technologies point towards defragmentation of services, standardization between the ANSPs, and introduce dynamic allocation of resources. Specifically, the virtual center, initially offered by the Swiss ANSP, Skyguide, aims to first merge the services between their Zurich and Geneva Area Control Centres (ACCs). This micro-problem of Skyguide is a role model for the macro-size problem of fragmentation on the European scale.

Currently, the SESAR solutions and the European ANSPs closely follow the virtual center implementation and are moving towards the adoption of the technology. This would mean that Europe-wide, there will be possibility of creating a virtualized layer on top of the physical layer, which would enable virtual service provision. In this configuration, a new entity called ATM Data Service Provider is defined both in the SESAR framework and the Airspace Architecture Study (AAC) from SESAR-Joint Undertaking (SESAR-JU). Of course, in order to support such a drastic change, regulatory and legal modifications should be required to support the operations of the virtual center.

6 Conclusion—The Next Step for Turkish ATM

Over time, Turkish ATM has adapted to the modern requirements of civil air traffic. This can be seen in terms of the milestones for the liberalization of airlines, airports, industry, and ground services. Turkey is one of the most important air traffic hubs in the world, connecting continents, cultures, and people. In that regard, being a regional player, Turkey establishes connections and collaborations with neighboring States through bilateral or supranational agreements. Given the fast-paced technological environment, radical changes are affecting the actors in the ATM landscape. This chapter has focused on the latest innovations and technologies and how they could be incorporated within Turkish ATMs.

Following these milestones, the chapter explains how the Turkish ATM liberalized airlines, airports, and ground services in a progressive fashion. Initially, public organizations organized and operated the main functions of air traffic. Following the first liberalization of ground services, airlines and airports are privatized in a progressive fashion. This allowed the entrance of private companies, competing for the provision for the services for air traffic in Turkey. By introducing low-cost competition from private airlines, such as Pegasus Airlines, ticket prices fell and air traffic becomes accessible for the majority of the Turkish population. With the rise of air traffic, the number of airports and airlines increased throughout Turkey. In order to accommodate the capacity and performance of air traffic, new airports were built, especially in Istanbul.

Turkey is a key ATM partner of the supranational actors in the region, such as EUROCONTROL. Closely following the SESAR implementation, Turkey implemented new solutions to support SESAR concepts in line with the European roadmap. The move towards a new communication system, VoIP, and implementation of free-route show that Turkish ATM has closely followed the technological developments in the global ATM. In the future, if these technologies are coupled with virtualization, these would be important enablers for the future of the Turkish ATM. The adoption of new technologies is important for the actors in ATM, as being an early adopter of new technologies creates a significant advantage.

Global technologies affect the ATM in three main ways. The first is the digitalization and virtualization that currently affect all the network industries, as in the case of air traffic. The second is the paradigm shift from legacy systems towards service-oriented architectures. The third is the rising importance of dynamic resource allocation. Concretely, new technologies address all three pillars by enabling new ways of service provision in ATM. In that regard, technology becomes more than a technical artifact and affects the legal, regulatory, political, and social dimensions. Thus, it becomes strategically important for the actors to align their interests towards adoption of new technologies. Furthermore, regulation should also follow the paradigm shift that is currently happening in ATM.

This multi-dimensional effect, of course, has substantial reflections on the current regulation of ATM and ANS and ATC services. In the future, disruptive technologies could alter the current roles of the actors that are relevant in the ATM, which could require substantial changes in the regulatory side. Looking to the latest technologies, location-independent service provision is one of the possible disruptions. In this way, new actors could emerge, focusing on the collection, maintenance, and sharing of ATM data. This will, in turn, require regulatory actors to define new ways of regulating the data and service provision surrounding it.

The next step for Turkish regulations is to closely observe the technological developments in the region, especially the European ATM, to which Turkey has already established close links in the scope of the SESAR program. Similar technologies may alter the landscape of the Turkish ATM. For the first time, the liberalization for the provision of services would be possible with the new technologies. This would alter the roles and business models of the actors, while also bringing substantial changes in the regulatory and legal aspects of air navigation services. Furthermore, these technologies enable both physical and virtual defragmentation, which will lead to regulatory changes, specifically in the provision of air navigation and control services in the future. These technologies lead to remote provision and the sharing of infrastructure, hardware, and software between ANSPs. Higher coordination for cross-border operations would be required to implement these new technologies.

References

Acar, A. Z., & Karabulak, S. (2015). Competition between full-service network carriers and low-cost carriers in Turkish airline market. *Procedia—Social and Behavioral Sciences, 207*, 642–651. https://doi.org/10.1016/j.sbspro.2015.10.134

Christidis, P. (2016). Four shades of Open Skies: European Union and four main external partners. *Journal of Transport Geography, 50*, 105–114. https://doi.org/10.1016/j.jtrangeo.2015.04.005

Çetin, T., & Benk, S. (2011). Regulation, deregulation, and competition in the Turkish airline industry. In T. Çetin, & F. Oğuz (Eds.), The political economy of regulation in Turkey (pp. 193–214). Springer, New York. https://doi.org/10.1007/978-1-4419-7750-2_9

Düzgün, M., & Tanyaş, M. (2014, October 30–31). The importance of Istanbul grand airport (IGA) for Turkey and its influence on widely regional air traffic around. In 12th International Logistics and Supply Chain Congress, Istanbul.

EASA. (2020). Turkey|EASA. [Online]. Retrieved September 17, 2020, from https://www.easa.europa.eu/domains/international-cooperation/easa-by-country/countries/turkey

EUROCONTROL. (2015). *2015 Comparison of air traffic management-related operational performance: U.S./Europe, 116*.

EUROCONTROL. (2019). *LSSIP 2019—Turkey local single sky implementation*. Brussels, Belgium.

Gemici, E., & Alpkan, L. (2015). an application of disruptive innovation theory to create a competitive strategy in Turkish air transportation Industry. *Procedia—Social and Behavioral Sciences, 207*, 797–806. https://doi.org/10.1016/j.sbspro.2015.10.169

Gerede, E. (2010). The evolution of Turkish air transport industry: Significant developments and the impacts of 1983 liberalization. Celal Bayar Universitesi IIBF, Manisa, *17*(2), 63–91.

Saldiraner, Y. (1992). *Sivil Havacılık Faaliyetleri ve Türk Sivil Havacılık Otoritesi İçin Organizasyon Yapısı Önerisi [Civil aviation activities and A proposal of organizational structure for Turkish civil aviation authority]*. Eskişehir. Anadolu Üniversitesi.

Saldıraner, Y. (2014). The new airport in Istanbul: Expectations and opportunities. *Journal of Case Research in Business and Economics, 5*, 1–11. http://www.aabri.com/manuscripts/131548.pdf.

Servantie, D. (2015). *A comparative analysis of EU and Turkish aviation policies, 13*. IKV Brief, Economic Development Foundation.

Yaman, K., Oktal, H., & Altan, M. (2015). Structural analysis of Turkish airspace by using GIS. *Journal of Air Transport Studies, 6*(1), 53–64. https://doi.org/10.38008/jats.v6i1.64

Engin Zeki has received his Ph.D. in Management of Technology from Ecole Polytechnique Federale de Lausanne presenting his thesis, titled 'Assessing Technology Adoption for the European Air Traffic Management: The Cases of the Virtual Centre and Flight-Centric Operations'. He holds a Master's Degree and a Bachelor of Science Degree, both in electrical and electronics engineering, focusing on signal processing and business intelligence. His interests include understanding and evaluating how new technologies come into existence and evolve. In parallel with academia, he gathers professional working experience mainly in public and private network industries, including air traffic, energy, and telecommunications, examining how new technologies could help these industries to elevate their performance and reach their key targets.

Regulation of Air Carriers

Ömer Faruk Erol

Abstract The civil aviation industry, and air transport activities, in particular, are regulated by both national and international regulations. The existence of this complementary structure reveals the need to evaluate international agreements and bilateral air transportation agreements together with national regulations. With the Turkish Civil Aviation Law number 2920, adopted on October 14, 1983, a new era has started in Turkish civil aviation from a legal perspective. The important innovations regulated by the Law No. 2920 are the opening of the civil aviation market to a private enterprise and the establishment of private airline companies in addition to allowing them to operate. However, it took almost 20 years for this regulation to be put into practice and reach significance in practice. In 2003, the rule that allowed Turkish private airline companies to operate on domestic flights started being used effectively in practice. There are currently many regulations and instructions for regulating the activities of air carriers, mainly Turkish Civil Aviation Law No. 2920 and the Commercial Air Transport Enterprises Regulation (SHY-6A). This study examines the relationship between air carriers and the administration within the scope of the regulation of their activities. First, the administrative organization for the regulation of civil aviation will be explained; afterward, the intervention of the administration in the sector will be examined in the context of both economic and safety-security issues.

Keywords Administrative Law · Aviation Law · Regulation · Civil aviation · Air transport · Airlines

This chapter is the updated, revised, and summarized version of the author's doctoral dissertation. For the expanded and Turkish version of this study, see: Ömer Faruk Erol, İdarenin Sivil Havacılık Alanında Hava Taşımacılığına İlişkin Faaliyetleri, On İki Levha Yayıncılık. (Erol, 2018).

Ö. F. Erol (✉)
School of Law, Ibn Haldun University, Istanbul, Turkey
e-mail: omer.erol@ihu.edu.tr

M. Eroğlu and M. Finger (eds.), *The Regulation of Turkish Network Industries*,
https://doi.org/10.1007/978-3-030-81720-6_11

1 Introduction

States' approach to aviation was at first negative, and the first legislative arrangements were to control the aviation activities instead of encouraging them. Over time, upon the understanding of the importance of aviation activities in the economy, the states made incentive arrangements this time. (Çağa, 1963, p. 16).

We can say that the regulation of the civil aviation sector has both national and international character. In other words, due to the fact that air transport activity is beyond the borders of the country, air carriers must comply with the regulations of the two countries in which they operate. Sometimes, even if there is a third country to carry out air transport activity, it is necessary to act in accordance with the regulations of all three countries. Therefore, both national regulations and international regulations can be implemented due to this transboundary nature of the civil aviation industry.

Both the bilateral air transport agreements signed between the two countries involved and the open sky agreements in which many states participate regulate the civil aviation sector, particularly the air transport activity, just as the state regulates a particular sector in its own country. The existence of this dual structure (both national and international regulations) reveals the need to evaluate the national regulations and regulations included in the bilateral air transport agreements together.

To examine the regulation of commercial air transport between the two countries, both the bilateral air transport agreement between these two countries and the national regulations of both countries should be considered. However, it is necessary to specify that our study will examine the scope of the regulation of air transport in Turkey, so an assessment will be made on the basis of Turkish legislation.

As a result of the evaluation of regulations on air transportation, we can examine the regulations under the two main topics: safety–security and economic order. In this context, the administrations that is in charge of the regulation of air transport will be examined first, followed by a discussion of the concepts and regulations of safety and security, and an examination of the concept of economic intervention and the issues regulated within the scope of economic intervention.

2 Administrations that are in Charge of the Regulation of Air Carriers

This section examines the relations between the authorities in charge and authorized for the regulation and supervision of Turkish air transport and their positions in the administrative organization, and the duties and powers of these administrations.

The Ministry of Transport and Infrastructure comes first among the administrations regulating air transportation. The Ministry is responsible for harmony between the other transportation models (road, rail, air, and sea) and the sustainable operation of a corresponding transport network in Turkey. In this direction, the Ministry ensures that highway and railway activities are carried out in a fair and sustainable competitive environment, depending on commercial, economic, social needs, and technical developments and in a manner that protects the public interest. However, these duties and authorities regarding the airway belong to the Directorate General of Civil Aviation (DGCA).

The duties and powers of the Directorate General of Civil Aviation were set out in Presidential Decree No. 4 (Presidential Decree About the Organization of Institutions Related to Ministries and Other Institutions and Organizations). The DGCA is an administrative authority, but not in the form of an Independent Administrative Authority, which has an example in the Turkish administrative organization, although its close regulation of a sector with its duties and powers brings it closer to the nature of an independent administrative authority. According to Presidential Decree No. 4 art. 437, the important duties and powers of the DGCA are as follows:

- To implement and follow up policies in order to ensure that civil aviation activities are organized and developed in line with the public interest, economic and social developments, and national security objectives.
- To regulate the principles that will ensure the sustainability and development of civil aviation activities in line with the international civil aviation rules and standards.
- To follow developments in the field of international civil aviation; to make and implement the necessary regulations for the implementation of rules and standards accepted by the international organizations of which Turkey is a member.
- To take the necessary precautions to prevent acts against the rules regarding civil aviation activities.
- To make arrangements for air traffic management services to be followed by civil aircraft using Turkish airspace.
- To provide recommendations on state policies on civil air transportation and to participate in the work of bilateral and multilateral agreements.
- To take the necessary precautions and make inspections to carry out safe flight operations with civil aircraft.
- To determine and implement the rules regarding administrative and technical sanctions to be applied to persons operating in the field of civil aviation and acting against the rules.
- To make regulations on technical infrastructure related to civil aviation and to ensure their implementation.
- To take measures to ensure the development of the civil aviation sector and to make recommendations to the related organizations on the necessary regulations.

The last administration discussed here is the General Directorate Of State Airports Authority (GDSAA). The GDSAA is a public economic enterprise established in accordance with the provisions of the Decree Law No. 233 on State Economic Enterprises dated June 8, 1984, having a legal entity, and being the relevant institution of the Ministry of Transport and Infrastructure. The objectives of the GDSAA are to operate airports, perform airport ground services, serve air traffic control services, and establish and operate navigation systems and facilities.

In summary, the most active administration regulating air transport is the Directorate General of Civil Aviation. The DGCA fulfills this task under the Ministry and acts in line with the Ministry's policies. On the other hand, the Ministry coordinates civil aviation regulation from above, directs air transportation and provides airport services and air traffic control services through the GDSAA.

3 Regulations of Safety and Security

The purpose of Turkish Civil Aviation Law No. 2920 is to ensure that civil aviation activities that show continuous and rapid development, apply advanced technology, and where speed and safety/security factors are of great importance, are organized in accordance with Turkish national interests and international relations.

In Turkish law, safety and security concepts are used without a clear distinction in the legal regulations related to civil aviation. These concepts are sometimes used interchangeably and sometimes in different laws to mean the same thing. For this reason, it is difficult to clearly separate the safety and security regulations in the national legislation. On the other hand, this situation causes confusion and difference in interpretation of legal regulations.

In Turkish law, the main issues related to safety and security in civil aviation are regulated in Turkish Civil Aviation Law No. 2920 and Presidential Decree No. 4 (Presidential Decree About the Organization of Institutions Related to Ministries and Other Institutions and Organizations). These two laws, along with regulations regarding the safety and security in the Turkish Criminal Code No. 5237, will also be examined.

3.1 Turkish Civil Aviation Law No. 2920

In Turkish Civil Aviation Law No. 2920, the concepts of safety and security were used together and sometimes interchangeably. In the law, concepts such as flight security, national security, life, and property security, as well as flight safety and life and property safety are used. Law No. 2920 does not make a complete distinction between the concepts of safety and security.

The regulations regarding safety and security in the Turkish Civil Aviation Law No. 2920 are as follows:

Article 7, entitled Flight Prohibitions and Restrictions regulates that the president may prohibit or limit the use of all or part of Turkish airspace with public order and safety considerations upon the positive opinion of the General Staff. In order to ensure flight security, the restrictions to be made by the General Staff and the Ministry of Transport and Infrastructure will be applied by the Ministry without being subject to this article.

In Article 27 regarding the inspection, it is regulated that those who operate with Turkish civil aircraft are subject to technical supervision to be performed or to have performed by the Directorate General of Civil Aviation in order to ensure the security of air navigation. Accordingly, in these inspections, if the operation poses a danger to life and property safety, the operation was suspended. Thus, both safety and security concepts were used to express the same phenomenon within the same article. However, the main purpose of the technical inspection is to ensure the safety of the air transport activity.

It is stated in Article 40, where aviation security is regulated, that specially trained armed security officers can work in Turkish registered civil aircraft. The Ministry of Interior is authorized to allow the presence of an armed security officer in foreign registered civil aircraft flying to Turkey on a reciprocal basis. The duties, powers, and responsibilities of the armed security officer are regulated by the regulation issued by the Ministry of Interior. The Regulation on the Duties, Powers, and Responsibilities of the Specially Trained Armed Security Officers on Turkish Registered Civil Aircraft were issued by the Ministry of Interior. The 100th article, regarding the powers and responsibilities of the pilot, regulates that the responsible captain pilot is authorized to take measures to ensure safety and order in the aircraft and to give orders and instructions to passengers and crew for this purpose and to remove them from the aircraft when necessary. In accordance with these regulations, "air police" will start working on Turkish registered civil aircraft in early 2021. However, because the relationship between the air police and the pilot is not clearly regulated, disputes may arise in practice.

3.2 Presidential Decree No. 4

In Presidential Decree No. 4, which included the Organization and Duties of the Directorate General of Civil Aviation, issues related to aviation security and safety are regulated (Articles 434–447).

The Directorate implements and follows up policies to ensure that civil aviation activities are organized and developed in line with the public interest, economic and social developments, and national security objectives. In addition, the directorate is responsible for taking the necessary measures and conducting inspections for safe flight operations with civil aircraft.

The Aviation Security Department was established as a service unit within the Directorate to fulfill its duties regarding security. The security-related duties of this department are to take necessary measures to prevent illegal interventions and other threats that threaten civil aviation security; to take necessary measures for the implementation of international security standards in the civil aviation sector; to monitor and control these measures; to ensure that "basic aviation security" training are given periodically in accordance with international standards and to follow their implementation; and to make regulations and supervise the transportation of illegal passengers.

As a result, Presidential Decree No. 4 Assigned important tasks to the Directorate for ensuring civil aviation security. To this end, the Directorate should make the necessary regulations, provide training and carry out inspections.

3.3 Safety/Security-Related Crimes in the Turkish Criminal Code No. 5237

A strict regulatory regime has been established to ensure safety and security in civil aviation. In civil aviation security regulations, the regime aims to prevent deliberate damage to the activities. Indeed, this deliberate damage is considered a crime by law. For this reason, in addition to making the regulations on security in civil aviation by law, criminal acts are also specified in the penal code. The behaviors related to risk and elimination of safety and security are regulated as crimes in the Turkish Criminal Code No. 5237.

Article 179, which concerns endangering of traffic security, makes it a crime to cause danger in terms of life, health, or assets of others through certain behaviors. In Article 223, another crime entitled "Hijacking or Seizure of Transport Vehicles" is regulated: Preventing the movement of a road transport vehicle, stopping it while it is moving or taking it to a place other than its destination by use of threat or violence or any other illegal means, is a crime.

As a result, we can say that the legal legislation contains strict regulations on ensuring safety and security in civil aviation. In the event that the safety and security in civil aviation is compromised by a movement contrary to these regulations, the relevant articles regulated in Law No. 5237 can be applied.

3.4 Regulations Regarding Safety and Security

Although Turkish Civil Aviation Law No. 2920 is not regulated in accordance with the safety and security distinction made by ICAO, this distinction is complied within regulations issued by DGCA.

Civil Aviation Safety Management System Regulation (SHY-SMS), Article 4 defines safety as taking necessary measures to eliminate unacceptable risk damage. The safety management system organized in SHY-SMS aims to operate the enterprises at the acceptable operational safety level and to create a "just and fair culture" and "safety culture" (Art. 5).

In the Commercial Air Enterprises Regulation (SHY-6A), which is the most important regulation for air carriers, the concepts of safety and security are used together, but with different meanings. SHY-6A in Article 14 regulates the measures to be taken by the commercial airline for the establishment of flight safety and security. These obligations are as follows. (1) The Air Carrier must employ a sufficient number of flight, technical and administrative personnel in accordance with the nature and scope of the operation to be carried out. (2) The air carrier must provide training and control to the personnel it employs and ensure the minimum qualifications of its personnel. (3) The air carrier should provide the minimum number of aircraft that should be in its fleet according to the nature of the license it owns and should keep these aircraft suitable for maintenance and flight. Also, aircraft in its fleet must be flightworthy. (4) The air carrier is responsible for the operations to be carried out in accordance with the minimum requirements of the national and international civil aviation legislation.

The administration has issued numerous regulations, instructions, circulars, and directives to enforce the provisions regarding safety and security in civil aviation, which are regulated both in legislation related to civil aviation.

These legal regulations are: Regulation on Providing Security, Execution of Duties and Services at Civil Airports, Ports and Border Gates; Regulation on Procedures and Principles Regarding Airline Carrier Liabilities; Regulation on Safety Assessments of Domestic and Foreign Aircraft (SHY-RAMP); Regulation on Reporting and Evaluation of Safety Events Related to Air Traffic Services (SHY 65–02); Regulation on Safety Surveillance in Air Traffic Management (SHY-GÖZETİM); and Regulation on Safety Management System in Civil Aviation (SHY-SMS).

Instructions on safety and security issued by the Directorate include the following: Airport Safety Standards Instruction (SHT-HES); Instruction for Safety Assessment for Domestic and Foreign Aircraft (SHT-RAMP); Instruction for Reporting of Civil Aviation Safety Incidents (SHT-OLAY); Instruction for the Implementation of the Safety Management System at Airports (SHT-SMS/HAD); Instruction on the Implementation of the Safety Management System in Commercial Air Transport Companies, Flight Training and Maintenance Organizations (SHT-SMS); Air Cargo and Mail Security Instructions (SHT-17.6); Aviation Organizations Security Management System (SeMS) Instruction (SHT-17.3); and Civil Aviation Security Training and Certification Instruction (SHT-17.2).

Safety and security are two different sides of the same coin. The purpose of the regulation of both is to prevent damage to people and property and to ensure that there is no loss of life. While safety regulations aim to prevent accidental harm; security regulations aim to prevent intentional harm (Dempsey, 2008, p. 67).

In order to prevent damage to people and goods in the air transport activity, civil aviation regulations should be harmonized with international regulations. This is because ensuring harmonization and coordination, especially in international air transport activities under the same sky will ensure that the aviation activities are carried out safely and securely. However, the absence of a distinction regarding the concepts of safety and security in the existing legal regulations and incompatibility with international regulations in this respect may cause legislative disputes, especially in the realization of international air transport activities. For this reason, safety and security distinctions should be made in national legislation and legislation should be updated by making legal changes regarding this issue.

4 Economic Regulations

The concept of intervention is used both in public service and law enforcement in the Turkish Administrative Law Doctrine. The concept of economic intervention can be confronted with both the state's ability to intervene in the economy through economic public services and intervention in fundamental rights and freedoms through law enforcement activities. (Yıldırım et al., 2018, pp. 431, 471) Considering the legal nature of the air transport activity, we can say that the economic intervention in the regulations for commercial air enterprises are the intervention in fundamental rights and freedoms through law enforcement activities.

Turkish Legal Doctrine states that both the 1961 Constitution and the 1982 Constitution include the articles on the regulation of the economy with the principle of social state, allowing the state to intervene in the economy (Yıldırım, 1991, p. 21). The state's intervention in the economy can basically take two forms: the regulation of the market by operating the state-owned enterprises; and drawing the general rules of the economic system with general regulatory processes. (Aslan, 1996, pp. 2–3).

The general regulatory processes for commercial air enterprises and the economic regulation of air transport activity are examined in three groups (Dempsey & Gesell, 2013, p. 62).

Entry and exit: It is possible for a company to enter a new market; that is, to start operating on a new route and to end this activity only with administrative permission. Turkish air carriers can operate if they obtain a "permit" and an "operating license".

Tariff of fares: In accordance with Article 25 of the Turkish Civil Aviation Law No. 2920 in Turkey, the tariff of fares shall be determined in accordance with the financial and economic conditions.

Competition (Antitrust): Participation and mergers of airline companies, company takeovers, and mutual commercial relations should be supervised. The aim of this supervision should be to make it suitable for public interest and to ensure that there are no anti-competitive activities in the market.

On the other hand, not every issue in the sector is regulated. Capacity, frequency, and degree of service are at the airline's sole discretion and depend on their commercial preferences. For example, it is up to the airline to choose to operate on a new route, to determine its price by its own decision, and to purchase a new aircraft (Dempsey & Gesell, 2013, p. 63). In Turkish civil aviation, on the other hand, it is preferred that every activity of air carriers is not regulated; some issues are regulated, and some issues (content of the service, delivery, route preferences, plane preferences, etc.) are deregulated over time.

4.1 *Permits and Operating Licenses*

According to the Turkish Civil Aviation Law No. 2920, airlines need to obtain permits and operating licenses to establish and operate. In addition, those companies that wish to carry out air transportation activities must meet certain conditions in advance and have these conditions during the activity. Both the permit conditions (national security and public order and not being convicted of crime) and the issues must be examined in the evaluation of the license application in relation to the safe and secure execution of the activity.

Turkish Civil Aviation Law No. 2920 and the Commercial Air Operations Regulation (SHY-6A) regulate the conditions that commercial air enterprises must fulfill in order to be able to carry out air transport activities, permits, and licenses to be obtained, the structure of the operator and the organization of the company, and the qualifications of the personnel to be employed.

In SHY-6A, Article 27, the steps to grant operating license are arranged as follows: (1) application phase; (2) pre-authorization phase; (3) document compliance phase; (4) main license phase; (5) inspection phase; (6) evaluation phase; and (7) issuing operating license phase. The first three of these phases correspond to the permit phase specified in Law No. 2920, and the remaining four to the operating license phase specified in Law No. 2920.

SHY-6A contains a regulation regarding the legal structure of the commercial air business. According to SHY-6A, Art. 6, enterprises that will engage in commercial air transportation activities with aircraft of fewer than 20 seats must be joint stock or limited liability companies; enterprises that will operate cargo transportation activities or aircraft with a seating capacity of 20 or more should be joint stock companies. These operators cannot change the legal structure of the company during the operation.

In SHY-6A, Articles 7–8 and 9 provide detailed regulations regarding the corporate action, the registered office, and the shareholders of the company. Accordingly, a working area other than aviation and aviation related works will not be included in the founding charter of commercial air enterprises that will carry out cargo transportation or passenger transportation activities only with aircraft with a seating capacity of 20 or more. The registered office of an air carrier must be found in the borders of the Republic of Turkey and capital structure ratios should be as follows: At least 51 percent of the shares must be registered; the majority shareholders and board members with voting rights and control of the majority of stakeholders must belong to Turkish citizens. On the other hand, the founder or co-founders and their representatives authorized to represent and bind should not have been found guilty of crimes determined by Article 18 of the Turkish Civil Aviation Law No. 2920, and there should not be an adjudication in bankruptcy or arrangement of bankruptcy for them.

A party that meets all these conditions must obtain a permit and an operating license in order to become an air carrier and to carry out air transportation activities. For this reason, the stages, conditions, and legal nature of the license and operating license are important.

4.1.1 Permit

According to Article 18 of the Turkish Civil Aviation Law, real and legal persons must obtain permission from the Ministry of Transport and Infrastructure to carry passengers or cargo or passengers and cargo for commercial purposes. The same article states that the permit required by commercial air enterprises can be bound to certain conditions and times. Article 18 of Law No. 2920 regulates that the license to be obtained in order to become a commercial air enterprise will not be given in case of certain conditions. These conditions are public order, national security, and being sentenced for specific offenses.

It is stated in Turkish law doctrine that public order will be maintained by ensuring that individuals live in health, well-being, and security in public places. Well-being is expressed as the absence of any irregularity and confusion that may adversely affect the ordinary course of life (Günday, 2013, p. 292). Events that violate the condition of well-being may vary according to the cultural level of the society and the place and time of the event (Yayla, 2009, p. 40). Health condition is defined as the protection of the society from infectious and widespread diseases and continuing their life under health conditions. Finally, the security factor is explained as ensuring that individuals live and move around in public places without being exposed to attacks, accidents, obstructions and without worrying about their lives and properties (Günday, 2013, p. 292).

On the other hand, the concept of "economic public order" emerged simultaneously with the economic activities and interventions of the administration (Özdemir, 2012, p. 101). With the concept of economic public order, the limitations imposed on freedom of contract and labor are generally discussed (Tekinsoy, 2011, p. 149). The doctrine states that the concept of public order is wide enough to include the concept of economic public order (Yıldırım, 1991, p. 30). Thus, the state's regulatory function regarding the economy has introduced the concept of "economic law enforcement" as a special type of law enforcement, and the purpose of the economic law enforcement is to establish the economic public order (Tan, 1990, pp. 166–167).

For all of the above-mentioned reasons, the application for permits should also be examined in terms of the economic public order within the scope of the public order. While all of these examinations are being carried out by the Administration, a decision should be made to realize the economic public interest and maintain the economic public order, in line with the economic model preferred by the state. Finally, examination and evaluation should be made in terms of public order within the scope of the general principles of law enforcement activities in a way that does not restrain the freedom of enterprise.

According to the Turkish Civil Aviation Law No. 2920, in the evaluation of the permit application made to the administration within the scope of national security, if it is concluded that the application may endanger the internal and external security of the State, and weaken the State against spying, sabotage, and destructive activities, the application for such permission may be rejected.

Finally, the last condition regulated in Article 18 of Law No. 2920 concerns being convicted of certain crimes. Real persons or the founding partners of legal persons and their authorized officials shall not get permission if they are convicted of these crimes.

Consequently, the administration will examine the application in terms of permit conditions, and will be able to grant permission if the information and documents presented are sufficient and not convicted for certain crimes and there is no contradiction to the national security or public order. The administration's assessment in terms of national security and public order can be evaluated within the scope of its discretion. This discretion will be reviewed by the administrative judiciary in terms of public interest and compliance with public service requirements. On the other hand, in the case of convictions for certain crimes, we can say that the administration has a bound authority (*compétence liée*) and cannot allow it.

4.1.2 Operating Licenses

According to Article 19 of the Turkish Civil Aviation Law, real and legal persons who will carry passengers and/or cargo on certain lines for commercial purposes by aircraft must also obtain an operating license after obtaining permission from the Ministry. Requirements for the operating license are regulated distinctly by the Commercial Air Operations Regulation (SHY-6A) and Turkish Civil Aviation Law.

Applications made for the operating license are examined by the Ministry in terms of eligibility and country benefits, in accordance with Article 21 of Law No. 2920. In this review, the enterprise to be established is evaluated in terms of the country's economy, national security, national transportation policy, and planned development principles.

According to Article 22 of Law No. 2920, real and legal persons who obtain permission from the Ministry should apply to the Ministry to obtain the operating license after they carry out their projects attached to the permit application. If it is determined that the project has been made after the necessary examinations carried out by the Ministry, an operating license is issued.

According to SHY-6A Art. 34, in the inspection phase, "effectiveness of the activities planned by the applicant," "adequacy of the facilities and equipment," "compliance with legal regulations and safe operational requirements" are examined. The organization of the enterprise and flight operation, ground management, technical, training, trade, financing, quality, safety, performance, flight operations (dispatch), team planning, and cabin services units are assessed during the inspection.

According to SHY-6A Art. 35, during the evaluation phase, the operation license request file and the inspection report prepared as a result of the inspection are evaluated together. If the evaluation result is affirmative, the operating license is issued.

The operating license is not transferable and the commercial title of the license holder, the type of transportation, the region or lines to be operated, and the limitations and restrictions indicated in the permit document is specified in the license.

According to Law No. 2920 and SHY-6A, having an operating license brings certain obligations to air carriers. The first concerns flight and fare schedules.

According to Art. 25, license holders can determine the tariffs and the date of implementation in accordance with commercial, financial, and economic conditions and announces to third parties three days before they come into effect. The flight schedules take effect seven days after they are determined by the license holder and approved by the Ministry of Transport and Infrastructure. According to SHY-6A art. 36, the air carrier should start its flight activities within six months after the operating license is issued. In addition, the carrier will not be able to interrupt flight activities for more than three months, provided that the conditions in SHY-6A are met. For this reason, the air carrier is obliged to continue the flight activity to ensure the continuity of the activity after the flight schedule is approved.

The air carrier also has other obligations. According to Article 26 of Law No. 2920, it must carry mail on every scheduled flight if requested. According to Article 27 of Law No. 2920, following the regulations put into force by DGCA and following the instructions given by DGCA regarding these regulations, it must provide all kinds of information and documents requested by DGCA and ensure that DGCA can always access the desired equipment. According to Article 28 of Law No. 2920, it must provide all technical and economic information requested for inspection. In accordance with the Regulation on the Rights of Passengers Traveling by Airline (SHY-YOLCU), the air carrier must provide a certain quality of service to the passengers or pay compensation if it does not.

4.2 Audit

Article 27 of the Turkish Civil Aviation Law regulates issues related to auditing. All public institutions and organizations and real and private legal persons operating with Turkish civil aircraft are subject to the technical supervision that DGCA does or will undertake to ensure air navigation safety. Audit principles, selection of audit staff, powers, and responsibilities will be determined by the regulation to be prepared by DGCA. Various regulations and instructions regarding the audit have been made by DGCA: Regulation on Duties, Authorities, Responsibilities, Working Procedures and Principles of Technical Auditors of DGCA (SHY TD01); Regulation on Authorized Audit, Technical Audit and Consultancy Institutions (SHY-YDTGMK); Instruction on the Inspection and Evaluation of Pilots and Air Vehicles Registered to Turkish Registration, Instruction on Principles and Procedures of Inspections During Flight for Cabin; Instruction on the Application Principles of Civil Aviation Enterprises Audits; Instruction on the Liability Insurance of Civil Aviation Enterprises Authorized Audit Firms; and Instruction for the Chief Auditors and Auditors to Take Part in Authorized Audit Firms.

In addition to Art. 27 of Law No. 2920 regarding the audit, Art. 28 of the Law regulates that the license holders are obliged to provide all the technical and economic information required for the inspection. We also see that, in Art. 44 of SHY-6A, issues related to technical inspection are regulated under the audit heading and in Art. 45, financial inspections are arranged under the financial audit heading. Therefore, we can say that the air carrier is subject to both technical and financial inspection.

4.2.1 Technical Inspection

Article 27 of the Turkish Civil Aviation Law regulates that air carriers are subject to maintain the technical inspection to ensure the safety of air navigation. The text of the article clearly states that the audit could be carried out by DGCA itself or authorized institution. According to the Regulation on Authorized Audit, Technical Audit, and Consultancy Institutions (SHY-YDTGMK), the subject of the audit to be carried out by the independent authorized audit firm, which will conduct audits on behalf of DGCA, varies according to the type of authorization to be taken.

The principles of the audit will be determined by DGCA. In these technical inspections, the technical inspectors assigned by DGCA investigate the aircraft's airworthiness, the crew's qualifications required for that aircraft type, the physical or mental capacity to carry out the flight in terms of safety and flight safety. In addition, inspections can be made with or without notice, during which the activities of the air carrier will be examined essentially in compliance with the standards set in national and international legislation.

4.2.2 Financial Inspection

Article 45 of the Commercial Air Operations Regulation (SHY-6A) contains regulations regarding the financial audit of the license holder under the title of Financial Audit. According to this, DGCA oversees the financial status of the air carrier and issues corrective warnings when deemed necessary. With the Instruction on Auditing of Commercial Air Enterprises in Financial and Fiscal Matters (SHT-6AF), the minimum requirements that commercial air enterprises should have regarding financial and fiscal requirements and the procedures and principles of the audits for meeting those requirements have been determined.

According to SHT-6AF Art. 5, the purpose of the audit is to evaluate whether the financial and fiscal conditions of the enterprises are sufficient to ensure the highest level of flight safety and to ensure that the necessary measures can be taken by the enterprises in any negative situation. According to SHT-6AF Art. 11, in case of negativity as a result of the evaluation made by DGCA, the directorate warns the enterprise, and the enterprise is requested to take necessary actions to correct its own financial situation.

Moreover, Article 11 of SHT-6AF regulates sanctions to be applied in the case of failure to comply with the evaluations made as a result of financial auditing. Therefore, the following requests of the air carrier may be rejected: requests to add aircraft to the fleet, new flight line requests, change requests that extend the license limits, rental requests from domestic and foreign companies. In addition to the rejection of these requests, additional letter of guarantee can be requested up to twice the existing letter of guarantee given under SHY-6A. All of these sanctions are aimed to improve the financial situation of the business and prevent the domino effect by showing an economic negative result in the civil aviation industry.

4.3 Administrative Sanctions

Law No. 2920 states that administrative sanctions will be applied in case of violation of the regulations of the civil aviation authority and civil aviation laws.

According to Article 27, DGCA is authorized to stop the flight operation (of an operator or workers requiring proficiency) and/or the operation of the operator if the following situations arise during the inspections: (a) The aircraft is not suitable for flight; (b) the crew does not have the required qualifications for that aircraft type or does not have the physical or mental capacity to carry out the flight; (c) operation poses a risk to life and property safety; or (d) The identified insufficiencies or outcomes directly concern life and property safety and flight safety.

Article 143 of Law No. 2920, entitled misdemeanor, imposes various administrative fines in case of violation of various articles of the law. In this context, administrative fines to be imposed in case of acts contrary to the Turkish Civil Aviation Law No. 2920 and regulations are regulated by the Regulation on Administrative Fines to be imposed by the Directorate General of Civil Aviation (SHY-İPC).

According to Article 30 ("decertification"), if the license holder acts contrary to the provisions in the third section of the Law; upon the notification to be made by the Ministry for the elimination of the violation, the license holder should eliminate the violations within the given period. If it does not, the permit for some or all of the lines it transports can be temporarily withdrawn or the license will be canceled immediately.

The main result of the decertification is the termination of the operator's air transport activity and to be subject to restrictions if the operator wants to carry out air transportation again. According to SHY-6A Art.13, in case of cancelation of the operating license of the carrier, which is operating in commercial air transportation and has a seat capacity of 20 or more, persons who are directly or indirectly a shareholder, member of the board of directors, general manager or responsible manager on the date of suspension or cancelation cannot take place as a stakeholder and board member, general manager or responsible manager in a new company requesting an operating license.

Another important result of the decertification concerns the letter of guarantee. According to SHY-6A Art.19/1-d, the letter of guarantee, which must be submitted at the licensing stage and be kept complete and valid during the operation, is converted into cash by the DGCA, without the need for a judgment, regardless of all debts and obligations of the operator. Upon an announcement made by DGCA due to the failure to fulfill the transport commitment because of the cancelation of the operating license, another airline may carry out the transportation within two months at the latest. In this case, transportation costs are paid from the letter of guarantee turned into cash to the airline company that carries out the transportation. Any remaining amount from encashment will be added to the treasury as income.

As a result, in case of decertification, the persons causing this result are prevented from performing air transport activities again. The intention is that those who do not conduct their air transport activities appropriately and do not comply with their license obligations will not be allowed to perform air transport activities again. On the other hand, in order to prevent passengers suffering due to the cancelation of the license, the continuity of the transportation activity is ensured by encashment of the letter of guarantee for two months.

4.4 Intervention to Tariff of Fares

Air carriers are subject to some regulations while determining the tariff of fares they will charge for these transportation activities. The regulation of the tariff of fares affects both the competition among other enterprises in the sector and the tax that the government will receive from this activity. In addition to the competition and tax issues, the tariff of fares directly affects the passengers. Therefore, the determination of the fee schedule means economic intervention in the market.

According to Presidential Decree No. 4 Art.44, it has been regulated that the DGCA is in charge of determining and supervising the implementation of floor and ceiling fees in civil aviation services, including transportation, to improve civil aviation activities and to provide a free, fair, and sustainable competitive environment in cooperation with the relevant service units and organizations of the Ministry.

We can say that the government has different approaches in different periods regarding the air carriers' tariff. Until 2001, the fee tariff had to be approved, while the 2001 amendment was lifted and the notification basis was introduced. In 2011, the ceiling–floor price practice was clearly regulated in the law, as the competition among the air carriers between 2011 and 2001 had different dimensions and sometimes became an obstacle to the public interest.

The state's regulatory function for the economy has introduced the concept of "economic law enforcement" as a special type of law enforcement, and the purpose of the economic law enforcement is to establish the economic public order (Tan, 1990, pp. 166–167).

As an economic law enforcement activity, ceiling–floor price practice provides the protection of the consumer by determining the ceiling price and prevents the competition from being destructive among the enterprises by determining the floor price. However, it should be noted that the floor price practice also eliminates the ability of air carriers to offer tickets at a competitive price and may result in passengers not being able to access lower fare tickets. On the other hand, while the ceiling price aims to protect the consumer, it may cause the quality of the service to decrease because enriching the service content and putting additional services will only be limited to the ceiling price.

Therefore, the administration should be active while determining the ceiling and floor prices and should not be left to "gentlemen's agreements" between the airline companies. It would be appropriate to have a base price that will make the activity sustainable for air carriers considering the costs of the air transport activity and to set a ceiling price in comparison to the profit rates of airlines worldwide.

5 Legal Regime of Air Transport for Foreigners in Turkey

The ability of foreigners to carry out activities related to air transportation in another country can be examined from two different perspectives: activities of foreign air carriers within the borders of Turkey and being stakeholders as foreign investors in the Turkish airline company.

5.1 Cabotage is a Limitation of the Activities of Foreign Air Carriers

Cabotage is defined as a ship's business between the ports of a country (www.sozluk.gov.tr, 20.08.2020). Although the source of the concept of cabotage is maritime law, it is also used in transportation law, especially in road and airline transportation (Mendes de Leon, 1992, p. XI). Cabotage in air law is defined as the transportation of passengers, cargo, or mail between two different places within a state (Dempsey, 2008, p. 608).

Cabotage practices are the biggest impediment for foreign airline companies to operate in another country. According to Article 31 of the Turkish Civil Aviation Law, entitled Cabotage, the transportation of passengers, mail, and cargo between two points by airlines within the borders of the Republic of Turkey can only be performed with Turkish aircraft.

The reason for cabotage is to protect domestic carriers, workers, and the industry against troubles caused by foreign carriers (Dempsey, 2008, p. 614). The purpose of the cabotage is to ensure public interest. In the regulations on cabotage, the public interest is provided by supporting the domestic carrier, ensuring the development of the country's capital, and preventing the domestic carrier from falling into a difficult situation against foreign carriers in domestic transportation.

In Article 7 of the Chicago Convention, cabotage is defined as follows: "*Each contracting State shall have the right to refuse permission to the aircraft of other contracting States to take on in its territory passengers, mail and cargo carried for remuneration or hire and destined for another point within its territory. Each contracting State undertakes not to enter into any arrangements which specifically grant any such privilege on an exclusive basis to any other State or an airlineof any other State, and not to obtain any such exclusive privilege from any other State.*"

States that are party to the Chicago Convention have the right to enforce cabotage. However, a state may renounce the right to enforce cabotage in favor of another state and allow that state to transport within its own country. In this case, according to Article 7, it must also grant permission to another state that meets the requirements for this permit.

Both security and economic reasons are the basis of cabotage practice. Foreign air companies carrying out air transportation activities between the two points in the country will put national security at risk because if the aircraft belonging to a foreign airline company is ready to park in any part of the country, it will pose a risk in case of a war with that country. Foreign airline companies can easily carry out intelligence activities since they operate all over the country. Finally, domestic operators may suffer economic losses because foreign airline companies compete with domestic operators in the domestic market.

Absolute cabotage practice may prevent obtaining the right to cabotage from other states. Therefore, in case of reciprocity and economic gains, the practice of cabotage should be stretched for the public interest.

In Turkish law, cabotage should be evaluated with a holistic approach, along with the policies of bilateral air transport agreements and practical applications which are known as back-door of cabotage. Instead of an absolute prohibition in the current situation, an approach that allows the right to cabotage to be recognized without compromising national security should be adopted. To this end, by making new legal arrangements, DGCA can be given discretion and DGCA can use this authority as a bargaining factor to enable domestic carriers to operate in foreign markets, which may be more profitable.

5.2 Establishment and Operation of Air Carriers of Foreigners

Launching the commercial aircraft business of foreigners in Turkey can be viewed from two perspectives: establishing a company for foreigners in terms of Turkish Commercial Code and sharing gates of companies that can perform air transportation.

Numerous amendments were made to the Turkish Commercial Code after 2011 regarding the establishment of foreigners and the appointment of foreigners to the board of directors. Today, a person who is not a Turkish citizen will be able to establish a joint stock company alone and create the entire board of directors from foreigners. Therefore, there is no longer any impediment for foreigners to establish a company in Turkey. However, this does not mean that companies established by foreigners can operate in every sector. Special regulations were introduced in Law No. 2920 regarding commercial air enterprises that will carry out air transportation activities.

According to Article 33 of the Turkish Civil Aviation Law, matters concerning foreign businesses conducting air transport between foreign countries and Turkey will be determined by the Ministry in accordance with the principle of reciprocity and bilateral and multilateral agreements to which Turkey is a party.

In the second part of the Regulation on Commercial Air Enterprises (SHY-6A), detailed arrangements have been made regarding the structure and share rates of the operators. According to this, the commercial headquarters of a company that will conduct air transport should be found within the borders of the Republic of Turkey. Moreover, the majority shareholders of these companies and board members with voting rights and control of the majority of stakeholders should be citizens of the Republic of Turkey. According to these regulations, foreign investors in Turkey, cannot own more than 49 percent of a company engaged in air transport operations and cannot have control of the company.

6 Deregulation of Air Transport in Turkey

Until the mid-1980s, only Turkish Airlines had operated in Turkish domestic airline transportation. After the Turkish Civil Aviation Law No. 2920 came into force, domestic airline transportation was liberalized and private enterprises were allowed to enter the sector (Gerede, 2015, pp. 168–169). However, these companies were not allowed to operate freely in domestic lines until the 1996 decision of DGCA. After this decision, private airline companies were able to fly freely to places where Turkish Airlines never flies. On the other hand, these companies could fly to the places where Turkish Airlines flies on days when Turkish Airlines does not have flight and on the days when Turkish Airlines does have a flight as long as Turkish Airlines is unable to meet demand (Gerede, 2015, p. 176). In 2003, when this restrictive decision was lifted, Turkish domestic airline transportation entered the period of liberalization, slot allocation was taken from Turkish Airlines, and seven private airline companies entered domestic transportation in seven years (Gerede, 2015, p. 186). The year 2003 was a turning point in liberalization and the environment of liberalization and competition started after this period, after which there may be a new concept for full opening of the market to competition (Kesici, 2011, p. 182).

On the other hand, the important liberalization is about the fee schedule. Before the amendment, the legislator regulated that both the fee schedule can be put into effect with the Ministry's approval. With the 2001 amendment, the license holder is not obliged to obtain Ministry approval when determining the fee schedule. The reason for this amendment appears at the first stage as removing the impediment to privatization of Turkish Airlines. In addition, a free market economy is provided in the civil aviation sector and the companies' wage policy can be freely determined by itself. As a result, the state will intervene less in the civil aviation market economically.

The results of the deregulation practices in Turkey can be summarized into two parts: a free way to enter the domestic market for domestic carriers and the elimination of the approval of the tariffs.

The ability of domestic carriers to freely enter the domestic market has brought a competitive environment. In particular, the fact that more than one carrier can operate instead of the presence of a single carrier in domestic lines has enabled the transportation opportunities offered to the passengers to increase, and the competition has led to cheaper and higher quality in air transportation activities. Therefore, the fact that more than one domestic carrier has been operating in the domestic market and the emergence of competition have been positive aspects of this deregulation practice.

In the deregulation of the fee schedule, the tide situation was experienced. The legal obligation regarding the approval of the fee tariff was removed first, and upon the emergence of unbalanced pricing in the market, legal arrangements were made and the ceiling–floor price determination authority was introduced. Therefore, since the deregulation regarding the fee schedule does not give the desired results, a new order was adopted instead of giving this up completely. The result obtained from this deregulation application and the new regulation made against them is an example of finding a balance between re-regulation and deregulation.

7 Conclusion

In Turkey, the role of air transport for the administration was amended with the privatization of Turkish Airlines. As a result of the privatization, the state has now moved to a position that does not offer air transport, but regulates it. In Turkish Law, the issues regulated by the administration in the legal regime for air carriers are related to safety and economic issues. While the most important role in safety and security regulation belongs to DGCA, economic intervention is carried out by both the Ministry of Transport and Infrastructure and DGCA.

The administration that will regulate the air transport activity must be fully authorized and independent in the regulation of the relevant issues. Because of the lack of full authority in safety and security issues while regulating the economic issues of air transport activity may cause confusion among authorities and bureaucratic difficulties in matters where safety and economic issues are interrelated, the sector cannot be guided as a whole. Therefore, it would be appropriate for the DGCA to expand its current duties and powers and regulate the entire civil aviation sector from a single source and independently.

The regulation of all aspects of the civil aviation sector by expanding the duties and powers of the DGCA should not be carried out by one person from a single source. The characteristics of civil aviation activities are different and concern different areas of expertise, so the sector must be addressed from multiple perspectives. For this reason, a board of directors should be established within the DGCA and this board should be in charge and authorized specially for the economic dimension of civil aviation. Security-related issues, on the other hand, may remain under the duty and authority of the general manager of the DGCA due to the emergence of issues that require immediate decision.

The intensity of the regulation for commercial air enterprises carried out by the DGCA should be reconsidered. In other words, a balance must be established between the regulation and deregulation of commercial air enterprises. In order to protect and maintain competition among commercial airline companies, competition must be controlled by the administration and a healthy competition environment must be established in the civil aviation sector. Therefore, it would be more appropriate to make arrangements in which discretionary power is granted to the administration when deregulating. It is not preferable to completely abolish the duties and authorities of the administration regarding the matters where commercial air enterprises are regulated. Instead, the duties and powers that can be used by the administration when deemed necessary in the presence of certain conditions should be regulated.

The Turkish air transportation system is currently regulated by the administration in accordance with international regulations and operates efficiently. It would be appropriate to make these changes in both organic and functional terms in order to develop Turkish air transport and to have a high competitive capacity in the international arena.

References

Aslan, Z. (1996). Devletin Ekonomiye Müdahalesi ve İktisadi Kolluk Faaliyeti İçinde Rekabetin Korunması. *İstanbul Üniversitesi Siyasal Bilgiler Fakültesi Dergisi, 14.*

Çağa, T. (1963). *Hava Hukuku Cild I: Genel.* İstanbul: Doğan Kardeş Matbaacılık.

de Leon, P. M. (1992). *Cabotage in air transport regulation.* Martinos Nijhoff Publishers.

Dempsey, P. S. (2008). *Public international air law.* McGill University.

Dempsey, P. S., & Gesell, L. E. (2013). *Public policy and the regulation of commercial aviation.* Coast Aire Publications.

Erol, Ö. F. (2018). *İdarenin Sivil Havacılık Alanında Hava Taşımacılığına İlişkin Faaliyetleri.* İstanbul: On İki Levha Yayıncılık.

Gerede, E. (2015). *Havayolu Taşımacılığı ve Ekonomik Düzenlemeler Teori ve Türkiye Uygulaması.* Ankara: Sivil Havacılık Genel Müdürlüğü Yayınları.

Günday, M. (2013). *İdare Hukuku* (10th ed.). İmaj Yayınevi.

Kesici, B. (2011). Sivil Havacılık Sektöründe Serbetleşme, Rekabet ve Rekabet Hukuku. In *Ulaştırma Sektöründe Serbestleşme, Rekabet ve Rekabet Hukuku Sempozyumu.* İzmir.

Özdemir, H. E. (2012). *Nükleer Güç Santrallarının Kurulmasına İlişkin Hukuki Esaslar.* İstanbul: On İki Levha Yayıncılık.

Tan, T. (1990). Anayasal Ekonomik Düzen. *Anayasa Yargısı Dergisi, 7*, 159–178.

Tekinsoy, Ö. O. (2011). *İdare Hukukunda Kamu Düzeni Kavramı.* İstanbul: XII Levha Yayıncılık.

Yayla, Y. (2009). *İdare Hukuku.* İstanbul: Beta Basım Yayım Dağıtım.

Yıldırım, T. (1991). *Türk İhracat Rejimi ve İlgili Mevzuat.* İstanbul: Kazancı Kitap Ticaret.

Yıldırım, T., Yasin, M., Kaman, N., Özdemir, H. E., Üstün, G., & Okay Tekinsoy, Ö. (2018). *İdare Hukuku* (7th ed.). On İki Levha Yayıncılık.

Ömer Faruk Erol is an assistant professor at Ibn Haldun University. He graduated from Istanbul University Faculty of Law in 2010. He completed his master's degree in 2012 with his thesis entitled "Environmental Impact Assessment According to the Environmental Law" and completed his doctorate in 2017 with his thesis titled "Civil Aviation Activities of Administration" at Marmara University. He completed his studies as a visiting researcher at Harvard University with the YÖK scholarship for master thesis and as a graduate research trainee at McGill University with the

TUBITAK scholarship for doctoral thesis. From the year he graduated, he worked as a research assistant at Yalova University, Istanbul Medeniyet University, and Marmara University respectively. He is currently serving as an assistant professor at the Faculty of Law at Ibn Haldun University. He lectures on Administrative Law and Aviation Law.

Regulation of Rail Transport

Şahin Ardıyok and Evren Sesli

Abstract The rail industry is a sector that is heavily regulated all over the world. Historically, railways constitute a good example of a natural monopoly due to their (infrastructure-based) high-cost structure, the presence of indivisibilities in their services, and economies of scale. Therefore, public control or ownership of railway infrastructure and public management of railways has become the most generally accepted approach. During the second half of the 1980s, there were increasing calls to improve regulations and privatize railways, mainly because of a decreasing trend in the quality of railways, the declining market share, and worsening financial performance. Various countries implemented a variety of regulation and liberalization methods. In this regard, we have witnessed a wide range of approaches to introduce competition into rail transport services. Against this backdrop, this chapter aims to review the Turkish railways' experience of liberalization, including some brief explanations on the main features of rail transport, the need for regulation in the railway industry, alternative models of regulation, and the trade-off between competition and regulation.

Keywords Regulation · Liberalization · Deregulation · Railways · TCDD · TCDD Taşımacılık

1 Introduction

Railways remain a crucial component of transportation networks (Warren, 2019, p. 3). However, after being the dominant transportation means for more than 100 years, railways have experienced a dramatic decline in global economic relevance during the last few decades (Cantos & Campos, 2005, p. 22). The decline of railways partially stem from the mismatch between railways' services and the expectations of customers

Ş. Ardıyok (✉) · E. Sesli
Balcıoğlu Selçuk Ardıyok Keki Attorney Partnership, Istanbul, Turkey
e-mail: sardiyok@baseak.com

E. Sesli
e-mail: esesli@baseak.com

M. Eroğlu and M. Finger (eds.), *The Regulation of Turkish Network Industries*,
https://doi.org/10.1007/978-3-030-81720-6_12

(Moyer & Thompson, 1992, p. 1). The other main reasons are public involvement in railways and the widespread effects of an obsolete regulatory framework (Cantos & Campos, 2005, p. 22).

Adapting to a changing environment is a necessity, as experience has shown that a modern, high-quality railway is a significant contributor to the development of a country's competitive economic structure (Ordover & Pittman, 1994, p. 1). Accordingly, countries have adapted various methods to regulate and modernize railways to overcome the sharp decline of the industry. Although the vertically integrated public service provider model, which is heavily protected from competition, has been the dominant strategy in railways, there are different experiences in both managerial and jurisdictional aspects (Warren, 2019, p. 1; Cantos & Campos, 2005, p. 23). Turkey's experience in the regulation of railways has similarities and differences from other countries. Turkey's unique geographical position, long history of railway transportation since the Ottoman Empire, and newly adapted liberalization policies following the need to renew the railways and harmonize policies with the EU provide instructive lessons.

For these reasons, the main aim of this chapter is to explain the milestones of Turkish railways' experience on the way to liberalization, with some brief explanations of the main features of rail transport, the need for regulation in the railways, and the trade-off between competition and regulation.

2 Main Features of Railways that Have Shaped Regulatory Preferences

Railways have unique characteristics that cause complicated issues, both in the determination of management policies of railways and its regulation or deregulation (Button, 1993; OECD Swiss, 2006, p. 22; Transport and ICT, 2017, p. 32;):

- Multiproduct nature of the activity
- Particular cost structure
- Public service characteristic
- Existence of indivisibilities
- Role of infrastructure[1]
- Externalities.[2]

Furthermore, the technology has changed slowly, economies of density[3] are high, the data availability is scarce, and the coordination of operations is more complex in railways than in other transportation modes (OECD Swiss, 2006, p. 22). This section describes some of these specific features that serve to explain railway framework and the main determinants of railway regulation.

2.1 Multiproduct Structure of Railways and Its Relation with Cost Structure

Railway is a multiproduct since operators generally supply freight and passenger services simultaneously (UN ESCAP, p. 99). Railways can be classified based on services (that is, passengers or freight), origin, and destination of service. In some cases, rail services that are self-funding (such as urban commuter services) are distinguished from rail services that are primarily funded from government funds (OECD, 2005, p. 25–26). For the designation of market structure and regulation-related issues, diversification based on freight and passenger services seems more relevant. The main activities in freight and passenger rail transportation are listed in Table 1 (Campos & Cantos, 1999, p. 4).

Several implications stem from the multiproduct nature of railways (UN ESCAP, p. 99). First, every railway activity faces different levels of competition from other transportation modes depending on the geographic, demographic, and economic characteristics of different countries[4] (OECD, 2005, p. 26). Second, it often becomes difficult to allocate total operating costs[5] among different services offered because of the existence of joint or common expenses to several rail users[6] (UN ESCAP, p. 99). As cost interdependence requires simultaneous decisions on prices and services, it becomes more complicated to implement regulatory tasks and frameworks in railways (Campos & Cantos, 1999, p. 4). Third, the sub-additivity[7] is another essential aspect of the effects of the multiproduct feature of railways at the cost level (Campos & Cantos, 1999, p. 4).

These implications are crucial for regulators and policymakers when designing the regulatory framework. The first question that the regulators and policymakers need to answer is whether it is more efficient for a single firm, rather than two separate firms, to supply both infrastructure and transport operation services. The second question is whether, if the infrastructure and services are separated, the supply of such services becomes more efficient within the context of a monopoly rather than the existence of competing firms (UN ESCAP, p. 99; Campos & Cantos, 1999, p. 4–5).

Table 1 Activities in freight and passenger rail transportation

Type of rail transportation	Activities
Freight	Transportation of bulk freight, supply of complete cargo trains, supply of postal services, other services of intermodal transportation
Passenger	Local services (including suburban and commuter trains), regional services, long-distance traffic, and high-speed trains on certain routes

Source (Campos & Cantos, 1999, p. 4)

2.2 Railways as a Public Service and Indivisibilities

Public administration plays a significant role in the organization and supervision of railway activities. Apart from the technical and economic features of the railways, providing transport services for the needs of customers is a politically sensitive issue (Transport and ICT, 2017, p. 108). Hence, railways are regarded as public or social services irrespective of their financial viability (UN ESCAP, p. 101).

The main reasons for defining railways as public services are interlinked with the integrative function of railways to lessen[8] geographical barriers, and the facilitative role of transportation to transmit economic and social aid to different regions of a country[9] (Campos & Cantos, 1999, p. 7; UN ESCAP, p. 83–84). Furthermore, it is crucial to fulfill the public service obligations (PSO) for particular geographical regions and/or a specific segment of the population in the form of compulsory provision, even for the unprofitable routes, timetables, and services (Campos & Cantos, 1999, p. 7; UN ESCAP, p. 101).

Another significant feature of railways is indivisibility of outputs and inputs. Rolling stocks, tracks, and stations, which constitute the capital inputs of railways, can be incremented only in discrete and indivisible units. However, it is challenging to examine similar fluctuations on the demand side (Campos & Cantos, 1999, p. 6). Furthermore, the lumpiness of railways due to location-specific and physically fixed infrastructure, accompanied by economies of density, have critical implications for investment and pricing decisions in railways (Campos & Cantos, 1999, p. 6; Transport and ICT, 2017, p. 32).

First, these features endow the railways with elements of a natural monopoly (Transport and ICT, 2017, p. 32). Although different railway regulation methods (such as vertical separation) have the potential to decrease the side effects that may stem from natural monopoly, they do not alter the capital-intensive structure of railways, which also has other indivisibilities within its production process. Second, differences in supply and demand of services cause excess capacity in rolling stock and infrastructure. Third, the significance of transportation costs increases at the limit of full capacity utilization.[10] Fourth, the existence of indivisibilities affects the presence of innovation and infrastructure improvement projects. Relatedly, public authorities usually defer innovation and infrastructure improvement projects and prefer carrying out investments in small discrete amounts[11] (Campos & Cantos, 1999, p. 6).

Given the above, the main objective for railways is to sustain an optimal level of service quality and variety while ensuring a high level of productive efficiency (Tzanakis, 2019, p. 5). The key factor to achieve this ultimate goal is preservation of the economies of scale and scope in infrastructure while introducing market elements and openness to competition (OECD Swiss, 2006, p. 22).

2.3 *Main Problems of Implementing Traditional Policies in Railways*

The main problems associated with the traditional policies on railways have been: (i) increasing financial losses, (ii) a high degree of inefficiency in management, (iii) a business activity oriented exclusively toward production targets rather than commercial targets, and (iv) a lack of dynamism (UN ESCAP, p. 84–85; Campos & Cantos, 1999, p. 9). These inefficiencies did not stem from an artificial reduction in the range of services, nor from excessively high fares, but from an unplanned increase in supply, which causes huge public subsidies (Campos & Cantos, 1999, p. 9). Therefore, goals set for the successful restructuring of railways, with or without the inclusion of the private sector, are (Ordover & Pittman, 1994, p. 1):

- Improving the utilization of railroads
- Providing incentives for investments in infrastructure and innovation in services
- Prevention of monopoly abuse
- Reducing the budgetary burden.

Policy-makers in many countries have agreed that the solution to this problem is to create competitive market-based railways that can align consumer needs with the provision of railway services in a financially stable way (UN ESCAP, p. 101–102). Such reforms have also emphasized the need for effective regulation and regulatory institutions instead of eliminating them (UN ESCAP, p. 83).

3 Regulation—Competition Trade-off in Railways

The healthy and efficient functioning of railways is correlated to regulatory framework and competition. Different regulations are implemented to obtain the efficient management of railways and to establish a nondiscriminatory access pricing system for managerial concerns (OECD Swiss, 2006, p. 24). Competition also emphasizes the importance of operation design, marketing strategies, and pricing policies to maintain and increase capacity utilization (Transport and ICT, 2017, p. 32). Whenever sector-specific regulations exist, the question is obvious: Would regulatory provisions be complementary to general competition law or should the implementation of competition law be restricted in certain conducts (Staebe, 2011, p. 1)? In this section, we explain the main requirements of a well-functioning sector-specific regulator, types of regulation, and competition policy concerns in railway transportation.

3.1 The Role of Sector-Specific Regulators and Types of Regulation

The design of regulatory institutions has certain requirements to establish a sound regulatory framework (Campos and Cantos, p. 56–58; UN ESCAP, p. 122; Tzanakis, 2019, p. 20):

- **Independence of regulators**: A regulator that is independent from government, train operators, infrastructure providers, and corporate and individual customers.
- **Judicial review**: Determine necessary reporting and audit requirements.
- **The scope and jurisdiction of regulator**: Define the scope of works.
- **Number of regulators and their appointment procedure**: Decide whether regulatory powers will be given to a single executive or in a board.
- **Accountability of regulator**: Work streams of regulators need to be open and transparent.
- **Transparency and consistency of regulatory rules**: Stakeholders should have knowledge and a common understanding about regulatory principles, and regulators need to be consistent in the implementation of these rules.

Designation of regulations with regard to the industry structure is crucial since this determines what needs be regulated (namely maintaining fair competition among rail operators, setting appropriate access charges and prices to end-users) (Transport and ICT, 2017, p. 132). To fulfill these needs, there are four types of regulation, which are connected to each other; these are summarized in Table 2 (Tzanakis, 2019, p. 28; Transport and ICT, 2017, p. 135; 140–142).

Table 2 Main objectives of each regulation

Type of regulation	Duties
Economic	• Determine of access charges • Maintain infrastructure investment • Prevent discriminatory access • Enhance competition • Regulate tariffs
Safety	• Establish safety standards • Supervise stakeholders • Give necessary permissions/authorizations
Environmental	• Issue national legislation for environmental protection • Undertake environmental impact analysis
Technical	• Set and supervise safety, environmental, and operational standards

Source (Tzanakis, 2019, p. 28; Transport and ICT, 2017, p. 135; 14–142)

3.2 *Current Competition Policy Concerns in Railways*

The introduction of competition to railways is worthy of discussion since designing railways solely from a sectoral regulatory perspective would not meet the needs of different stakeholders (Finger et al., 2017, p. 1). Initially, regulatory mechanisms entail changes in the role of regulator overtime to foster competition and market mechanisms whenever possible, while simultaneously providing a stable legal and institutional framework for economic activity (UN ESCAP, p. 123). Later, rather than seeing competition purely as a sectoral regulatory aim, other relevant competition dimensions become apparent (such as tendering of public services obligation contracts, state aid issues), and they are more difficult to approach from a purely sectoral point of view (Finger et al., 2017, p. 1).

Type of competition is another determinant factor. In inter-module competition, where railways already face competition from other modes of transportation, cooperation between regulators and competition authorities is crucial to establish competitive markets that will diminish the need to regulate tariffs. As part of this cooperation, the regulator should monitor the level of competition in the transportation sector and may intervene actively to promote competition, sometimes in cooperation with the competition authority (Tzanakis, 2019, 29). In the case of market competition, third-party access to railway infrastructure can lead to lower prices, increased innovation, and the development of new markets due to competition among different rail operators. However, in this scenario, state-owned incumbent railways have usually complained that new market entrants compete only for the most profitable markets ("cream skimming"), leaving the incumbent to serve the least profitable markets, which it may be under an obligation to serve (Tzanakis, 2019, 29; Finger et al., 2017, p. 4). The results of inefficient implementation of competition in railways are (Tzanakis, 2019, 29):

- Reduced investment in infrastructure
- Increased need for government subsidies
- Continuing financial losses
- Closure of loss-making services.

The regulator has a duty to prevent cream skimming, and maintain competition[12] that is fair between market players and of benefit to customers.[13] Finding a good balance between open access competition and competitively tendered services is important. Regulators should consider the difficulty for commercial operators to compete against services that are subsidized (Finger et al., 2017, p. 4). A duopoly model for the transition period with efficient implementation of a regulatory framework has been said to allow the incumbent rail operator to adapt to the new market structure by controlling its costs and ending internal cross-subsidies (Montero et al., 2016, p. 3). However, the effect of competition on total welfare and the issue of how to regulate the market and conduct capacity allocation to maximize welfare remains unclear (Broman & Eliasson, 2016, p. 232). The economics literature suggests that competition improves welfare if it generates enough new passengers. Otherwise, price competition with homogenous services causes a price war (Montero & Melero, 2020, p. 10).

Country examples indicate that railways have continued to have a monopoly situation, sometimes complemented by smaller niche actors (Broman & Eliasson, 2016, p. 233). Mainly, the routes with the highest usage and best infrastructure attract the interest of newcomers. Consequently, there is growing concern that efficient widespread competition might not emerge in railways as it has emerged in other liberalized network industries (Montero, 2019, p. 185).

4 Turkey's Experience of Railway Regulation

The history of the railways, within the current national borders of Turkey, dates back to 1856. After the railway-dominant period that continued until the beginning of the 1950s, Turkey experienced a sharp decline in railway transportation, with huge financial losses that prevented the government from maintaining necessary investments in railways and providing efficient transportation services. Hence, Turkey inevitably followed global trends in railway liberalization policies and started to discuss alternative models of deregulation.

As Turkey proceeded in its negotiations with the EU for membership, legal harmonization of railways was one of the crucial targets on the agenda in the 2000s. Hence, the process of harmonization with the rail policy of the EU was one of the driving forces that accelerated the deregulation process. The Act regarding the Liberalization of Railway Transportation No. 6461 (Law No. 6461) entered into force on May 1, 2013, and it anticipated a gradual transition period for some railway activities until December 2018.

This section briefly examines the history of Turkish railways, the need for reform, the crucial events during the liberalization process, and the current situation after a few years of deregulation experience.

4.1 Historical Background of Turkish Railways

The Ottoman Empire implemented a railway construction project in the 1850s (Kulat, 2018, p. 4). Although the Empire considered railways as a solution to transportation problems in the early years, later railways turned out to be the critical means of achieving economic, military, and political goals (Yıldırım, 2002, p. 316). However, since the Empire did not have sufficient financial and technical resources for railway construction, it granted privileges and concessions to British, German, and French companies to build and expand railway network (Engin, 1993, p. 43–44; Shakibaei & Alpkokin, 2017, p. 1303) and carried out railway management of only a few lines (Rota, 2009).

Almost half[14] of the railway tracks that were built during the Ottoman era were left within Turkish national borders after the establishment of the Republic of Turkey in 1923 (TCDD Taşımacılık A.Ş., 2017, p. 28). To take over the operations of railways, a Turkish state company called the "General Directorate of Anatolian Baghdad Railways"[15] was founded in 1924 and renamed as the "General Administration of State Railways and Ports"[16] in 1927[17] (TCDD Taşımacılık A.Ş., 2017b, p. 28; Arıak, 2009, p. 31).

The years between 1923 and 1940 are seen as an "expansion period" (Arıak, 2009, p. 31) or a "railway-dominant period" (Shakibaei & Alpkokin, 2017, p. 1303) in which the Republic of Turkey purchased and nationalized all railways that were operated by foreign companies (TCDD Taşımacılık A.Ş., 2017b, p. 28; Arıak, 2009, p. 31). Turkey invested in railways and built an average of 200 km every year (Servantie, 2015, p. 8), with the main aim of creating an indispensable component of a self-sufficient economy[18] (Shakibaei & Alpkokin, 2017, p. 1303).

In the 1950s, Turkey—like the rest of the world—changed its transportation priorities from railways to road and "the golden age of road transportation" started in the country,[19] lasting until the beginning of the 2000s (Arıak, 2009, p. 31). Turkey slowed down its railway investment and built only 945 km of railways between 1951 and 2003 (Ministry of Transportation and Infrastructure, 2018, p. 226). The only exception to the road-intensive transportation policies in that period was "1983–1993 Transportation Interim Plan," which aimed to decrease the share of roads in transportation (Arıak, 2009, p. 32).

The diminishing importance of railways continued until the beginning of the 2000s. With the goal of integrating Turkey into the EU, the government allocated more funds and accelerated investments in railways (Shakibaei & Alpkokin, 2017, p. 1303). Turkey used the EU's legal framework as a base for its own regulatory reforms and shaped its own policy through development plans. It set the target of providing competitive, high-quality, and sustainable railway services for consumers at affordable prices, and gradually liberalized railways. Investment projects were activated and 1983 km of railways were built between 2004 and 2018, at an average of 138 km per year, and the construction of a further 4015 km of railway is planned (Ministry of Transportation and Infrastructure, 2018, p. 226).

4.2 *The Reasons for Reform in Turkish Railways*

The Ministry of Transportation governed intercity rail transport with three governmental bodies as shown in Table 3.

TCDD played a crucial role in railway operations[20] and acted as a legal vertically integrated monopoly for the management of infrastructure and provision of railway services (Arıak, 2009, p. 33). TCDD used conventional trains for passenger transportation until 2009 (Solak, 2018, p. 14) and rail lines, which did not cover some large cities of Turkey, did not constitute a suitable network for efficient transportation, especially for freight activities (Arıak, 2009, p. 33). On the other hand, Turkey

Table 3 Responsibilities of governmental bodies in the pre-deregulation period

Governmental bodies	Responsibility
General Directorate of railways, harbors, and airports construction	Infrastructure investments of transportation modes
General Directorate of railroad transportation	Planning of railway transportation in accordance with economical, technical, social, and national security policies
TCDD	Operating, expanding, and renovating the railways and ports

Source (Karamanoğlu, 2012, p. 56–57)

had a relatively extensive road network of approximately 427,000 km, with a large number of private trucking companies (Arıak, 2009, p. 36).

Although transportation market for both passenger and freight traffic grew substantially in Turkey, railways continued to lose market share against road transportation. The share of railways in freight transportation decreased from 68.2 to 5.4% between 1950 and 2000, while the share of railways in passenger traffic decreased from 42.2 to 2.2% (Shakibaei & Alpkokin, 2017, p. 1303). Prioritization of road-intensive transportation policies, rather than the modernization and enhancement of railways since the 1950s, has led to the railway infrastructure becoming outdated and hence, causing deficiencies in both the efficiency and profitability of railways. The primary reasons why the railways could not meet the needs of customers, and hence the sharp decline in the share of railways, are listed below (Arıak, 2009, p. 33–34):

- Relatively aged portfolio of locomotives and rolling stock[21]
- Insufficient length of tracks
- Lack of sufficient double lines
- Limited length of electrified lines
- Problems related to gradients, axle load, and sleeper types.

Under these circumstances, modernization of infrastructure was necessary to improve the quality of railways, to meet the expectations of customers, and to increase the efficiency and profitability of the railways. Besides, the government had a duty to provide railways to its citizens at affordable prices with an optimal level of service quality and variety as a public good (OECD, 2005, p. 27). Furthermore, the public provision of railways has importance in social policies like regional development and employment opportunities (Karamanoğlu, 2012, p. 12).

Turkey initially rehabilitated rail tracks to meet customer expectations and accelerate the railways' modernization process. The first project, which was the rehabilitation of Ankara–İstanbul passenger lines for high-speed trains, was introduced in 1996 and started operations in 2004. Turkey benefitted from Spanish government funds for that project, which aimed to reduce travel time between two important cities of Turkey and constituted a reasonable alternative to road transport. However, a severe accident on that track shifted the government policy toward building a brand new track compatible with high-speed trains instead of rehabilitation policies (Aslan,

2012, p. 29). Accordingly, Turkey started constructing a double-track high-speed line between Ankara and Istanbul in 2005. The first phase of the project that connects Ankara to Eskişehir was opened for passenger transportation in March 2009, which changed the competitive landscape on behalf of railways against road transportation. Furthermore, another high-speed line connecting Ankara to Konya was also started its passenger operations in 2011 (Aslan, 2012, p. 30). Even though there have been improvements in passenger operations, railways still did not become a prominent and stable alternative to road transportation. Besides, building new tracks with expensive technology required higher budgets that increased the long-lasting financial losses of TCDD.

TCDD's losses increased from €197.2 million in 2000 to €508.1 million in 2013 (Solak, 2018, p. 16; TCDD Statistics, 2014, p. 91). Although profit maximization has never been the main target of public railways, operational revenues are indispensable for accomplishing social policies while carrying out railway operations. The increasing subsidy needs of TCDD, the shortage of resources for new investments, and the decreasing market share of railways against other transportation modes forced public bodies to look for structural policy changes (Kabasakal & Solak, 2009, p. 27; Kaynak, 2002, p. 28; Özcan, 2006, p. 1058).

In addition, accession negotiations with the EU accelerated in the first years of the millennium and Turkey decided to focus on the deregulation process to accomplish the requirements of EU rail policy. The EU and World Bank provided technical assistance for the liberalization of the Turkish railways with the aim of EU integration and decreasing the financial losses of TCDD (UNECE, 2018 as cited in Koning, 2012). Accordingly, Turkey stated that its priorities for railways were to increase the share of railways in freight transportation and to invest in some essential railway connections in the ninth Development Plan (Arıak, 2009, p. 46).

4.3 The Liberalization Period

Until the 2000s, TCDD had an active role in the network infrastructure, infrastructure management, and rail operations in a vertically integrated structure with high levels of financial losses (Shakibaei & Alpkokin, 2017, p. 1304, as cited in Knieps, 2004). As the railway network would continue to consume a large part of future investments for maintenance and modernization without providing the necessary revenues itself or becoming efficient, Turkey needed to choose an appropriate deregulation scheme for the liberalization of its railways.[22]

One of the concerns during the deregulation period was the integration process with the EU, and Turkey initiated accession negotiations with the EU in 2005 on 33 chapters, including transportation policy. Against this background, the Ministry of Transport accelerated the works for the restructuring of the Turkish State railways to achieve the liberalization of the railway market in Turkey according to Europe's experience after the European Commission's legislation packages aimed to introduce competition in the railway market and improve efficiency. Although there have been general improvements and investments regarding the railway infrastructure and railways since 1995 to increase rail transportation efficiency, it has been difficult to examine any significant structural change. Therefore, according to the European Commission's progress report on Turkey in 2010, railways were the only transport sector in which no remarkable progress was observed. The 2012 progress report also mentioned the failure to adopt "a more comprehensive railway law" as one of the obstacles regarding railway liberalization policies (Aslan, 2012, p. 35).

Accordingly, Turkey inevitably started to examine the experience of EU countries during its liberalization efforts (UNECE, 2018, p. vi), which requires having separate infrastructure and operations in railways, at least in terms of accounting with nondiscriminatory infrastructure charges and slot allocation. Accordingly, vertical separation became a prominent strategy for the deregulation of railways (Karamanoğlu, 2012, p. 13–14; UNECE, 2018, p. 5). Turkey adopted the German model of vertical separation, which aims to provide infrastructure services and railway operations through two separate legal entities affiliated to a holding company (Solak, 2018, p. 8). In other words, Turkey opened up the market for competition in both freight and passenger transportation services but did not implement horizontal separation as Great Britain had done (Aslan, 2012, p. 35).

The legal foundation of the deregulation was Law No. 6461,[23, 24] which entered into force on May 1, 2013, replacing the state monopoly in railways. According to Law No. 6461, railways were divided into infrastructure management and railway management. TCDD was restructured into an infrastructure provider and operator under the Ministry of Transportation, while a new publicly owned company was established under the name of TCDD Taşımacılık A.Ş. as a train operator.

One of the main aims of Law No. 6461[25] was to restructure Turkish railways and strengthen its role in the transportation system to create competitive and economically and socially sustainable railways that are compatible with EU legislation. The enactment of the liberalization law did not mean the privatization of the railway sector (Karakülah, 2018, p. 165–166) but instead meant the enhancement of the competitive market landscape. In this regard, the goal is for the private sector to provide freight and passenger transportation services as well as railway infrastructure.

Although Law No. 6461 was enacted in 2013, its implementation and the adoption of secondary legislation have actually been spread over time. Moreover, a gradual transition period was foreseen for some railway activities, which covered a five-year period starting from the coming into force of Law No. 6461, to financially support the PSOs and other activities of TCDD and TCDD Taşımacılık A.Ş. While TCDD received direct public support from the central budget, the losses of TCDD Taşımacılık A.Ş. were compensated through allocations received from TCDD via installments on the paid capital of the company. The main aim of state funding during that transition period was to support TCDD with its infrastructure expenditures, since market growth rates and transaction volumes would not be sufficient at the beginning of the deregulation period. The secondary legislation was largely completed as of 2018 (Solak, 2018, p. 18).

4.4 *Organizational and Regulatory Framework in the Turkish Railways*

Law No. 6461 discharges TCDD from its duties and responsibilities concerning railway operations. TCDD continues its activities as an infrastructure provider and operator under the Ministry of Transportation. TCDD[26] has also has the monopoly power to direct railway traffic in exchange for fees levied on the both public and private operating companies.

TCDD Taşımacılık started operations on January 1, 2017, and has executed all operational activities of railways that were previously under the responsibility of TCDD. As a requirement of the unbundling process, TCDD Taşımacılık has its own financial, legal, and human resources that are allocated to the operations of the railways (TCDD Taşımacılık A.Ş., 2017b, p. 29). TCDD Taşımacılık's organizational structure consists of three main departments, which are responsible for (i) logistics, (ii) passenger transportation, and (iii) vehicle maintenance, respectively (Solak, 2018, p. 20). TCDD Taşımacılık currently operates high-speed trains, conventional trains, intracity urban trains, as well as Marmaray[27] and Başkentray.[28] Due to the five-year period for obligatory financial support, the central government covered the budgetary deficits of both TCDD and TCDD Taşımacılık until December 2018 (Solak, 2018, p. 26). Since the transition period ended, only high-speed trains, construction of new lines and modernization, or renewal of existing rail lines have been publicly financed as a part of infrastructure investment (UNECE, 2018, p. 14).

Furthermore, regulatory and supervisory public bodies are structured to guarantee technical safety rules and competition between market players. The General Directorate of Railway Regulations acts as a national security authority for all railway operations (railway security, interoperability, licensing, and standards of vehicles). The Directorate coordinates the PSOs in addition to its duties in supervising competition in the market. It is responsible for the free, transparent, and nondiscriminatory functioning of the railways, and thus has the right to declare minimum and maximum prices (Solak, 2018, p. 19). The Railway Accident Investigation Board was established on November 2011 for examining major accidents. Figure 1 shows the current organizational structure of railway transportation in Turkey.

4.5 *Level of Competition and Market Figures in the Post-deregulation Period*

The deregulation period was described as an opening of railways to competition rather than the privatization of railways. In this respect, Law No. 6461 has given private entrepreneurs the opportunity to invest in railway operations. Involvement of the private sector in railway transportation services is crucial for the successful realization of the liberalization process. Three freight and three private passenger train operators and one agency have been authorized for railway operations (Ministry of

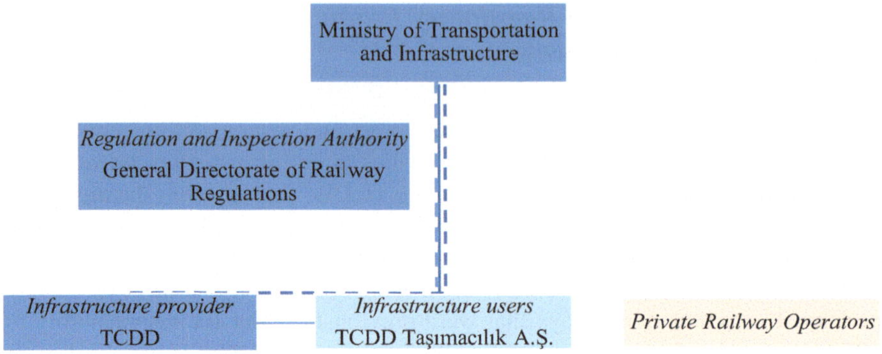

Fig. 1 Current organizational structure of Turkish railways. *Source* Report of Turkish Ministry of Transportation and Infrastructure—"Ulaşan ve Erişen Türkiye 2018", p. 234

Table 4 Number of railway coaches

Type of coaches	TCDD	TCDD Taşımacılık	İzban	Private	Total
Freight	–	16,609	–	3973	20,582
Passenger	–	990	–	–	990
Locomotive	–	669	–	–	669
High-speed trains	–	19	–	–	19
Multiple unit	–	210	24	–	234
Track overhaul car	319	–	–	5	324
Total	319	18,497	24	3,978	22,818

Source (Akbayır & Öztürk, 2018, p. 172)

Transportation and Infrastructure, 2018, p. 234). However, due to the need for large investments in railway operations, only two private companies currently have freight transportation services: Körfez Ulaştırma[29] and Omsan Lojistik,[30] which operate with 30 and 45 train operators, respectively (Akbayır & Öztürk, 2018, p. 171). For passenger transportation, private entrepreneurs have still not launched railway operations. Only two municipal companies, in İzmir and Kayseri, have operations in urban lines. Currently, only TCDD Taşımacılık operates intercity lines. To ensure freedom of movement in passenger transportation by train, it is essential to guarantee affordable prices for railway services for passengers. Therefore, Law No. 6461 allows the Ministry of Transportation to delegate these PSOs to public and private train operators. Currently, four high-speed lines, 15 main lines, and 29 regional lines are obliged to operate as a public service (Karakülah, 2018, p. 173). Table 4 shows the number of railway coaches that private entrepreneurs and TCDD Taşımacılık have in their rail operations.

A lack of investment and the need for further development of Turkish railways have been the most important motives behind the deregulation of the railways. Before deregulation, the national railways system was approximately 8500 km. Conventional lines increased by only 3.9% in the post-deregulation period and reached 9023 km in 2017 (TCDD Statistics, 2014, p. 12). However, a large increase occurred in the high-speed lines from 397 km in 2009 to 1213 km in 2017. Similar figures can be seen for the number of high-speed train passengers (TCDD Taşımacılık A.Ş., 2017a, p.12). While high-speed trains carried 4.2 million passengers in 2013, the number of passengers increased by 92.8% in 2018 and reached 8.1 million passengers (TCDD Taşımacılık A.Ş., 2017b, p. 47; Ministry of Transportation, 2019, p. 331).

Maintenance of the railway tracks increased from 1015 to 1276 km in 2014 but started to decrease rapidly between 2015 and 2018 (Ulaşan ve Erişen Türkiye, 2018, p. 229). Improvements in the railway tracks at the beginning of the post-deregulation period had a positive effect on the share of railways. While the share of railways in total transportation was steady at around 5% between 2006 and 2013, it increased to 7.1% in 2018 (10th Development Plan, 2013, p. 109). In 2018, railways carried 31.6 million tons of freight, an increase of 18.8% compared to 2013, and a 98.7% increase compared to 2003 (Ministry of Transportation, 2019, p. 328).

5 What to Expect

Railways have long acted as monolithic organizations that control their infrastructure, operations, and managerial functions while unilaterally determining what services to supply to an often-captive market, with little incentive for productivity gains (Crozet, 2016, p. 22; Moyer & Thompson, 1992, p. i). Nevertheless, as other transportation modes have continuously weakened the traditional market for railways, railway reforms are required to remedy to this situation (Crozet, 2016, p. 22). Reforms aims to achieve better services at lower prices and, hence, reach positive welfare effects compared to a vertically integrated public monopoly. Despite the fact that deregulation period is quite new for many countries, the outcome seems to be a monopoly situation, sometimes complemented by a few niche market players (Broman & Eliasson, 2016, p. 233). There continues to be debate on the issues listed below (Broman & Eliasson, 2016, p. 232; Crozet, 2016, p. 23; OECD Swiss, 2006, p. 24):

- How to reach on-rail competition that gives customers several options on a single rail line
- How two or more rail operators can achieve effective competition
- How to design mechanisms for eliminating incentives on the access rights of different operators.

Whether railways stay with a monopolistic structure or head toward an oligopolistic structure, regulators and competition authorities continue to have a prominent role. For the efficient functioning of railways, a particular regulation format needs to be created that considers competition concerns. For example, while designing the liberalization procedure, the winning railway operators could be subject to pre-determined commitments for a specified period in terms of the routes they will operate, their frequencies, timetables, stations, stops, etc. (Montero & Melero, 2020, p. 12). However, regulators should also avoid a long period of commitment that turns out to be an obstacle. Otherwise, it might risk the viability of these railway operators in the long-run (Montero & Melero, 2020, p. 12). Regarding competition concerns, it is essential to monitor closely concentrations in the market, the market shares of the various operators, and hence, help the survival of smaller operators in the case of an oligopolistic market structure (Broman & Eliasson, 2016, p. 233). Also, integrating competition rules and railway-specific regulation is usually more easily said than done (Finger et al., 2017, p. 2).

Turkey has experienced similar steps in the regulation of railways. After the rail-intensive era between 1923 and 1950, Turkey had difficulty ensuring essential investments for maintenance and development of the railways due to a high level of financial losses. Inefficiencies in railways with the presence of EU negotiations accelerated the transformation policies. Law No. 6461 entered into force in May 2018 after many years of discussion and preparation. As part of the liberalization of the railways, TCDD, which was a legal monopoly in railway operations and infrastructure under the supervision of the Ministry of Transport, was transformed into a public company that is responsible for the infrastructure of railways, and a new subsidiary was established for railway operations. Deregulation also allowed the involvement of the private sector in passenger and freight transportation. Liberalization aimed to sustain a certain level of competition in freight and passenger transportation operations on the railways and to make necessary investments for the maintenance and enhancement of railway lines.

Since the deregulation envisaged a gradual transitional period for opening up the market, with financial subsidies for TCDD Taşımacılık until the end of 2018, it is difficult to reach definitive conclusions on the level of competition in the market. Considering freight transportation, two private companies currently have operations only for their own freight, and they carried 17.47%[31] of total freights in 2019. However, Turkey still struggles with insufficient connections to production centers and insufficient port connections of the railway network. This situation limits the participation of private sector and prevents reaching the expected benefits of deregulation in freight transportation (Solak, 2018, p. 15). On the other hand, two private companies as newcomers rented their locomotives from TCDD Taşımacılık, and they benefited from the idle capacities, but this could create barriers of entry or increase costs for latecomers. For passenger[32] transportation, there has been almost no progress since deregulation. There are four main reasons for this situation: (i) inadequate public investments in the routes with highest usage or demand (such as İstanbul–Ankara and İzmir–Ankara); (ii) conflict of interest with other transportation modes in the same routes; (iii) deferment or reluctance in private sector involvement and investments due to uncertainty in economic conditions; and (iv) uncertainty in liberalization schedule (e.g., which lines will be privatized, when it is scheduled, and how tender process will design). Finally, public subsidies on behalf of a public company in a capital-intensive market and uncertainty over the requirements of these subsidies have the potential to reduce the investments of new entrants in both passenger and freight market. Although, according to the 11th Development Plan, railway lines with PSOs should be offered to private companies through public tenders (11th Development Plan, 2019, p. 125), it is difficult to say that the level of competition between public and private railway operators has been sufficient to achieve the expected benefits of liberalization in the market since the end of 2018.[33]

Moreover, although the current level of competition is far from a competitive market structure, there are already issues regarding the delegation of supervisory powers in competition issues. The General Directorate of Railway Regulations has the power to maintain a level of competition in the market to ensure a competitive market structure after deregulation. However, it is difficult to describe an independent regulatory landscape since the Directorate works as a branch of the Ministry of Transportation. The Turkish Competition Authority is also responsible for the implementation of competition legislation without any sector-specific restriction, including railways. Therefore, the debate over the clarification of roles of the Turkish Competition Authority and the General Directorate of Railway Regulations remains ongoing, as it is in many other countries. Accordingly, removing uncertainties in public subsidies and clarifying the supervisory authorities for competition issues could be the next steps toward the ultimate aim of creating efficient and cost-effective railways. Relatedly, it might be expected that Turkey could decrease its operational costs due to low traffic density since, as a general expectation, it is argued that vertical separation reduces costs as long as the traffic density is low (Aslan, 2012, p. 36 with reference to Mizutani & Uranishi, 2011). However, the current landscape has been far from achieving any cost savings concerning the ambiguity in regulators' roles. In other words, the effects of a strong regulator and well-designed incentives to improve capacity and efficiency seem to be the missing parts that are crucially important in reducing costs. Although a strong regulator with extended powers and a high degree of independence should be the priority, safety and market regulation will be conducted within the Ministry of Transport. There might be concerns about the independence of safety and regulatory authority that works under the Ministry of Transport and ensures nondiscrimination among all railway operators. Accordingly, a more independent market regulator seems to be a better option to provide well-functioning railways and to reach liberalization goals on efficiency.

Table 5 Recommendations for a sustainable and successful deregulation period

Category	Recommendations
Regulatory	• An independent regulator from the Ministry of Transportation could prevent political intervention • Maintaining the independence of the Railway Accident Investigation Board has been an important step in sustaining the confidence of citizens regarding the prevention of accidents on railways • Active role of Turkish Competition Authority to accelerate liberalization in railways (e.g., sector inquiries)
Policy design	• Policies should consider interoperability between rail infrastructure and train operations while designing tenders, investment alternatives and train operations • Strategic planning in allocation of freight transportation (e.g., increasing share of private operators by enhancing
Operational activities	• It is vital to take effective action for the construction of the secondary lines of boarding points, especially for freight transportation (such as outsourcing the project to the private sector in return for a deduction from its service fee, as happened during the liberalization of the electricity market) • It is essential to invest infrastructure to connect railways to seaports and logistic centers • Conversion of the main routes to double-tracks instead of single-track lines, along with electrification and signaling • Consolidation of dispersed logistics centers and activation of current logistic centers as soon as possible
Finance	• Since it is difficult for railways to self-finance and governments continue to subsidize both infrastructure and train operations, faster and more predictable public finance could accelerate the private sector investments in railways • Consultation with banks and financial institutions in terms of financing private sector investments
Competition	• Development of active cooperation between sector-specific regulators and the competition authority is crucial to accelerate the effectiveness of liberalization process • As the liberalization proceeds, digitalization started to affect the competition among different transportation modes. In this regard, antitrust issues in railways need to consider the developments in adjacent markets or alternative transportation modes (such as long-distance bus and long-distance car-sharing markets)

Table 5 lists recommendations[34] that could be useful for achieving a successful deregulation period based on other countries' liberalization experience.

Notes

1. From the regulatory point of view, infrastructure can be regarded either as a natural monopoly or as any other competitive economic activity that could be provided by multiple operators or by a single firm under some sort of concession or license arrangement (Campos & Cantos, 1999, p. 6).

2. The policy goal of PSO is often supported by the idea that railways contribute less to the rise of negative externalities than other transportation modes. Externalities are concerned in the implication of socially optimum prices and the intermodal implication in terms of the economic consequences of regulation (Campos & Cantos, 1999, p. 7–8).
3. Economies of density exist when the total cost to transport units of freight from their points of departure to their intended destinations decreases by increasing the utilisation of the existing vehicle fleet and infrastructure capacity within a market area (Pienaar et al., 2012, p. 44).
4. For example, passenger services have a smaller share of passenger transport market in countries with lower population densities and more considerable distances between major urban cities (such as the USA, Canada, and Australia). By contrast, there is a higher market share for passenger services in Europe, where population densities are higher and roads tend to be more congested (OECD, 2005, p. 26).
5. There are four categories of railways costs: (i) train working costs, (ii) track and signalling costs, (iii) terminal and station costs, and (iv) administration costs (Campos & Cantos, 1999, p. 5 with reference to Waters, 1985).
6. It is important to make a clear distinction between avoidable costs and other costs for regulatory considerations (UN ESCAP, p. 99; Campos & Cantos, 1999, p. 4).
7. A cost function is sub-additive when the provision of services by a single firm is more efficient than the same production carried out by two or more companies (Campos & Cantos, 1999, p. 4, with reference to Baumol, 1977).
8. Due to the low rolling resistance of steel wheels on railways, railways were extremely fuel-efficient and relatively cheap. Some form of public control was obligatory in many countries for military and industrial reasons (Campos & Cantos, 1999, p. 7).
9. The achievement of economic growth and poverty reduction requires good physical access to resources and markets, while the quality of life is generally dependent on the quality of physical access to health services, education, etc. (UN ESCAP, p. 83–84).
10. For example, transportation costs of an additional unit of traffic may be insignificant when there is idle capacity. Conversely, transportation cost may be substantial when the capital is at the limit (Campos & Cantos, 1999, p. 6).
11. Instead of dramatic changes in current railway network, public authorities prefer partial renovations that often introduce technical asymmetries between tracks (Campos & Cantos, 1999, p. 6 with reference to Boyer 1997).
12. Since it is more challenging to develop competition for passenger services than for freight, regulations often varies for each (Tzanakis, 2019, 30).
13. The rail regulator in Great Britain has developed the Not Primarily Abstractive test to prevent new open access operators merely cherry picking franchised services. The EU has developed an Economic Equilibrium Test for the same concerns (Finger et al., 2017, p. 4).

14. 4136 km of the total 8619 km of railway tracks (TCDD Taşımacılık A.Ş., 2017a, p. 28).
15. The Directorate operated under the Ministry of Public Works (TCDD Taşımacılık A.Ş., 2017a, p. 28; Arıak, 2009, p. 31).
16. It was converted into a state economic enterprise under the name "Republic of Turkey General Directorate of State Railways Administration" (TCDD) in 1953 (TCDD Taşımacılık A.Ş., 2017, p. 28; Arıak, 2009, p. 31).
17. The Directorate has operated under the Ministry of Transportation since 1939 (TCDD Taşımacılık A.Ş., 2017, p. 28; Arıak, 2009, p. 31).
18. In 1950, the length of the railways was 3578, 3208 km of which had been completed by 1940 (Karamanoğlu, 2012, p. 54).
19. The main reasons for the shift are (i) an industrialization process based on agriculture and consumer goods, (ii) change in government policies on behalf of railroads under the impact of the Marshall Aid Plan, (iii) difficulties in the allocation of financial resources to capital-intensive railway investments, (iv) developments in automobile industry, (v) flexibility of road transportation, and (vi) consumers' preference for door-to-door transportation (Kabasakal & Solak, 2009, p. 28; Karamanoğlu, 2012, p. 54).
20. TCDD also managed ports (İzmir, Mersin, and İskenderun) and ferry lines in Van Lake (Solak, 2018, p. 14).
21. Approximately 62% of the stock of TCDD was 20 years or older (Arıak, 2009, p. 34).
22. EU railway liberalization evolved in three major areas: (i) the separation of infrastructure management and train operation services, (ii) market opening or liberalization of railway services, and (iii) enhancement of interoperability and technical harmonization to ensure an integrated rail network (UNECE, 2018, p. vi).
23. Published in the Official Gazette No. 28634 and dated 1 May 2013.
24. Even though Law No. 6461 is described as the first step in the deregulation period, the Governmental Decree concerning the Organization and Duties of the Ministry of Transport, Maritime Affairs and Communication No. 655 was actually the first step taken to achieve the liberalization of the railway sector.
25. Law No. 6461 covers railways only at the national level and does not include regulations regarding the activities of subways, trams and other intracity railway systems.
26. TCDD has seven regional directorates (Haydarpaşa, Ankara, İzmir, Sivas, Malatya, Adana, and Afyonkarahisar), (http://www.tcdd.gov.tr/content/67, (last access: 2.8.2020)) and one directorate of high-speed trains (YHT Bölge Müdürlüğü). TCDD has only one subsidiary, which is the railway operator company TCDD Taşımacılık A.Ş. (http://www.tcdd.gov.tr/content/34, (last access: 2.8.2020)). On March 4th, 2020, three subsidiaries of TCDD (TUVASAS—Türkiye Vagon Sanayi A.Ş., TUDEMSAS—Türkiye Demiryolları Makinaları Sanayi A.Ş. and TULOMSAS—Türkiye Lokomotif ve Motor Sanayi A.Ş.) were merged and Türkiye Raylı Sistem Araçları Sanayi A.Ş. (TÜRASAŞ) was established with a presidential decree (http://www.tcdd.gov.

tr/content/34 (last access: 2.8.2020)). TCDD also continues its seaport operations at the ports of Haydarpaşa and İzmir. In the previously privatized ports (İskenderun, Derince, Samsun, Mersin and Bandırma) TCDD has supervisory duties, http://www.tcdd.gov.tr/content/68, (Last access: 2.8.2020)).

27. Marmaray is the 76.6-km-long rail line in Istanbul, which comprises a rail tunnel under the Bosphorus strait.
28. Başkentray is the 39-km-long rail line in Ankara.
29. Körfez Ulaştırma, which is a wholly owned subsidiary of Tüpraş, carries fuel oil between the Kırıkkale and İzmit refineries of Tüpraş with five locomotives rented from TCDD Taşımacılık and 491 self-owned rolling stocks since 2017.
30. Omsan Lojistik carries freight transports of a cement company (Solak, 2018, p. 25) via 15 rented locomotives from TCDD Taşımacılık and 512 rolling stocks. It has railway operations between Sivas-İskenderun, İzmir/Aliağa-Kayseri, and Hatay/Payas–Kayseri.
31. Körfez Ulaştırma: 1.83 million ton (Tüpraş Annual Report, 2019, p. 36), Omsan Lojistik: 3.3 million ton (http://www.omsan.com/lojistik/hizmetlerimiz/demiryolu-tasimaciligi, Last access: 7.8.2020), TCDD: 24.2 million (2015–2019 Annual Report, p. 55).
32. Only two municipalities (İzmir and Ankara) have licenses for passenger transportation, but they have used them for intracity passenger transportation. In this regard, the market has not yet been opened up to competition.
33. The share of private railway operators in freight transport is expected to reach 40% in 2023 (Railways Engineers Association, Interview with the General Manager of TCDD Taşımacılık A.Ş. p. 29).
34. For some of these recommendations please consider Finger et al., 2017, p. 2–3.

References

Akbayır, Ö., & Öztürk, M. M. (2018, October). *Türkiye Demiryolu Sektöründe Serbestleşmedeki Mevcut Durum*. Paper presented at Anadolu Üniversitesi.

Arıak, A. (2009). *Deregulation of Turkish railways for freight transportation*. Master's Thesis, Bilkent University.

Aslan, Ö. (2012). *European railway reform: A study for Britain, Germany, France and the case of Turkey*. Master Thesis, Institute for Transport Studies The University of Leeds.

Baumol, W. J. (1977). On the proper cost tests for natural monopoly in a multiproduct industry. *American Economic Review, 65*(5), 810–822.

Boyer, K. D. (1997). *Principles of transportation economics*. Addison-Wesley.

Broman, E., & Eliasson, J. (2016). Market dynamics in on-rail competition. 19th Europe working group on transportation meeting, 5–7 September 2016, Istanbul Turkey. *Transportation Research Procedia, 22*, 232–244.

Button, K. J. (1993). *Transport economics*. Edwar Elgar Publishing Limited.

Campos J., & Cantos, P. (1999). Rail transport regulation, prepared for the Economic Development Institute of the World Bank.

Cantos, P., & Campos, J. (2005). Recent changes in the global rail industry: Evaluating the new regulatory instruments. *European Transport/Transporti Europei, 29*, 22–45.

Crozet, Y. (2016, October). *Introducing competition in the European rail sector. Insight for a holistic regulatory assessment. Assessing regulatory changes in the transport sector*. Roundtable, ITF/OECD, Stockholm, Sweden.

Eleventh Development Plan (2019). *Presidency of Turkish Republic*. Department of Strategy and Budget for the period between 2019 and 2023, Ankara.

Engin, V. (1993). *Rumeli Demiryolları*. Eren Yayıncılık, İstanbul.

Finger, M., Bert, N., Bouchard, K., & Kupfer, D. (2017, Spetember 4). *Competition in passenger railways in Europe*. European University Institute, Robert Schuman Centre for Advanced Studies, Policy Brief, Issue: 2017.

Governmental Decree Concerning the Organization and Duties of the Ministry of Transport. *Maritime affairs and communication No. 655 published in the Official Gazette No. 28102* and dated November 1, 2011.

Kabasakal, A., & Solak, A. O. (2009). Demiryolu Sektörünün Rekabete Açılması. *Dumlupınar Üniversitesi Sosyal Bilimler Dergisi, 24*, 27–34.

Karakülah, T. (2018). *Avrupa Birliği Demiryolları Sektöründe Kamu Hizmet Yükümlülüğü ve Türkiye Durum Analizi*. Avrupa Birliği Uzmanlığı Tezi. Ulaştırma, Denizcilik ve Haberleşme Bakanlığı, Dış İlişkiler ve Avrupa Birliği Genel Müdürlüğü, Ankara.

Karamanoğlu, C. (2012). *Demiryolu Sektöründe Yapısal Reformlar ve Rekabet: Serbestleşme Öncesi Türkiye için Öneriler. Rekabet Kurumu Uzmanlık Tezleri*, Ankara.

Kaynak, M. (2002). Yeni Demiryolu Çağı Yüksek Hızlı Trenler ve Türkiye. *Ekonomik Yaklaşım, 13*(42–43), 23–53.

Knieps, G. (2004). *Privatisation of network industries in Germany: A disagreed approach*. CESifo Working Papers, No. 1188.

Koning, P. (2012). UK rail reform, an insider's perspective. *AECOM*.

Kulat, M. V. (2018). *Türkiye'de Demiryollarının Serbestleştirilmesi*. Ph.D. Thesis, Yeditepe University, İstanbul.

Law No. 6461. *The act regarding the liberalization of railway transportation published in the Official Gazette No. 2863* and dated May 1, 2013.

Ministry of Transportation and Infrastructure. (2018). *Report: Ulaşan ve Erişen Türkiye 2018*, 219–354.

Ministry of Transportation and Infrastructure. (2019). *Report: Ulaşan ve Erişen Türkiye 2019*, 223–385.

Mizutani, F., & Uranishi, S. (2011). Does vertical separation reduce cost? An empirical analysis of the rail industry in OECD countries. *Journal of Regulatory Economics*.

Montero, J. J. (2019). Asymmetric regulation for competition in European railways. *Competition and Regulation in Network Industries, 20*(2), 184–201.

Montero, J., Ramos, R., & Giuricin, A. (2016). Open with care: The duopoly model for the transition to competition in long-distance passenger railway transportation. *Competition and Regulation in Network Industries, 17*, 3–4.

Montero, J. J., & Melero, R. R. (2020*). Competitive tendering for rail track capacity: The liberalization of rail services in Spain*. European Institute Working Paper RSCAS 2020/27.

Moyer, N. E., & Thompson, L. S. (1992). *Options for reshaping the railway*. Infrastructure and Urban Development Department of the World Bank. Policy Research Working Papers (Transport).

OECD. (2005). *Policy roundtables, structural reform in the rail industry, No. DAF/COMP* (2005) 46.

OECD. (2006). *OECD Review of regulatory reform in Switzerland*. Regulatory authorities for air transport, railways, telecommunications and postal services.

Official website of Körfez Ulaştırma A.Ş. Retrieved September 3, 2020, from http://korfezulastirma.com/

Official website of Omsan Lojistik. Retrieved September 3, 2020, from http://www.omsan.com/lojistik/hizmetlerimiz/demiryolu-tasimaciligi

Official website of TCDD. Retrieved September 3, 2020, from http://www.tcdd.gov.tr

Ordover, J., & Pittman, R. (1994).*Restructuring the railway for competition*. Antimonopoly Law Handbook.
Özcan, İ. Ç. (2006). Demiryolu Ulaştırmasında Kamu-Özel İşbirliği Modeli ve Türk Tecrübesi. *Uluslararası Demiryolu Sempozyumu Bildiriler Kitabı* (Vol. 2). Ankara.
Pienaar, W. J., de Bod, A., & Havenga, J. H. (2012). Economies of density in rail freight transport: Potential for utilisation in Southern Africa. *Journal of Governance and Regulation, 1*(3), 44–49.
Rota, Y. (2009). Türk Demiryolculuğunun Oluşumu: Behiç (Erkin) Bey – 1. *Transport.*
Servantie, D. (2015). *İktisadi Kalkınma Vakfı Değerlendirme Notu: AB ve Türkiye Demiryolu Politikalarının Karşılaştırmalı Analizi.*
Shakibaei, S., & Alpkokin, P. (2017). Deregulation of Turkish state railways based on public-private partnership approaches. *World Academy of Science, Engineering and Technology International Journal of Transport and Vehicle Engineering, 11*(9), 1303–1309.
Solak, A. O. (2018). Türkiye Demiryollarında Serbestleşme. *Rekabet Dergisi*, 4–37.
Staebe, E. (2011). *Railways between sector specific and competition regulation—A view from Deutsche Bahn*. 3rd European rail transport regulation forum, 5 December 2011, Florence, Italy.
TCDD Taşımacılık A.Ş. (2017a). *Annual report*. Retrieved May 11, 2020 from http://www.tcddtasimacilik.gov.tr/files/3/Strateji/Faaliyet_Raporlari/faaliyet_raporu_2017.pdf
TCDD Taşımacılık A.Ş. (2017b). *Statistics report 2017*. Retrieved from May 11, 2020, from http://www.tcddtasimacilik.gov.tr/uploads/tcdd-t-2017-yillik-baski-final.pdf
TCDD. (2014). *2009–2013 Annual statistics*. Retrieved May 11, 2020, from http://www.tcdd.gov.tr/files/istatistik/20092013yillik.pdf
Tenth Annual Development Plan. (2013). *Ministry of development for the period between 2014 and 2018*, Ankara.
Transport and ICT. (2017).*Railway reform: A toolkit for improving rail sector performance*. Washington, DC: World Bank, License: Creative Commons Attribution CC by 3.0.
Tüpraş. (2019). *Annual report*. Retrieved from August 7, 2020, from https://tprstaticfilessa.blob.core.windows.net/assets/uploads/faaliyet-raporlari/TUPRAS_YILLIK_2019.pdf
Tzanakis, K. (2019). The Regulation of the Railway Sector "Targeting an Optimal Level of Service Quality and Efficiency".
United Nations, ESCAP (2001). *Regulation in transport: The practice*. Bangkok, Thailand.
United Nations, UNECE. (2018). *Railway reform in the ECE region*. New York and Geneva.
Warren, M. J. (2019). *Rail transport 2020*. Lexology, Law Business Research Ltd, London.
Waters, W. G. (1985). Rail cost analysis. In K. J. Button, & D. E. Pittfield (Eds.), *International railway economics*. Gower, Aldershot.
Yıldırım, İ. (2002). Osmanlı Demiryolu Politikasına Bir Bakış. *Fırat Üniversitesi Sosyal Bilimler Dergisi, 12*(1), 311–324.

Şahin Ardiyok is a partner at Balcıoğlu Selçuk Ardıyok Keki Attorney Partnership and the head of Competition Law, Public Policy, and Regulation Team. He represents clients, government institutions, NGOs, and multinational companies in antitrust investigations and merger/acquisition notifications, regulatory authority filings, appeal cases, antidumping investigations (across Turkey, the EU, the USA, and Canada), and PPP projects. Before joining in the private practice, he worked at the Turkish Competition Authority for five years and conducted investigations in sectors including telecommunications, media and broadcasting, information technology and electronic appliances. He graduated from Business Administration Department of Turkish Army Academy in 1991 and Faculty of Law of Ankara University in 1997. He also holds an MBA from the Ankara University and an LL.M. from the University of Chicago. Previously, he taught the courses on "Telecommunications Law" and "Pharmaceutical Law" at Istanbul Bilgi University and "Economic Regulation and the Law" (in English) and "Energy Law and Policy" at Bilkent University which is located in Ankara. Moreover, he is the advisory board member of the Competition Law Center of Istanbul Bilgi University, and he instructs various certificate programs at various universities, Bar Associations, and other institutions. Recognized both nationally and internationally for his competition law and regulatory practices, he is an active delegate of the International Chamber of Commerce (ICC) National Committee in the ICC's Competition Commission and serves as an NGA for the various working groups of International Competition Network (ICN). He frequently speaks at conferences and symposia on competition law and regulated industries; he frequently writes on competition law, regulation and law and economics. His latest publications include a book on Regulation Law that provides an end-to-end coverage of all aspects in the field of regulatory practice. Legal500 and Chambers & Partners consistently rank him as a leading individual in the field of competition and compliance.

Evren Sesli is working in the area of competition law and economics since 2005. She worked at the Turkish Competition Authority for thirteen years as a chief case handler and an economic analyst. Along with her professional activities at the Turkish Competition Authority, she presented Turkey's contributions at the OECD, International Competition Network and Association of Competition Economists. She got her undergraduate degree from the Economics Department of Bogazici University (İstanbul). She achieved academic excellence during her graduate studies in Economics and Social Sciences at the Bocconi University (Milan), where she graduated with her thesis on "Outstanding Competition Policy Concerns in Online Advertising." Since September 2017, she works as a Counsel in BASEAK. She provides consultancy and advice on competition law to national and international companies operating in different sectors.

Regulation of Motorways and Main Land Roads

Faruk Cirit and Volkan Recai Cetin

Abstract Over the last two decades, transport sector infrastructure investments in Turkey have increased significantly, particularly in the road sub-sector, and the country has also witnessed a crucial paradigm shift in road transport sector. The Turkish Government made medium- and long-term plans to upgrade the quality of the national road network by building almost 20,000 km of dual-carriage roads. In addition, while new road investments and maintenance of existing roads were financed through the national budget, over the last 10 years the government has utilized the public-private partnership model as a means of creating an alternative financing method for major motorway projects, by which around 750 kms of motorway projects were completed and another 600 kms are under construction. These developments have led to a rapid increase in freight and passenger transport volumes. In parallel, the government put into effect new regulations to protect road infrastructure and increase traffic safety and security on the road network and improve the overall efficiency of the transport system. In this chapter, with reference to the legal structure and institutional responsibilities, the main regulations implemented for motorways and the main land roads in Turkey are discussed, and some regulation recommendations are then proposed to improve the quality and efficiency of transport service provision.

Keywords Motorways · State roads · Road transport · Passenger transport · Freight transport · Transport regulation

The views and opinions expressed in this chapter are those of the authors and do not necessarily reflect the views of the Presidency of Turkey Strategy and Budgeting Office.

F. Cirit (✉) · V. R. Cetin
Presidency of Turkey, Strategy and Budgeting Office, Ankara, Turkey
e-mail: faruk.cirit@sbb.gov.tr

V. R. Cetin
e-mail: volkan.cetin@sbb.gov.tr

M. Eroğlu and M. Finger (eds.), *The Regulation of Turkish Network Industries*,
https://doi.org/10.1007/978-3-030-81720-6_13

1 Introduction

Roads perform an important function as they lie at the intersection of many sectors and social and economic activities. Road transport is composed of the interaction between users, motor vehicles, transport services, and road infrastructure, such as urban roads, main land roads,[1] and motorways.

Since road transport has door-to-door accessibility advantage and has been supported politically for more than half a century, road transport dominates Turkey's transport system to such an extent that its modal share is around 88.5% of both passenger and freight transport. However, rail and air transport have been given more importance in recent years, increasing their share in the transport system and making the system more efficient and sustainable.

Road transport services are categorized as passenger and freight transport services, which have different characteristics in terms of competition. As in any other network industry, economies of scale and scope are critical factors affecting the level and scope of competition and regulations in road transport as well. When the demand for timeliness is high and volumes of traffic between any two points are relatively low compared to the efficient size of a transport vehicle, economies of scale and scope are said to be stronger (OECD, 2000).

Passenger transport can be categorized into three parts: urban, local, and intercity passenger transport. Urban passenger transport is regulated by local authorities and is not covered in this chapter. Local passenger transport, which stands for transport activities between districts, is subject to certain limitations defined by each local authority on the basis of quotas and terms of service. Intercity passenger transport is open to competition, although the Ministry of Transport and Infrastructure (MoTI) may determine floors and ceilings for passenger transport tariffs according to economic and social conditions.

In intercity passenger transport, timeliness is very important and the volume of traffic is relatively low compared to size of a bus. Therefore, economies of scale are important in passenger transport, especially in intercity domestic market (OECD, 2000). Nevertheless, liberalization of intercity bus services has generally been successful throughout the world, although there is a certain level of risk in terms of concentration.

On the other hand, the freight market can be categorized as domestic and international in terms of operational coverage. While domestic freight transport in Turkey is open to competition, international freight is subject to restrictive bilateral and multilateral agreements, especially in terms of quotas, quantities, and service standards.

In the full-truckload segment of freight transport, timeliness is relatively unimportant and the volumes of freight are relatively large compared to truck size, and economies of scale are not as significant as in the time sensitive and/or less than truckload segment, including express packages delivery services (OECD, 2000).

---

[1] In Turkey, main land roads are classified as state and provincial roads. In this chapter, these two road types are combined as state roads.

Unlike vertically integrated railway industry or airway industry, in road transport, infrastructure and motor vehicle ownership are totally separated. Therefore, financing of infrastructure in road transport differs significantly from rail and air industries. In addition, since road transport has historically been open to competition, regulations are generally concentrated on technical aspects, operational safety, protection of environment and users, traffic management, and infrastructure financing.

In this chapter, the main objective is to give an overall idea of how road transport evolved and how it is regulated in terms of management, planning, investments, organizational duties, and responsibilities. With this in mind, the remainder of this chapter is organized as follows.

Section 2 summarizes the background of road transport in Turkey, excluding the urban road network. Section 3 will explain the regulatory framework of state roads and motorways in Turkey, both in terms of passenger and freight transport perspectives. Finally, the last part is devoted to expectations about the future regulatory requirements in a constantly evolving economic, social, and political environment.

2 Background of Road Transport in Turkey

This part summarizes the evolution of the road transport sector in Turkey with reference to the main policies, important dates, institutional structure, and statistical data.

2.1 Road Transport Policies Between 1923 and 1960: From Rail to Road

In the first decades of the Turkish Republic, which was founded in 1923, importance was given to railway infrastructure investments. Since the Republic took over railway lines constructed during the days of the Ottoman Empire, the most developed means of land transport was the railway transport.

In the two Five-Year Industrialization Plans prepared in 1932 and 1936, priority was given to iron-steel, coal, and machinery industries and to have economic transport options for the products of these industries; therefore, railway sector investments dominated transport investments (Turkish Chamber of Civil Engineers, 2006). On the other hand, although the Roads and Bridges Institute was founded in 1929 as the first institutional body responsible for managing and maintaining road transport infrastructure, the road network grew slower than the railway network (DGH, n.d.-a). In 1923, railway and road network lengths were 1378 and 18,335 km, respectively. From 1923 to 1940, while the annual growth rate of railways was 10%, that of roads' was just 4.9%. Therefore, in 1940, railway network length reached 6947 km, whereas the road network increased to 41,582 km (Turkish State Railways [TSR], (n.d.); DGH, personal communication, July 21, 2020).

After World War II, road transport investments gained importance and priority over the railway sector. In 1947, a group of experts led by H. G. Hilts, who was the deputy director of US Federal Highway Administration, conducted technical studies regarding the need for road network improvements. In 1948 they presented a report entitled "Status of Roads in Turkey", which recommended building 35,000 km of road network and the establishment of a new institution to manage and coordinate these investments (Emiroglu & Uzmay, 2013). Meanwhile, by the approval of the parliament on July 8, 1948, Turkey was included in the Marshall Plan (Guler, 2004), which would give Turkey an opportunity to benefit from financial aid. Constructing new roads and promoting road transport was among the main policies imposed by the Marshall Plan. Following the Hilts' report and the Marshall Plan, on October 8, 1948, the Council of Ministers passed a decree for a "Nine-Year Road Plan", which targeted the construction of 23,054 km of new roads. In addition, on February 11, 1950, the Directorate General of Highways (DG Highways [DGH]) was founded (Emiroglu & Uzmay, 2013).

Based on the Nine-Year Road Plan and the funds received under the Marshall Plan, emphasis was shifted from railway to road transport infrastructure. Between 1940 and 1960, the average annual growth rate of road network (2%) was triple that of the railway network (0.6%). By the end of 1960, while the railway network was 7895 km, the road network reached 61,542 km (TSR, n.d.; DGH, personal communication, July 21, 2020).

Regarding the modal split between rail and road transport, data reveals how fast the transformation process took place in the transport system. Until 1950, passenger transport was carried out by roads and railways, whereas freight transport was maintained by rail and maritime transport. For domestic passenger transport, the shares of the road, rail, and maritime sectors were 49.9%, 42.2%, and 7.5%, respectively. On the freight transport side, the corresponding figures were 17.1%, 55.1%, and 27.8% (Chamber of Mechanical Engineers, 2012). In 1960, however, the share of road transport in both passenger and freight transport went up rapidly, reaching 67% and 37% respectively (State Planning Organization [SPO], 1963).

As the supply in road infrastructure increased, so did the demand for motor vehicles leading to increased motor vehicle imports. In 1959, the first automotive factory (Otosan) was founded under the American–Turkish partnership. In the 1960s, other new factories were also founded to produce other types of vehicles. These developments also supported the dominance of road transport in Turkey.

2.2 *Road Transport Policies Between 1960 and 2000: From Road to Motorway*

Beginning in the 1960s, rather than constructing new roads, DG Highways focused on upgrading the standards of the existing road network and began implementing asphalt surface pavement works. From 1960 to 1970, the total length of paved roads rose by 172% and reached 19,226 km. In the same period, share of paved roads jumped from 11.5% to 32.3% (DGH, personal communication, July 21, 2020).

In the 1970s, in addition to upgrading the standards of existing roads, new types of road projects such as beltways and motorways came into the foreground due to the increased population, income level, and vehicle traffic in and between cities. In 1973, the first motorway project was implemented in Istanbul, including the construction of the first bridge on the Bosphorus Strait (DGH, n.d.-a).

Beginning in the 1980s, the construction of motorways was the main policy for the DG Highways, and several motorways projects began during this period. In 1988, the second motorway project in Istanbul was opened, including the second bridge over the Bosphorus (DGH, n.d.-b). Following these developments, motorway investments gained momentum in the 1990s.

In 1980, the total length of state roads was 60,761 km (DGH, personal communication, July 21, 2020), and the length of motorway network was just 24 km (Cetin, et al., 2011). However, with intensive investments, the motorway network grew to 241 km in 1990 and 1674 km in 2000 (Turkish Statistical Institute [TURKSTAT], n.d.). During the same period, the state road network only increased slightly, to 61,090 km. In addition, the share of paved roads (including motorways) in Turkey climbed from 59% in 1980 to 81% in 1990 and eventually to 92% in 2000 (DGH, personal communication, July 21, 2020).

In parallel with these developments, modal shares also kept changing dramatically in favor of road transport. From 1960 to 1985, the modal share of road transport rose from 67 to 95% in passenger transport, and from 37 to 80% in freight transport. This trend continued in the following years and the figures by 2000 were 96.0% and 90.2%, respectively (TSR, 2005).

2.3 *Road Transport Policies in the 2000s: Divided Roads and Public-Private Partnerships*

In the millennium age, another paradigm shift took place regarding road transport investments with the election of a new government in November 3, 2002.

In the "Urgent Action Plan" published on January 3, 2003, the government gave priority to upgrading existing state roads to divided roads with a target of reaching 15,000 km (SPO, 2003). The rationale behind this policy was stated as the lack of capacity, national and international network integrity, and traffic safety on the road network (DGH, 2019).

As of the end of 2002, the total length of divided roads (including motorways) was 6040 km and the target set by the Urgent Action Plan was achieved in 2009. However, the government decided to maintain this policy and network length reached 18,863 km in 2010, 23,107 km in 2015, and 27,301 km in July 2020 (TURKSTAT, n.d.; DGH, personal communication, July 21, 2020). DG Highways further plans to achieve 29,514 km by 2023 (DGH, 2019).

As a result of these developments, the share of road transport investments increased in total public investment expenditures. While that figure was 15% in 2002, it doubled in 8 years and reached 30% in 2010. During this period, although 483 km of new motorway sections were opened to service, the majority of the expenditures originated from construction of divided roads and upgrading existing state roads. At the end of 2010, the divided road network was 18,863 km, including 2197 km of motorways (SBO, personal communication, July 21, 2020; DGH, n.d.-c; DGH, personal communication, July 21, 2020).

Beginning in 2010, the government decided to benefit from public-private partnerships and to tender motorway projects under Build-Operate-Transfer (BOT) scheme, with the aims of developing the motorway network faster, not creating an additional burden on public investment budget, and benefit from the efficiency and operating capabilities of the private sector.

Between 2010 and 2020, five BOT motorway projects were tendered, with a total length of 1366 km. As of July 2020, two of these projects were completed (525 km), one has been partially completed (224 km completed, 186 km under construction), and two are under construction (431 km) (SBO, personal communication, July 20, 2020). Including the opened sections, the total length of motorways in operation was 3148 km in July 2020 and the target for 2023 is set as 3779 km (DGH, 2019).

In the last decade, in addition to intensive road investments, the government has also supported rail and air transport sectors to balance the dominance of road transport. By means of such policies, the modal share of road transport slightly decreased and, by the end of 2017 it realized as 88.9% for passenger transport and 88.4% in freight transport (SBO, 2019).

From the management perspective, all state roads as well as motorways realized by traditional procurement methods were built, operated, and maintained by the DG Highways on behalf of the MoTI. On the other hand, motorways constructed under the BOT scheme are operated and maintained by the contractor company.

BOT projects are tendered on the basis of annual guaranteed traffic and tolls. In this scheme, the government mainly takes demand and exchange rate risks, whereas private parties only assume the construction risk.

BOT motorway projects aimed to improve the transport network in a shorter time period and make more fiscal room for divided roads funded by public budget. However, after the motorways began operation, the government needed to pay a significant amount due to the fact that realized traffic was below the guaranteed levels. The main reason behind low traffic is the high tolls compared to relatively cheap state motorways, and free divided state roads. In order to increase traffic in these projects, further steps were taken, such as lowering the tolls for users and subsidizing the difference, as well as diverting heavy vehicle traffic to these motorways on certain parts of projects.

3 Regulation of State Roads and Motorways

This part of the chapter explains the regulation of roads in Turkey, first by describing the organizational and legal structure and then by giving regulatory details for accessing the market, traffic safety and security, and fiscal and environmental regulations.

3.1 Main Stakeholders and Institutional Structure

As roads play an important role in many sectors and social and economic activities, various authorities and organizations are involved in the regulation process of state roads and motorways. Such responsibilities range from the construction and maintenance of roads to the management of freight and passenger operations.

The MoTI and the Ministry of Interior (MoI) take the lead in regulating state roads and motorways. However, other authorities including the Ministry of Industry and Technology, the Ministry of Education, the Ministry of Family, Work, and Social Services, the Ministry of Environment and Urbanization, and the Ministry of Treasury and Finance have indirect responsibilities.

Major duties of the ministries and their related institutions are defined in the presidential decrees. While Presidential Decree No. 1 includes the organizational structure of the government system of the Republic of Turkey and the institutional laws of ministries and their central directorates, Presidential Decree No. 4 encompasses laws of related institutions to the ministries.

In addition to presidential decrees, other legislation regulates the rules and operations on state roads and motorways. Among these, the two major laws are Highway Traffic Law No. 2918, and Highway Transport Law No. 4925, which assign additional duties for ministries and their central and related directorates.

Some other ministries also have duties related to road transport and motor vehicles. For instance:

- The Ministry of Industry and Technology determines and inspects technical specifications and standards of motor vehicles to allow vehicles to enter the market.
- The Ministry of Education is responsible for training driver candidates and making their examinations, authorizing and inspecting driving courses, and providing certificate to successful candidates.
- The Ministry of Environment and Urbanization regulates all kinds of activities that may have an effect on environment, and accordingly has responsibility for regulating standards for pollutants from the transport sector. In addition, it is the

coordinating organization for local authorities (municipalities) and indirectly have responsibility for the urban road network.

- The Ministry of Family, Work and Social Services and its related institution Vocational Qualifications Authority have responsibility for setting standards and managing the procedures for professional qualifications.
- The Ministry of Treasury and Finance has the authority to impose taxes and fees related to motor vehicles, fuels, and transport business and services.

After a brief introduction, the duties and responsibilities of two key ministries—MoTI and MoI—will be explained in detail by referencing underlying legislation.

3.1.1 Ministry of Transport and Infrastructure

The main duties of MoTI are regulated by the 474th article of Presidential Decree No. 1. In that article, MoTI is assigned to:

- Plan, establish, operate, and develop infrastructure, networks, and systems related to transport, maritime, communication, and postal business and services;
- Ensure that such business and services are economic, fast, safe and secure, quality, environmentally friendly, and provided in a fair and sustainable competitive environment for the best interests of the public.

Since the provision of such duties is so complex and has different characteristics, the ministry established central and related directorates that are structured to perform duties in specific sub-sectors, such as road, maritime, air and railway transport.

Among these directorates, DG Highways and DG Regulation for Transport Services are the two units responsible for performing duties related to state roads and motorways. The duties of these directorates are further discussed in the following sections.

DG Highways

DG Highways is a related directorate to the MoTI and is the main responsible body for planning, constructing, and maintaining state roads and motorways. Being a related institution, Chap. 17 of Presidential Decree No. 4 is reserved to institutional law of DG Highways. According to the 211th article of that decree, the duties and responsibilities of DG Highways are defined as:

- Preparing plans for state roads and motorway routes;
- Constructing state roads and motorways, keeping them under constant maintenance so as to be used safely;
- Determining technical characteristics and preparing specifications for the construction and maintenance of the road network that is under their responsibility;
- Determining, publishing, and controlling road sign standards, taking international standards into account;

- Determining roads signs required for road network and placing them in appropriate locations;
- Conducting works for any required plans, maps, surveys, and making research and development studies;
- If needed, being partners with companies authorized to construct motorways under public private partnerships.

Moreover, Highway Traffic Law No. 2918 also imposes some other duties on the DG Highways, such as:

- Changing speed limits (by the permission of MoI) originally determined in the Highway Traffic Bylaw;
- Placing roads signs on state roads and motorways in line with speed limits;
- Giving opinions on traffic and vehicle techniques, examining and approving projects related to road safety;
- On behalf of the MoTI, carrying out inspections at roadside control stations.

Finally, pursuant to the Services of Directorate General of Highways Law No. 6001, DG Highways is in charge of determining road charges for state roads and motorways.

DG Regulationfor Transport Services

DG Regulation for Transport Services is a central directorate of the MoTI and its duties are defined in the 477th article of Presidential Decree No. 1. According to that article, DG Regulation for Transport Services is mainly assigned to:

- Ensure that highway and railway transport activities and combined transport services are maintained in an economical, fast, safe and secure, quality, environmentally friendly, and integrated way and are provided in a fair and sustainable competitive environment for the best interests of the public;
- Ensure that transport of dangerous goods are carried out in accordance with international conventions, standards, and legislation;
- Determine and control the minimum qualifications for parcel and passenger terminals, stations, and storage facilities that are used for road and railway transport activities, as well as transportation of dangerous goods;
- Determine the terms of professional qualification for drivers, machinists, carriers, agencies, logistics, and terminal operators operating in the field of road and railway transport and transport of dangerous goods, and authorize them by arranging training programs;
- Determine the terms of service, financial qualification, and professional dignity of carriers, agencies, logistics, and terminal operators operating in the field of road and railway transport;
- Determine minimum qualifications for all kinds of vehicles used in road transport activities, in terms of type, capacity, ownership, and age, including vehicles carrying dangerous goods;

- Determine the procedures and principles regarding the periodic technical inspections of road vehicles and the minimum qualifications of those companies that will be authorized to perform inspections, and control these authorized companies;
- Determine the floor and ceiling prices (when necessary) and supervise their application regarding all kinds of charges related to road transport activities and services.

In addition, by the 8th article of the Highway Traffic Law No. 2918, DG Regulation for Transport Services (on behalf of the MoTI) is assigned to conduct vehicle and emission inspections, and control inspection stations.

3.1.2 Ministry of Interior

MoI has a large organizational structure with many directorates and institutions. Among these, the Directorate of General Security (DG Security [DGS]-police) and General Command of Gendarmerie have responsibilities for regulating road transport.

DG Security is a related directorate to the ministry and carries out the main duties regarding management of traffic on state roads and motorways. The directorate has special security units based on their field of work, such as traffic police and motorway police.

General Command of Gendarmerie is another related institution to the ministry and is responsible for carrying out duties related to traffic management and security on rural roads, which are also included in state roads.

The ministry plays a crucial role in the safety, security, planning, operation, and management of passenger and freight traffic. The main duty of the ministry is regulated by the 254th article of Presidential Decree No. 1 and is defined to ensure and control traffic organization on all kinds of roads.

In addition to Presidential Decree No. 1, Highway Traffic Law No. 2918 also assigns MoI to:

- Regulate and manage traffic;
- Register vehicles and manage procedures for license plates, registration documents, and driver licenses;
- Control vehicles, documents, and equipment that need to be kept in vehicles;
- Control drivers and the documents they should travel with;
- Prepare traffic accident reports.

3.2 *Market Access and Authorization Regulations*

This part is designed specifically for market access and authorization regulations; however, most of technical regulations and standards will be covered in this part as they are closely interrelated with market regulations.

3.2.1 Vehicle Registration and Compulsory Liability Insurance

As mentioned in the previous part, any vehicle that accesses the transport market must be registered by the MoI. On behalf of the ministry, the DG Security carries out this duty. However, with reference to Bylaw No. 24356, notaries are authorized to perform the sale and transfer of vehicles and all other kinds of registration procedures on the directorate's behalf. Upon registering the vehicles, notaries hand in license plates along with registration documents.

According to Highway Traffic Law No. 2918, another important requirement for operating a vehicle is to have compulsory liability insurance. This insurance is also compulsory for international drivers. International drivers can insure their vehicles upon entering the country at the customs office. Vehicles without this insurance are prohibited from driving in Turkey.

Finally, with reference to Communique No. 20752, which defines the general terms and conditions for liability insurance, clauses that are covered by insurance are financial losses, health expenses, permanent disability, and death.

3.2.2 Vehicle and Emission Inspections

With reference to Highway Traffic Law No. 2918, all kinds of motor vehicles must pass inspection at specific time intervals. The duty of regulation and management of vehicle and emission inspections are maintained by the DG Regulation for Transport Services (Presidential Decree No. 1, Highway Traffic Law No. 2918 and Highway Traffic Bylaw No. 8182 Bylaw). DG Regulation for Transport Services delegated this authority to a private company for 20 years by tendering in 2007 (TUVTURK, 2007).

Bylaw No. 7182 regulates the details for inspection stations and vehicle inspection. According to the bylaw, private cars, tractors, and vehicles with two and three wheelers must pass an inspection in their third year and then every two years thereafter; all other vehicles must pass inspection every year.

In addition, vehicles are also required to pass emission inspections regulated by Exhaust Emission Control Bylaw No. 23410. These inspections are provided by both TUVTURK and other private entities authorized by the Ministry of Environment and Urbanization.

Emission inspections are carried out for cars (vehicles with up to nine seats, including the driver) in their third year and then every two years thereafter. All other vehicles have to pass these emission tests every year.

3.2.3 Pricing and Road User Charges

With reference to Law No. 6001, which concerns the "Services of Directorate General of Highways", the responsibility for determining road charges for state roads and motorways is given to the DG Highways. As mentioned in previous parts, state roads are free of charge. However, some engineering structures part of these roads such as bridges and tunnels can be subject to tolls. For instance, beltways in Istanbul are used free of charge, but two state-owned bridges on the Bosphorus are tolled. On the other hand, all motorways (except free beltway sections) are toll roads.

Charges for state-owned motorways and other related infrastructures are determined by the DG Highways. Tolls on BOT motorways are determined by the contract and are subject to changes by contract clauses.

Toll collection systems were first used in 1973 with the construction of the first bridge on the Bosphorus. In that system, tolls were collected manually in cash by staff located at toll booths. Beginning in 1999, in addition to manual toll collection, a new automatic system called "Automatic Passing System" (OGS) was put into use. In this system, payment was made directly from drivers' bank accounts through communication between a special device attached to the vehicles and the OGS gantries. In 2005, an alternative system to OGS, the "Card Passing System" (KGS) was implemented, in which drivers stopped by booths and made their payments by swiping/tapping their cards.

Between 2005 and 2011, all three toll collections systems were in use together; however, on the 14th of March, 2011, the manual toll collection system was discontinued, and OGS and KGS were left the only alternative payment methods. Later, at the end of 2013, the KGS system was replaced by "Fast Passing System" (HGS) in which, using RFID technology, payments are collected by gantries from a sticker attached to the windshield without having the vehicle stopped.

As of July 2020, 2 million vehicles use the OGS system, and more than 13 million use the HGS system (DGH, personal communication, July 21, 2020).

International vehicles using motorways and bridges are also subject to toll charges. Upon entering Turkey, the drivers of these vehicles must have either OGS or HGS devices to use toll roads and bridges.

Finally, with respect to the pricing policies and applications, it is observed that the tolls on the BOT motorways are almost five times higher than those operated by DG Highways. Therefore, tolls should be compatible with the income level or willingness to pay of the region or country; otherwise, the potential economic and social benefits may not be realized.

3.2.4 Authorization and Rules for Passenger and Freight Transport Operations

Passenger and freight transport involves many parties and requires detailed and comprehensive legislation. Numerous laws and bylaws refer to activities, liabilities, special conditions in passenger, and freight transport, and they include responsibilities, and rules that both private companies and government agencies must abide by to gain access to the market. Regarding passenger and freight transport activities, the primary pieces of legislation that must be referred to are the Highway Transport Law No. 4925 and the Highway Transport Bylaw No. 24299. While the Highway Transport Law defines the main duties, responsibilities and procedures, the Bylaw goes into detail and lists specific conditions, figures, and other critical information. In this part, most important articles of the legislation will be explained.

First, to carry out activities in passenger and freight transport, transport brokerage, and run warehouse and parcel post businesses, it is obligatory to obtain a certificate of authorization from the MoTI.

In order to apply for a certificate of authorization, both legal and natural persons to work in this field must satisfy necessary requirements. Legal persons must have professional dignity, financial, and professional qualifications. On the other side, natural persons must hold a professional qualification certificate.

According to Bylaw No. 24299:

- Professional dignity is defined as having a good reputation in the field of operation, abiding by the rules imposed by the authorities regarding road transport activities, not being convicted and imprisoned for crimes such as smuggling, theft, bribery, fraudulent bankruptcy, or drug, weapon, and human trafficking.
- Financial qualification is defined as having adequate financial resources for starting, running, and managing a company related to road transport activities.
- Professional qualification is defined as being equipped with required training, knowledge, and skills to perform road transport activities. Standards and legislation for professional qualification are regulated by the Vocational Qualifications Authority. Being a related institution to the Ministry of Family, Work, and Social Services, the Vocational Qualifications Authority determines national qualifications in all professional fields and authorizes companies to conduct training and examination for acquiring certificates of professional qualification.

A certificate of authorization is given with an annex called a "Vehicle Paper," which includes information about the size of vehicle fleet and types of vehicles that the certificate owner can operate. In addition, each vehicle in this paper is given a specific "Vehicle Card" displaying the owner of the vehicle and the certificate of authorization. The certificate and attached papers are valid for a period of five years and cannot be sold or handed over to any other parties.

The aforementioned bylaw defines many types of certificates of authorization with respect to types of vehicles operated and businesses carried out. Each certificate type has its own specific terms and conditions, such as the number of vehicles to possess, the capacity of related vehicles, equipment, or warehouse, and the minimum financial ratios to run their businesses. These very specific details are not included in this chapter, but certificate types given to natural and legal persons are summarized below with the corresponding types of vehicles and operations.

- Type A: Passenger transport by cars in domestic or international operations.
- Type B: Passenger transport by buses in domestic and/or international operations
- Type C: Freight transport in domestic and/or international operations
- Type D: Passenger transport by buses in domestic operations
- Type F: Run passenger transport agency in domestic and/or international operations

- Type G: Run freight and parcel transport agency in domestic and/or international operations
- Type H: Work as a broker for freight transport in domestic and/or international operations
- Type K: Freight transport in domestic operations
- Type L: Run logistics business in domestic and/or international operations
- Type M: Run parcel post business in domestic and/or international operations
- Type N: Run transport store in domestic operations
- Type P: Run distribution business in domestic operations
- Type R: Run transport organization business in domestic and/or international operations
- Type T: Run passenger and freight terminals.

Regarding passenger transport operations, market actors are quite institutionalized and market is well regulated. On the other hand, domestic freight transport market is quite accessible for natural and legal persons. However, the market structure is dominated by small noninstitutional family enterprises, which means there is an excess supply of freight operators in the market. As a result, almost 35% of trucks in the traffic are operated unladen, which causes both economic inefficiencies and higher traffic accident risk. Therefore, regulatory rules should be adapted to allocate resources more efficiently without creating an entry barrier to market.

3.2.5 Operational Time Restrictions for Drivers

In addition to having professional qualification requirements to operate in the market, drivers working in passenger and freight transport must satisfy some additional terms and obey rules regulating their working and resting hours.

Qualifications and terms that drivers must hold are defined in the Highway Transport Bylaw No. 24299. According to the 34th article:

- Drivers must be less than 66 years old;
- Drivers operating buses with capacity of more than 16 passengers must be at least 26 years old;
- Drivers must receive a health report showing that they are healthy in terms of body and psycho-technic, and renew that report every five years;
- Drivers operating vehicles carrying dangerous goods must have training documents for that specific goods;
- Drivers must not have been convicted of any crime related to drugs, weapons, human and customs smuggling, or terrorist acts.

Regarding the rules and terms for working and resting hours of the drivers, Section 6.3 of the Highway Traffic Bylaw No. 8182 can be referred to. According to Article 98 of that bylaw:

- Drivers of commercial vehicles that transport freight with a maximum weight exceeding 3.5 tons and those transporting more than nine passengers (including the driver) must not drive more than nine hours in total and more than 4.5 h in a single trip within any 24-h time period. These drivers must rest for at least 11 h during a 24-h period and must take one day break (at least 24 h) after a maximum of six consecutive days of driving.

- International passenger transport drivers can drive for 12 days, after which they must take a two-day break. Total hours of driving in two weeks' time cannot exceed 90 h.

Last but not least, operators running passenger and freight transport business must have tachograph devices equipped on their buses, trucks, and tractor trucks. The tachograph registries must be kept for one month in the vehicle and five years in the operator's office. Both operators and drivers are responsible for ensuring the tachographs are working properly and must keep their registries available for required time periods. These devices are vital for tracking whether drivers and operators are following the rules regarding working and resting hours, as well as other traffic rules such as speed limits.

3.2.6 Roadside Gauge and Weight Controls

As regulated by Highway Transport Law No. 2918 and Highway Transport Bylaw 8182, roadside gauge and weight controls are assigned to the MoTI and on behalf of the ministry; these controls are performed by DG Highways.

As of July 2020, the DG Highways had a total of 98 stations in 39 cities, and an additional 100 stations are planned to open by 2024 (DGH, personal communication, July 18, 2020).

The above-mentioned legislation also determines gauge and weight limits by vehicle type. According to the legislation, the maximum height of motor vehicle must not exceed 4 m, and the maximum width is 2.55 m, except for vehicles with refrigerators and trolleybuses, which are allowed up to 2.6 and 2.65 m, respectively. The maximum length allowed for motor vehicles depends on the vehicle type and number of trailers and varies from 12 to 22 m.

On the other hand, vehicle weight limits are regulated in two categories; one is maximum load on axles and axle groups, and the other one is total load on the vehicle.

For the first category, the maximum allowable load on a single axle is 10 tons for a non-drive axle and 11.5 tons for a drive axle. If the vehicle is two-axle then, depending on the distance between the axles, the maximum allowed load varies between 11.5 and 18 tons. Similarly, depending on the distance between axles, the maximum load allowed on axle group for two-axle trailers and semitrailers ranges from 11 to 20 tons, whereas it varies between 21 and 24 tons for axle group of 3 or more axles.

For the second category, the total load allowed varies between 18 and 44 tons depending on the vehicle type and number of trailers. The total load must be less than or equal to:

- 18 tons for two-axle vehicles and trailers

- 25 tons for three-axle vehicles
- 28 tons for three-axle vehicles with semitrailers and articulated buses
- 32 tons for four-axle vehicles
- 36 tons for four-axle vehicles with trailers and semitrailers
- 40 tons for five (or more)-axle vehicles with trailers or semitrailers
- 44 tons for vehicles with a three-axle tractor and a two or three-axle semitrailer carrying a forty-feet ISO container.

Passenger and freight transport operators must load their vehicles according to the gauge and weight limits. However, in the case of specific circumstances, by taking special permission from the Ministry of Transport and Infrastructure and paying a special transport fee, vehicles can be operated with loads exceeding the limits. Although the amount of the fee is fixed as of July 2020, it is planned to charge a fee depending on the distance covered for the transport of the goods.

3.3 *Traffic Safety and Security Regulations*

This part discusses safety and security regulations with respect to speed limits, intelligent transport systems (ITS) applications (ITS), and transport of dangerous goods.

3.3.1 Speed Limits

Speed limits are determined in Highway Traffic Law No. 2918 and Highway Traffic Bylaw No. 8182. The maximum speed allowed for cars is 120 km/h on motorways, 110 km/h on divided roads, and 90 km/h on single carriageways. The minimum speed limits are 15 km/h for both divided roads and single carriageways, and 40 km/h for motorways.

Speed limits are given in further detail in the Highway Traffic Bylaw. According to the 100th article, unless otherwise posted/stated, speed limits for some specific types of vehicles (without a trailer) are as follows:

- Cars can reach a maximum speed of 50 km/h in residential areas, 90 km/h on single carriageways located in non-residential areas, 110 km/h on divided roads located in non-residential areas, and 120 km/h on motorways.
- Minibuses, buses, and motorbikes can speed up to 50 km/h in residential areas, 80 km/h on single carriageways located in non-residential areas, 90 km/h on divided roads located in non-residential areas, and 100 km/h on motorways.
- Trucks, tractor trucks, and pickups may drive no faster than 50 km/h in residential areas, 80 km/h on single carriageways located in non-residential areas, 85 km/h on divided roads located in non-residential areas, and 90 km/h (95 km/h for pickups) on motorways,

- For vehicles with dangerous goods, the speed limits are 30 km/h in residential areas, 50 km/h on single carriageways located in non-residential areas, 60 km/h on divided roads located in non-residential areas, and 70 km/h on motorways.

In 2010, amendments to the Highway Traffic Law and related secondary legislation increased speed limits on divided roads in Turkey. In a cost-benefit analysis study covering a heavy road section, Cetin et al. (2018) asserted that the new regulation is not an acceptable policy decision for the objective of improving road safety. As most of the divided roads were designed and constructed before the above-mentioned amendments, these new speed limits were not compatible with the geometric and technical standards of current infrastructure. Since divided roads do not allow increased speed limits in most sections, this regulation needs to be reviewed and revised according to the physical standards of the roads.

Finally, as stated in the Highway Traffic Law No. 2918, it is forbidden to possess, produce, import or sell any kind of devices used to detect or avoid speed radar and other devices used by security forces.

3.3.2 Intelligent Transport Systems Applications

ITS policy is among the top priorities of the MoTI and many studies are being conducted in terms of both policies and investments in ITS. For instance, National Intelligent Transport Systems Strategy Document (2014–2023) was published in 2014. That document refers to many application fields for ITS, including passenger information, traffic management, freight transport, urban transport, and electronic toll collection systems. This part of the chapter summarizes ITS applications that are aimed at improving traffic safety and security.

First of all, in order to provide safety and security on roads, in 2011 DG Security began using "Electronic Control Systems (EDS)" (named Traffic Electronic Control System-TEDES until 2016), which were equipped with cameras and other surveillance equipment. These systems aimed to detect speed limit violations and other traffic safety and security rules. Since 2017, they have been implemented with required technology to detect average speeds and to ticket vehicles exceeding speed limits. As of July 2020, average speed controls have been implemented in 34 corridors on state roads in various cities. In addition, average speeds are detected between entry and exit points of state-owned motorways (DGS, personal communication, July 21, 2020).

Regarding the traffic safety and security issues in road transport, "Traffic Management Systems" are among the most efficient tools. These systems receive traffic data from sensors and cameras and convey them to traffic management centers. Traffic management centers pass these data on to information and instantly inform drivers about accidents, road, and weather conditions, and other emergency conditions through such means as radio broadcasts, variable traffic, or message signs (National Intelligent Transport Systems Strategy Document (2014–2023), 2014). DG Highways plans to have 18 "Traffic Management Centers", located in 17 regional centers and in the DG Highways headquarters (DGH, personal communication, July 18, 2020).

Traffic management systems in operation as of July 2020 are managed within their local management traffic centers and are not interconnected through an ITS framework. After the completion of these centers, all existing and new intelligent transport systems applications will be connected to these centers, and an integrated transport management system will be achieved. Data collected from one sensor or camera will be accessible from the main center in the headquarters or by other users. Moreover, all independent sub-systems used in road transport will be gathered and controlled within the same traffic management system. The communication between data receivers, regional and main centers, and drivers will be maintained through proper means such as fiber-optic and wireless technologies. In order to provide safe and fast information technologies infrastructure, a total of 15,000 km of fiber-optic cables is planned to be laid primarily along the main highway routes (DGH, personal communication, July 18, 2020).

Among existing traffic management systems, the variable message sign system is a widely used method of informing drivers. Messages conveyed through these systems include road and weather conditions such as construction activities, lane closures, accidents, precipitation, fog, and ice (DGH, personal communication, July 18, 2020).

In addition, being an integral part of road transport, signalization systems used on the road network are mainly fixed-time applications. In recent years, however, adaptive traffic signal systems started to be implemented, in which vehicle flows and congestion are taken into account and signal frequencies are determined automatically.

According to a pilot project implemented in a junction located in Polatlı county of Ankara, before-and-after studies revealed that the project led to reduction in waiting times by 36%, fuel consumption by 73,500 L, and CO_2 emissions by 178 tons (DGH, personal communication, July 18, 2020).

Another ITS application is the “Passenger Information System”, created and operated by the DG Highways. This system can be used through the directorate’s website and as a mobile application. It includes information regarding congestion levels on the national road network, weather conditions, tourist places, open and closed roads, and roads under construction. Users can also utilize a navigation system that allows them to find the fastest and safest route to their destination.

3.3.3 Transport of Dangerous Goods

Another important element in traffic safety is the transport of dangerous goods. As regulated by the Highway Traffic Bylaw No. 8182, physical and chemical substances that are explosive, flammable, easy to inflame, caustic, toxic, and radioactive substances are considered dangerous and must be transported in accordance with international and national rules and procedures. According to Bylaw, international rules and standards are defined in the “European Agreement Concerning the

International Carriage of Dangerous Goods by Road (ADR)". Turkey is a party to this agreement and is represented by DG Regulation for Transport Services as the competent authority.

On the national side, rules for transport of dangerous goods are regulated by Transport of Dangerous Goods on Highways Bylaw No. 31450. According to Bylaw, in compliance with ADR provisions, dangerous goods on roads must be transported in a safe and environmentally friendly way with minimum negative impact on human health. Dangerous goods prohibited by ADR cannot be transported on the roads of Turkey.

Companies transporting dangerous goods must possess a Dangerous Goods Operation Certificate related to their field. This certificate is given by the MoTI and is required for operators involved in activities for transporting, filling, packing, loading, unloading, sending, and receiving dangerous goods, as well as for tank-container/portable tank operations.

Among these operators, those working in sending, transporting, packing, filling, loading, unloading dangerous goods must also employ a dangerous goods safety advisor or attend training programs certified by the Institution of Dangerous Goods Safety Advisory, which is authorized by the MoTI.

In addition, vehicles carrying dangerous goods with fixed or removable tanks over 1 m^3 or with transportation equipment such as tank-container, portable tank, and multi-element gas containers over 3 m^3 should have an ADR/Vehicle Conformity Certificate. Vehicles carrying dangerous goods in bulk or packaged (except for explosive materials) are exempt from this certificate.

On the other hand, drivers of vehicles carrying dangerous goods must possess an additional professional qualification document called a Certificate of Driver Training for Dangerous Goods Transport. The MoTI regulates the rules, procedures, and standards about this certificate and authorizes institutions to conduct training programs and examinations.

3.4 Fiscal and Environmental Regulations

This part of the chapter explains taxation policies for motor vehicles and fuel types in order to provide insights into how they are utilized to regulate the transport market.

3.4.1 Motor Vehicle Taxation

Operators of all motor vehicles are obliged to pay an annual motor vehicle tax. The terms and conditions for this tax are regulated by the Motor Vehicles Tax Law No. 197. In addition to the law, to determine tax amounts and other additional rules, a general communique for motor vehicle taxes is published annually.

Until the end of 2017, these taxes were determined with respect to vehicle type, age, and engine size. However, beginning on January 1, 2018, the government added vehicle values to the criteria set.

Motor vehicle taxing criteria are not well designed according to environmental concerns. As motor vehicles with higher engine sizes are subject to higher taxes than vehicles with smaller engines, younger vehicles, which are more environmentally friendly than older ones, are subject to higher taxes. Table 1 illustrates the methodology used in calculating motor vehicle taxes. For convenience, the taxation system is presented for cars only. The vehicle values in the table are gathered from Motor Vehicles Tax General Communique No. 52, which determines taxes for 2020.

These tax values are determined for cars with internal combustion engines with respect to age, motor size, and vehicle value. However, cars with electric motors have an exception regarding their tax amount. According to Communique No. 52, such vehicles are subject to 75% discount from the taxes determined for a specific tax group. For instance, a one-year-old electric car with motor power less than 70 kW and with vehicle value less than 51,800 Turkish Lira would pay only 241 (one forth of 964) Lira.

Finally, for vehicles other than cars, motor vehicle taxes are determined with respect to vehicle age for minibuses, age, and engine size for panel vans, age, and passenger capacity for buses, age and maximum weight for pickups, and trucks.

3.4.2 Fuel Taxations

Fuel taxation is one of the most controversial topics in Turkey. Ignoring its market regulatory function, the government sees fuel taxes as a means of creating revenue and the tax levels have always been a matter of debate in society. While some argue that taxes should be determined in line with environmental impacts or polluter pays policy, the government usually adjusts taxes and fuel prices according to budgetary needs and constraints.

In general, fuel prices in Turkey are calculated by adding three types of taxes—special consumption tax, value-added tax (VAT), and commission for Energy Market Regulatory Authority (EMRA)—as well as dealer margins on top of refinery prices. While calculating the fuel price, the special consumption tax and the EMRA commission were added on refinery entry prices to find refinery exit price before VAT. Therefore, in an unusual way, the VAT base includes special consumption tax (Kantarci, 2018).

According to Turkish Petroleum Market Report 2019 and Liquefied Petroleum Gas (LPG) Market Report 2019, costs incurred by taxes and other commissions made up, on average, 50.22% of gasoline sales price, 42.98% of diesel sales price, and 41.28% of LPG prices. As of the end of 2018, diesel vehicles represented 50.7% of the total, followed by gasoline (27.9%) and LPG (20.9%) vehicles. The rest (0.4%) were electric cars or vehicles with missing information. Therefore, although the number of high-polluting vehicles comprises the majority of the vehicle stock, the tax regime is designed in favor of these vehicles.

Table 1 Motor vehicle taxes for cars in 2020

ICE Size (S) in cm^3 Electric motor power (P) in kW	Vehicle values before taxes (V) in Turkish Liras	Group	Vehicle age				
			1–3 years	4–6 years	7–11 years	12–15 years	Above 16 years
$S \leq 1300$ $P \leq 70$	$V \leq 51,800$	1	964	672	376	284	100
	$51,800 < V \leq 90,800$	2	1059	739	413	313	110
	$V > 90,800$	3	1156	806	451	341	120
$1301 \leq S \leq 1600$ $70 < P \leq 85$	$V \leq 51,800$	4	1678	1258	730	516	198
	$51,800 < V \leq 90,800$	5	1846	1384	803	566	217
	$V > 90,800$	6	2014	1510	875	618	237
$1601 \leq S \leq 1800$ $85 < P \leq 105$	$V \leq 129,800$	7	3260	2549	1499	915	355
	$V > 129,800$	8	3557	2779	1637	999	387
$1801 \leq S \leq 2000$ $105 < P \leq 120$	$V \leq 129,800$	9	5136	3955	2325	1384	545
	$V > 129,800$	10	5603	4316	2536	1510	594
$2001 \leq S \leq 2500$ $120 < P \leq 150$	$V \leq 162,100$	11	7704	5593	3494	2087	826
	$V > 162,100$	12	8405	6101	3811	2278	901
$2501 \leq S \leq 3000$ $150 < P \leq 180$	$V \leq 324,400$	13	10,741	9345	5838	3140	1152
	$V > 324,400$	14	11,719	10,194	6369	3426	1257
$3001 \leq S \leq 3500$ $180 < P \leq 210$	$V \leq 324,400$	15	16,358	14,720	8867	4426	1622
	$V > 324,400$	16	17,847	16,057	9672	4828	1771
$3501 \leq S \leq 4000$ $210 < P \leq 240$	$V \leq 519,200$	17	25,720	22,210	13,080	5838	2325
	$V > 519,200$	18	28,060	24,228	14,271	6369	2536
$S \geq 4001$ $P > 240$	$V \leq 616,500$	19	42,097	31,568	18,696	8403	3260
	$V > 616,500$	20	45,924	34,438	20,396	9166	3557

Apart from their classical fiscal purpose, taxes are also important instruments for various objectives, such as those that affect consumers' spending behavior and internalize negative externalities. In this context, a fuel taxation scheme should be revised according to environmental concerns and the principle of user or polluter pays.

4 Conclusion and Future Regulatory Requirements

After intensive railway-oriented policies and investments for the first three decades of the young Republic of Turkey, it is clear that funds received under the Marshall Plan played a very significant role in shifting from rail to road transport-oriented policies.

Until the 1970s, the main objective was to expand the road network and improve accessibility, which led to increased motorization and urbanization. Especially after the 1980s, increased urbanization and interaction between metropolitan cities required high quality and uninterrupted transport facilities, so motorway-oriented policies were adopted. Another milestone in road transport history was paved after the elections in 2002. The new government targeted the replacement of state roads, which were mostly single carriageways, with divided roads, and adopted to finance new motorway projects via the public-private partnership method. With these developments, both passenger and freight transport in Turkey have dominated by road transport. While the share of passenger transport doubled in the past 70 years, that of freight has quintupled in the same period.

From the institutional perspective, DG Highways on behalf of MoTI, and DG Security on behalf of MoI came forward as the key authorities in constructing and maintaining road infrastructure, as well as managing traffic operations.

The evolution of road transport policies also affected the transformation in economic and social conditions in Turkey. From the economic perspective, importing motor vehicles and oil had a negative effect on balance of payments and manufacturing industry as the production structure was not ready for such a rapid transformation. In addition, investing in road infrastructure, which was relatively poor compared to the railway infrastructure, and allowing passenger and freight transport operations to shift to the road network definitely created an economic opportunity cost for the country. Especially in freight operations, using trucks instead of railway led to inefficiencies and higher unit costs.

From the social perspective, the development of road transport led to a perception of availability and accessibility for people due to enhanced door-to-door transport opportunities. In addition, many entrepreneurship emerged in passenger and freight transport operations, creating new jobs and additional employment. However, increasing traffic on roads raised traffic safety issues and losses from traffic accidents in Turkey continue to be a major socio-economic problem. Therefore, in parallel to the changing paradigm in the world, including developments in ITS applications and human-centered policies, an institutional and regulatory framework based on safe system approach should be adopted.

Except for international freight transport, road transport has historically been open to competition, so regulations are generally concentrated on technical aspects, operational safety, protection of environment and users, traffic management, and infrastructure financing. Although access to market is relatively easy due to competitive environment, operating in the market requires strict rules and procedures. In addition, to maximize efficiency in freight transport operations, instead of maintaining a small family enterprise-based market structure, a more institutionalized structure should be incentivized.

Moreover, there are out-of-date provisions and fines in the laws, which should be updated according to the changing economic and social conditions, and monetary provisions should be regulated essentially in the secondary legislation. Confusing and even contradictory provisions should be eliminated; a clear, straightforward, and predictable regulatory framework should be established for all parties involved in the system.

Regarding pricing policies, the price gap between state-owned and privately owned motorways should be reviewed to allow more users to benefit from the existing infrastructure, thereby decreasing the financial burden on the public budget and maximizing economic and social benefits.

With respect to motor vehicle and fuel tax policies, taking environmental concerns and the pay-as-you-pollute principle into account, a fairer taxing scheme should be designed.

Finally, a new transport system paradigm is evolving from being car-centered to people-centered, and policymakers have, for some time, been forced to address road transport-related issues. This has been putting increasingly unbearable burdens on society, such as lives lost, congestion, economic losses, and pollution. Moreover, key game changers such as automation, digitalization, and mobility sharing are expected to shape the future of road transport and, thus, the transport system as a whole. Therefore, technological upheaval, new business models, and societal needs are expected to turn the transport sector upside-down. However, to harness the potential of new technologies and business models and to better meet the needs of society, public authorities must define and coordinate all actors for the public interest and establish efficient and equitable regulatory framework for the new system. Standardization, data governance, and data security have also become more critical than ever before, especially in an environment that is becoming increasingly digitalized and integrated.

References

Cetin, B., Baris, S., & Saroglu, S. (2011). Türkiye'de Karayollarının Gelişimine Tarihsel Bir Bakış [A historical overview of the development of highways in Turkey]. *Cankiri Karatekin University Journal of the Faculty of Economics and Administrative Sciences, 1*(1), 123–150. http://www.acarindex.com/dosyalar/makale/acarindex-1423874310.pdf.

Cetin, V. R., Yilmaz, H. H., & Erkan, V. (2018). Baris S., & Saroglu, S. (2011). The impact of increasing speed limit in Turkey: The case of Ankara-Sivrihisar road section. *Case Studies on Transport Policy, 6*(1), 72–80. https://doi.org/10.1016/j.cstp.2017.11.004

Chamber of Mechanical Engineers. (Eds.). (2012). *Ulaşım ve Trafik Politikalarında Planlama Gerekliliği* [*The need for planning in transport and traffic policies*]. Türkiye Mühendis ve Mimar Odaları Birliği. https://www.mmo.org.tr/sites/default/files/bb5a4b976e6b652_ek_0.pdf

DGH. (2019). *2019–2023 Stratejik Planı* [*2019–2023 strategic plan*]. https://www.kgm.gov.tr/SiteCollectionDocuments/KGMdocuments/Kurumsal/StratejikPlan/strateji(2019-2023).pdf

DGH. (2020). *Trafik ve Ulaşım Bilgileri 2019* [*Traffic and transport information 2019*]. https://www.kgm.gov.tr/SiteCollectionDocuments/KGMdocuments/Istatistikler/TrafikveUlasimBilgileri/19TrafikUlasimBilgileri.pdf

DGH. (n.d.-a). *Tarihçe* [*History*]. https://www.kgm.gov.tr/Sayfalar/KGM/SiteTr/Kurumsal/Tarihce.aspx

DGH. (n.d.-b). *Dünden Bugüne Karayolları* [*Roads from past to present*]. https://www.kgm.gov.tr/SiteCollectionImages/KGMimages/Gorseller/DundenBuguneKarayollari/tarihce.pdf

DGH. (n.d.-c). *Yıllar İtibariyle Ödenek ve Harcamalar* [*Budget allocations and spendings by year*]. https://www.kgm.gov.tr/SiteCollectionDocuments/KGMdocuments/Istatistikler/ButceOdenekHarcamaGelir/YilItibariOdenekHarcamaTR.pdf

Emiroglu, K., & Uzmay, Ü. (2013). *Demiryolu Ansiklopedisi* [*Railway encyclopedia*]. TSR.

EMRA. (2020a). *Turkish liquefied petroleum gas (LPG) market report 2019.* https://www.epdk.gov.tr/Detay/Icerik/3-0-108/yillik-sektor-raporu

EMRA. (2020b). *Turkish petroleum market report 2019.* https://www.epdk.gov.tr/Detay/Icerik/3-0-107/yillik-sektor-raporu

Guler, Y. (2004). II. Dünya Harbi Sonrası Türk–Amerikan İlişkileri (1945–1950). *Gazi Üniversitesi Kırşehir Eğitim Fakültesi, 5*(2), 209–224. http://kefad.ahievran.edu.tr/InstitutionArchiveFiles/f44778c7-ad4a-e711-80ef-00224d68272d/d1a3a581-af4a-e711-80ef-00224d68272d/Cilt5Sayi2/JKEF_5_2_2004_209_224.pdf

Kantarci, H. B. (2018). Türkiye'de Akaryakıt Üzerinden Alınan Vergilerin OECD ve AB Ülkeleri ile Karşılaştırılması. *Sosyoekonomi, 26*(35), 229–247.

OECD. (2000). *Policy roundtables: Competition issues in road transport.* https://www.oecd.org/daf/competition/sectors/2379173.pdf

SBO. (2019). *2020 Yılı Programı* [*2020 annual program*]. Strateji ve Bütçe Başkanlığı. http://www.sbb.gov.tr/wp-content/uploads/2019/11/2020_Yili_Cumhurbaskanligi_Yillik_Programi.pdf/

SPO. (1963). *Birinci Beş Yıllık Kalkınma Planı (1963–1967)* [*The first five-year development plan (1963–1967)*]. Devlet Planlama Teşkilatı Müsteşarlığı. http://www.sbb.gov.tr/kalkinma-planlari/

SPO. (2003). *Acil Eylem Planı* [*Urgent action plan*], Devlet Planlama Teşkilatı Müsteşarlığı.

TSR. (2005). *Turkish state railways annual statistics (2000–2004).* http://www.tcddtasimacilik.gov.tr/sayfa/istatistikler/

TSR. (n.d.). *İstatistikler (1923–2005)* [*Statistics (1923–2005)*]. http://www.tcddtasimacilik.gov.tr/sayfa/istatistikler/

Turkish Chamber of Civil Engineers. (2006). 1923–1940 Dönemi Demiryolları [Railways between 1923 and 1940]. *Türkiye Mühendislik Haberleri, 2*(3), 442–443. http://www.imo.org.tr/resimler/ekutuphane/pdf/21.pdf

TURKSTAT. (n.d). *Transportation statistics: Road lengths.* http://www.turkstat.gov.tr/PreIstatistikTablo.do?istab_id=1585

TÜVTURK 20 Yıllığına Araç Muayene İstasyonlarını İşletecek. (2007, August 15). TUVTURK. https://www.tuvturk.com.tr/basin-bultenleri.aspx?url=basinbultenleri/tuvturk-20-yilligina-arac-muayene-istasyonlarini-isletecek.aspx

Ulusal Akıllı Ulaşım Sistemleri Strateji Belgesi (2014–2023) [*National intelligent transport systems strategy document (2014–2023)*]. (2014). https://www.resmigazete.gov.tr/eskiler/2014/10/20141025-21-1.pdf

List of Legislation

Araç Muayene İstasyonlarının Açılması, İşletilmesi ve Araç Muayenesi Hakkında Yönetmelik [Vehicle Inspection and Opening and Operating Vehicle Inspection Stations]. Public Bylaw 7182. (2004). https://www.mevzuat.gov.tr/mevzuat?MevzuatNo=7182&MevzuatTur=7&MevzuatTertip=5

Araçların Satış, Devir ve Tescil Hizmetlerinin Yürütülmesi Hakkında Yönetmelik [Conduct of Services for Sale, Transfer and Registration of Vehicles]. Public Bylaw No. 24356. (2018). https://www.mevzuat.gov.tr/mevzuat?MevzuatNo=24356&MevzuatTur=7&MevzuatTertip=5

Bakanlıklara Bağlı, İlgili, İlişkili Kurum ve Kuruluşlar ile Diğer Kurum ve Kuruluşların Teşkilatı Hakkında Cumhurbaşkanlığı Kararnamesi [The Presidential Decree about the Organizational Structure of Related Institutions of the Ministries and Other Organizations]. Presidential Decree No. 4. (2018). https://www.mevzuat.gov.tr/mevzuat?MevzuatNo=4&MevzuatTur=19&MevzuatTertip=5

Cumhurbaşkanlığı Teşkilatı Hakkında Cumhurbaşkanlığı Kararnamesi [The Presidential Decree about the Organizational Structure of the Presidency]. Presidential Decree No. 1. (2018). https://www.mevzuat.gov.tr/mevzuat?MevzuatNo=1&MevzuatTur=19&MevzuatTertip=5

Egzoz Gazı Emisyonu Kontrolü Yönetmeliği [Exhaust Emission Control]. Public Bylaw No. 23410. (2017). https://www.mevzuat.gov.tr/mevzuat?MevzuatNo=23410&MevzuatTur=7&MevzuatTertip=5

Karayolları Motorlu Araçlar Zorunlu Mali Sorumluluk Sigortası Genel Şartları [General Terms for Compulsory Motor Vehicles Liability Insurance]. Public Communique No. 20752. (2015). https://www.mevzuat.gov.tr/mevzuat?MevzuatNo=20752&MevzuatTur=9&MevzuatTertip=5

Karayolları Trafik Kanunu [Highway Traffic Law]. Public Law No 2918. (1983). https://www.mevzuat.gov.tr/mevzuat?MevzuatNo=2918&MevzuatTur=1&MevzuatTertip=5

Karayolu Taşıma Kanunu [Highway Transport Law]. Public Law No. 4925. (2003). https://www.mevzuat.gov.tr/mevzuat?MevzuatNo=4925&MevzuatTur=1&MevzuatTertip=5

Karayolu Taşıma Yönetmeliği [Highway Transport Bylaw]. Public Bylaw No. 24299. (2018). https://www.mevzuat.gov.tr/mevzuat?MevzuatNo=24299&MevzuatTur=7&MevzuatTertip=5

Karayolu Trafik Yönetmeliği [Highway Traffic Bylaw]. Public Bylaw No 8182. (1997). https://www.mevzuat.gov.tr/mevzuat?MevzuatNo=8182&MevzuatTur=7&MevzuatTertip=5

Motorlu Taşıtlar Vergisi Kanunu [Motor Vehicles Tax Law]. Public Law No. 197. (1963). https://www.mevzuat.gov.tr/mevzuat?MevzuatNo=197&MevzuatTur=1&MevzuatTertip=5

Motorlu Taşıtlar Vergisi Genel Tebliği [General Communique of Motor Vehicle Tex]. Public Communique No. 52. (2019). https://www.mevzuat.gov.tr/mevzuat?MevzuatNo=34104&MevzuatTur=9&MevzuatTertip=5

Tehlikeli Maddelerin Karayoluyla Taşınması Hakkında Yönetmelik [Transport of Dangerpus Goods on Highways Bylaw]. Public Bylaw No. 31450. (2019). https://www.mevzuat.gov.tr/mevzuat?MevzuatNo=31450&MevzuatTur=7&MevzuatTertip=5

Faruk Cirit currently works as a senior transport specialist at the Department of Transport and Logistics at the Presidency of Turkey Strategy and Budgeting Office, which is in charge of national transport policies, transport planning, and infrastructure investments. He earned his Bachelor of Science degree in industrial engineering from Istanbul Technical University (ITU), Master of Science degree in industrial engineering from Bogazici University, and Master of Arts degree in economics from Duke University. He is also a Ph.D. candidate in industrial engineering. His research areas are infrastructure project appraisal, transport policies and regulations, and sustainable urban transportation.

Volkan Recai Cetin currently works as a consultant at the World Bank Ankara Office. Prior to this position, he served as the Head of Department of Transport and Logistics at the Presidency of Turkey Strategy and Budgeting Office and Ministry of Development, and as a senior transport specialist at TR Prime Ministry State Planning Organization. He holds a B.Sc. in civil engineering, MPA degree from Cornell University Cornell Institute for Public Affairs, and Ph.D. degree in Public Economics from Ankara University. He conducted research as a visiting scholar at University College London Centre for Transport Studies and Quantitative and Applied Spatial Economic Research (QASER) laboratory. He is also a part-time instructor at Istanbul Commerce University and Yıldırım Beyazıt University. His research areas focus on transport economics and planning; infrastructure investments, policy, and regulation; policy analysis, and project management.

Regulation of Urban Transport Infrastructures and Services—The Case of Istanbul

Umut Alkım Tuncer

Abstract Cities are centers of economic activity, creativity, and innovation. They are complex socio-technical systems and transport is one of the key elements that facilitates their growth. However, the fast pace of urbanization has certain consequences. Limited resources have made increasing transport demand a challenge for urban dwellers, who may lose time in congested traffic, be unable to access a service because there are not enough transport options, or have difficulty expanding their business because moving goods is costly. Regulation of urban transport comes into play here, affecting performance or leading to efficiency gains, and defining the rules of movement in such systems. This chapter focuses on urban transport regulations and identifies global problems. It introduces the actors and transport-related regulations in Turkey, specifically the city of Istanbul, which has undergone rapid urbanization in recent decades. The current market structure in terms of urban transport infrastructures and public transport is also reviewed. The chapter then discusses the current situation in the market and problems pertaining to regulations in line with the findings from a literature review.

Keywords Transport regulation · Urban transport · Public transport · Istanbul · Privatization · PPPs

1 Introduction

The trend toward urbanization has meant that urban transport is steadily gaining importance as a research area. The global urban population recently exceeded the rural population and it is projected that 68 percent of the world's population will be living in urban areas by 2050 (UN, 2018). Transport demand will also increase exponentially. From this angle, urban transport market(s) and how they are regulated become crucial research topics.

U. A. Tuncer (✉)
EPFL ME, Ras Al Khaimah, UAE
e-mail: umut.tuncer@epfl.ae

M. Eroğlu and M. Finger (eds.), *The Regulation of Turkish Network Industries*,
https://doi.org/10.1007/978-3-030-81720-6_14

Urban transport has different characteristics than intercity or international travel. Trips are usually short in distance and involve certain vehicles, such as taxis, trams, cable cars, and monorail systems. The high passenger volume means that these systems and vehicles have evolved to meet the ever-increasing travel demand. Public transport started to become essential for cities to function due to limited road space and the inefficiencies of private car use. Therefore, regulation of "public" transport is also relevant within an urban context and requires the analysis of local actors.

This chapter focuses on regulation of urban transport in Istanbul, the most populated city in the Republic of Turkey. Like other countries, Turkey has experienced rapid urbanization in recent decades, with Istanbul being the main destination. The city has struggled to find space and provide transport services to this growing population. Other major cities in Turkey have had similar experiences. In order to analyze Istanbul, we first look at global practices in urban transport regulation and its concepts and then discuss Istanbul, with a broader look at Turkey.

2 Theoretical Background—Transport in Urban Areas, Its Governance and Regulation

The objectives of regulations are to prevent market failures, decrease their external effects and, in doing so, serve the public interest (ECMT, 1983, p. 14). Unregulated markets can create monopolies, limiting the competition and creating negative effects for the public. On the other hand, heavy regulation is also undesirable as it can increase transactions costs, hindering innovation and competition (Button, 2005; Macário, 2011; Meakin, 2004, p. 92).

The transport sector is subject to regulations at the local, national, and international levels. Regulations, along with institutions and markets, define roles in transport operations and related infrastructure systems (Button, 2005, p. 102). They also help keep competition in markets alive, which is believed to bring benefits to users, such as low prices, safe and better quality services (Meakin, 2004; Sohail et al., 2006). It makes more sense when we consider the current privatization trend in the sector, especially in transport infrastructure. Increasing transport demand necessitates infrastructure investments, but they are usually very costly. Public–private partnerships (PPPs) come into play here as they give governments an option to finance such investments without incurring a substantial amount of spending or debt in the short term (OECD/International Transport Forum, 2013a, 2013b, p. 71). The same trend has also been observed in existing infrastructure and former public monopolies such as rail are increasingly privatized, which has raised calls for better regulations (Button, 2005, p. 102).

Regulation instruments in the transport sector are "price regulations, market access and capacity regulation, conditions on the form of supply, control over investment and taxes and subsidies" (ECMT, 1983, p. 7). These instruments can be used by national and sub-national regulators (Button, 2005, p. 104). Urban transport is subject to regulations at the national and local levels, as cities are almost always part of greater transport systems that connect them with other cities and rural areas. Highways and rail networks are some examples. Moreover, transport in urban areas can be affected by taxes and subsidies, which are defined by higher levels of government, such as ministries (Button, 2005, p. 73). Turning to urban context, it can be said that the first regulatory framework that needs to be analyzed is the one for public transport systems, as it is the second most popular mode of choice for urban dwellers after private cars. When we consider urban public transport, prevalent modes are heavy rail and light rail systems, public buses, taxis and paratransit such as minibuses and shared taxis. Public transport markets can be closed, meaning that all services are provided by a single operator, or there can be a competition in the market where operators compete for passengers. Competition can also be "for" the market and this time operators compete for the exclusive right to provide public transport services. Another type is a deregulated, open market where any qualified operator can provide services (Meakin, 2004, pp. 93–95).

Urban public transport used to be dominated by public monopolies, as in the case of national railways. It was usually local public enterprises that provided bus and rail services, and modes such as taxis were under local concession. With liberalization and an attempt to bring competition, terms such as tendering for services, PPPs and licensing started to become more popular, in line with the similar developments in the overall transport market (Sohail et al., 2006, p. 177). However, many examples of public monopolies still exist—it has certain advantages, such as easy government control on services and fares—but is usually the case that there are heavy subsidies and inefficiencies in the management because decisions are usually made on a political basis, unlike private operators (Meakin, 2004, pp. 25–26). Public operations in transport also means that regulators and operators are on the same side and this is one of the challenges, together with the current general absence of transport regulators at the urban level.

Apart from liberalization, another development in urban transport with a disruptive nature has been digitalization, which has brought new subjects such as integration of passenger information systems, ticketing, regulation of passenger data, and transport network companies like Uber (Audouin & Finger, 2019, p. 250). From a regulatory point of view, it is likely that most of the future challenges will be technology-driven, as in other sectors, and on-time revisions of existing regulations will be needed in order to have a balanced modal distribution and a good functioning transport system.

3 Different Global Approaches to Urban Transport Regulation

Transport characteristics differ greatly between "developed" cities and "developing" cities. Following the 1980s liberalization movement, privatization started to become the norm for developing cities, as it had been in developed ones (Sohail et al., 2006,

p. 179). However, some clear distinctions remained. The general pattern for developed cities is high car ownership, "for the market" competition in public transport, and high subsidies, as operators do not rely solely on fares to cover their expenses. There is usually integration of fares and planning because systems are user-oriented. Also, market entry conditions and regulator's roles are clear. On the other hand, developing cities see low car ownership due to lower income levels and operators often compete in the market, mostly relying solely on fare revenues to cover their expenses (Meakin, 2004, pp. 97–98). Operators competing for passengers and less costly paratransit modes become more common in developing cities, which also leads to fragmented ownership of vehicles and routes, making the regulation and integration more challenging.

The liberalization of transport infrastructure and operations has produced relatively different results in industrialized countries. In the rail sector, for example, there has been a complete privatization in the US. Most of the EU countries have also seen forms of decentralization, but the public sector still plays a role in countries such as France, Italy, and Spain (ECMT, 2006a, 2006b, p. 21). It is believed that UK's Private Finance Initiative (PFI) in 1997 became a model for such movements in other developed countries and the UK's experience was a means of testing the performance of different forms of PPPs in infrastructure and operations (Button, 2005; ECMT, 2006a, 2006b). There has been a similar trend in public transport and public monopolies are becoming scarce, with countries and cities adopting service procurement through competitive tendering (Meakin, 2004, pp. 26–28). In most of the cases, there is a public transport authority (PTA) that grants the right to operate either after operators apply or through a tendering process initiated by the authority, as in Germany's public bus framework (Beck, 2012, p. 3). This process allows authorities and regulators to give concessions to operate on the basis of the lowest subsidy, as there are usually subsidies involved in such services (Merkert & Hensher, 2012), or other contractual obligations such as service quality (Albalate et al., 2012, p. 87). Some studies on Polish and London public transport markets have noted that tendering of services resulted in efficiency gains and cost savings (Beck, 2012; Meakin, 2004). On the other hand, tendering and contracting can be a costly process in which every aspect of the service has to be defined in detail. Even so, there can still be a risk of incompleteness and, at this point, regulators have to be in a position to renegotiate to avoid a form of misconduct (ECMT, 2006a, 2006b, p. 74). The same risks prevail in transport infrastructure projects carried out in the form of PPPs. These are usually long-term projects and failure to transfer risks in contractual agreements can later become a burden for governments, regulators, and ultimately the taxpayers themselves, especially if there are guarantees provided by public sector to realize this type of project (OECD/International Transport Forum, 2013a, 2013b; Zegras, 2002).

Finally, as mentioned earlier, cities—and, more specifically, urban transport—are facing challenges in regulating services created by digitalization such as platforms. Usually, existing actors are not in favor of such players entering the market because it means that they will be losing market share. One example is New York City's taxis. After Uber's entry to the market, the value of the limited number of taxi plates ("medallions") started to decline (Puche, 2019, p. 38). Apart from platforms, digitalization has also paved the way for new transport modes like shared e-scooters, which has affected modal share and markets, especially for short trips.

4 Regulation of Urban Transport in Turkey

The main body responsible for transport in Turkey is the Ministry of Transport and Infrastructure (MoTI). There are various institutions under the Ministry to provide infrastructure for and regulate road, rail, maritime and air transport modes. Relevant ones for urban transport are the Directorate General of Highways, which, apart from other roads, constructs and maintain state roads that also run through cities; and Turkish State Railways (TCDD), the public monopoly in rail infrastructure and operations, which also provides service in some cities in the form of suburban railway systems. With Legislative Decree Nr. 655 dated November 1, 2011, urban railway projects such as underground metro and trams are subject to the review and approval of General Directorate of Infrastructure Investments. Approved project proposals are then delivered to the Ministers' Council for final approval. Roadside inspections, traffic fines and rules are enforced by the Ministry of Interior (Togan, 2016, pp. 37–38).

Turkey has a unitary form of administration and the central government is represented by appointed governors in 81 provinces that include cities. These provinces also have a local elected government called municipalities (Union of Municipalities of Turkey, 2020). Cities with a population of more than 750,000 have metropolitan municipalities. With Law Nr. 6360 dated November 12, 2012, the number of metropolitan municipalities increased to 30 and their jurisdictional areas were extended to provincial borders, abolishing a system in which a central government authority called provincial special administration existed. In addition, villages and small-scale municipalities called "belde" lost their legal status in these provinces and became neighborhoods under district and metropolitan municipalities with elected mayors (Savaş-Yavuzçehre, 2016).

Municipality Law Nr. 5393 dated July 13, 2005, gives municipalities the power to "designate the numbers, fare and tariffs, timing and routes of any type of service and public transport vehicles that are operated on land, sea, water and railways, together with taxis; designate and operate or cause to operate or lease the stops and vehicle parking spots on motorways, roads, avenues, streets, squares and similar places; carry out all works of traffic." The same powers are also given to metropolitan municipalities through Metropolitan Municipalities Law Nr. 5216 dated July 23, 2004, but metropolitan municipalities are also required to prepare transport master plans, establish infrastructure coordination centers (ICC), and transport coordination centers (TCC). On the other hand, powers given with the same Law to metropolitan municipalities transfer powers of "traffic planning, coordination and routing, the designation of taxi, public minibus and service-bus stopping places and parking areas and the determination of the number of such vehicles" to TCCs. With a further by-law entitled "Regulations on Coordination Centers of Metropolitan Municipalities," issued by the Ministry of Interior on June 15, 2006, fare regulation and licensing powers given to metropolitan municipalities by Metropolitan Municipalities Law were transferred to TCCs. The same regulation defines the members and decision-making procedure of TCCs. Chaired by mayors, TCCs have a permanent membership structure and members, apart from metropolitan municipality representatives, involve a set of central government bodies such as the Ministry of National Education, the Ministry of National Defense, Turkish State Railways, and the General Directorate of Highways. Through a recent amendment, central government bodies have more members than metropolitan municipalities and decisions in TCCs are made with absolute majority rule.

Public transport provision is under the responsibility of municipalities or metropolitan municipalities in accordance with the aforementioned laws. Municipality Law Nr. 5393 states that municipalities "Provide public transport, and to this end, establish or cause to establish and operate or cause to operate public transport systems of all sorts, including buses, maritime and waterway vessels, underground systems and rail systems." Similarly, Metropolitan Municipalities Law Nr. 5216 gives metropolitan municipalities the power and responsibility to "Provide metropolitan public transport services, and to this end, establish or cause to establish and operate or cause to operate such facilities, and issue licenses for public transport vehicles, including taxis and service buses, on land and sea within the metropolitan boundaries." As can be inferred from these subclauses, municipalities are also in a position to outsource public transport infrastructure and operations.

Separate urban transport authorities are not the case for cities in Turkey as they are in some global examples. However, there have been attempts in the most populous cities: Istanbul, Ankara and İzmir. The General Directorate of Istanbul Electricity, Tramway and Tunnel Enterprises (İETT), the General Directorate of Ankara Electricity, Gas and Bus Enterprises (EGO), and the General Directorate of Electricity, Water, Gas and Trolleybus (ESHOT) are examples that were established through central government decrees in 1939, 1942, and 1943 respectively. These institutions are a part of metropolitan municipalities in the cities where they currently provide service and their role as an authority and operations has been reduced to some specific mode of public transport until now. İETT, for example, only provides public bus and BRT services, with some legacy tram lines that are still under operation. It also acts as an authority for private bus operators and carries out inspections for service quality. EGO operates public buses, metro network and cable-car lines in Ankara and ESHOT provides bus services in İzmir. Over time, some metropolitan municipalities decided to provide public transport services through private companies established as subsidiaries. For example, Metro Istanbul company, which was established in 1988 by the Istanbul Metropolitan Municipality (IMM), currently operates tram, metro, light rail, funicular, and cable-car lines in the city (Tuncer, 2017, p. 39).

5 Case of Istanbul—Market and Regulatory Challenges

Istanbul is a megacity in Turkey with a population of 15,519,267 as of 2019 (Turkstat, 2020). The population growth rate is very high, considering that it was 11,076,840 in 2000 and 1,166,477 in 1950 (IMM, 2020a, 2020b, 2020c; Turkstat, 2020). The city is located on the northwestern part of Turkey, with the Bosphorus Strait that connects the Mediterranean to the Black Sea cutting across and dividing Istanbul into two parts, known as the Asian Side and the European Side. Even though it is not the capital city, Istanbul is important for the overall economy of the country as it accounts for approximately half of central government's tax revenues and one-fifth of country's GDP (COMCEC, 2015; Tuncer, 2016).

5.1 Recent Developments and Regulatory Challenges in Urban Transport Infrastructure

The main urban transport infrastructures in Istanbul consist of roads, railways, ports, and bridges that connect the two sides of the city. During the last decade, there have been ambitious programs to extend the railway network and implement large-scale projects involving bridges and highways in order to meet the growing transport demand. The length of railway network increased from 28 km in 1994 to 233 km in 2020 and there are 17 different ongoing construction projects with an additional total length of 221 km (IMM, 2020a, 2020b, 2020c). The railway network comprises metro, tram, cable-car lines and funicular systems, and investments are made either by the IMM or MoTI. Other noteworthy developments in the infrastructure are the Marmaray Tunnel, the Eurasian Tunnel, the İstanbul–İzmir Motorway project, and the Third Bosphorus Bridge and the North Marmara Motorway project. Apart from the Marmaray Tunnel, these projects were all realized under recent and popular PPP schemes by the central government and met with criticism from the public and opposition parties in the Parliament because of the Treasury guarantees for daily toll revenues. Marmaray is a railway tunnel under the Bosphorus and is part of a suburban rail line between Gebze and Halkalı. Designed to be used for passenger and freight transport, the tunnel project was completed by TCDD in 2013 at a cost of US$4.2 billion, and has been operated by the same institution since then. The Japanese International Cooperation Agency (JICA) and several European development banks invested in the project (Togan, 2016, p. 52). The Eurasian Tunnel followed a similar strategy, to connect two sides of the city with an immersed tube, and was completed in 2016. This is a highway tunnel that was built and has been operated through a build–operate–transfer (BOT) model developed by MoTI. The investment cost is US$1.25 billion and the consortium that won the bid, formed by Yapı Merkezi of Turkey and SK E&C of South Korea, has the operational rights for 25 years (ATAŞ, 2020). The İstanbul–İzmir Motorway BOT project involves the construction and operation of highways and a suspension bridge between Gebze and Orhangazi. The consortium of Nurol, Özaltın, Makyol, and Göçay of Turkey and Astaldi of Italy has the operational rights for 22 years and 4 months and the project was completed in 2019. With a total investment cost of $7.2 billion including the suspension bridge, it is Turkey's first privately operated highway (Nurol, 2017; Otoyol, 2016). The North Marmara Motorway consists of motorways, connecting roads and a bridge over the Bosphorus Strait, which connects the two sides of city along with the 15th of July Martyrs' Bridge and Fatih Sultan Mehmet Bridge, which were built in 1973 and 1988, respectively. IC İçtaş of Turkey and Astaldi of Italy has the operational rights for 10 years and 2 months for the bridge that was completed in 2016 with an investment cost of $2.5 billion (Togan, 2016, p. 44). When recent data for utilization is analyzed, it can be seen that daily toll collection in Eurasian Tunnel is close to guaranteed figures, but the Yavuz Sultan Selim Bridge of the North Marmara Motorway and the Osmangazi Bridge of the İstanbul–İzmir Motorway are not still at the desired levels and the central government had to compensate approximately $550 million in 2017 for these three infrastructures due to Treasury guarantees (TBMM, 2018). Fares in these recent infrastructure PPP projects are defined by MoTI in negotiation with consortiums through periodic reviews (Şevik & Eser, 2016). Sharp fare increases in short time periods, especially in the Eurasia Tunnel, have been a subject of public debate (AA, 2020; UAB, 2019).

5.2 Modal Share and Public Transport Market

The main modes of transport in Istanbul, apart from private cars, are public buses, metro, tram, cable-car, funicular lines, minibuses, taxis, shuttles (minibuses and buses that carry students and public/private sector personnel from home to school/work), and a BRT line. In line with population growth and increasing demand, overall ridership figures are on the rise, especially in road and railway modes. Maritime modes and their limited share in the market have been relatively steady in recent years (Fig. 1).

The main operator in rail systems is Metro İstanbul, which is a subsidiary company of IMM. It provides services on 13 lines and a network of 154.25 km (Metro İstanbul, 2020). The other important operator is TCDD that operates Marmaray, a 76.6 km suburban line stretching from east of the city to its west and uses the Marmaray tunnel for the Bosphorus Strait crossing (Marmaray, 2020). In road transport, public bus services are provided by İstanbul Otobüs A.Ş. (İOAŞ), which is again a subsidiary of the IMM and was established recently, privately owned buses called Özel Halk Otobüsleri (ÖHO), and İETT, which is also the operator of the BRT line. İETT has recently attempted to privatize its services through service procurement. This takes the form of competitive bidding for the purchase of new buses, which includes services such as maintenance, cleaning, supply of drivers, and parking inside bus depots (Tuncer, 2017, p. 49). Minibuses, taxis, and shuttles are other modes and are operated privately. The main operators in maritime passenger transport are Şehir Hatları A.Ş., a subsidiary of the IMM, İDO A.Ş., and private boats. İDO A.Ş. was

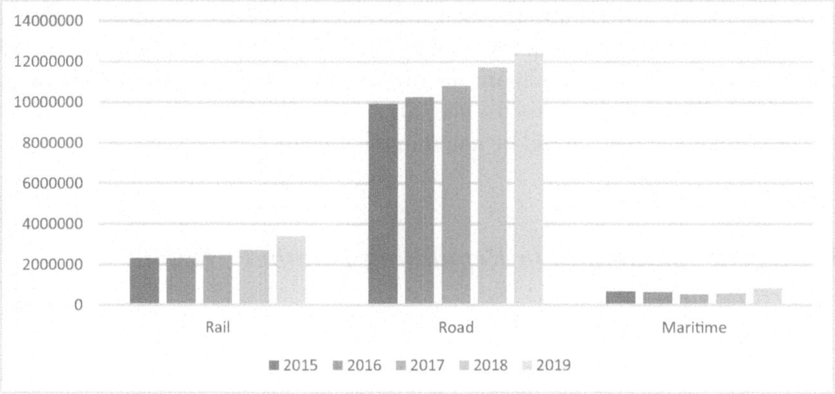

Fig. 1 Maximum daily ridership figures in Istanbul's public transport system between 2015 and 2019. *Source* Prepared by the author using IMM statistics (IMM, 2020a, 2020b, 2020c)

established in 1987 to provide mainly intercity ferry services between İstanbul and other coastal neighboring cities in the Marmara region. It was privatized in 2011 through a tender, which was won by a consortium that includes international partners, such as the Stagecoach Group of Scotland (Tuncer, 2017, p. 88) (Fig. 2).

The current public transport market structure in Istanbul was generally shaped in the early twentieth century when there was a period of a relatively deregulated market and paratransit modes operated by the private sector became the main mode of public transport. Advanced systems such as underground metro did not come onto the local agenda until the 1980s when existing modes started to become inefficient and insufficient. However, there have been strong investments in public transport, especially after 1989 when the first metro line was opened (Tuncer, 2016, p. 29). This has also reflected itself in the application of modern technologies. For example, an advanced feature of Istanbul's public transport system is that there has been a common payment system for fares since 1995. This application started with a

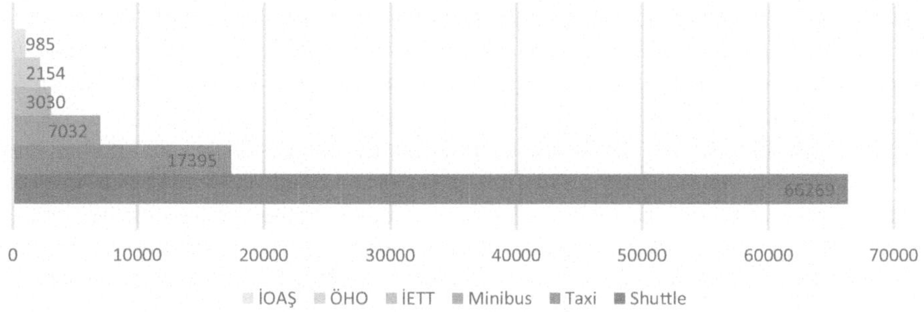

Fig. 2 Number of vehicles operated in road transportation in Istanbul. *Source* Prepared by the author using IMM statistics (TUHİM, 2020)

touch-on-memory (TOM) button, which was replaced in 2019 with the İstanbulkart smart card (Tuncer, 2018). This card can be used to pay fares in public buses, BRT line, metro system, Marmaray, maritime modes, and some taxis (BELBİM, 2020). Opening of the first driverless metro line 2017, Uskudar–Çekmekoy, is also one the recent developments that should be noted in this regard.

5.3 Recent Challenges in Regulating Istanbul's Public Transport

As noted, the TCC is the regulatory body in Istanbul that has the power to define fares, market access, capacity, and service conditions. However, there are still some components in the overall transport system that are regulated at the central government level, such as bridges and state roads.

It can be said that most of the regulatory challenges in Istanbul are due to the strong historical presence of paratransit modes such as taxis, shuttles, and minibuses. These markets have historically been regulated only through market entry restrictions and fare regulations in major cities of Turkey. This has resulted in problems in terms of passenger safety and a monopolistic market structure. Moreover, there have been practices to restrict vehicle medallions, creating a situation where there are incumbent individual operators or some individuals who own a certain number of medallions, as in the case of Istanbul. With an entry regulation applied by the IMM in 1991, the number of medallions was restricted to 18,000 and the amount has not changed since then, which has led to ever-increasing medallion prices—one is currently valued at approximately US$300,000, as the transfer of medallions is not restricted (Çetin & Eryiğit, 2011, p. 478). With the entry regulation and increasing medallion prices, incumbent medallion owners have developed relatively strong lobbying power over the IMM in years and even the central government. This has manifested itself in the banning of Uber's entry to the market in 2019, before which there were street demonstrations by taxicab drivers, sharp increases in taxi fares over short time periods, and an objection to the IMM, which recently started a study to increase the number of medallions. The entry regulation applied in 1991 caused a low supply of services, because the population of the city has been more than doubled since then, which has created a market for illegal taxis, of which there are estimated to be 50,000 (Çetin & Deakin, 2019).

Similar problems have also been observed in minibuses, shuttles, and private bus operators due to market entry regulation and operating rights. Minibuses also have a medallion restriction and have the exclusive rights to operate on specific routes. Private bus operators (ÖHO) emerged after an attempt to liberalize the bus market in the 1980s as İETT acted as a public monopoly that provided these services until then, even if there were such services as early as 1920 without a license. However, with the Ministers' Council Decision Nr. 86/10553, issued in 1986, these operators, most of which are individuals, were given the concession to operate on certain routes, mostly the profitable ones. This has created difficulties on İETT's side, which is the public bus authority in Istanbul along with its roles as an operator, in regulating their services in the form of planning and service quality (Canıtez et al., 2019). Moreover, there has been competition "in the market" for taxis, minibuses and (until recently) private bus operators, as fares are the only source of income, unlike modes operated by IMM in which there are subsidies if there is loss-making (Tuncer, 2017, p. 48). As a result, these modes of transport all compete for passengers during the day and there are usually overcrowding problems. In addition, there is a continuous demand for fare increases. A recent regulatory intervention by IMM will likely improve the situation with private bus operators. With IMM Council Decision Nr. 906 dated September 17, 2020, all private bus operations have been transferred to İETT and there will be a combination of fixed and performance-based subsidies. The latest development for paratransit was the medallion restriction for shuttles in 2019 through TCC decision Nr. 2019/6–1 dated June 11, 2019. Shuttle owners had been lobbying for the restriction for some time and the decision came before the last local elections (IMM, 2019). On the other hand, there have been recent quality and driver regulations to improve paratransit services. The TCC's decision Nr. 2016/9–7 dated December 28, 2016 defines the age limitations for commercial paratransit services. In line with this decision, the minimum age for minibuses is defined as 10 years old, taxis as five, shuttles carrying students as 12, and shuttles carrying personnel as 15. Similarly, all public buses need to be low-floor, accessible, and the age limit is 15, as per decision Nr. 2015/7–5 dated October 26, 2015. There have been also regulations to increase passenger safety and better control all road public transport modes. With TCC's decision Nr. 2013/11–16 dated December 26, 2013, all public transport vehicles need to be equipped with GPS and cameras and GPS IDs need to be submitted to IMM. On the driver side, through the TCC decision Nr. 2018/3–6 dated May 9, 2018, a program named Public Transport Service Quality Evaluation System (PTSQES) was enacted. Under this program, all bus, minibus, taxi, and shuttle drivers have to obtain a "public transport vehicle operator license" that must be renewed every two years. License requirements include elements in terms of not having certain types of road traffic accidents and substance abuse history, and successful completion of regular trainings. It also brings in a scorecard system for drivers, in which those who have the license start with 100 points and lose points with each violation, with a possibility of license suspension.

6 Analysis

Urban transport and its regulations are still at the development stage in Turkey and Istanbul. This is very much in line with the overall ongoing decentralization and liberalization trend in the country and its transport market. In the form of administration, Turkey has traditionally been on a centralist course and it can be argued that the Metropolitan Municipalities Law was an attempt to decentralize and give more control to local elected authorities within provincial boundaries. However, recent changes in the membership structure of TCCs have been a sign of stepping back because there are now more members with voting power from the central government side in these urban transport regulation bodies present in major cities of Turkey, which have metropolitan municipalities and such bodies through the respective law. Moreover, the centralist approach continues in the urban railway sector, as project proposals are subject to review and approval of the central government.

Our review of the roles, powers, and responsibilities of TCCs shows that they are clearly defined by the Metropolitan Municipalities Law, and the establishment of such separate urban transport regulation bodies in major cities of Turkey is clearly a reform and a way forward for Turkish cities. On the negative side, however, membership structure and voting system have their drawbacks because all members with voting power are either from the local government or central government. With the respective by-law, representatives of NGOs and the academia can be invited to TCC meetings in order to be consulted, but there is no such permanent member with the power to vote, meaning that public participation in urban transport regulation remains relatively weak. The other downside is the existence of critical transport infrastructures, bridges, and state roads, which are not regulated at the local level even if there has been an inclination toward such regulations. However, considering the whole urban transport system, externalities that can occur as a result of such fragmentation are rather limited.

On the market side, there is a mixed structure in infrastructure and operations. There is a current trend of liberalization, most notably at the central government level, and in construction and operation of transport infrastructure. This is very much in line with the ongoing liberalization process in the national railway sector that started with Law Nr. 6461 dated May 1, 2013, and is expected to be completed by 2023 (Togan, 2016, p. 54). In the context of Istanbul, new bridges, tunnels, and highways have been built and started to provide services in the last decade under PPP schemes. The same tendencies are also present in the public transport market. Examples include the privatization of İDO A.Ş., a new service procurement model of İETT, and the establishment of İOAŞ by IMM to provide bus services. Railway services are in a state of public monopoly as the only service providers are Metro İstanbul, a subsidiary of IMM, and TCDD of the central government. The situation is the same with the BRT line, where the only operator is İETT. Paratransit modes have historically been operated by the private sector and the abovementioned medallion restrictions and exclusive rights to operate on certain routes have created a monopolistic structure.

It is observed that, even if there have been important developments recently, urban transport regulation in Istanbul has its challenges. Starting with the infrastructure, it is evident that regulating the fares and tariffs in recent large-scale PPP projects observing public interest will not be easy for the central government because of guarantees provided. However, these infrastructures are still very young and there can be an increase in demand, which can make the negotiations easier for all the actors involved. When it comes to public transport, most of the challenges stem from the monopolistic structure in modes operated by the private sector. As indicated earlier, until recently, regulations in paratransit were limited only to market entry and fare regulations. The situation was the same for buses operated by the private sector.

Coupled with the strong lobbying power developed by incumbent operators in time, quality, driver, and even fare regulations have historically been challenging. On the other hand, recent regulations for GPS tracking, cameras, vehicle age limitations, and development of PTSQES for all drivers prove that there is a reverse trend and there has been a change in regulation policies.

Based on the current developments, it can be understood that transport in Turkey and Istanbul will continue going through a liberalization process and the question of how to regulate such infrastructures and services will come up regularly. Modal share and market structure in Istanbul will likely change because of the investments to railway systems and ongoing projects. This is also demonstrated in the ridership figures for the past five years. With an extended rail network that provides service with more routes, the road transport sector may lose some of its current market share. Maritime transport services' share has historically been limited and this situation will not change unless there is a change in local policies and investments to this mode. Current regulatory challenges with incumbent operators in paratransit will presumably remain the same if the situation with operating rights and entry restrictions does not change. Recent medallion restriction for shuttles indicates that similar practices may continue and such a market structure can be preserved in the future. On the other hand, a shift toward rail transport will naturally decrease the long-time dominance of such services and there can be a decrease in the lobbying power of their operators. It is also likely that safety and quality concerns with these modes will decrease with the recent regulations of the TCC, even if the market structure does not change.

Banning Uber from entering the market indicates that TNCs may continue to be restricted from entering the market for a while. However, it does not mean that alternative modes and services will not develop. Investments in bicycle roads by IMM are increasing and shared e-scooters have recently grown in popularity in Istanbul and other major cities in Turkey. It has been announced by MoTI that there will be specific regulations for e-scooter services.

7 Conclusion

Urban transport infrastructures in Turkey have traditionally been under public provision and public monopoly, but there has been a shift in policies during recent decades. Public transport, on the other hand, went through a relatively deregulated period, in which paratransit modes became the main means of traveling in urban areas and, to balance the market and make sure such services exist, public investments have been made into railway systems and bus services. There have also been several privatization attempts in different forms, such as service procurement. Reviewing the findings, it is clear that developments and inclination in urban transport sector of Istanbul are consistent with Sohail et al. (2006) observations at the global scale in terms of privatization. However, it can be said that general risks described by Zegras (2002) and ECMT (2006a, 2006b) are also valid for the infrastructure PPP projects in Istanbul that have been implemented recently. If the risks are not transferred to the contractors, costly renegotiations are a possibility and a "privatize now, regulate later" approach may prevail. The experience with these PPP projects will define how such infrastructures will be constructed and operated in the future and contracting models can be revised after the current trend.

The public transport market has elements from both developed and developing city characteristics described by Meakin (2004). As in developed cities, market entry conditions and regulators' roles are clear and there is fare integration. Moreover, there can be subsidies involved in major modes such as rail and bus. IMM Council Decision, Nr. 906 dated September 17, 2020, brought all bus operators under the roof of İETT and provides guaranteed payments to existing private operators; this demonstrates that the local government is adopting and will be testing a more user-friendly approach at the expense of increased spending. The decision states that all private bus routes will be reorganized by taking the rail network into account, and buses will be operated by İETT. It means that there will be more of a "competition for the market" approach and problems with racing for passengers, planning, and safety can be reduced. On the other hand, a substantial amount of paratransit operators still compete in the market and rely on fare revenues to cover their expenses. They have a fragmented ownership structure and regulating them in the way of private bus operators can be more challenging. However, investments into rail transport indicate that it will be the main mode of urban transport in the near future, which will gradually decrease the demand for such services.

Istanbul's practices in paratransit modes and infrastructure provision are, to some extent, generalizable to other cities under rapid urbanization. Increasing transport demand can result in service provision that is not very well planned and harder to regulate later. In terms of public transport, this means that services can still be provided with fewer subsidies, but the trade-offs for cities are not being able to have a good control over the form of supply and having less user-friendly systems. Istanbul's experience with taxis and minibuses and with private bus operators set a good example in this regard. Moreover, as in Istanbul, cities can refer to PPPs for the necessary costly infrastructure to meet increasing transport demand. It is very early to conduct a good analysis for the PPP projects in Istanbul, but the establishment of a TCC, its recent quality and safety regulations for paratransit, the decision to adopt a guaranteed payment model for private bus operators, fare integration, and ambitious railway investments that have the potential to balance the market can be a pattern to follow for comparable cities moving toward a more developed market and services.

References

AA. (2020, February 04). *EKONOMI haberleri: Anadolu ajansi.* Retrieved from Anadolu Ajansı Websitesi: https://www.aa.com.tr/tr/ekonomi/ulastirma-ve-altyapi-bakani-turhan-kanal-istanbulun-imar-plani-onaylandi/1723588

Albalate, D., Bel, G., & Calzada, J. (2012, October 20). Governance and regulation of urban bus transportation: Using partial privatization to achieve the better of two worlds. *Regulation & Governance*, 83–100.

ATAŞ. (2020, 08 31). *Avrasya tüneli anonim şirketi.* Retrieved from Avrasya Tüneli websitesi: https://www.avrasyatuneli.com/en/corporate/how/finance

Audouin, M., & Finger, M. (2019). Conclusion. In: M. Finger, & M. Audouin (Eds.), *The governance of smart transportation systems* (pp. 249–255). Springer Nature Switzerland AG.

Beck, A. (2012). *Competition for public transport services—Institutional framework and empirical evidence of bus services in Germany.* Physica-Verlag.

BELBİM. (2020, 08 31). *Üye işyerleri: Belbim.* Retrieved from BELBİM Websitesi: https://www.belbim.istanbul/belbim-uyelik-uye-isyerleri

Button, K. (2005). Myths and taboos in transport policy. In P. Rietveld & R. Stough (Eds.), *Barriers to sustainable transport* (pp. 37–53). Spon Press.

Canıtez, F., Çelebi, D., & Beyazıt, E. (2019). Establishing a metropolitan transport authority in Istanbul: A new institutional economics framework for institutional change in urban transport. *Case Studies on Transport Policy*, 562–573.

Cetin, T., & Deakin, E. (2019). Regulation of taxis and the rise of ridesharing. *Transport Policy*, 149–158.

COMCEC. (2015). *Urban transport in the OIC megacities.* COMCEC Coordination Office.

Çetin, T., & Eryiğit, K. (2011). Estimating the effects of entry regulation in the Istanbul taxicab market. *Transport Research Part A*, 476–484.

ECMT. (1983). *Possibilities and limits of regulation in Transport Policy.* OECD Publications Office.

ECMT. (2006a). *Transport and decentralization.* OECD Publications Service.

ECMT. (2006b). *Transport services: The limits of (De)regulation* . OECD Publications Service.

IMM. (2019, 06 14). *News: IMM.* Retrieved from Istanbul Metropolitan Municipality Website: https://www.ibb.istanbul/News/Detail/35494

IMM. (2020a, 08 31). *Istanbul metropolitan municipality.* Retrieved from Istanbul Metropolitan Municipality website: http://www.ibb.gov.tr/tr-TR/BilgiHizmetleri/Istatistikler/Documents/demografi/t211.pdf.

IMM. (2020b, 06 19). *Istanbul metropolitan municipality.* Retrieved from Istanbul Metropolitan Municipality website: https://www.ibb.istanbul/News/Detail/35514

IMM (2020c, 08 31). *Yolculuk türü bazinda yolculuk sayisı: İBB açik veri portali.* Retrieved from İBB Açık Veri Portalı: https://data.ibb.gov.tr/dataset/yolculuk-turu-bazinda-yolcu-sayisi.

Macário, R. (2011). *Managing urban mobility systems.* Emerald Group Publishing Limited.

Marmaray (2020, 08 31). *Marmaray Hakkinda: Marmaray.* Retrieved from Marmaray: http://marmaray.gov.tr/marmaray-hakkinda/.

Meakin, R. (2004). *Training course: Bus regulation and planning—bus sector reform.* GTZ.

Merkert, R., & Hensher, D. (2012, July 12). Regulation, trust and contractual incentives around transport contracts—Is there anything bus operators can learn from public air service contracts? *Research in Transportation Economics*, 67–78.

Metro İstanbul. (2020, 08 31). *Operation services: Metro İstanbul.* Retrieved from Metro İstanbul Website: https://www.metro.istanbul/en/content/i%C5%9Fletmehizmetleri.

Nurol. (2017, 01 01). *Nurol holding A.Ş.* Retrieved from Nurol Holding A.Ş. websitesi: https://www.nurol.com.tr/otoyol-yatirim-ve-isletme-a-s.

OECD/International Transport Forum. (2013a). *Better regulation of public-private partnerships for transport infrastructure.* OECD Publishing/ITF.

OECD/International Transport Forum. (2013b). *Understanding the value of transport infrastructure.* OECD/ITF.

Otoyol. (2016, 01 01). *Otoyol yatirim ve İşletme A.Ş.* Retrieved from Otoyol Yatırım ve İşletme A.Ş. website: https://isletme.otoyolas.com.tr/?page_id=4037.

Puche, M. L. (2019). Regulation of TNCs in Latin America: The case of uber regulation in Mexico City and Bogota. In: M. Finger, & M. Audouin (Eds.), *The governance of smart transportation systems.* Springer Nature Switzerland AG.

Savaş Yavuzçehre, P. (2016). The effects of the law no. 6360 on metropolitan municipality system in Turkey. *European Scientific Journal*, 291–303.

Sohail, M., Maunder, D., & Cavill, S. (2006, January 4). Effective regulation for sustainable public transport in developing countries. *Transport Policy*, 177–190.
Şevik, S., & Eser, L. (2016). Road pricing and privatization implementation in Turkey: A model proposal. *The International Journal of Economic and Social Research*, 89–108.
TBMM. (2018, 11 16). *The grand national assembly of Turkey*. Retrieved from The Grand National Assembly of Turkey website: https://www2.tbmm.gov.tr/d27/7/7-3678sgc.pdf.
Togan, S. (2016). *The liberalization of transportation services in the EU and Turkey* . Oxford University Press.
TUHİM. (2020, 08 31). *Mevcut toplu ulaşim sayilari—İBB toplu ulaşim hizmetleri müdürlüğü*. Retrieved from İBB Toplu Ulaşım Hizmetleri Müdürlüğü: https://tuhim.ibb.gov.tr/İstatistiksel-bilgiler/mevcut-toplu-ulaşım-araç-sayıları.
Tuncer, U. (2018, October 15). The role of technology in public transport integration and governance—Smart card use in Istanbul and Mexico City BRT systems. *IGLUS Quarterly*, 7–13.
Tuncer, U. A. (2016). *Constraints for urban public transport authorities in implementing BRT projects* (Unpublished master's thesis). IGLUS—EPFL.
Tuncer, U. A. (2017). *İstanbul 'da toplu ulaşimin organizasyonel yönetimi ve finansmani* (Unpublished master's thesis). Bahçeşehir Üniversitesi.
Turkstat. (2020, 08 31). *Turkish statistical institute*. Retrieved from Turkish Statistical Institute website: http://www.turkstat.gov.tr/UstMenu.do?metod=temelist.
UAB. (2019, January 31). *Basın açiklamalari: UAB*. Retrieved from Ulaştırma ve Altyapı Bakanlığı Websitesi: https://www.uab.gov.tr/basin-aciklamalari/avrasya-tuneli-gecis-ucretleri-hakkinda.
UN (2018, May 16). *News: Department of economic and social affairs*. Retrieved from United Nations Web Site: https://www.un.org/development/desa/en/news/population/2018-revision-of-world-urbanization-prospects.html.
Union of Municipalities of Turkey. (2020, August 31). *Local authorities*. Retrieved from Union of Municipalities of Turkey Website: https://www.tbb.gov.tr/en/local-authorities/.
Zegras, C. (2002). *Private sector participation in urban transport infrastructure provision*. GTZ.

Umut Alkım Tuncer is the Manager of Innovative Governance of Large Urban Systems Program (IGLUS) at Ecole Polytechnique Fédérale Lausanne (EPFL). He holds bachelor's degrees in English language teaching, translation and interpretation from Hacettepe University. He also has master's degrees from Istanbul Bahçeşehir University and EPFL in urban governance and transportation management. His academic research is mainly focused on urban mobility.

Regulation of Turkish Intercity Coach Industry

Ali Osman Solak

Abstract In Turkey, coach transport is the most common mode of intercity passenger transportation and coach services have been carried out by the private sector in free-market conditions since 1926. This chapter focuses on intercity coach market regulations and the results of these regulations. Due to regulations and increased intermodal competition, the number of operators in the market has decreased. We can predict that intermodal competition will increase in the coming years, which will lead to a further decline in the number of operators and increased collaboration among operators. Based on Turkey's experience, we can say that it is possible to carry out public transportation without government subsidies, and that intermodal and potential competition are significant for intercity passenger transportation.

Keywords Turkey · Intercity coach industry · Regulation · Competition

1 Introduction

In many countries, especially in Europe, as a way of protecting rail transport, coach services were not allowed to operate on the intercity routes where rail services exist for many years or have not developed due to strict regulations. During recent decades, many of these countries have liberalized their intercity coach markets for commercial services. In Turkey, the intercity coach market has not been exposed to any limitations or regulations to protect railways. Since 1926, transport services have been carried out by the private sector in free-market conditions. In the early years, coach operations mostly carried out on routes where there were no railways. Coach usage and the number of companies increased in intercity passenger transportation over the years, in parallel with the developing road network, growing population, and growing economy. Coach transport became the most widely used mode in intercity public passenger transportation, compared with low-service quality, state-operated railway

A. O. Solak (✉)
Department of Public Finance, Faculty of Economics and Administrative Sciences, Bolu Abant Izzet Baysal University, Bolu, Turkey

M. Eroğlu and M. Finger (eds.), *The Regulation of Turkish Network Industries*,
https://doi.org/10.1007/978-3-030-81720-6_15

transport and uncommon and expensive air transport. Since the 1970s, coaches have reached almost every city in Turkey, including the center where there are rail lines. However, the lack of regulation and growing demand led to haphazard development. The market consisted of many companies that could benefit from the cost advantages of scale and density economy, could operate efficiently, and were not institutionalized (Solak, 2020, p. 16). Regulations were issued for the market starting from 1990, a law was enacted for the first time in 2003, and subsequent regulations have been made within the framework of this law.

This chapter focuses on the regulation of the Turkish intercity coach industry. The following section deals with the theoretical background and country practices. The third section presents a brief history of coach transport and the fourth section about the regulatory institution. The fifth section discusses the regulations before and after 2003 and their effects. The sixth section provides a current outlook and a future projection of the industry, before the final section presents an overview and includes some general lessons learned from Turkey's experience.

2 Theoretical Background and Country Practices

According to the public economic theory, perfectly competitive markets ensure efficient use of resources without any government intervention. Situations requiring government intervention are limited to market failures and unfair welfare distribution of the market. In these situations, governments either produce goods and services directly themselves or regulate the private sector. These regulations are market entry-exit regulations, price and quantity regulations, and performance and quality regulations.

Government intervention in the transportation services is usually due to the existence of natural monopolies, network industries, and externalities. Government intervention is comparatively strong in passenger transportation since the universal service dimension of passenger transportation services and the market does not yield satisfying outcomes in sociopolitical terms (Van de Velde, 2004, p. 1–8).

The market structure and the government's role in intercity coach services vary by country. Some countries liberalized their markets a long time ago, while intercity coach services have not been carried out for many years or not developed in others because of strict regulations. The USA and Britain both liberalized their markets in 1980, with Britain being the first European country to do so. In the following years, other countries have also implemented coach transportation reforms to provide a low-cost alternative to railways. The liberalization process in Poland started in 1990, while Sweden and Norway opened up their markets in 1998 and 1998–2003, respectively. Japan liberalized its market in 2002 and Italy started the liberalization process in 2007 and completed it in 2014. Germany and France opened up their markets later than other countries, in 2013 and 2015, respectively, to protect railways. The institutional structure and regulatory institutions for the liberalized intercity coach transportation market differ among these countries.

Studies on liberalization, regulations, and competition in the coach transport market show us that the effects of liberalization and regulations vary by country (Aarhaug & Fearnley, 2016; Aarhaug et al., 2018; Alexandersson et al., 2010; Augustin et al., 2014; Beria et al., 2018; Blayac & Bougette, 2017; Crozet & Guihéry, 2018; Dunmore, 2016; Dürr et al., 2016; Grimaldi et al., 2017; Knorr & Lueg-Arndt, 2016; Leiren and Fearnley, 2008; Sakai & Takahashi, 2013; Taylor & Ciechanski, 2008; Van de Velde, 2004, 2013; White & Robbins, 2012). In general, the results of liberalization are positive in terms of the number of customers, new entries to the market, the number of routes and frequency, prices, and service innovation. Markets grow quickly after liberalization and then a small number of companies dominate the market. Route and frequency regulations are often observed. Coach operators in many countries are under intense intermodal competition pressure, especially from railways and apps such as BlaBlaCar developed in recent years.

3 Historical Developments

The Republic of Turkey, founded in 1923, had a low-standard 18,335 km road network and 4136 km of railway lines inherited from the Ottoman Empire (KGM, 2019a, p. 3; TCDD, 2019a, p. 34). In the early years of the Republic (1923–1950), limited investment resources were mainly used for railway construction, and passenger and freight transport was mostly carried out by railways. However, road construction has continued, albeit at low standards. The length of the road network reached 41,582 km in 1940 and 47,080 km in 1950. Since 1950, with the effect of the financial and technical aid of the USA in 1948 (known as Marshall aid and the Marshall Plan), road construction was prioritized. For this purpose, the General Directorate of Highways (KGM) was founded in 1950.[1] By 1960, the road network reached 61,452 km; apart from some small settlements, all parts of Turkey were accessible by road. Since 1960, the focus shifted to raise the standards of the existing road network (KGM, 2019a, p. 3; Tekeli & İlkin, 2004, p. 382).

In Turkey, coach transport began in 1926 with KâmilKoc, founded in Bursa, a privately owned company. The developing highway network, increasing population, and growing economy also increased passenger transportation on highways and, in parallel, coach use in intercity passenger transportation also increased. In the early years, coach transport was mostly carried out on routes that had no railways. Contrary to examples in European countries, governments did not prevent the development of coach transport for protecting railways. They favored coach transport because rail lines did not reach all cities. In following years, the increase in the quality of the road network, the expansion of the road network, the development of coach technology, the increase in domestic and international tourism, and a lack of investment in railways all increased the demand for coach services. Since the 1970s, coach transport has

[1] KGM has continued its activities as the institution responsible for the construction, maintenance, and repair of highways until today.

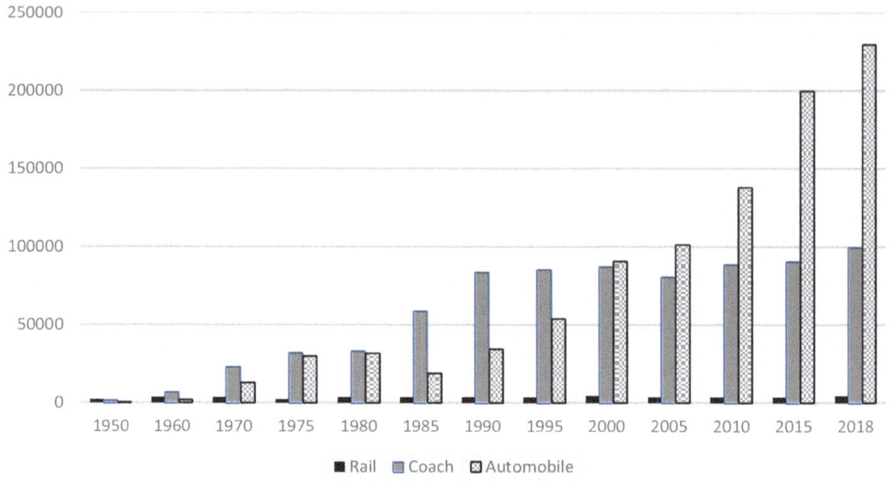

Fig. 1 Intercity passenger transportation. *Source* Prepared by the author using KGM (2019b) and TCDD (2019b) statistics

operated in almost every city in Turkey, including cities where there were railway lines (Solak, 2020, p. 16). In this sense, we can say coach transport played a significant role in terms of linking different geographical parts of the country and social mobility. Coach transport used much more than rail transport. The share of the scheduled and unscheduled coaches in passenger transportation grew from 32.9% in 1950 to 61.4% in 1985. With the increase in automobile use, this rate declined in the following years (see Fig. 1). The number of operators and coaches in the industry also increased over the years. As of the end of 1997, 430 companies with 7305 coaches operated in the scheduled coach market (Ministry of Transportation, 1998, p. 58).

Since its early development, scheduled intercity coach services were conducted by privately owned companies under free-market conditions without receiving any direct government subsidies. Until 1990, the market was not subjected to significant economic regulation. There was not even a regulation for companies to obtain an operating license to enter the market. Similarly, there was no considerable intervention on prices, using contracted coach, routes, schedules, and frequencies.

The absence of any restrictions on entering the market and using contracted coaches has facilitated the increase in the number of operators and the growth of the market. By using contracted coaches instead of their own coaches, the companies shared the risk/income with the coach owners and did not have to use resources to buy coaches. In practice, companies received commissions from contracted coach owners operating under their business titles, based on the number of passengers. In this sense, companies acted as umbrella organizations of coach owners. Even individuals who had only one coach was able to enter the market by contracting with companies.

The growth of intercity passenger transportation demand and the increase in the number of companies also intensified the competition. Companies improved their service quality and kept prices low in the face of increasing competition in many routes, which made coach transport much more advanced than railways operated by the state as a monopoly with low-service quality and air transport that is not common and affordable. However, the lack of regulation and the growing demand led to haphazard development. The market consisted of many companies that could not benefit from the cost advantages of scale and density economy, could not operate efficiently, and were not institutionalized. This market structure was mostly detrimental to individual contracted coach owners. The fact that coach owners are generally low-educated people who only calculate their costs and profits by considering their fuel costs and ignore costs such as depreciation, maintenance, and inflation, contributed to this situation.

4 Regulatory Institution

In Turkey, the Ministry of Transportation was established by Law No. 3613 in 1939. The first institution to regulate road (freight and passenger) transportation was the Land Transport Department, which was founded in the same year as the central organization of the Ministry. This institution was named as the Directorate of Highways Transportation Department in 1945 through Law No. 4770, and later became the General Directorate of Land Transportation as a result of the reorganization carried out in 1973. In 1983, with Decree-Law No. 182, the General Directorate was transformed into the Directorate of Land Transportation Department. With Law No. 3348, enacted in 1987, the General Directorate of Land Transportation was re-founded. The institution served under this name for many years, and in 2011 was renamed the General Directorate of Highway Regulation (GDHR) via Decree-Law No. 655. The regulatory institution served as a central organization directly linked to the Ministry of Transportation since its early years and has never been autonomous. A change made in January 2020 abolished the GDHR and the highway and rail transportation regulators gathered under one roof as the General Directorate of Transportation Services Regulation (GDTSR).

The regulatory institution had not been effective in intercity passenger transportation until 1990, as it was mainly concerned with the regulation of international freight and passenger transportation. The institution started to be active in intercity passenger services with the Regulation on Intercity Passenger Transportation on Road issued in 1990. The institution has been more efficient with the Road Transport Law (No. 4925) enacted in 2003 and the Road Transport Regulations (RTR) issued in 2004, 2009, and 2018.

The GDTSR has seven principal duties regarding coach transport. (1) To ensure that the transport activities are carried out economically, serially, safely, with high quality, environmentally, and following the public interest in a free, fair and sustainable competitive environment. (2) To determine service conditions, financial capability, and professional reputation of the carrier, ticket agency, terminal operator, etc., and to authorize and supervise them. (3) To determine the professional competence conditions of the people working in the industry, especially the drivers, and to provide training, to make an examination related to this. (4) To determine the procedures and principles regarding the public service obligation. (5) To define minimum standards for passenger terminals and to supervise them. (6) To determine minimum standards for all kinds of vehicles used in transportation activities in terms of type, capacity, ownership, age, etc. (7) If necessary, to set floor and ceiling prices regarding all kinds of road, bridge and tunnel fees, terminal usage fees, and transport activities with the sole aim to provide a free, fair, and sustainable competitive environment.

The regulatory institution monitors transport activities, including coach transport, on highways through 13 Regional Directorates spread out across the country. The duties of the regional directorates are the issuance of entry license, supervision of operators, renewal of vehicle cards, the permission of lines and routes, and approval of timetables and price tariffs.

5 Regulations for the Market

5.1 Regulations Before the Road Transport Law (2003)

In Turkey, there was no law or regulation directly regulating long-distance highway transportation until 1990. The first legal regulation was the Regulation on Intercity Passenger Transportation on Roads, issued in 1990. According to the Regulation, companies must obtain a D1 license for scheduled coach services, for which they require a minimum of three coaches and some business capital. The Regulation also introduced an age limit for coaches and the obligation to approve timetables and price tariffs for companies. In 1994, the Regulation on Intercity Passenger Transportation on Road was renewed. The minimum coach requirement was increased from three to seven, and the age limit for coaches was lowered. However, revisions made in the following years reintroduced the minimum of three coaches and the previous age limit. This regulation remained in force, with minor changes made on different dates, until 2004, when the Road Transport Regulation (RTR) was issued.

The 1990 and 1994 Regulations had no direct barriers to entry to the market, and the companies that had D1 licenses had the opportunity to use an unlimited number of contracted coaches. There was no significant intervention regarding the price, time, route, and frequency. The regional directorates approved timetables and price tariffs planned by the operators without changing them. The principal purpose of these initial regulations was to register the industry. They also aimed to increase the institutionalization of the companies with the condition of a minimum number of coaches and to reduce idle capacity with an age restriction for coaches. Although some of these regulations were withdrawn—with the reasons being that the industry was not ready, the financial structure of the companies did not allow them, and the partnership culture of the companies was not developed—the regulations have been significant in terms of controlling the transport services that carried out for a long time, both unregulated and unreported.

In the regulations between 1990 and 2003, excess supply from the previous years in the market continued and companies and coaches operated more than necessary due to the lack of entry barriers and the fact that they could use contracted coaches without limitation. In 2003, there were 461 operators in the market with D1 licenses. The absence of sunk cost risk was also efficient in excess supply in the industry. The coach transport industry does not require any substantial fixed investment. Even though self-owned coaches must be financed totally by the operator, any operator leaving the market can quickly sell those coaches on secondary markets. The presence of many local–regional and non-institutional companies reduced the load factor and efficiency. The lack of sufficient statistics about the industry makes it impossible to present this case in figures. However, the fact that operators were operating with low load factors, except during religious holidays/seasons, has been revealed by the industry representatives and relevant public authorities on various platforms (Aydın, 2011; DPT, 2006; Erdoğan, 2011; ITU, 2005).

The low load factor in the industry sometimes led to predatory price wars between operators in the absence of price floor regulations. Operators that served in the same routes were engaged in predatory price wars to take others out of the market, which prevented the development of a healthy competitive environment. This feature was also expressed in the decisions of the Competition Authority.[2]

The level of competition can be described as generally high due to the high number of companies and the absence of any barrier to market entry. This competition was also reflected in prices. Although uncompetitive pricing was seen occasionally in some lines where only a few companies operated, prices generally reflected a highly competitive market structure. However, the competitive structure that had emerged as a result of having too many companies did not increase companies' efficiency. On the contrary, it led to inefficient outcomes because of the industry's network structure, where the economy of scope and density have a significant role.

5.2 *Regulations After the Road Transport Law (2003)*

After the enactment of the RTL in 2003, freight and passenger transport activities on the highways were carried out within the framework of this Law. The RTL was prepared considering the aim of adjusting with the European Union and represented

[2] See Competition Authority Decision No. 10–68/1445–545 of 28.10.2010 and Decision No. 09–27/576–136 of June 11, 2009.

a significant turning point for the coach transport industry.[3] For the implementation of the RTL, the Road Transport Regulation (RTR) entered into force in 2004. The RTR was renewed in 2009 and 2018 and several changes were made on different dates. While the 1990 and 1994 Regulations only regulated the intercity passenger transportation market, the RTR also regulates the freight transportation market.

The RTL and the RTR had four aims regarding the scheduled coach market (Ministry of Transportation, 2007, p. 151). (1) To reduce the existing idle capacity and eliminate it completely at the end of a process. (2) To improve corporate transportation instead of individual transportation. (3) To increase the number of companies that are well organized and perform efficiently, with a large number of self-owned coaches, instead of companies that cannot operate efficiently, with a large number of contracted and a small number of self-owned coaches. (4) To have a transportation industry, in accordance with the European Union acquis, that has a professional reputation, professional competence, financial capability, higher service quality, operate orderly, safely, serial, economic, and is integrated with other transport modes.[4]

5.2.1 Entry, Route, and Tariff Regulations

Companies must obtain a D1 license to operate in the scheduled coach market. In order to get a D1 license, companies must have self-owned coaches with a total capacity of at least 150 seats (in the 2004 Regulation, the seat capacity requirement was 200, this number changed later) and business capital of 60,000 TL. The license fee was 30,000 TL in 2004 and was reassessed every year. For example, it was 38,828 TL in 2009, 61,372 TL in 2016, and 30,000 TL in 2018. In 2020, the license fee increased excessively to 200,000 TL. Licenses need to be renewed every five years and the renewal fee is 15% of the first fee.

According to entry regulations, there is no direct barrier to entry to the coach market. The license fee, which was increased excessively in 2020, may constitute an indirect entry barrier by increasing the entry cost for new companies. However, considering the equity and/or asset sizes required for the companies to enter the market, the fee is not a deterrent and is affordable for the enterprises.

According to the 2004 and 2009 Regulations, the operators were able to take one line for approximately every fifty self-owned seats. In practice, the applications made by the operators to take lines were approved by the regional directorates regardless of the demand structure on the line. There were no daily service (departure) frequency restrictions on these lines. With the 2018 Regulation, this line allocation

[3] Since starting accession negotiations with the EU on October 3, 2005, Turkey has been seeking to revise its legislation to compatible with EU acquis.

[4] Professional reputation refers to acting in accordance with the codes of conduct of commerce and to complying with the rules related to road transportation activities. Professional competency refers to having the education, knowledge, skill, and equipment required to carry out the road transportation activities. Financial capability refers to having the necessary fiscal sources to establish a company engaged in transportation activities, to manage such a company efficiently and suitably. See Council Directive 96/26/EC and Council Directive 98/76/EC.

regulation changed, and operators were able to take lines according to the number of coaches. Accordingly, operators use lines for four months and regional directorates take the load factor and effectiveness into account when increasing or decreasing the daily service frequency on these lines. This line allocation regulation aims to increase the load factor and enable operators to carry out more efficient transport services. However, regional directorates must monitor load factors to implement this regulation. With the 2018 Regulation, the Transportation Electronic Tracking and Control System (U-ETDS) was established within the Ministry to keep data about the activities of the companies, including load factors. Accordingly, companies have to report their daily departure, passenger, personnel, and vehicle information through the system. This system gives the Ministry the opportunity to access the data in the market in real-time and accurately. The implementation of the system was postponed to January 2020.

According to the line and tariff regulations, companies with a D1 certificate have no entry barrier to a new line and no restriction for daily service frequency. However, the line regulation introduced in 2018 may constitute an entry barrier. At this point, it will be necessary to see the practices before making any evaluation.

Although the Regulations expressed that limitations and/or temporary arrangements may be imposed on the issuance and/or number of the licenses, the number and/or capacities of vehicles, and transport lines and/or routes, no restriction have been imposed so far.

In the 2004 Regulation, the permission to use unlimited contracted coaches, which had previously been given to operators, was changed to a maximum of twice the number of self-owned coaches. This change aimed to increase the number of institutionalized companies with a large number of self-owned coaches. This arrangement was revised by the 2018 Regulation, as the contracted coach permission tripled the number of self-owned coaches. The latest revision aimed to increase consolidation and efficiency by enabling big companies to work with smaller ones as subcontractors in the market on a larger geographical scale, considering the collaborations and franchise agreements between companies in recent years.

The regulations made after 2003 exerted significant impacts on the market. Entry regulations, especially those limiting the number of contracted coaches, have reduced the number of operators in the market. The number of operators in 2003 was 461, but decreased over the years and fell to 338 in 2019. Insufficient microdata, such as load factors and daily departures, means we cannot show before and after the regulations comparatively in terms of reducing idle capacity and increasing efficiency, which were aims of the arrangements.

Besides the impact of entry regulation, the increase in the automobile and airline traffic and new high-speed trains on some lines reduced the number of operators in the market, as discussed in the following section. In the face of increasing intermodal competition, companies consolidated in order to reduce costs, and various companies that could not keep up with intense competition had to leave the market.

Despite the decline in the number of operators, there was generally no monopolistic pricing due to intermodal competition pressure and potential competitive threat.[5] In fact, the regulatory institution has not set any price ceiling since it has not been concerned with excessive prices. Between 2004 and 2015, the Competition Authority fined seven operators and issued warnings to eight operators in the industry on the grounds of competitive violations, such as price setting, market sharing, and supply setting (Solak, 2014).

5.2.2 Price Regulations

Regarding prices, the Regulations state that price floors and/or ceilings may be imposed for a specified period if market prices act to the detriment of the national economy and/or public interest and/or unreasonable prices are set and/or the competitive environment deteriorates. In practice, the regulatory institution sets price floors according to the length of lines, but it does not set price ceilings. As previously mentioned, the competitive pressure has kept prices within reasonable limits for consumers, so price ceiling implementation is not required. The price floor implementation aims to protect the operators from predatory price settings, which are common. In this sense, it has reached the aim of regulation. Operators that provide services on the same lines set prices that are very close to each other due to the "mutual dependency" seen in oligopoly markets, and turn to service quality competition instead of price competition.

The regional directorates do not intervene with the upper limit of the prices provided by the operators, and solely check whether the price tariffs comply with price floors in the approval process. Operators cannot charge more than approved price tariffs and cannot make more than a 30% discount. Prices have to be the same on departure and return on the same route. Operators regularly use their 30% discount rights and apply ticket prices at a discount, but they remove these discounts during periods of high demand, such as New Year holidays, semester holidays, and summer holidays.

With the amendment made in 2012, operators can—with the permission of the regional directorate—use the special discount once a year for up to four months for a number of seats that will not exceed 10% of the total number of seats for the coach, without being subject to a 30% discount limit. This amendment opened the way for flexible/dynamic pricing in coach services as in air transportation and other industries. However, it is not used much by operators.

The regulatory institution set price ceilings for coach terminals, which have had some indirect impacts on the market. It also removed terminal fees for coaches that enter the terminal only to drop passengers. The purpose of this regulation is the concern that terminal operators may charge excessive prices because the terminals serve in a monopoly position. Since terminal charges increase transport costs, and

[5] Similar conditions are observed in the coach services on some other countries. In Norway, for example, despite market consolidation and collaboration between companies, there were no monopoly prices due to potential competition (Leiren and Fearnley, 2008).

thus ticket fares, keeping these fees at a reasonable level will benefit both companies and consumers.

5.2.3 Coach Drivers and Company Managers' Regulations

Before 2004, drivers who had a bus driving license could use intercity coaches regardless of their education and age. The 2004 Regulation introduced education and age restrictions for coach drivers. Accordingly, coach drivers have to be at least high school graduates and between the ages of 26 and 65. Additionally, coach drivers have to own a driver professional competency certificate (DPC) and a bus driving license, and obtain a psycho-technical health report. These safety regulations aimed to reduce traffic accidents involving coaches.

The 2004 Regulation also introduced regulations for company managers. Companies with a D1 license are required to employ at least two people, one with a senior manager (SM) certificate and another one with a mid-level manager (MM) certificate. This regulation is mainly aimed to increase the institutionalization of companies. MM, SM, and DPC certificates are given within the scope of the "Road Transport Activities Professional Competency Training Regulation" issued in 2004.

5.2.4 Quality Regulations

The age limitation for coaches, introduced with the 1990 Regulation, has changed many times over the years. According to the latest changes, the age limit for self-owning coaches, which provide the minimum capacity, is 10 for the first application and during the activity period, and the age limit for the other coaches is 19. Coaches have to have at least 25 seats, including the driver. In addition to these arrangements related to the coaches, the following regulations were also introduced: departures and arrivals made from a terminal; no passenger downing-loading, except for terminals and intermediate stops; and the companies establish an active Web site that includes line, route, timetables and price tariffs, branch, agency, their communication, and similar information. Besides, regulations regarding the physical standards of coach terminals were introduced starting with the 2009 RTR.

5.2.5 Public Service Obligation

With the 2004 Regulation, public service obligations and public service contracts entered the highway transport legislation for the first time. Accordingly, if necessary, the regulatory institution may make a tender regarding transport services within the scope of the public service contract. The contractual provisions may not have consequences against or in favor of the other operating license owners who provide the same service. In practice, no public service contracts have been signed since 2004.

6 Current Outlook of the Industry and Projections for the Future

6.1 Current Outlook and Market Structure

In Turkey, approximately 27% of intercity passenger transport was carried out by scheduled and unscheduled coaches in 2018[6] (see Table 1). Generally, older people, low-income groups, and students prefer scheduled coaches. The market structure is changing rapidly, adapting to seasons such as semester holidays and summer holidays. Scheduled coach transport has a network structure that reaches all cities in Turkey. Ankara is a natural hub for the center and eastern regions of the country and Istanbul for the western regions. Coach services are carried out from these hubs to almost all other cities.

As of the end of 2019, 338 companies operated in the market, with Metro Tourism, KâmilKoc, Nilufer Tourism, and Pamukkale Tourism dominating the market. Many operators are operating on a local–regional scale and only operate between two or three cities, but carry the conditions of market entry in terms of the number of self-owned coaches.

The service quality of the companies is high due to the intramodal and intermodal competition. Companies offer services such as introducing extra comfortable, express coach services, offering in-vehicle food and beverage services, providing intracity and airport shuttle services, and selling tickets from the call center and online. Besides the companies' own Web sites, applications such as Obilet.com and Biletall.com also offer coach ticket sales services over the Internet. The increase in ticket sales through the call center and the Internet have enabled the companies to reduce their costs by reducing their ticket sales branches as well as customer satisfaction.

KâmilKoc, one of the biggest operators in the market, was acquired by the German company Flixbus one of the most significant transporters in Europe in 2020. While it is too early to evaluate the effects of Flixbus in the market, it can start having meaningful impacts on the market. KâmilKoc has already removed catering, assistants, and hosts/hostesses to reduce the costs. Preferring big coaches to carry more passengers, the company also will remove coaches with 2 + 1 seats and will use 54-seat coaches for all distances. Flixbus, the leading firm in affordable transportation, has brought in the system it has implemented in Europe and America to Turkey. Removing the services that have been standard for years will have a significant cost advantage. The effect of these changes for the whole industry will be determined by how customers will react to these changes and whether they are adopted by other operators. In this context, we see that some other operators have realized the same system.

[6] There is no data on scheduled and unscheduled services.

Table 1 Intercity passenger transportation according to transport modes (million passenger-km)

Years	Automobile	%	Coach	%	Airline	%	HST	%	Conven Rail	%	Maritime	%
2000	90,560	–	87,391	–	3555	–	–	–	4240	–	n.a	–
2005	101,227	53.0	80,925	42.4	3992	2.1	–	–	3602	1.9	1241	0.6
2010	137,857	55.8	89,056	36.0	15,159	6.1	476	0.2	3017	1.2	1570	0.6
2015	199,895	61.3	90,839	27.9	29,790	9.1	1847	0.6	1812	0.6	1836	0.6
2016	213,853	63.3	86,999	25.7	31,730	9.4	1871	0.6	1397	0.4	2059	0.6
2017	221,241	62.5	93,493	26.4	34,018	9.6	2218	0.6	1426	0.4	1833	0.5
2018	229,439	61.9	99,924	27.0	34,996	9.4	2551	0.7	1762	0.5	2101	0.6

Source KGM (2019b); TCDD (2019b)

6.2 Intermodal Competition

In Turkey, the number of passengers transported by coach has increased in recent years, but the share of coach transport is declining due to intermodal competition (see Table 1). Coaches face competitive pressure from the other transport modes on most transport lines. The increasing demand for private automobile use and air transport is particularly challenging for coach operators.

On medium and long distances, air transport is a significant competitor for coaches. With the opening up of the airline market to competition in 2003, new carriers have entered the market and ticket prices have become cheaper. Furthermore, many new airports have been built in recent years and the number of transport lines has increased. Domestic flights flew to 26 points from two centers in 2003, but this increased to 56 points from seven centers in 2019 (UAB, 2019, p. 406). Increasing the number of lines and cheaper ticket prices significantly increased the competitive pressure from air transport. Besides, government implementations such as the exemption of airline companies from the excise tax on fuel prices and the absence of a price ceiling for air carriers—which caused plane tickets to become even cheaper than coach tickets on some lines—have also affected the coach services. As can be seen from Table 1, the share of air transport increased from 2.1% in 2005 to 9.4% in 2018. Many of these airline passengers used to prefer coach transport for travel in the past.

Private automobile ownership has increased over the years in Turkey. The number of automobiles per 100 persons increased from 6.8 in 2000 to 15.1 in 2018 (KGM, 2019a, p. 11). Parallel to automobile ownership, automobile use/mobility between cities has also increased. While 53% of intercity transportation was made by automobiles in 2005, this ratio increased to 61.9% in 2018 (see Table 1). This increase has mostly affected coach transport. Besides automobile ownership, BlaBlaCar (a well-known app that supports ride-sharing private automobiles) has also become critical for intermodal competition in Turkey.

In Turkey, high-speed trains (HST), which started to operate on the newly built railway line in 2009, constitute a considerable alternative to coaches. As of the end of 2018, the HST line length was 1213 km (TCDD, 2019c, p. 30). While HST fares in the available lines are slightly more expensive than coach fares, passengers prefer to travel with them as they provide relatively comfortable and faster travel opportunities. Coach transport on these lines has decreased significantly. On the Ankara-Eskisehir line, for example, an average of 572 passengers per day are carried by conventional trains before HST, while this number reached 6000 on weekdays and 7500 on weekends after HST. The railways' share (HST + conventional trains) increased from 8 to 72% with HST (TCDD, 2017, p. 53). Coach operators affected by this increase have reduced the number of daily departures on this line, and some have even left the line completely.

6.3 Consolidation and Collaborations

In the past 10–15 years, the number of operators in the market has decreased and considerable concentration has occurred. While the consolidation was partly due to new regulations, it was also a result of intermodal and intramodal competition. Among the companies that withdrew from the market were large and well-established companies as well as small ones. Two examples are Varan Tourism, which was founded in 1946 and went bankrupt in 2016, and Ulusoy Travel, which was founded in 1937 and went bankrupt in 2017. With Varan Tourism and Ulusoy Travel withdrawing from the market, the market shares of the remaining leading companies—Metro Tourism, KâmilKoc, and Nilufer Tourism—have increased. Although the number of companies has decreased, the number of coaches has not. While 461 operators had 7600 coaches in 2003, 338 operators had 8024 coaches in 2019. Most of the coaches in the closing operators contracted or sold with existing companies in the market. Although this was not exactly a company merger or acquisition, there was an indirect merger.

In addition to consolidation, collaborations that are not in violation of competition law have been observed among companies. The rapid growth period in the industry between 1950 and 1995 is now over. Issues such as cost-effectiveness and productivity, which companies ignored during the growth period, have gained importance. Companies have turned to collaborations in order to reduce their costs and increase their competitiveness. These collaborations are mostly related to sales networks, ground services, shuttle services, and private terminals. For example, Pamukkale Tourism and Anadolu Tourism have collaborated to connect their ticket sales networks to reduce their costs and sell their tickets at more points. In addition to these kinds of collaborations, large companies such as KâmilKoc and Metro Tourism made franchise agreements with small companies to use their brands, broad service network, and web infrastructure. In addition, in some cities such as Bartın and Kastamonu, all companies have completely removed local urban shuttle services to reduce their costs (Solak, 2020, p. 17).

Consolidation and collaboration in the industry may constitute a contradiction with the Competition Law. At this point, the Competition Authority's evaluation of the relevant market, potential competition, intermodal competition, and the cost structure of the market become significant.[7] The Competition Authority has only one judgment regarding mergers and takeovers in the industry.[8] In the decision of the Competition Authority regarding Ulusoy Travel's acquisition of Varan Tourism in 2012 stated that there are no competition concerns due to reasons such as the presence of both national and local companies in the market, the absence of market entry barriers, the presence of alternative vehicles (such as airline, railway), and the fact

[7] Examples observed in Norway and Sweden. In Sweden, the National Competition Authority did not favor collaborations. In Norway, despite the National Competition Authority's opposition, collaborations were supported by the government due to the presence of intermodal competition and increased efficiency (Aarhaug and Fearnley, 2016; Alexandersson et al., 2010).

[8] See Competition Authority Decision No. 12–42/1264–415, dated 28.08.2012.

that market share rates may change significantly even within one year. The Competition Authority has only one judgment regarding collaboration too.[9] In the decision about the collaboration agreement signed between Pamukkale Tourism and Anadolu Transport, the Competition Authority concluded that this agreement met the conditions of exemption upon the justifications that the deal would yield a considerable reduction in the costs of the companies, the opportunity to sell their tickets at more booking offices, and passengers having the chance to get tickets with discount prices at more locations.

6.4 Future Insights

Looking at the intercity passenger transportation market in general reveals that the state supports air and rail transport to create a balanced transportation market by increasing the share of air and rail transport and invests in this direction.[10] In air transport, many airports have developed in recent years and new airports are emerging. These airports will increase the number of airline flight destinations when they are completed, meaning that air transport will increase even more. On the other hand, in rail transport also, HST lines have been built and some lines are developing now. We can expect that these lines receive a significant demand when becoming operational as the completed HST lines. Besides, automobile ownership and usage are increasing, as is BlaBlaCar usage. All these developments show us that, in the coming years, the intermodal competition will increase in intercity passenger transportation.

A significant factor that will affect the intermodal competition, and thus coach transport, will be per capita income. Individuals with greater income prefer to travel by aircraft, HST, or automobile instead of coach transport.

Against increasing intermodal competition, coach operators need to reduce their costs and bring innovative solutions to maintain their competitiveness. At this point, we can predict that consolidation and collaboration will continue in the industry in two ways.

The first way is with the industry's internal dynamics. Examples of this have been seen in the industry. While companies used to serve as an umbrella for individual coach owners, the new trend is that a small number of national companies are umbrellas for small local companies. These national companies will technically support subcontractors and will be a kind of platform. The amendment to increase the contracted coach permission with the 2018 Regulation aims to open the way for this. Flixbus, which is one of the largest coach transport companies in the world and entered the Turkey market by acquiring KâmilKoc, has a business model based on collaboration with regional companies. According to this business model, local partners are responsible for the daily operations of the coaches, while Flixbus conducts commercial permits, network planning, marketing, pricing, quality management,

[9] See Competition Authority Decision No. 12–06/189–51, dated 09.02.2012.

[10] For airport and HTS projects that are under construction, see UAB (2019).

and customer service activities. Flixbus has, in a sense, digitized the industry and established a transportation network/technology platform on the Internet. It offers a cost-effective service with smart network planning and a dynamic pricing system. Flixbus charges a specific commission for each ticket. Thanks to this business model, the company has achieved significant growth in a short time. It dominates the European market and has become the market leader in many European countries such as Germany, Italy, France, Benelux countries, and Poland.

The second way is the transport line regulation introduced with the 2018 Regulation. This regulation has the potential to increase consolidation and collaboration. It will also consolidate coach transportation on the lines, affect load factors, reflect on costs and prices, and increase competitiveness if it is implemented effectively by considering supply–demand balance in the lines and without harming intramodal competition.

In the coming years, there will also be developments in favor of coach transport. In recent years, a significant length of highway was built and some are still under construction. Although these are toll highways, they have shortened travel times and their high-speed limits and travel comfort have made them more comfortable. Accordingly, we can expect that coach transport will increase on the lines passing these new highways. Besides, increasing environmental sensitivity may cause passengers to prefer coaches instead of automobiles in the coming years. Governments may provide subsidies for coach services or make regulations to direct people to coach transport.

7 Conclusion

Coach transport is the most used mode in intercity public passenger transportation in Turkey. Coach services have been carried out by privately owned companies in free-market conditions without any subsidies from governments since the first development. In the initial years, the number of companies and coaches increased due to the absence of regulations and entry restrictions, and due to rapidly growing demand. These companies are mostly small, local, and non-institutionalized. Many local–regional and non-institutional companies reduced the load factor and efficiency because the industry has the nature of the network. The market has been regulated since 1990. The principal purposes of these regulations are to register the industry, reduce the idle capacity, increase efficiency, and turn the companies into institutionalized companies that have a professional reputation, professional competence, and financial capability. A law was enacted for the first time in 2003 and the subsequent regulations were made within the framework of this law.

Over time, with the effect of the regulations and the increased intermodal competition, the number of operators reduced, and partial concentration occurred in the market. Despite the consolidation in the market, monopolistic pricing was generally not seen due to intermodal competition pressure and potential competitive threat.

Price floor implementation was introduced to protect the companies against predatory price practices, frequently seen in the past, which prevented predatory price wars in the market. In this sense, the implementation has achieved its aim. With the effect of regulations for coach drivers and company managers, companies have become more institutionalized and the staff quality has also improved.

Recent developments point to the following future trends: Intermodal competition will increase further in the coming years. Along with the increase in intermodal competition, the share of coach transport will decline proportionally, though not necessarily numerically. The increasing intermodal competition will further consolidate the market and improve collaborations. Foreign investors that have just entered the market can further increase the institutionalization of other companies with the new dynamism and business model they bring to the industry.

The lessons learned from Turkey's experience for coach services are as follows:

- It is possible to carry out the intercity transportation without government subsidies. In Turkey, with coaches, the privately owned companies provide passenger transportation services to all cities without using any public funding and without any public service obligation contract for years.
- Having a free transportation market is significant in terms of market development. In addition to many other factors,[11] the fact that transportation activities are carried out under free-market conditions is important, given that coach transportation is the most commonly used public transportation mode.
- Intermodal competition and potential competition are significant factors in passenger transportation and should be taken into account while evaluating the transportation market. A decline in the number of companies in any transport line does not always lead to monopolistic behavior. The existence of intermodal competition and potential competition affects firm behavior.

References

Aarhaug, J., & Fearnley, N. (2016). Deregulation of the Norwegian long-distance express coach market. *Transport Policy, 46*, 1–6.

Aarhaug, J., Farstad, E., Fearnley, N., & Halse, A. H. (2018). Express coaches: An up-hill battle after liberalization? *Research in Transportation Economics, 72*, 82–91.

Alexandersson, G., Hultén, S., Fearnley, N., & Longva, F. (2010). Impact of regulation on the performances of long-distance transport services: A comparison of the different approaches in Sweden and Norway. *Research in Transportation Economics, 29*, 212–218.

[11] We can say the leading factors are as follows: the road network has reached all cities and flexible nature of road transportation; the railway lines have not reached all centers and have used conventional trains for years without investment; and there were a limited number of airports and flight fees were not accessible for many years.

Augustin, K., Gerike, R., Sanchez, M. J. M., & Ayala, C. (2014). Analysis of intercity bus markets on long distances in an established and a young market: The example of the US and Germany. *Research in Transportation Economics, 48*, 245–254.

Aydın, T. (2011). Kara ulaştırması ve karayolları. Rekabet Kurumu (Der.), Ulaştırma sektöründe serbestleştirme, rekabet ve rekabet hukuku sempozyumu kitabı içinde (s. 47–69).

Beria, P., Nistri, D., & Laurino, A. (2018). Intercity coach liberalisation in Italy: Fares determinants in an evolving market. *Research in Transportation Economics, 69*, 260–269.

Blayac, T., & Bougette, P. (2017). Should I go by bus? The liberalization of the long-distance bus industry in France. *Transport Policy, 56*, 50–62.

Crozet, Y., & Guihéry, L. (2018). Deregulation of long-distance coach services in France. *Research in Transportation Economics, 69*, 284–289.

DPT (State Planning Organization) (2006). *9. Kalkınma planı—karayolu ulaşımı özel ihtisas komisyonu raporu.*

Dunmore, D. (2016). *Comprehensive study on passenger transport by coach in Europe.* Final report, No. MOVE/D3/2014–261. Steer Davies Gleave.

Dürr, N. S., Heim, S., & Hüschelrath, K. (2016). Deregulation, competition, and consolidation the case of the German interurban bus industry. *Journal of Transport Economics and Policy, 50*(2), 164–188.

Erdoğan, M. (2011). Kara ulaştırması ve karayolları. Rekabet Kurumu (Der.), Ulaştırma sektöründe serbestleştirme, rekabet ve rekabet hukuku sempozyumu kitabı içinde (s. 71–74).

Grimaldi, R., Augustin, K., & Beria, P. (2017). Intercity coach liberalisation. The cases of Germany and Italy. *Transportation Research Procedia, 25*, 474–490.

ITU (Istanbul Technical University). (2005). *Ulaştırma ana planı stratejisi sonuç raporu.*

KGM (General Directorate of Highways). (2019a). *Karayolu ulaşım istatistikleri 2018.* https://www.kgm.gov.tr/SiteCollectionDocuments/KGMdocuments/Yayinlar/YayinPdf/KarayoluUlasimIstatistikleri2018.pdf, (Accessed in April 11, 2020).

KGM (General Directorate of Highways). (2019b). *Karayolu ulaşım istatistikleri.* https://www.kgm.gov.tr/Sayfalar/KGM/SiteTr/Yayinlar/Yayinlar.aspx, (Accessed in April 11, 2020).

Knorr, A., & Lueg-Arndt, A. (2016). Intercity bus deregulation in Germany–Intramodal and intermodal effects after two years. *Research in Transportation Economics*, 59, 323–329.

Leiren, M. D., & Fearnley, N. (2008). *Express coaches—the story behind a public transport success.* Paper presented at the 2008 European Transport Conference, held in Leeuwenhorst Conference Centre, October 6–8, 2008.

Ministry of Transportation. (1998). *9. Ulaştırma şurası karayolu ulaştırması komisyon raporu.*

Ministry of Transportation. (2007). *Ulaşımdan iletişime kalkınan Türkiye 2003–2007.*

Sakai, H., & Takahashi, Y. (2013). Ten years after bus deregulation in Japan: An analysis of institutional changes and cost efficiency. *Research in Transportation Economics, 39*(1), 215–225.

Solak, A. O. (2014). Türkiye otobüs ile şehirlerarası tarifeli yolcu taşımacılığı sektörünün rekabet düzeyi. *Rekabet Dergisi, 15*(4), 40–77.

Solak, A. O. (2020). Turkish intercity coach industry and COVID-19: Lessons and future insights. *Network Industries Quarterly Turkey, 1*(1), 16–20.

Taylor, Z., & Ciechański, A. (2008). What happened to the national road carrier in a post-communist country? The case of Poland's state road transport. *Transport Reviews, 28*(5), 619–640.

TCDD (Republic of Turkey State Railways). (2017). *Demiryolu sektör raporu 2016.*

TCDD (Republic of Turkey State Railways). (2019a). *Demiryolu sektör raporu 2018.*

TCDD (Republic of Turkey State Railways). (2019b). *İstatistik yıllıkları.* http://www.tcdd.gov.tr/content/35, (Accessed in April 15, 2020).

TCDD (Republic of Turkey State Railways). (2019c). *Annual statistics 2014–2018.* http://www.tcdd.gov.tr/files/istatistik//20142018yillik.pdf, (Accessed in April 13, 2020).

Tekeli, İ., & İlkin, S. (2004). *Cumhuriyetin harcı 3: Modernitenin altyapısı oluşurken.* İstanbul Bilgi Üniversitesi Yayınları.

UAB (Ministry of Transportation and Infrastructure). (2019). *Ulaşan ve erişen Türkiye 2019.*

Van de Velde, D. (2004). *Reference framework for analyzing targeted competitive tendering in public transport*. TransportøkonomiskInstitutt.
Van de Velde, D. (2013). Long-distance coach services in Europe. In M. Finger & T. Holvad (Eds.), *Regulating transport in Europe* (pp. 115–139). Edward Elgar.
White, P., & Robbins, D. (2012). Long-term development of express coach services in Britain. *Research in Transportation Economics, 36*, 30–38.

Ali Osman Solak is an associate professor of the Department of Public Finance at Bolu Abant Izzet Baysal University. He lectures on Microeconomics, Public Economics, Mathematical Economics, Statistics, and Public Choice. Solak earned his Ph.D. in Economics from Sakarya University (2011). He has an M.Sc. degree in Economics from Selcuk University (2006) and a B.Eng. degree in Civil Engineering from Cukurova University (2002). Between 2005 and 2012, he served as a civil engineer in the Ministry of Transport. His main research interests are public economics, transportation economics, and housing economics.

Regulation of Postal Sector in Turkey

Volkan Recai Cetin

Abstract The postal sector has undergone a structural change while letter post volumes have decreased and parcel and express markets have grown rapidly as a result of ever-increasing e-commerce and developing technology. This transformation has raised the importance of deregulation and effective re-regulation to ensure sustainable universal service provision while creating a competitive environment to increase quality of service for all types of postal services. This chapter discusses the Turkish postal sector and its regulation. Mainly regulated by the Postal Services Law enacted in 2013, postal liberalization has not been completed in Turkey and the incumbent universal service provider (PTT) is granted quite a large reserved area along with the resources funded by the operators. The lack of competition and uncertainty in regulation remain as major problems that are hindering improvements in the quality of service and increases in the efficiency of service provision. On the other hand, steadily growing demand for parcel and express services and the new information and communication technologies that help reduce costs and provide value-added services are emerging as opportunities for all operators, as long as an effective and sound regulatory environment is introduced and sustained.

Keywords Postal services · Regulation · Universal service · Competition · Efficiency · Turkey

1 Introduction

The postal industry served as the world's main communication method for decades, but has been one of the sectors most affected by digitalization and has experienced a structural evolution. Thanks to alternative communication technologies, letter post volumes have decreased steeply, whereas parcel and express mail volumes have surged as a result of growing e-commerce. According to the International Post Corporation (2019), global letter volume decreased by 5.9% in 2018, whereas parcel volume

V. R. Cetin (✉)
The World Bank, Washington, USA
e-mail: volkan.cetin@sbb.gov.tr

M. Eroğlu and M. Finger (eds.), *The Regulation of Turkish Network Industries*,
https://doi.org/10.1007/978-3-030-81720-6_16

grew by 9.1% and only one-third of the revenues in the postal industry resulted from the letter post segment. While the letter-to-parcel volume ratio was 13:1 in 2005, it decreased to 4:1 in 2015 and is further expected to decrease to 1:1 in 2025; the corresponding ratios for a typical incumbent are 30:1, 10:1, and 3:1, respectively (McKinsey & Company, 2019). In European countries,[1] between 2013 and 2016 period, letter post volumes declined by 4.2% annually, on average, whereas parcel and express volumes grew by an average of 13% each year (Copenhagen Economics, 2018a).

Incumbent operators have tried to react to this irreversible evolution by making operational adjustments, reorganizing offices, and optimizing logistics routes according to the changes in composition and location of demand. They have also attempted to reorganize labor and minimize competitive disadvantages of public servant status of employees by implementing flexible employment schemes and outsourcing. Having relied heavily on revenues from letter posts, they are now trying to increase revenues by penetrating into parcel and express segments and proliferating services such as banking activities and e-government services.

On the other hand, these operators still enjoy the advantages of their already established large networks that help them to benefit from economies of scale and scope, varying level of reserved areas, reputation, and loyalty advantages.

The structural change in the industry and better understanding of the dynamics of postal economics have changed the regulatory approach and enhanced the importance of effective regulation policies to ensure universal service and better quality postal services. In many countries, the postal sector is one of the gradually liberalized network industries. Postal services markets are completely or partially liberalized to achieve higher productivity, lower prices, and better quality of service, and ultimately to increase the welfare of the society.

In Turkey, as in many other countries, postal sector deregulation and re-regulation policies lag behind other network industries such as energy and electronic communications. There are two likely reasons for this delay. First, governments generally consider postal offices of these public enterprises and their ubiquitous networks as an essential channel of their relations with the society. Second, the postal industry does not exhibit monopolistic bottlenecks or requires heavy investments or large sunk costs, contrary to the electronic communications or energy sectors, and competition can develop in some segments within market dynamics.

During the continuous structural evolution period in the postal sector, as summarized above, Turkey has not yet completed the basic regulation efforts. Even though the Postal Services Law was enacted as far back as 2013, with a vision of gradual liberalization and ensuring competition, the very fundamental regulation of drawing the boundaries of the reserved area granted to the incumbent has not been endorsed yet and this gap in the regulation leads to uncertainty and failure to reach the desired level of competition.

In the second section of this chapter, fundamental issues in postal services regulation as a network industry will be discussed briefly. Nevertheless, the objective

[1] European Union Member States, European Economic Area Member States, and Switzerland.

of this chapter is to inform the readers about the postal sector and its regulation in Turkey; therefore, the economic justifications of the regulatory approaches are beyond the scope of this chapter. The third section will explain the legal and institutional framework in the Turkish postal sector and present a detailed discussion of universal service regulation, competition regulation, and other regulatory issues in Turkey. The fourth section will briefly explain the rather competitive parcel and express services segment, the final section is devoted to expectations about the future regulatory requirements of postal sector in a constantly evolving communication technologies environment.

2 Fundamentals of Postal Services Regulation as a Network Industry

Network industries such as telecommunications, postal services, energy, and transport have historically been the sectors in which states intervene substantially. Governments intervene in these sectors to ensure that these services are accessible by all users regarding their essential role in economic development as inputs of many industries and indispensable role in social inclusion. Even though governments have gradually withdrawn from production facilities in many countries after privatization of state-owned enterprises, they still play a crucial role by setting polices and strategies, regulation, subsidization, and financing where necessary.

Different types of market failures such as externalities, scale and scope economies, public good characteristics, and imperfect markets, and the need for large investments and sunk costs associated with these investments constitute common economic justifications of intervention in network industries. The postal sector is similar to other network industries in terms of having economies of scale, scope, and density characteristics. On the other hand, it differs from them in terms of its labor-intensive structure, lower investment needs, and much lower sunk costs.

Given that many alternative technologies are widely used for communications, for provision of postal services by public enterprises, and for other types of government intervention to be rationale and acceptable, there must be market failures that cannot be solved by competitive means. Even though postal markets have been liberalized in many countries, with a few exceptions such as the Netherlands, Germany, and Portugal in Europe, incumbent operators are still public administrations or publicly owned companies acting subject to private law (Copenhagen Economics, 2018b).

The main pillars of the regulation facilities in the postal industry are universal service and competition regulations. Governments regulate price and quality of universal services as well to make them accessible for all users at affordable prices and at a sufficient quality to fulfill its role in economic and social development. Cremer et al. (2000) asserted that the need for universal postal service provision lies in its role as a public service, a redistribution mechanism based on uniform pricing throughout the country, a solution to network externalities, and a tool for regional development.

Although there are many alternatives to letter post for communicating, postal services remain an important medium for communication between citizens and governments and a necessary infrastructure for trade. In this context, universal postal service are accepted as a crucial tool for business activities, economic development, and social inclusion. Meanwhile, governments' interventions to ensure universal postal service necessitate regulations to establish and protect competition in a liberalized market.

The literature about postal economics includes many studies questioning the validity of the economic justifications behind the public provision of postal services, exclusive rights granted to incumbents, access, and universal service regulations (Bradley & Colvin, 1994; Cohen & Chu, 1997; Irmak, 2008; NERA, 2004; OECD, 1999; Panzar, 1991). This chapter does not focus on these arguments or question the economic justifications in the field of postal industry, but it does examine current postal sector regulation dynamics in Turkey.

3 Postal Sector Regulation in Turkey

This section starts by explaining the legal framework for postal regulation in Turkey and the stakeholders in Turkish postal sector, followed by a discussion of universal service regulation, competition regulation, and other regulatory issues such as tariffs and access regulations in the sector.

3.1 Legal and Institutional Framework

Turkish postal sector is mainly regulated by Law no. 6475, the Postal Services Law, which was enacted in May 2013. The Law replaced the former Postal Law, Law no. 5584, which dated back to 1950 when postal services were the basic form of communication and the backbone of the relationship among governments, citizens, and businesses and were considered an essential service provided by a public institution.

A related law is the Road Transport Law, Law No. 4925, which regulates commercial passenger and freight road transport, as it regulates the transportation facilities of the postal operators. Furthermore, there are provisions related to the postal sector regulation in the laws regulating the establishment and operations of the stakeholders of the sector, such as the Ministry of Transport and Infrastructure and Information and Communication Technologies Authority.

There are also several secondary pieces of legislation regulating different aspects of the postal sector in Turkey.

In this section, as the main sectoral law that designs the regulatory framework of Turkish postal sector, the scope of the Postal Services Law will be summarized briefly and regulations will be explained in detail in further sections about the stakeholders in the sector, universal service regulation, competition regulation and other regulatory issues.

3.1.1 Postal Services Law

The objective of the Postal Services Law is defined as liberalizing the postal services market to ensure sustainable provision of high-quality and affordable postal services for all users in a competitive environment, creating a transparent and financially strong sector and setting the procedures and principles regarding the provision of postal services and effective regulation.

The Law is composed of two main parts. The first part sets the regulatory framework for the postal sector while the second part, which was substantially amended by Presidential Decree No. 4 about the establishment of public institutions in 2018, covers provisions regarding the corporatization process and operations of the incumbent operator, namely the Turkish Postal Administration (PTT/Turkish Post).

Postal items covered by the Law are defined as all items of correspondence and their special services to be conveyed and delivered to the address on the item indicated by the sender or indicated by the sender's order, or taken to the place according to the demand of the sender, books, catalogues, newspapers, periodicals, and materials for blind and partially sighted people and postal packages (parcels), which have a maximum weight of 5 kg or maximum volume of 50 dm^3 and contain merchandise with or without commercial value.

A postal package (postal parcel) is defined as any item sent by a postal service provider that does not contain materials for correspondence and has a maximum weight of 30 kg or a maximum volume of 300 dm^3. In this context, the Law covers all postal items and postal packages up to a weight of 30 kg and volume of 300 dm^3. However, these two limits do not imply similar scales in postal sector. A package that has a volume of 3 dm^3 is accepted as 1 kg as approximation in collection; therefore, the scope of the Law practically covers packages up to 100 kg, which is far beyond the generally accepted limits for a postal package.

The Law brought regulations in four main interrelated areas.

Firstly, to separate policy setting, regulation, and operation activities and thereby eliminate conflict of interest in the sector and to increase regulatory capacity. The Information and Communication Technologies Authority (ICTA), which has been the electronic communications sector regulator since 2000, was assigned as the national regulatory authority (NRA) of postal sector as well.

Secondly, the first step toward liberalization of the postal services market was taken by defining the reserved area granted to the incumbent. Meanwhile, the quantitative measures of the exclusive rights, namely the weight and/or price of a letter, was not determined at the Law and the president (formerly the Council of Ministers) was given the right to set the limits of the monopoly, upon the recommendation of the ICTA and the proposal of the Ministry of Transport and Infrastructure. However, the presidential decree has not been endorsed yet, so the large legally reserved area granted to the incumbent remains.

Thirdly, the principles, scope, and financing of the universal postal service are regulated as a prerequisite for a liberalized market to ensure that every user can access postal services of a certain quality at an affordable price.

Finally, the law also regulates the corporatization of DG Turkish Postal Administration as a publicly owned company that operates subject to private law provisions.

In addition to these regulations, the Law contains provisions on authorization, tariffs, consumer protection rules, confidentiality, and sanctions to ensure a properly functioning market.

After the enactment of the Law, the NRA issued three by-laws about the provision of postal services, authorization of service providers and administrative sanctions to design the procedures and principles in the application of the Postal Services Law.

3.2 Functional Separation in the Sector and Institutional Framework

To eliminate the conflict of interest in the sector and to increase the regulatory capacity, the Postal Services Law separated policy setting, regulation, and operation activities. The main public institutions in the sector are the Ministry of Transport and Infrastructure, the Information and Communication Technologies Authority, and the Turkish Postal Administration.

3.2.1 Ministry of Transport and Infrastructure

The Ministry of Transport and Infrastructure (MoT) is the line ministry that is responsible for designing policies and strategies in the field of postal sector for the sustainable provision of economical, accessible, safe, high-quality services in an environmentally friendly, equitable, and competitive environment.

The MoT is responsible for designing policies in the field of universal service provision in line with economic, social, cultural, and technological developments and monitoring their application. The MoT can take necessary measures to ensure universal service provision, including planning, construction, and operation of required infrastructure. The net cost of universal service provision is estimated and universal service providers are monitored and compensated if necessary by the Ministry.

3.2.2 The National Regulatory Authority: Information and Communication Technologies Authority

As mentioned above, the ICTA is the NRA for the electronic communications sector and the postal sector. In many countries, the postal sector is regulated by the electronic communications sector NRA with the aim of benefiting from existing regulatory capacity on the grounds that regulation needs are similar in these two network industries. However, such a decision creates a risk that the importance of the postal sector may be underestimated given the size and complexity of the electronic communications sector.

The ICTA's main responsibility is defined in the Law as making the necessary regulations to ensure the provision of safe and sustainable postal services to the users at an affordable rate and in accordance with the policy and strategies determined by the MoT.

The ICTA makes all postal sector-specific secondary regulations except the ones about universal service standards and its financing, which are at the discretion of the MoT. The Authority is responsible for taking all measures for establishing and protecting the competitive environment in the market and ensuring the provision of high-quality services. Regulated areas include authorization procedures, monitoring of authorized service providers, tariffs, consumer protection, dispute resolution, and administrative sanctions.

3.2.3 The Incumbent: Turkish Postal Administration

The incumbent operator, the Turkish Postal Administration (PTT/Turkish Post), has been the universal service provider since its establishment in 1840. It was corporatized in 2013 by the Postal Services Law to eliminate the inefficiencies arising from its public enterprise status and prepare the firm for the prospective competitive environment after the liberalization of the market. PTT is a 100% publicly owned capital firm, its shares are owned by Turkish Wealth Fund and its operations are subject to private law. PTT operates and provides universal service under the terms of the Service Contract signed with the ICTA.

PTT has a large network, with around 5000 offices throughout the country. As the letter market shrinks gradually, PTT intends to have a growing share of parcels in its revenues. Moreover, using the advantage of its ubiquitous network, PTT has diversified its services by growing into banking activities with many services provided either on its own or in collaboration with commercial banks. As of September 2019, 1305 out of the PTT's 5091 offices were located in provinces that do not have any bank branches and PTT is the only entity that can handle financial transactions as an infrastructure for trade and the relationship between the government and citizens (Strategy and Budgeting Office, 2019).

PTT stopped publishing periodical statistics about its operations, revenues, network, employment, etc. after being restructured as a company subject to private law. The ICTA does not publish sectoral statistics in the above-mentioned fields either. The lack of data remains as a problem for transparency and predictability in the sector.

3.2.4 Private Postal Operators

As of September 2020, 35 private operators are authorized by the ICTA to provide postal services outside the scope of the legally reserved area. Of these 35 service providers, 19 firms are authorized to operate at national level, while a further 12 are granted authorization to operate only in the province of Istanbul, which has the largest population in Turkey and is the center of commerce.

As of December 2019, private licensed operators in Turkey had approximately 12,000 offices and 120,000 employees (KARID, 2019). Thanks to the steadily growing e-commerce in B2C markets, the demand for parcel and express services of private operators is expected to rise.

3.3 Universal Postal Service Regulation

In a liberalized postal sector, for commercial reasons, new entrants that are not forced to provide universal service are naturally expected to operate only in profitable areas, such as urban and industrialized regions of a country and profitable consumer groups. Anticipated cream-skimming operations of new competitors raise the risk of eroding the funds of the incumbent, which are reserved for cross-subsidizing the non-commercial services. In this context, in liberalized markets, a need emerges for an effective universal service regulation to fulfill the policy objective of sustainably providing postal services of certain quality for all users at affordable prices with economic and social considerations.

Even though the universal service obligation (USO) is justified by social considerations, the funding mechanism of the universal service can itself have distortionary effects on social welfare (Cetin, 2007). Therefore, it is essential to regulate universal service effectively and to comply with the generally accepted regulation principles such as ensuring efficiency, sustainability, transparency, accountability, and proportionality.

Setting universal service regulation parameters such as the scope, funding mechanism, universal service provider (USP) designation, and estimating the net costs of universal service provision should be based on sound analyses that take initial market structure, demographic conditions, urbanization level, geographic properties, tax systems, etc. into account. However, as funding mechanisms have their own advantages and disadvantages when compared to each other, and as there is a trade-off in complying with the above-mentioned principles, universal service regulation policies are the results of political decisions that depend on redistribution policies of the governments in achieving economic and social welfare goals.

As the first step in universal service regulation, governments should define the services (and their quality) that will be accessible by all consumers at affordable prices. After defining the scope of universal service in parallel with the social and economic objectives, to apply sound universal service regulation, governments should make effective decisions regarding three areas: the funding mechanism of universal service, designation of the universal service provider(s), and calculation of the net cost of universal service provision.

3.3.1 Scope of Universal Service

The scope of universal service should ideally cover services that are indispensable for social inclusion and increase the welfare of the society despite the intervention in the market (Cetin & Irmak, 2012). Irmak (2008) suggested two additional conditions that should be met for a service to be accepted as universal service. Firstly, the service must be already in use by the vast majority of consumers. Secondly, it would not possible for the rest of the population to access that service at affordable prices because of high costs unless it is regulated.

According to the Postal Services Law, the scope of the universal postal service covers the acceptance, collection, sorting, dispatch, distribution, and delivery of postal items weighing up to two kilograms, postal parcels weighing up to twenty kilograms, printed material, and materials for blind and partially sighted people. Universal postal services must be accessible by all users at affordable prices every working day, not less than five working days a week, except the special conditions deemed by the MoT.[2] The scope of the universal postal service may be amended by the president, taking the geographical, social, cultural, economic, commercial conditions into account upon the proposal of the MoT.

The main universal service principles explicitly mentioned in the Law are providing services without any form of discrimination (arising from language, nationality, gender, ideological considerations, religion, etc.), taking measures regarding the special needs of people in need of social protection, providing services related to registered and insured items, and taking responsibility in case of lost, theft, and damage of postal items and offering uninterrupted services except in case of *force majeure*.

[2] By-law on Universal Postal Services defines the exceptions in rural areas without established postal offices. In rural areas, in counties with a population of fewer than 2,000 people, postal items within the scope of universal service must be collected and delivered at least one working day per week. Delivery services must be provided at least two working days per week in counties with a population of 2,001–5,000 people and at least three working days a week in counties with a population of more than 5,000 people. The net cost of universal service is calculated by taking the costs and revenues in these areas into account.

3.3.2 Funding of the Universal Service

Universal service in postal market has been traditionally financed by state-owned incumbent operators that have been granted exclusive monopoly rights to finance non-commercial activities with the profits of services provided in low-cost regions. Since the 1980s, monopolies have been extensively criticized on the grounds that they failed to achieve economic efficiency in service provision or to create incentives for the incumbents to provide services of good quality at lower prices due to the lack of competition.

In line with the deregulation and re-regulation policies incorporating liberalization of network industries, reserved areas were completely or partially abolished in many countries and a need for alternative funding mechanisms emerged to ensure sustainability of service provision for all users.

In addition to monopoly rights, commonly used funding mechanisms are (i) establishment of a compensation fund contributed by postal operators (including pay-or-play schemes) and (ii) compensation from general budget. These three mechanisms have their own pros and cons in terms of ensuring efficiency in service provision, providing incentives for productivity, lower prices and higher quality of service, lower administrative costs and applicability, and fulfilling necessary universal service regulation principles of sustainability, transparency, neutrality, and proportionality.

Regardless of the funding mechanism, the costs of universal service provision will eventually be borne by consumers. However, the consumer groups that will carry the financial burden and the burden reflected on each consumer will be determined by the funding method (Cetin & Irmak, 2012).

To ensure the provision of services in line with the agreed scope, if a mechanism is insufficient to compensate the net costs of universal service provision with tolerable inefficiencies in the market and there is a need for additional financing, different mechanisms can be utilized in a complementary way. However, when two alternative mechanisms are applied together without a sound quantitative assessment, which is hardly computable, the advantages of each method may disappear, whereas their disadvantages persist.

Before the enactment of the Postal Services Law, the legally reserved area in Turkey covered all open and closed letters and postcards sent as items of correspondence. The Postal Services Law was prepared with the intention of gradually liberalizing the letter post segment in which there was already de facto competition after the entrance of many parcel and express service providers. The Law contained provisions about both postal monopoly –which was planned to be only partially abolished, a decision that already lagged behind market dynamics—and about the establishment of a compensation fund contributed to by the service providers.

The reserved area granted to PTT covers (i) acceptance, collection, sorting, dispatch, distribution, and delivery of domestic and international items of correspondence that either weigh or are priced below the limit set by the president upon the proposal of the MoT; (ii) acceptance, collection, sorting, dispatch, distribution, and delivery of all legal notifications and electronic notifications without prejudice to Taxation Procedures Law; (iii) postal services of the Turkish Armed Forces during peacetime; and (iv) the issuing of and selling of postal stamps, personalized stamps, commemorative stamps, envelopes, and postcards stating the price to be charged.

Although the Law was enforced in 2013, the presidential decree setting the weight or price limit that will precisely define the size of the postal monopoly has not yet been endorsed. Therefore, the postal monopoly involving all items of correspondence without a quantitative limit is still in force. The consequence of the large legally reserved area is not only the lack of competition and efficiency losses in the market, but also the uncertainty faced by the operators, which affects the operational and investment decisions of both already licensed firms and potential entrants.

In addition to the postal monopoly, a sectoral fund was established to compensate the net costs of universal service provision. The majority of the funds are generated by a certain share of revenues of operators, which acts as an additional tax, and a smaller amount is provided by a certain share of administrative fines in postal sector. All licensed postal operators, including PTT, contribute to the fund and any operator that is authorized as a universal service provider can claim funds for the compensation of net costs.

Employing two alternative financing mechanisms simultaneously hinders the monitoring of the effectiveness of regulation in the market and its effects on the stakeholders, and thereby the net effect on social welfare. The main advantage of a reserved area is its applicability and lower administrative costs. However, when it is applied with a compensation fund in a complementary way, it is essential to conduct sound quantitative analysis so as to avoid creating an unfair financial burden either on the incumbent universal service provider or on the rival operators. Firstly, net cost of the universal service provision must be calculated. Secondly, it is essential to estimate the size of the funds available for the incumbent that results from the granted reserved area. The difference between two estimations could justify the need for additional financing from the compensation fund. The need for these complicated analyses and the efforts to monitor the violations of monopoly in the market eliminate the applicability benefits of granting a reserved area, whereas the commonly criticized downsides of a monopoly—such as the allocative and distributional inefficiencies, lower productivity, and the lack of incentive to innovate and provide higher-quality service—persist.

If the incumbent is over-compensated due to the lack of accurate above-mentioned analyses, payments of the operators to the compensation fund take the form of an additional proportional tax, which will eventually be borne by consumers and will have a negative effect on their welfare.

3.3.3 Designation of the Universal Service Provider

USP designation mechanism is another important policy step in universal service regulation as the designation procedure defines the results of the selected method of financing. As long as a postal monopoly is applied to finance universal service obligations, the incumbent is taken inherently as the universal service provider. On the other hand, when a sectoral fund or general taxes are used for financing, there are four ways to designate USP: (i) the incumbent can be designated as the universal service provider; (ii) designation can be done based on the results of an auction in which the operator demanding the lowest compensation for a pre-determined quality of service is assumed to operate the most efficiently; (iii) all operators can be required to provide universal service; or (iv) no designation is made and provision is left to be accomplished by the market dynamics.

In Turkey, in line with the dual financing mechanisms, USP designation is applied as a hybrid policy as well. The incumbent, PTT, which is granted a reserved area, is designated as the universal service provider by the Postal Services Law and operates as a USP according to the terms defined in the Service Contract signed with the ICTA.

Additionally, all postal service providers can be designated as USPs upon their own request if they fulfill the terms and conditions in the authorization contract that is signed with the MoT and can claim compensation from the universal service fund. However, this provision is clearly non-functional and, given the substantial advantages of PTT, such as the already established large network of offices, large scale, and monopoly rights, no private operator is expected to demand designation under current conditions.

3.4 Competition Regulation

In a network industry with market failures and reserved areas, to establish and sustain a competitive environment, the NRA must effectively regulate competition and eliminate practices in which the incumbent uses the funds that are reserved or given as compensation for funding the net cost of universal service to cross-subsidize its services in non-reserved areas. By using the funds, the incumbent may set low and/or distortionary prices in competitive segments of the market and deter entry into market or urge rivals that operate more efficiently to exit the market. As incumbent postal operators are usually publicly owned, their main objective is not profit maximization; therefore, they may keep the prices under the marginal cost of the services long enough to keep the competitors outside the market by cross-subsidizing.

In addition to cross-subsidization, the incumbent may also distort competition in different ways such as volume discounts, bundling and tying practices, and blocking access to its network (Cetin & Irmak, 2012).

It is essential to monitor and audit the accounts of the universal service provider in a sector with any size of reserved area and compensation funds. The ICTA is responsible for monitoring the accounts of the postal operators and preventing cross-subsidization to protect competition. To eliminate cross-subsidization, accounting separation is critical. According to the Postal Services Law, all service providers, including the incumbent universal service provider, must keep separate accounts for each service licensed by the ICTA and the incumbent must keep separate accounts for services within the reserved area and outside the reserved area as well.

However, according to Cetin (2007), separating accounts in the above-mentioned way may not generally be enough to prevent cross-subsidization because (i) the border between competitive and uncompetitive segments is not clearly defined in dynamic markets with ever-changing technology and consumer preferences, and (ii) it is not easy to separate operational and administrative costs accurately in an organization that provides many services and benefits from economies of scope. The existence of two simultaneous compensation mechanisms makes the control of funds even more complicated and requires strict rules, transparency, high regulatory capacity, and effective monitoring.

3.5 Other Regulatory Issues: Access, Tariffs, Quality of Service Regulations

In addition to universal service and competition regulations, in order to establish and sustain competition in the postal sector after deregulation and ensure universal service, NRAs regulate other closely linked parameters such as access to infrastructure, tariffs, and quality of service. Regulatory principles and the procedures of these parameters are provided in detail in the By-Law on Provision of Postal Services issued by the ICTA.

Access to network is a critical issue in achieving a competitive environment. Depending on the alternative views about the characteristics of postal infrastructure and the results of the debate about whether all, none, or a part of the incumbent's network should be considered as an essential facility for the provision services, the NRA may ex-ante require the incumbent or an operator that has significant market power to allow access to its network under certain conditions or leave the agreement completely to market dynamics, or apply a hybrid policy, as is the case in Turkey.

In Turkish postal regulation, network access is not ex-ante regulated. However, if an operator demands access to certain postal infrastructure of another and two operators cannot agree on the terms of an access contract, they can request mediation from the ICTA.

Service providers are free to set their prices within market dynamics in a transparent and non-discriminatory way without prejudice to the Competition Law. However, the ICTA may regulate tariffs of postal service providers that have significant market power to ensure competition and protect consumers. Tariffs may be regulated to eliminate distortionary pricing and cross-subsidization.

Tariffs for services in the scope of universal service must be set in line with the base tariff declared by the ICTA and must be affordable, non-discriminatory, and transparent. Tariffs may be regulated by the ICTA if deemed necessary. Geographically uniform pricing must be employed, except for bulk mail discounts, which must reflect costs and for tariffs for users in need of social protection.

Regarding quality of service, the ICTA is entitled to impose certain quality of service standards to operators; however, there are currently no requirements such as a certain percentage of letters to be delivered in specified number of days after collection.

4 Parcel and Express Services

Historically, parcel and express services have usually been excluded from the reserved area. In Turkey, they are mostly provided by private operators that also operate subject to the Postal Services Law and are licensed by the ICTA. Additionally, their transport facilities are subject to specific provisions in the Road Transport Law and they need to be authorized by the MoT to provide services.

The Road Transport Law and its related by-law involve provisions about the collection, transport, and delivery of postal packages. Contrary to the Postal Services Law, the Road Transport Law considers these services as road freight transport facilities similar to other logistics operations instead of components of a communication service provided throughout a well-defined network.

In Turkey, private operators started to operate in the postal market long before the legal framework was prepared as a result of the increasing need for delivery of parcels in line with ICT and e-commerce developments and users' demand for better quality of service. Before the enactment of the Postal Services Law, they provided their services subject to the Road Transport Law as a temporary regulation and were not authorized as postal service providers. As a challenge to the legal monopoly of the former Postal Law, which dated back to 1950 and represents de facto liberalization, the parcel delivery services segment in Turkey was developed in a similar way to value-added express services in developed countries with fast delivery practices, and the operators captured the market with their speed, reliability, and track-and-trace services. Currently, most of the standard parcels are delivered at D + 1 between major cities.

Basic parcel services such as acceptance, collection, sorting, dispatch, distribution, and delivery of parcels weighing up to 20 kg are included in the scope of the universal service in Turkey and are subject to conditions and requirements similar to items of correspondence that are included in the scope of universal service. Collection and delivery of parcels universally turned out to be more important with the rise in e-commerce, both in B2B operations and retail trade in the B2C segment. B2C e-commerce in Turkey exhibited an average annual growth rate of 35% during the 2015–2019 period and the growth rate in 2019 reached 39% (TUBISAD, 2020). As a result of the surging derived demand, the main parcel and express service providers that are authorized to operate at national level have gradually increased the number of their offices in rather small provinces; nevertheless, PTT is the sole universal service provider for parcels.

5 Future Expectations for Postal Services Market and Regulatory Requirements and Concluding Remarks

As the developments in ICT shape every sector, operators always move before regulatory practices and regulations emerge as a response to developments in markets. Significantly affected by digitalization, the postal sector is no exception.

The focus of regulation in the sector will continue to be on ensuring a universal postal service while providing high-quality services in a competitive environment. However, we are at a new stage of discussing the scope of the universal service in an era of ever-spreading digital services and the ways to finance USO at this age.

USO can be redesigned in several ways, such as reducing the frequency of delivery, decreasing the required speed of delivery, reducing the number of post offices, loosening the delivery at premises requirements, and applying varying requirements for urban and rural areas.

The role of postal services in communications has notably changed, on the contrary, there is no alternative for most of the physical deliveries. Although letter demand has plummeted, increasing demand for parcels, which is also a part of the USO, has created difficulties in terms of cutting the frequency and speed of deliveries in rural areas, particularly for businesses that rely on regular collections and deliveries.

On the other hand, budgetary restrictions have limited the sustainability of universal services for the incumbents. Despite efforts to increase market share in parcels, the loss of historical revenues from letters and competition with private operators has eroded the funds and created the need for additional compensation. Meanwhile, the effects of easing the USO requirements may create substantial reductions for the incumbents. For instance, Copenhagen Economics (2017) estimated that the costs of USO would almost be halved in Norway if delivery frequency was adjusted to every second week day.

Despite the strict regulatory approaches in providing services of general economic interest to facilitate internal markets, loosening decisions were observed for USO requirements recently, even in some European countries. In Denmark, for example, in response to a reduction in letter post volumes of more than 80%, letter post universal delivery requirements were decreased from five days a week to three in 2009 and one day a week in 2018 (McKinsey & Company, 2019). Meanwhile, parcels are still required to be delivered five days a week to facilitate trade in Denmark, in which 82% of the population is engaged in e-commerce, at a rate far above the EU averages (Copenhagen Economics, 2018b). In the same period, some countries reduced delivery frequency and some of them started to impose different requirements for urban and rural areas (Copenhagen Economics, 2018a).

In this context, regarding sustainability, it is essential to comprehensively analyze the potential effect on consumers and operators both in the short and long terms, by taking into account changes in the size and composition of demand, the ubiquity of alternative means of communication, rising demand for physical deliveries, and financial restrictions.

Regarding the changing role of postal services in the economy, it is expected that there will be a continuous need for sound and dynamic regulation in the postal industry in upcoming years to ensure the provision of high-quality services for all users.

References

Bradley, M. D. & Colvin, J. (1994). An econometric model of postal delivery. In M. A. Crew & P. R. Kleindorfer (Eds.), *Commercialization of postal and delivery services: National and international perspectives*. Kluwer Academic Publishers.

Cetin, V. R. & Irmak, I. (2012). *Posta hizmetlerinin ekonomisi ve regulasyonu*. Rakun.

Cetin, V. R. (2007). *Posta hizmetlerinin serbestlestirilmesi ve regulasyonu; Turkiye posta sektoru reformu için bir oneri*, Devlet Planlama Teskilati.

Cohen, R. H. & Chu, E. H. (1997). A measure of scale economies for postal systems. In M. A. Crew, & P. R. Kleindorfer (Eds.), *Managing change in the postal and delivery industries*. Kluwer Academic Publishers.

Copenhagen Economics. (2017). *Effects of changing the USO in Norway*. https://www.copenhageneconomics.com/publications/publication/effects-of-changing-the-uso-in-norway.

Copenhagen Economics. (2018a). *Main developments in the postal sector (2013–2016)*. Publications Office of the European Union

Copenhagen Economics. (2018b). *Main developments in the postal sector (2013–2016): Country fiches*. Publications Office of the European Union,

Cremer, H., Gasmi, F., Grimaud A., & Laffont J. J. (2000). The cost of universal service in the postal sector. In M. A. Crew, & P. R. Kleindorfer (Eds.), *Current directions in postal reform*. Kluwer Academic Publishers.

International Post Corporation. (2019). *Global postal industry report 2019: Key findings*. https://www.ipc.be/sector-data/postal-sector/key-findings.

Irmak, I. (2008). *Evrensel hizmet; Turkiye posta sektorunde evrensel hizmet regulasyonu için oneriler*, Devlet Planlama Teskilati.

KARID. (2019, Dec 18). *Koordinasyon toplantisi sunumu*, Turkiye Ulastirma ve Lojistik Meclisi Koordinasyon Toplantisi. http://www.karid.org.tr/2018-yili-tobb-ulastirma-meclisi-koordinasyon-toplantisi-yapildi/.

McKinsey & Company. (2019). *The endgame for postal networks: How to win in the age of e-commerce*. https://www.mckinsey.com/industries/travel-logistics-and-transport-infrastructure/our-insights/the-endgame-for-postal-networks-how-to-win-in-the-age-of-e-commerce.

NERA. (2004). *Economics of postal services: Final report to the European Commission DG-MARKT*. NERA Economic Consulting.

OECD. (1999). *Promoting competition in postal services*, OECD Publishing.

Panzar, J. C. (1991). Is postal service a natural monopoly? In M. A. Crew, & P. R. Kleindorfer (Eds.), *Competition, and innovation in postal services*. Kluwer Academic Publishers.

Strategy and Budgeting Office. (2019). *2020 yılı Cumhurbaşkanlığı yıllık programı*. Cumhurbaskanligi Strateji ve Butce Baskanligi.

TUBISAD. (2020). *Turkiye'de e-ticaret: 2019 pazar büyüklüğü.* http://www.tubisad.org.tr/tr/guncel/detay/TUBISAD-E-Ticaret-hacmi-2019da-yuzde-39-buyuyerek-83-1-milyar-TLye-ulasti-/58/2572/0.

Volkan Recai Cetin currently works as a consultant at the World Bank Ankara Office. Prior to this position, he served as the Head of Department of Transport and Logistics at the Presidency of Turkey Strategy and Budgeting Office and Ministry of Development, and as a senior transport specialist at TR Prime Ministry State Planning Organization. He holds a B.Sc. in civil engineering, MPA degree from Cornell University Cornell Institute for Public Affairs and Ph.D. degree in Public Economics from Ankara University. He conducted research as a visiting scholar at University College London Centre for Transport Studies and Quantitative and Applied Spatial Economic Research (QASER) laboratory. He is also a part-time instructor at Istanbul Commerce University and Yıldırım Beyazıt University. His research areas focuses on transport economics and planning; infrastructure investments, policy and regulation; policy analysis and project management.

Regulation of Ports in Turkey

Kazım Yeni and Soner Esmer

Abstract In global trade, the fact that 85% of cargo is transported by the maritime sector brings that sector to a critical point. Three elements are needed for sea transportation to take place. These are the cargo transported by sea, the ships carrying these cargoes, and the ports where these vessels are served. With the increasing globalization trend in the 1990s, the maritime sector reached global coverage in both ship management and port management as a network industry. As a newly industrialized country with high growth rates, Turkey's share in global trade is increasing every year. With the increasing foreign trade, the roles of the ports are getting stronger, which increases the interest of national and also international investors in Turkish ports. In fact, with privatization practices that started in 1997, the public sector has largely abandoned its port management identity and has taken the rule-setting and regulatory identity. Today in Turkey, there are approximately 180 port facilities, 80 of which are large-scale, mainly operated by the private sector. This study was conducted as part of a thorough assessment of port regulation in Turkey. The study includes not only current regulations regarding the port management, but also port management and cargo classification of businesses operating in Turkey, and also presents handling figures.

Keywords Ports · Regulations · Turkey

1 Introduction

Ports are places that are characterized by the essential function of exchanging cargo between ship and shore (Robinson, 2002). The critical nature of a port is a connection point. It is a platform that links sea and inland transportation, the local hinterland and overseas foreland, various shipping and transport service providers, as well as trade and the urban system where the port is located (Lam & Song, 2013). Depending on

K. Yeni · S. Esmer (✉)
Iskenderun Technical University, İskenderun, Turkey

S. Esmer
Dokuz Eylul University, İzmir, Turkey

M. Eroğlu and M. Finger (eds.), *The Regulation of Turkish Network Industries*,
https://doi.org/10.1007/978-3-030-81720-6_17

the developments in world trade, the traditional port-to-port transportation concept has turned into a door-to-door approach; as a result, the importance of the services provided in ports has increased. Ports are elements embedded in value-driven chain systems, in value chain constellations; they deliver value to shippers and other third-party service providers in the value-driven chain (Robinson, 2002).

In terms of collecting and sending cargoes subject to foreign trade to overseas markets, as well as ensuring that cargoes brought from overseas markets are distributed to the hinterlands, the ports are considered as logistics nodes and are service businesses of strategic importance. Developments in the global production and supply chain have effects on port management and port operations.

Turkey is located in a strategic position close to the main trade routes with serious economic objectives. As far as Turkey's land-bridge position in North–South and West–East transit transportation is considered, ports have a vital impact on the efficiency of logistics activities of Turkey (Yeni & Tuna, 2003). Furthermore, Turkey with its ports and straits, which are linked to the Aegean and the Black Sea, is known as a critical node in the sea transportation network, especially due to its strategic position on Far East-Europe and the Black Sea seaborne trade routes (Esmer & Duru, 2016). In this sense, ports are among the most important transportation infrastructures.

As with every business, developments and changes in the external business environment affect port business and operations. As a result, port business management must adapt to changing environmental conditions, be one step ahead of their competitors, and has the ability to think and see in the long run. In this context, this study defines and explains the legal environmental factors for port business with a general view of the port business in Turkey.

2 Overview of the Port Sector in Turkey

As of 2020, there are about 180 ports in the coast of Turkey, 128 of them are operated by private enterprise, while 23 of them are operated by municipalities. The remaining ports are operated by two public units: Turkish Maritime Facilities (TDI) and Turkish State Railways (TCDD). Most public ports are commercialized and operated by the private sector (Esmer & Duru, 2016).

Table 1 shows the total amount of handled cargo between 2015 and 2019 in ports in Turkey.

Based on Table 1, 72% of the cargo handled in ports in Turkey consists of foreign trade (imports and exports), 12% of cabotage, and the rest is made up of transit cargoes. Table 2 shows export and import figures of foreign trade loads for the same period.

Based on Table 2, only 37% of foreign trade cargoes consist of export cargoes, while 63% consists of import cargoes. However, the classification of cargoes handled in ports in Turkey between 2015 and 2019 is given in Table 3.

Given Table 3 above, cargo handled in ports in Turkey consists of bulk liquids (33%), solid bulk (28%), containerized (23%), general cargo (14%), and in-vehicle

Table 1 Total cargo handled in ports in Turkey (2015–2019)

Year	Foreign trade cargoes (Tons)	Cabotage cargoes (Tons)	Transit cargoes (Tons)	Total (Tons)
2015	300,478,930	52,472,668	63,085,097	416,036,695
2016	309,937,639	53,300,216	66,963,307	430,201,162
2017	347,348,092	60,396,079	63,429,725	471,173,896
2018	328,969,455	59,555,845	71,628,260	460,153,560
2019	353,081,390	56,112,724	74,974,298	484,168,412

Source Ministry of Transport and Infrastructure (2020) Republic of Turkey

Table 2 Export and import figures of Turkey by sea freight (2015–2019)

Year	Export (Tons)	Import (Tons)	Total (Tons)
2015	92,152,622	208,326,308	300,478,930
2016	94,805,120	215,132,519	309,937,639
2017	113,692,068	233,656,024	347,348,092
2018	110,424,635	218,544,820	328,969,455
2019	131,676,578	221,404,812	353,081,390

Source Ministry of Transport and Infrastructure (2020) Republic of Turkey

Table 3 Classification of cargo handled in ports in Turkey (2015–2019)

Year	Bulk liquids (Tons)	Solid bulk (Tons)	Container (Tons)	General cargo (Tons)	In-vehicle (Tons)	Total (Tons)
2015	146,554,192	110,701,538	87,025,857	63,505,731	8,249,377	416,036,695
2016	145,024,367	116,516,338	94,928,597	65,564,619	8,167,241	430,201,162
2017	152,897,347	128,842,715	107,917,908	72,147,186	9,368,740	471,173,896
2018	139,717,069	133,653,221	114,231,465	63,977,765	8,574,040	460,153,560
2019	155,253,914	150,344,563	118,768,010	52,672,991	7,128,934	484,168,412

Source Ministry of Transport and Infrastructure (2020) Republic of Turkey

cargo (2%). The bulk liquid cargo group has the highest value in handling because of petroleum and petroleum products (especially crude oil).

More than half of the ports in Turkey are located in the Istanbul, Kocaeli, Gemlik, Izmir, and Iskenderun regions. The others stretch across Turkey's coastal areas (Esmer & Duru, 2016).

There are 71 harbor masters (serving as Port Authorities) of the Ministry of Transport and Infrastructure (https://www.uab.gov.tr/liman-bas-kanliklari). Table 4 shows the top ten harbor masters in terms of maximum cargo handled in 2019.

Table 4 Top ten harbor masters in terms of maximum cargo handling (2019)

Rank	Harbor master	Total handled (Tons)	Share (%)
1	Kocaeli	72,196,415	14.91
2	Botaş	66,945,044	13.83
3	Aliağa	65,799,062	13.59
4	İskenderun	62,167,713	12.84
5	Mersin	36,373,703	7.51
6	Ambarlı	34,649,484	7.16
7	Tekirdağ	29,933,977	6.18
8	Gemlik	13,908,352	2.87
9	Karabiga	12,969,988	2.68
10	Zonguldak	11,960,291	2.47
Others		77,264,383	15.96
Total		484,168,412	100.0

Source Yeni (2020)

According to Table 4, 84% of all handling in Turkey takes place in the ten harbor masters' administrative areas. Table 5 shows the top ten harbor masters in terms of maximum container throughput in 2019. Table 5 also shows that the ports with most container handling are clustered in Marmara Region (Istanbul, Kocaeli and Gemlik), Mediterranean (Mersin and Iskenderun), and Aegean (Izmir and Aliaga).

Table 5 Top ten harbor masters in terms of container handling (2019)

Rank	Harbor master	Total handled (TEU)	Share (%)
1	Ambarlı	3,104,882	26.8
2	Mersin	1,854,312	16.0
3	Kocaeli	1,715,193	14.8
4	Tekirdağ	1,413,962	12.2
5	Aliağa	1,132,480	9.8
6	Gemlik	861,657	7.4
7	İskenderun	680,120	5.9
8	İzmir	541,679	4.7
9	Antalya	150,678	1.3
10	Samsun	67,426	0.6
Others		69,448	0.6
Total		11,591,838	100.0

Source Yeni (2020)

3 Port Governance Models in Turkey

After the Republican era in Turkey, commercial port operations were conducted by the public authorities. After 1980, in parallel with the developments in foreign trade, the total handling volume of the ports in Turkey increased. On the other hand, factories (industrial establishments) that brought their raw materials by sea and transport their products by ships located on the coastal side have naturally equipped their facilities with piers or port facilities. In the following years, as the public-operated ports could not meet the increasing needs, factory ports or piers provided port services for third parties since 1987 (TURKLIM, 2006). After this period, all major ports on the coasts of Turkey have started investments made by the private sector.

The first private port corporation was established in Turkey in 1987 and became operational in 1992 in Bursa. The fact that Bursa was an industrially developed city caused the port to be established within the boundaries of this city (Esmer & Duru, 2016). After this date, many other private entrepreneurs implemented their investments to operate ports in different regions. These ports, unlike factory ports, are mostly *exclusive* or *specialized ports* in the form of container terminals and have been created solely to provide port services to third parties.

The legal basis of the port privatization process in Turkey was created at the beginning of the 1990s (Esmer & Duru, 2016). The privatization of ports in Turkey is regulated by the Law on Privatization Practices—Law No. 4046 and by Law on Build–Operate–Transfer Projects, which were published in the Official Gazette in 1994, and all privatizations were carried out accordingly. However, under these laws, ports can be privatized provided that the property rights are reserved, and the transfer period of the operating right cannot exceed 49 years. The related institutions regarding privatization of ports are the Ministry of Treasury and Finance, Privatization Administration, Ministry of Transportation and Infrastructure, Competition Authority (Çetin, 2012).

The governance models of ports in Turkey can be categorized as follows.

3.1 Public Ports

Until 1992, there had been only public ports, which were publicly owned and operated by different state-owned enterprises and municipalities. In Turkey, the Port of Haydarpaşa was the first port purchased by public in 1924 and was operated with a special regime and transferred to TCDD in 1927 (TURKLIM, 2006). The best-known public institutions are the Turkish State Railways (TCDD) and Turkish Maritime Facilities (TDI). While almost all of the ports of these two institutions have been commercialized with the Devolution Operating Rights/Transfer of Operations Rights method, a few are still operated by these two public institutions. The Port of Haydarpaşa and Port of Izmir are still operated by TCDD. Lake Van is the largest body of water in Turkey and TCDD also operates Lake Van Ferries and jetties in Van and Tatvan. TDI still operates Sarayburnu, Kuruçeşme, Kabatepe, Gökçeada—Kuzu, Gökçeada—Uğurlu, Güllük, Paşalimanı, Taşucu. However, some of these ports (Port of Haydarpaşa, Port of Izmir, and Port of Güllük) are expected to be commercialized soon by Privatization Administration.

3.2 Private Ports: Commercialized Ports, Factory Ports, Ports Established by Private Entities

With the implementation of privatization, the market shares of public ports decreased and the place and importance of the private sector in the port sector increased. TDI, which has 23 ports, transferred the operations of all major ports it has to the private sector, mainly in the 30 years following privatization. The most important of these ports are Tekirdağ, Antalya, Trabzon, Galataporti Kuşadası, and Çeşme (TDI, 2020).

The ports operated by TCDD, the other state-owned enterprise, have larger capacities than TDI ports. These ports are also very close to industrial areas and have relatively large areas. TCDD ports have been privatized by the transfer of operating rights method, like TDI ports. TCDD privatized five of its seven ports: İskenderun, Derince, Bandırma, Mersin, and Samsun.

On the other hand, factory ports serve businesses operating on the coasts. The majority of factory ports and the ports established by private entities are of Turkish origin (Esmer et al., 2016). These ports are located mostly in the Aegean, Marmara, and Mediterranean regions. The biggest ports in Turkey in terms of dry bulk and general cargo handling are mainly factory ports and operated by private entrepreneurs. Based on those metrics, the top ten ports are Isdemir, Icdas, Erdemir, Eren, Port of Mersin (commercialized), MMK, Colakoglu, Yesilyurt, Batı Liman, and Akcansa (TURKLIM, 2019).

In addition to these ports, there are also independent of port operations in Turkey. These ports serve third parties as private ports. Some of these ports are Marport, Kumport, Mardaş, Socar, Nemport, and Yılport.

4 Regulations and Legal Elements Affecting the Port Sector

According to the hierarchy of norms, the Constitution is at the top, followed by laws, international conventions, decree laws, bylaws, regulations, directives, circulars, communiqués, and instructions. Decree laws, bylaws, regulations, directives, communiqués, and bulletins are enacted based on laws and also international conventions. International conventions (such as MARPOL—the International Convention for the Prevention of Pollution from Ships, and SOLAS—Safety of Lafe at Sea) become valid with the promulgation of the law regarding appropriate participation in related conventions after Turkey became a party to the convention. On the other

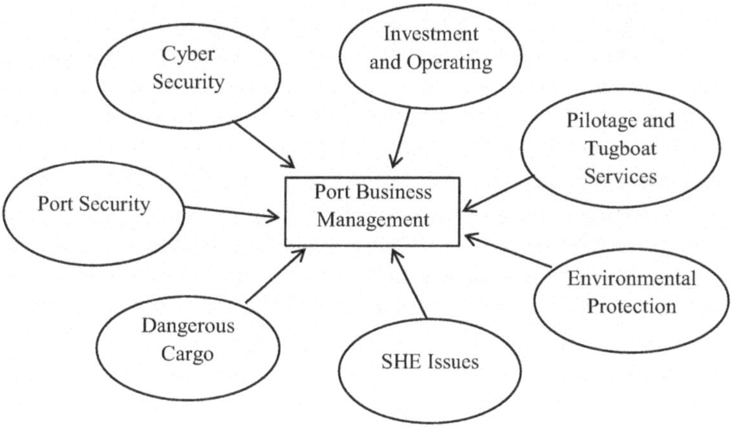

Fig. 1 Main legal factors affecting port management in Turkey

hand, the constitutional amendments in 2017 have led to significant changes to the structure of the executive branch and, accordingly, the powers of the president have been rearranged (Akyilmaz, 2019; İba & Söyler, 2019). With the transition to the new government system, presidential decrees and executive orders have emerged. While a presidential decree originates directly from the Constitution, an executive order can be put into force only based on the scope of laws.

The main legal factors, currently in force, that affect port management are shown in Fig. 1 in the aspects of investment and operating, pilotage and tugboat services, environmental protection, occupational health and safety (SHE), handling of dangerous goods, port security, and cybersecurity aspects.

4.1 Construction and Operating of a Port Facility

Regarding construction and operating of a port facility in Turkey, the main legal factors including laws, regulations, and communiqués are explained in this section.

The Law on Ports—Law No. 618 was enacted in 1925. According to this law, the government is responsible for administration, cleaning, deepening, widening, dredging, placing buoys, and keeping them in good condition. All ships coming in and out of the ports in Turkey are subject to the provisions of this law. The maritime coordinates of the borders of ports, jurisdiction areas of the ports, rules to be followed in navigating, berthing to docks and piers, mooring to and leaving buoys of the ships and other marine vessels, loading and unloading methods, places and times of commercial, explosive, flammable and similar dangerous cargoes, and available period for ships can stay in the port, the prevention of environmental pollution, and other issues are determined by the regulations to be issued by Ministry of Transport and Infrastructure with this Law. The Regulation on Ports was enacted in 2012 by

Ministry of Transportation and Infrastructure, mainly based on the Law on Ports to regulate to determine the administrative boundaries and area and anchorage areas of the harbor masters; to regulate the rules to be followed when sailing, anchoring, berthing, mooring, or leaving the coastal facilities of ships or sea vehicles in the port administrative areas, loading and unloading methods, place and time of all kinds of cargo and passengers, notifications for ships or sea vehicles, pilotage and towing requirements, other relevant issues regarding the safety of navigation, life, property and environment, and discipline in the administrative fields following the requirements of international conventions.

The Law on Ports Construction—Law No. 6237 was enacted in 1954 to meet the pier, dock, breakwater, shelter, port and related facilities, equipment, repair, dredging, and also to study the project needs of cities and towns on the country's coast. Prior to this law, the Law on Ports Construction, Extension, Improvement and Equipment—Law No. 5775 was enacted in 1951 (Yurtoğlu, 2019). Law No. 6237 is still in force. The two laws led to the formation of the national port infrastructure in Turkey.

Environmental Law—Law No. 2872 was enacted in 1983 to protect the environment in line with the principles of a sustainable environment and development. One of the definitions in this law is "Environmental Impact Assessment—EIF." EIF refers to the projects that are planned to be realized and their positive and negative effects, and measures to be taken to prevent the negative effects of these projects or to minimize them to the extent that they do not harm the environment. According to the Regulation on Environmental Impact Assessment—EIF, port investors who would like to construct jetty or terminal or any port operator who would like to make port extension project must prepare and present an "environmental impact assessment report" and a "project introduction file" to the Ministry of Environment and Urbanization.

The Law on Coasts—Law No. 3621 was enacted in 1990 to determine the principles of conservation and use of the public benefit, taking into account the natural and cultural characteristics of the sea, natural and artificial lakes and streams, coasts, and the coastal strips that are under the influence of these places and are their continuation. This law covers the regulation of the sea, natural and artificial lakes and shores of streams, and the coastal strips surrounding the coasts of seas and lakes, and describes the principles regarding the opportunities and conditions to benefit from these places for the public interest. The definition of coastal line, shore edge line, coast, coastal band, the coincidence of coastal line, and shore edge line. Piers, dock, breakwater, shelter, port and related facilities can be built, provided that a development plan is approved by the Ministry of Environment and Urbanization. The Regulation on the Implementation of Law on Coasts, enacted in 1990, covers the issues that clarify the implementation of Law on Coasts. Additionally, there is the Notification on Planning and Implementation Process in Coastal Buildings and Facilities, which determines the procedures and principles regarding the transactions to be carried out by the investors and public institutions, including information and documents (prior permission, feasibility report, modeling report, hydrographic and oceanographic report, hydraulic model experiment, geological and geotechnical survey report) to be requested regarding the buildings and facilities to be constructed.

The Regulation on Procedures and Principles Regarding Operation Permit to Coastal Facilities was enacted in 2007 by the Ministry of Transportation and Infrastructure based on the Law on Coasts to regulate the procedures and principles for operating permits by the Ministry for ports, cruise ports, marinas, passenger terminals, piers, docks, shelters, berthing areas, fuel/liquefied petroleum gas pipelines, buoy systems and similar coastal facilities, and other superstructure facilities for sea transportation. Shipyards, boat manufacturing, and boatyards are not covered by the regulation. Under this regulation, all public and private ports are obliged to obtain a Coastal Facilities Operation Permit from the Ministry. However, the criteria for permits differ according to the cargo handling types of coastal facilities. This permit has a maximum duration of five years. At the end of the term, the ports are inspected and the permits are renewed if deemed appropriate. During these audits and surveys, the requirements of international conventions are also questioned.

Turkey is an earthquake-prone country and has experienced many devastating earthquakes. Regarding the possible impacts of earthquakes on port structures, two regulations have been published, one of which is currently in force and the other of which will be in force in 2021. The Earthquake Technical Regulation for Construction of Coastal and Port Structures, Railways and Airports was enacted in 2007 and the design principles of coastal and port structures under the effect of different earthquake levels have determined following this regulation. Additionally, the Regulation on Earthquake for Coastal and Port Structures in Turkey was published in 2020 but will be enacted in October 2021 to determine the necessary rules and minimum conditions for the design of new coastal and port structures under the effect of earthquakes, evaluation of the performance of existing structures, and strengthening design. The entry into force of the new regulation is predicted to cause a change in the previous regulation or remove the previous regulation.

The Regulation on Dredging of Sea and Inland Waters was enacted in 2016 by the Ministry of Transport and Infrastructure. The purpose of this regulation is to determine the procedures and principles regarding dredging activities to be carried out in sea and inland waters. The dredging activities in territorial waters of Turkey and inland waters of the country require a ministerial permit. Dredging activities in Turkish waters can only be carried out by Turkish-flagged ships that are licensed by the Ministry of Transport and Infrastructure. The Regulation on Environmental Management of Bottom Dredging Materials was enacted in 2020 by the Ministry of Environment and Urbanization. The purpose of this regulation is to determine the procedures and principles regarding the environmental management of dredging activities carried out in the sea, coastal areas, and river mouths, the beneficial use of dredging materials resulting from dredging activities, and the discharge or disposal of them in the marine environment in a way that does not harm the environment or human health. The regulation covers the standards of dredging materials, the criteria of the sea areas where dredging materials will be poured, discharge methods, disposal and beneficial use possibilities, permits to be obtained, monitoring the effects of the activity on the marine environment.

The Law on Evaluation of Immovable Properties of the Treasury and Amendments to the Value Added Tax Law—Law No. 4706 was enacted in 2001 by the

Ministry of Treasury and Finance to immovable properties owned by the treasury to the economy in a shorter time. The part of this law that concerns the port sector is that all infrastructures to be constructed and built in the sea direction of the coastal line are considered as the private property of the treasury and under the rule and possession of the state; even those constructed and built by private entrepreneurs. The Easement Right is established between the General Directorate of National Estate, which is one of the bodies of Ministry of Treasury and Finance, and port investor to construct and build a port facility on immovable property (land) under private ownership of the treasury and rule and possession of the state. Basically, regarding the operating of ports by establishing the right of easement, a "pre-permit agreement" and then a "usage permit agreement" are drawn up between the Ministry of Treasury and Finance and the port investor. The duration of the usage permit agreement does not exceed 49 years and pursuant to this agreement, port operators regularly pay an "easement right or usage fee" to the Ministry of Treasury and Finance during the contract period. Moreover, in case of providing services to third parties, a "relative rent fee" is also paid from the income of these services by the port operator to the Ministry of Treasury and Finance. All public ports and commercialized ports are exempt from usage fee and relative rent fees.

The Law on the Protection of Competition—Law No. 4054 is the main regulatory element regarding competition and was enacted in 1994 to prevent agreements, decisions, and practices that prevent, distort and restrict competition in the goods and services markets and from abuse of their dominance by undertakings dominating the market, and to ensure the protection of competition by making necessary regulations and regulations and inspections. Even though ports are not involved with any specific national competition rules, the Turkish Competition Board tends to follow EU competition rules. For example, all commercialization (privatization) applications and mergers and acquisitions that exceed certain thresholds need approval by the Turkish Competition Board (Çamlıbel & Burnett, 2016). There is a new regulation regarding port fees and charges, namely the Communiqué on Determining and Application of Service Items Provided by Coastal Facilities and Base and Ceiling Charges of These Items, which was enacted in 2020 by the Ministry of Transport and Infrastructure. The purpose of this communiqué is to regulate the issues regarding the determination and application of the services provided in the coastal facilities operated for commercial cargo and/or passenger transportation, the construction, modification, repair, and maintenance of ships, and water vehicles and the ship recycling activities and determination and application of base and ceiling fees of these services by getting permission from the Ministry. According to this communiqué, port operators are obliged to determine service items they provide for cargo, passengers, and ships, including charges, and submit to the ministry. After receiving the approval of the Ministry, these fees and charges are announced on corporate websites and related platforms and put in the force.

The Customs Law—Law No 4458 was enacted in 1999 by the Ministry of Trade. The purpose of this law is to determine the customs rules for the good and vehicles entering the customs territory of the Republic of Turkey.

4.2 *Pilotage and Tugboat Services*

According to the Law on Cabotage—Law No. 815, enacted in 1926, only Turkish citizens and Turkish-flagged ships can make freight and passenger transport, pilotage, and towing services in Turkish territorial waters. The Regulation on Ports specifies the requirements of pilotage and tugboat services, including the principles. Thereafter, 500 GRT (gross tonnage) and larger tankers and ships and marine vessels carrying all kinds of dangerous goods, 1000 GRT and larger Turkish-flagged ships and marine vessels, 500 GRT and larger foreign-flagged ships and sea vehicles, and above-1000 GRT foreign-flagged commercial and private yachts must take a pilot when berthing and unberthing to coastal facilities. The minimum number of tugs and their minimum towing power for ship maneuvering is also determined by this regulation.

The Regulation for Pilotage and Tugboat Services was enacted in 2020, determines the procedures and principles regarding pilotage, specifies requirements of pilotage and tugboat services, and the qualifications required of those who will provide these services, and to perform the necessary authorization and inspections in Turkey's maritime jurisdiction, the Turkish Straits region, channel and inland waters. The pilotage services in Turkey are not open to competition. The pilotage and tugboat services at private ports are provided by authorized private enterprises, which are determined by the Ministry of Transport and Infrastructure. The commercialized ports and some of the private ports are also authorized to deliver these services. However, the tariff for pilotage and tugboat services is determined by the Ministry of Transport and Infrastructure.

5 Environmental Protection

According to the Regulation on Ports, coastal facility operators must take all necessary precautions to ensure that all kinds of waste and residual materials do not accumulate in the sea and the dock during and after cargo handling, do not spill into the sea during and after cleaning, and do not flow into the sea with rainwater over the quay.

The Law on Principles of Emergency Response and Compensation for Damages in Pollution of Marine Environment by Oil and Other Harmful Substances—Law No. 5312 was enacted in 2005 by the Ministry of Environment and Urbanization. The purpose of this law is to ensure maritime safety and to take into account the rights and obligations arising from international law (MARPOL—International Convention for the Prevention of Pollution from Ships) and national law regarding the preventing of maritime pollution and to determine the principles of interventions and preparedness to be applied in order to eliminate or reduce or limit the danger of pollution arising from the activities of ships and coastal facilities in emergencies, principles of determination, and compensation for damages resulting from such events. This law has an implementation regulation, as well as many circulars and communiqués.

According to the Regulation on the Principles on the Application of Law on Emergency Response and Compensation for Damages in Pollution of Marine Environment by Oil and Other Harmful Substances, emergency response management plan for the ports should be prepared, documented, and reported by an authorized company. Response procedures, drills, personnel, training, and equipment requirements should be clearly defined in the report, which should be built upon a broad risk evaluation. This plan is checked and approved by the Ministry of Environment and Urbanization in coordination with the Ministry of Transport and Infrastructure. After approval, the risk-owner is expected to supply and apply all the requirements defined in the plan.

The Regulation on Taking Wastes from the Ships and Waste Control was enacted in 2004 by the Ministry of Environment and Urbanization. This regulation was prepared in parallel with Environmental Law, as well as MARPOL Convention and the Directive of the European Parliament and the Council (Directive No. 2000/59/EC) on Port Waste Reception Facilities Used for Ship Wastes and Cargo Residues. According to this regulation, ports are obliged to establish and operate waste reception facilities that have sufficient capacity and technical equipment to serve the waste generated by ships and cargo residues and dispose of the wastes collected in their waste reception facilities following the provisions of the Environmental Law and the relevant legislation. The fees of waste reception from the ships are determined by the Ministry of Environment and Urbanization. Besides, there are many regulations, directives, circulars, and communiqué linked to Environmental Law.

6 Occupational Health and Safety

The Law on Occupational Health and Safety—Law No.6331 was enacted in 2012 to govern the health and safety standards adopted by employers in Turkey. The law aims to regulate the duties, powers, responsibilities, rights, and obligations of employers and employees to ensure occupational health and safety in workplaces and to improve existing health and safety conditions. This law covers all types of employment, work, and workplaces that belong to public and private sectors, owners, and/or employers of subject workplaces and representatives/agents of such employers, and all employees including apprentices, interns and trainees, regardless of the fields of activity in which they are involved (Voyvoda & Bicer, 2012). According to the Regulation on Procedures and Principles of Employee Health and Safety Training, employers are obliged to prepare training programs for their employees on occupational health and safety, to provide suitable places and tools and equipment for training, to ensure that employees participate in these programs, and to keep records of the training. The Regulation on Occupational Health and Safety Services was enacted in 2012 to regulate the establishment of workplace health and safety units, their duties, powers, responsibilities, authorization, cancelation of their licenses, working procedures, and principles. According to this regulation, port operators are obliged to appoint one or more workplace doctors and other medical personnel and occupational safety specialists. On the other hand, the Regulation on Duty, Authority, Responsibility and Training of Occupational Safety Specialists regulates the qualifications, training, and certification of occupational safety specialists, including their duties, powers, and responsibilities. These specialists, who will be appointed by the employers, must have a valid certificate. Similarly, there is Regulation on Duty, Authority, Responsibility and Training of Workplace Doctor and Other Medical Person for regulating their qualifications, training, and certification. According to the Regulation on Occupational Health and Safety Committees, port operators (since the number of employees at ports exceeds 50) must establish a committee to work on occupational health and safety. The Regulation on Occupational Health and Safety Risk Assessment was enacted in 2012. Under this regulation, port operators must conduct a risk assessment (or have one conducted) in terms of occupational health and safety to ensure, maintain, and improve the work environment and the health and safety of employees. The Regulation on Stopping Work in Workplaces states that work may be ceased/stopped in some or all of the workplace if there is a life-threatening matter for the employees in the working methods and forms or work equipment or if there is no risk assessment until the hazardous situation is eliminated. There are also many other regulations, directives, circulars, and communiqués regarding occupational health and safety in ports. Furthermore, Turkey ratified 59 of the ILO's (International Labor Organization) Conventions, 55 of which have been implemented. One of these is the Occupational Safety and Health (Dock Work) Convention 1979 (No. 152), for which the term "dock work" covers all and any part of the work of loading or unloading any ship, as well as any work incidental thereto; the definition of such work shall be established by national law or practice.

The Law on General Public Health—Law No. 1593 was enacted in 1930 by the Ministry of Health. This law was one of the most important laws of the newly founded Republic of Turkey. The law stipulates the establishment of public councils in provinces and districts within the framework of a public health organization. This law gives a special place to combating epidemic and communicable diseases. The By-Law on International Health was enacted in 2005 (latest version) by the World Health Organization (WHO), which Turkey is part of, and ship inspections are carried out by the Handbook of for Inspection of Ships and Issuance of Ship Sanitation Certificates of WHO. According to the Law on General Public Health, ships arriving from foreign country ports are first inspected by Coastal Health Inspection Centers (responsible for and authorized to take all kinds of health measures following the provisions of national and international legislation against infectious and epidemic diseases that may enter and leave the country through sea Gates) of the Directorate-General of Health Services for Borders and Coasts of Turkey and unless given Free Pratique, there is no contact with the ship and no any cargo handling can be done. The Regulation on Environmental Health Procedures to be Applied at International Entry Points was enacted in 2013 based on the Law on General Public Health and the By-Law on International Health. According to this regulation, port operators are obliged to continuously manage their areas for in a healthy and safe manner regarding vector, indoor air quality, and drinking and utility water, including their tanks, food, and waste. In this context, ports are audited annually by the related Coastal Health Inspection Center. According to the Directive on Disinfection Procedures to be Applied in International Operating Vessels and Entry Points, port operators are obliged to work with the companies that authorization certificates.

The Law on Social Insurance and General Health Insurance—Law No. 5510 was enacted in 2006 by the Ministry of Family, Labor, and Social Services. The purpose of this law is to secure people in terms of social insurance and general health insurance, to determine the persons who will benefit from these insurances, and the rights to be provided, the conditions, and the methods of financing.

The Law on Vocational Competency Institution—Law No. 5544 was enacted in 2006 by the Ministry of Family, Labor, and Social Services. The purpose of this law is to determine the principles of national qualifications in technical and vocational fields on the basis national and international occupational standards; establishment of the Vocational Qualifications Authority to establish and operate the necessary national qualification system to carry out activities related to auditing, assessment, evaluation, and certification. The Regulation on National Vocational and Qualification Standards was enacted in 2015. Based on this law and regulation, national vocational standards and qualifications for port workers (e.g., forklift operators) have been determined.

According to the Regulation on Ports, coastal facility operators must form firefighting teams equipped with necessary tools and equipment in case of fires and take all necessary precautions to ensure that all kinds of waste and residual materials do not accumulate in the sea and the dock during and after cargo handling, do not spill into the sea during and after cleaning, and do not flow into the sea with rainwater over the quay. They must have lighting, fenders, depth, light or lightless buoys, routing signs, audible and visual safety warning and similar elements for ships and marine vehicles to berth or to moor to shore facilities safely.

7 Handling of Dangerous Goods

Regarding regulations for the handling of dangerous goods at ports in Turkey, many pieces of legislation are based on the International Maritime Organization (IMO's) most foremost conventions (SOLAS and MARPOL), which deal with various aspects of maritime safety.

International Safety Guide for Oil Tankers and Terminals (ISGOTT) *was published* by the International Chamber of Shipping (ICS), the Oil Companies International Marine Forum (OCIMF), and the International Association of Ports and Harbors (AIPH). ISGOTT is a definitive guide to the safe carriage and handling of crude oil and petroleum products on tankers and at terminals and properly used by all oil terminals in Turkey. The Regulation on Transport of Dangerous Goods by Sea was enacted in 2015 (based on SOLAS, MARPOL Conventions, Law on Ports, Law on Protection of Life and Property at Sea—Law No. 4922) by the Ministry of Transportation and Infrastructure to ensure that dangerous goods transportation activities are carried out by sea as economically, rapidly, and safely with high quality, with a minimal negative effect on the environment, and in compliance with transportation activities. Within the scope of this regulation, port operators handling dangerous goods are obliged to employ a dangerous goods safety consultant or receive service from one in all their activities within the scope of dangerous goods transportation and obtain a Dangerous Goods Compliance Certificate. Port operators are also obliged to prepare a Dangerous Goods Guide that explains all procedures, responsibilities, and measures related to dangerous goods. This guide must be available to all relevant facility personnel, public authorities, and port users. The port operators must provide training for persons involved in loading, unloading, and handling of dangerous goods. Port operators ensure that emergency arrangements are made and all relevant parties are informed. The container terminals must create a suitable storage area for segregation and stowage rules for containers carrying dangerous cargo and take all necessary fire, environmental, and other safety measures in this area.

8 Port Security

In Turkey, regulations of port security are shaped by national regulations besides the ISPS (International Ship and Port Facility Security) Code, the aim of which is to detect security threats and take preventive measures against security incidents affecting ships and port facilities used in international trade.

The Regulation on International Ship and Port Facility Security Code was enacted in 2007 by the Ministry of Transportation and Infrastructure based on ISPS Code, Regulation (EC) No. 725/2004 of the European Parliament and of the Council of 31 March 2004 on enhancing ship and port facility security, Commission Regulation (EC) No. 884/2005 of 10 June 2005 laying down procedures for conducting Commission inspections in the fields of maritime security, and Directive 2005/65/EC of the European Parliament and of the Council of 26 October 2005 on enhancing port security. The purpose of this regulation is to ensure the effective implementation of ISPS Code on Turkish-flagged ships and in port facilities in Turkey. According to regulation, port operators are obliged to make a Security Assessment of the Port Facility and submit it to the Ministry of Transportation and Infrastructure. Pursuant to the Regulation, port operators must appoint a Port Facility Security Officer (PFSO) and have approval from the Ministry. Port operators have to prepare a Port Facility Security Plan and submit it to the Ministry. The Ministry then issues the Port Facility Security Certificate with a validity five years and port operators must then implement the port facility security plan. As a result of the audit by the Ministry, international ship traffic is not allowed in the port if the provisions of ISPS Code are not fulfilled and the operation permit of the port is suspended. Within the scope of the regulation, the powers and duties of RSOs (Recognized Security Organizations) that the port operators work are also defined. Drills and exercises should be carried out regularly by port operators to ensure that the requirements of port facility security plan are fulfilled effectively.

The Regulation on Security at Civil Airports, Ports and Border Gates, Execution of Duties and Services was enacted in 1997 by ministries' decree. This regulation has been prepared to apply the security measures determined by the International Civil Aviation Organization (ICAO) and the European Civil Aviation Conference (ECAC) at civil airports, ports, and border gates for the regular, continuous, and effective performance of security and other services under the management of a single authority (governor of the city, or deputy of the governor or district governor). In accordance with this regulation, entry to civil airports, ports, and border gates is carried out subject to control and permission (all personnel and vehicles affiliated with public and private organizations). The ship agents, customs brokers, stevedoring firms, etc. are obliged to receive permission from related *Customs Authority* to enter and do business in ports.

The Law on Private Security Services—Law No. 5188 was enacted in 2004 and evaluates private security services as complementary to public security. This law covers issues related to private security permits, licensing, and auditing of persons and organizations that will provide this service. The security services at ports in Turkey are provided by port operators' personnel or by authorized companies as subcontracting. These people have the authority to search for people and items that are trying to enter the areas they protect and secure, with various security systems such as X-rays, and they are even authorized to possess and carry firearms in some cases. Based on this law, the Regulation on the Implementation of Law on Private Security Services includes granting operating permits to private security companies, private training institutions, work permits to private security officers, qualification of private security training, curriculum, conditions to be sought in trainers and training centers, and qualifications as a result of training auditing security services. The provision of port security in Turkey (especially when the ISPS Code Level is 3), Customs Enforcement, Coast Guard, the police, and gendarmerie forces also play active roles.

9 Cybersecurity

Regarding cybersecurity, the port operators in Turkey, which handle a minimum of 10 million tons of cargo or a minimum of 500,000 TEUs of container per year, are required to establish an Institutional CIRT—Cyber-Incident Response Team by International Organization of Standardization (ISO) and International Electrotechnical Commission (IEC) 27001: Information Security Management System standards. For this requirement, the Ministry of Transportation and Infrastructure has issued a circular that is a guideline for the establishment of the Institutional CIRT based on International Maritime Organization (IMO's) Resolution MSC. 428(98) on Maritime Cyber-Risk Management in Safety Management Systems.

10 Assessment of Regulations and Legal Elements Affecting the Port Sector in Turkey

The previous section outlined the main legal factors directly affecting port management. The legal elements explained above have been amended many times according to the current needs of the sector, to improve the foreign trade of the country, working standards, and to meet the requirements of the international regulations. Although the Ministry of Transport and Infrastructure is the regulatory authority on a large scale for the port sector in Turkey, almost all other ministries and their related bodies also play a regulatory role for the port industry. Regarding the process of port investments, operations, and management in Turkey, there is no specific port authority as there is in Europe and the Far East. At this point, it is necessary to note that, regarding any coastal facility construction and building in Turkey, the views and approval of almost all ministries and also Turkish Armed Forces General Staff are required. This situation prolongs the time for approval of the development plans and realization of the investments, especially for private ports. The regulation on Dredging of Sea and Inland Waters has paved the way for companies to conduct dredging activities, and fleet expansion has been experienced. However, the existence of marine vehicles for dredging activities that the port sector needs or can meet all kinds of needs has not yet been fully established.

The reason for the Relative Rent Fee was the protection of Monopoly Rights of the ports operated by public institutions (TDI and TCDD). In this context, private port operators have been paying relative rent fees (apart from usage fees) from their income gained from port services to third parties. However, all public ports and commercialized ports are exempt from usage fee and relative rent fees. On other hand, many of the public ports have been commercialized and the remaining ones (especially the Port of Haydarpasa and the Port of Izmir) are predicted to be commercialized soon. Therefore, the reason for the occurrence of the relative rent fee practice no longer exists. The relative rent fee practice negatively affects the port operating cost and creates unfair competition among port operators, which have different levels of competition in different geographical locations, so it is considered that the abolition of this practice is considered as positive for the Turkish port operators.

Regarding port tariffs, there is newly enacted regulation—the Communiqué on Determining and Application of Service Items Provided by Coastal Facilities and Base and Ceiling Charges of these Items—which allows the Ministry of Transport and Infrastructure to intervene directly in port tariffs and has been widely discussed in the sector. Public port institutions (TDI and TCDD) have always been pioneers in terms of identifying items and charges of the port's services in the Turkish port sector. However, the criteria for determining the base and ceiling charges for the port services are not described or indicated in the communiqué. The base and ceiling charges that are intended to be brought with this communiqué will negatively affect the sustainable competitive environment. If there is a high rate of reduction in the charges of services provided by the port operators, the following scenarios will be possible. The port operators may have to focus more on operating costs and there will may a potential obstacle to the proper execution of their management for critical subjects such as employment and service quality. It is predicted that port operators may think more carefully about their future investment decisions; ultimately, there will be a possibility of experiencing the negativities experienced by the public port operators for a period. Conversely, if there is a high rate of increase in the charges of services, higher port costs for the carriers and Turkish BCOs (beneficial cargo owners) will return as an increase in freight rates and logistics costs, which may prevent Turkish foreign trade and adversely national competitiveness.

There are many regulations, directives, circulars, and communiqués linked to Customs Law, and they are constantly revised and amended to improve foreign trade of the country. For example, the Regulation on Customs, which was put into force in 1999 (on the same date as the law) to regulate the procedures and principles of customs transactions, has been amended many times since then. The port operators should closely follow the constantly revised and amended customs legislation because many ports in Turkey with limited port storage areas and cargo must remain at the ports' areas as soon as possible to ensure field efficiency and productivity.

According to the Law on Cabotage, only Turkish citizens and Turkish-flagged ships can conduct freight and passenger transport, pilotage, and towing services in Turkish territorial waters. According to this law, empty container transport between ports in Turkey is only conducted with Turkish-flagged vessels and this situation will negatively affect the positioning of empty containers in Turkey and directly affect the logistics costs and freight rates for Turkish BCOs. The empty containers can be considered as "mobile holds of the ships," not as a cargo, vehicle, or packages to be transported by a ship. Thus, carriage of empty containers by its carrier, which operates foreign-flagged ships, between Turkish ports, may be treated as an exceptional application under Cabotage Law.

The recent changes in the occupational health and safety legislation have led to positive changes in port management structures, organizations, working conditions, personnel recruitments, training, and business conduct. No Turkish ports have been quarantined yet during the COVID-19 epidemic because of strict application of Turkish Coastal Health Inspection Centers under the Law on General Public Health; Turkish port operators have kept their operations. Besides, in Turkey, many institutional port operators use the ISO (International Organization of Standardization) management system (ISO 14001: Environmental Management System, ISO 45001: Occupational Health and Safety Management System), which helps them comply with requirements of occupational health, safety, and environment protection legislation.

11 Conclusion

Turkey has long been an area of geostrategic importance and a key transit route between Asia and Europe. Turkey is also pivotal due to the natural transit route between the Black Sea and the Mediterranean. Turkey's seas have over 8000 km of coastline surrounding the three sides. Turkish ports located on both the Mediterranean and Black Sea coasts have played an active role in regional trade for many years. Ports in Turkey are mainly located in the Istanbul, Kocaeli, Gemlik, Izmir, and Iskenderun regions. More than half of all ports in Turkey are in these regions, and the others are spread across Turkey's coastal areas.

Since the 1920s, radical changes have been experienced in the management of Turkish ports. These changes are generally supported by legal regulations. The governance models of ports in Turkey can be mainly categorized as public and private ports. The best-known public port institutions are TCDD and TDI, while almost all the ports of these two institutions have been commercialized. Other public companies (BOTAS—TTK) operate their ports to handle only the company's goods. There are also municipality ports that generally take the form of ferry and passenger piers. Private ports in Turkey are commercialized ports, factory ports, and ports established by private entities. While the market share of public ports decreased, the private ports' market share increased in the last 30 years.

In some ways, the port authority model in Turkey varies from models known in other parts of the world. For example, there is no unified port authority in Turkey to oversee all port-related activities as there is in countries in Europe and the Far East. However, there are 71 harbor masters (as port authorities) of the Ministry of Transport and Infrastructure. Although the Ministry is the regulatory authority on a large scale for the port sector in Turkey, almost all other ministries and their related bodies also play a regulatory role for the port industry.

This chapter has explained the main legal factors affecting port management in the aspects of investment (construction, building) and operating, pilotage and tugboat services, environmental protection, occupational health and safety, handling of dangerous goods, port security, and cybersecurity aspects. All of the regulations mentioned in the section are updated according to current commercial conditions and are intended to provide public benefit. It should be noted that there are other legal elements (such as Turkish Commercial Law, Law on Obligations, etc.) that port operators must comply with or follow. Further studies could evaluate the effects of the other legislation not mentioned in this chapter.

References

Akyilmaz, C. (2019). 2017 Anayasa Değişikliğine Göre Cumhurbaşkanlığı Kararnameleri. *Ankara Hacıbayram Veli Üniversitesi, Hukuk Fakültesi Dergisi C, XXIII*(Sayı 1), 191–207.

Çamlibel, E., & Burnett, G. M. (2016). Ports & Terminals: 2016. *Turkey* (pp. 97–100). Law Business Research Ltd.

Çetin, K. (2012). *Limanlarda Örgütsel Değişim ve Değer Zinciri Sistemlerinde Etkililik Analizi.* İzmir: Dokuz Eylül Üniversitesi Yayınları.
Esmer, S., & Duru, O. (2016). Port governance in Turkey: The age of the global terminal operators. *Research in Transportation Business & Management, 22*, 214–223.
Esmer, S., Nguyen, H. O., Bandara, Y. M., & Yeni, K. (2016). Non-price competition in port sector: A case study of ports in Turkey. *The Asian Journal of Shipping and Logistics, 32*(1), 3–11.
İba, Ş., Söyler, Y. (2019). Yeni Hükümet Sisteminde Cumhurbaşkanlığı Kararnamesi ile Cumhurbaşkanı Kararının Nitelik Farkı ve Hukuki Sonuçları. *Anayasa Yargısı, Cilt, 36*(Sayı 1), 195–223.
Lam, J. S. L., & Song, D. W. (2013). Seaport network performance measurement in the context of global freight supply chains. *Polish Maritime Research, 20*, 47–54.
Ministry of Transport and Infrastructure. (2020). Republic of Turkey. https://denizcilikistatistikleri.uab.gov.tr/yuk-istatistikleri. 01 September, 2020.
Olivier, D. (2005). Private entry and emerging partnerships in container terminal operations: Evidence from Asia. *Maritime Economics & Logistics, 7*(2), 87–115.
Robinson, R. (2002). Ports as elements in value-driven chain systems: The new paradigm. *Maritime Policy and Management, 29*(3), 241–255.
TDI (Turkish Maritime Facilities). (2020). https://www.tdi.gov.tr/isletme-hakki-devir-yontemi-ile-ozellesen-limanlar/. 15 September, 2020.
TURKLIM (Port Operators Association of Turkey). (2006). *Turkey port sector report*. Istanbul.
TURKLIM (Port Operators Association of Turkey). (2019). *Turkey port sector report*. Istanbul.
Voyvoda, O. K., & Bicer, A. E. (2012). *Law on occupational health and safety*. Çakmak Avukatlık Bürosu.
Yeni, K., & Tuna, O. (2003). Logistics oriented developments in container ports: An analysis of Turkish Ports. *IAME Annual Conference* (pp. 736–743), 3–5 September, Busan.
Yeni, K. (2020). *Liman Hizmetlerinde Fiyatlandırma*. Ankara. Akademisyen Kitabevi.
Yurtoğlu, N. (2019). Cumhuriyet Döneminde Türkiye'de Liman ve İskele Politikaları. *Atatürk Araştırma Merkezi Dergisi. Fall 2019, 100*, 505–554.

Kazım Yeni is an assistant professor at Iskenderun Technical University, Barbaros Hayrettin Naval Architecture and Maritime Faculty; he lectures on the Management and Organization, Strategic Management, Maritime Business, Port Operations, and Management. He holds a Master of Science from Dokuz Eylul University, Institute of Social Sciences, Department of Maritime Business Administration in 2001, and a Ph.D. from Dokuz Eylul University, Institute of Social Sciences, Department of Maritime Business Administration in 2017. He worked at private companies' in Turkey in port divisions at the managerial level between 2006 and 2020. His academic research focuses on port management and strategic management. He published a book entitled "Pricing in Port Services." He has several publications in the field of logistics management, strategic port management both in English and Turkish. He provides consultancy to several companies regarding port management.

Soner Esmer is the Dean of Iskenderun Technical University, Barbaros Hayrettin Naval Architecture and Maritime Faculty, and a professor at Dokuz Eylul University Maritime Faculty. He obtained his undergraduate and postgraduate degrees in Maritime Business Administration and specialized in maritime and port management. In addition to these matters, he teaches lectures and supervises academic dissertations in logistics and supply chain management. His research's primary motivation sources are current developments, including discussions in logistics, maritime transport, and port management. He took part in many Public Works, including Transportation Coastal Structures and Tourism Coastal Structures Master Plan Studies, Development Plan Transportation Specialization Commissions, Integrated Coastal Plans. Dr. ESMER is currently working as an author, researcher, book and journal editor, columnist, and reviewer in scientific journals.

Regulation of Fixed Telecommunication Services

Bülent Gökdemir

Abstract Regulation policy for fixed telecommunication services in Turkey has been in line with that of the European Union. This policy involves enhancing retail competition by means of effective wholesale regulations. From this perspective, the incumbent monopoly's market power in fixed voice services has been reduced through mobile and VoIP substitution; increasing competition in the retail market caused by the deregulation and robust regulations imposed on the wholesale markets. As for fixed broadband services, retail competition has been improving through an increasing share of fiber and intense regulations for wholesale markets. However, apart from OECD countries, cable has not been able to contribute to competition to the extent necessary because of a lack of required investment. Strengthening and expanding cable infrastructure and imposing regulatory remedies on cable infrastructure owner will enhance retail competition in the broadband service market.

Keywords Fixed voice market · Fixed broadband market · Regulations · Retail competition · Infrastructure competition

1 Introduction

It is nearly two decades since the Turkish telecommunication market was liberalized. The state monopoly, Türk Telekom, has been privatized; the regulatory legislation has been enacted and a sectoral regulator—namely the Information and Communication Technologies Authority (ICTA)—has been established.

The Electronic Communication Law (ECL)[1] that was put into force in 2008 is the main legislation regulating telecommunication sector. The primary objectives of the law are to establish a competitive market environment, secure consumer rights, enable service delivery all over the country, ensure efficient resource allocation, and promote technological developments. Various complementary laws and secondary

[1] Electronic Communication Law. (Law No. 5809). *Official Gazette* dated November 10, 2008, and numbered 27050 (Turkish version).

B. Gökdemir (✉)
Bilkent University, 3 D3 Blok No: 14, Çankaya, Ankara, Turkey

M. Eroğlu and M. Finger (eds.), *The Regulation of Turkish Network Industries*,
https://doi.org/10.1007/978-3-030-81720-6_18

legislation have also been enacted to attain the said objectives, which seem to overlap with what are addressed by EU Directive 2018/1972[2] establishing the European Electronic Communications Code.

As per Directive 2009/140/EC,[3] the main purpose of the regulatory framework is to reduce ex-ante regulation in a phased manner as competition in markets develops and, ultimately, to govern the sector through competition law enforcement. It can be assumed that competition is a general prioritized solution that paves the way for realizing all other objectives by EU. From this point of view, the objective of ex-ante regulatory intervention is stated as generating benefits for end-users by making retail markets effectively competitive on a sustainable basis as Directive 2014/710/EU[4] addresses.

Unlike in the EU, Turkish legislation makes no direct expression that underlying ex-ante regulation should be reduced and retail competition should be enhanced as much as possible. However, the rationale of ECL, references in market analysis studies of the ICTA to EU legislation, and the practice of the ICTA imply the Turkish regulatory approach has been on the same path as the EU. Thus, competition can be taken as the main reference point in the discussion of telecommunication regulation.

There are two main options for competition in telecommunication sector: infrastructure competition or retail competition. Infrastructure competition concerns competition through different facilities or technological options, such as DSL, fiber, or cable infrastructures, whereas retail competition occurs when operators offer services to the final consumers by getting access to network. In this model, firms such as fixed telephone call providers or Internet service suppliers are able to purchase wholesale capacity or other elements of the network from the infrastructure provider.

Infrastructure competition has more to do with the amount of investment in alternative facilities or technologies. Thus, regulatory regime has less room to stimulate infrastructure competition than it does for retail competition. Turkey, like the EU, seems to prioritize retail competition. Hence, the role of regulation plays an important role in attaining the objective of establishing competition in the market.

Turkey's current regulation methodology is very similar to that of the EU as well. The ICTA has authorization to conduct in-depth market analysis in a participatory manner. The Turkish Competition Authority, NGOs, and the principal companies operating in the sector deliver contributions to market analysis study in which the

---

[2] EU Directive 2018/1972 of the European Parliament and of the Council of December 11, 2018, Establishing the European Electronic Communications Code. *Official Journal of the European Union* dated December 17, 2018, and numbered L 321/36.

[3] EU Directive 2009/140/EC of the European Parliament and of the Council of November 25, 2009, Amending Directives 2002/21/EC on a Common Regulatory Framework for Electronic Communications Networks and Services. *Official Journal of the European Union* dated December 18, 2009, and numbered L 337/37.

[4] EU Commission Recommendation 2014/710/EU of October 9, 2014, on Relevant Product and Service Markets within the Electronic Communications Sector Susceptible to ex-ante Regulation in accordance with Directive 2002/21/EC of the European Parliament and of the Council on a Common Regulatory Framework for Electronic Communications Networks and Services. *Official Journal of the European Union* dated October 11, 2014, and numbered L 295/79.

relevant markets are determined; the companies obtaining efficient market power are identified and regulatory obligations imposed on the companies with efficient market power are specified. Based on market analysis studies, the Information and Communication Technologies Board (ICTB) makes the formal decision to impose the obligations in question.

In this context, this chapter focuses on analysis of current regulatory regime for fixed telecom services from the perspective of retail competition. Regulations for infrastructure competition will also be touched upon where necessary. The analysis will be conducted on two main axes: fixed voice service and fixed broadband Internet service where the state of affairs will be put forward through delivering foremost statistics and comparative data. Determination of relevant markets by the ICTA will be discussed and regulatory framework for the relevant markets will be introduced, elaborated, and discussed.

2 State of Affairs in Fixed Telecommunication Services

Turkey's fixed telecommunication service market has been dominated by the incumbent monopoly, Türk Telekom, which offers broadband, mobile, fixed voice, and TV products and services to its subscribers. As of the end of 2019, Türk Telekom had 14.6 million fixed access lines, 11.4 million broadband, 3.5 million TV, and 22.9 million mobile subscribers (Türk Telekom, 2019).

As of 2019, Türk Telekom's consolidated revenue was 23.7 billion TL, of which mobile revenue constituted 38%, fixed broadband 28%, fixed voice 11%, corporate data 8%, and others (TV, interconnection, international sales, etc.) 15% (Türk Telekom, 2019).

Türk Telekom has a great number of small and large rivals, both in the fixed voice and broadband markets. According to ICTA statistics, there are currently 160 licensed fixed voice service providers and 293 Internet service providers in Turkey.[5]

In the fixed voice market, the main competitors of Türk Telekom are Türknet, Türksat, Superonline, İşnet, NetGSM, Vodafon Net, and Millenicom. In the fixed broadband market, the most prominent competitors are Superonline, Vodafon Net, D-Smart, Türksat, Türknet, and Millenicom.

As for retail competition, when Türk Telekom owns nearly 81%, telephone call providers (TCPs) have 19% of the market with respect to net sales from fixed telephone service as of December 2019.[6] For the last ten years, TCPs' market share of TCPs has increased by eight points, four of which occurred since the retail market was deregulated in 2014 (ICTA, 2010a; ICTA, 2014; ICTA, 2019a).

Considering voice traffic for the same year, the market shares of Türk Telekom and TCPs are, respectively, 82% and 18% in local calls, 46% and 54% in intercity

[5] https://yetkilendirme.btk.gov.tr/Yetkilendirme/index.xhtml;jsessionid=JYmlHCOclLaWrVDrIVa-ZZu2KpNLb3-hh1g3gvRX.dmzweb (last visited on 11.7.2020).

[6] The net sales of Türk Telekom include access, interconnection, and other network incomes.

calls, 40% and 60% in calls to mobiles, and 31% and 69% in international calls. In terms of total traffic originating from fixed lines, the portions of Türk Telekom and TCPs are 50.3 and 49.7, respectively (ICTA, 2019a).

The market shares of the TCPs have made significant progress over the last decade. The nearly tenfold increase in the market share of TCPs in respect of total traffic originated from the fixed line shows there has been a considerable improvement in retail competition, especially after deregulation (ICTA, 2010a; ICTA, 2019a).

Deregulation of the retail market has been shown to boost the performance of TCPs in some segments of the fixed voice market. Since deregulation, TCPs have gained 18 points of market share in intercity calls, 15 points in calls to mobile, and 25 points in international calls (ICTA, 2015; ICTA 2019a).

There are two main infrastructures for the fixed voice service: PSTN[7] and VoIP.[8] As the market share of PSTN that wholly belongs to Türk Telekom is 89%, VoIP's portion amounts to 11% in terms of the number of subscribers. From 2015 to 2019, the subscription number of VoIP increased from 706,000 to 1.147 million. This development shows that there is potential for infrastructure competition for the fixed voice market (ICTA, 2015; ICTA 2019a).

A prominent feature of Turkey's fixed voice market is its small share of total call traffic. Fixed voice calls constituted only 2.2% of total traffic at the end of 2019, with the remainder belonging to mobiles (ICTA, 2019a). According to the Turkish Competitive Telecom Operators Association (TELKODER), this number is small compared to European countries such as the UK and Germany, where fixed voice calls account for 31% and 52% of those countries' totals, respectively (TELKODER, 2020).

During the last decade, fixed voice has lost ground in terms of the number of subscribers and total traffic because it has been shifting to mobile. The number of fixed voice subscribers has lowered from 16.5 to 11.3 million in the last 10 years. However, the number of mobile subscribers has increased from 62.8 to 80.8 million over the same period. As of 2019, the share of fixed voice was approximately 18% (ICTA, 2010a; ICTA, 2019a).

Hence, Türk Telekom's market power has been fading through two channels; fixed voice has been substituted by the mobile and Türk Telekom's infrastructure, PSTN, has been losing market share to VoIP. This trend implies Türk Telekom's superiority has been weakening in the shrinking market. Furthermore, by taking the abovementioned outlook of competitive retail sub-segments, it can be alleged that Türk Telekom no longer has sufficient power to determine economic parameters by acting independently of its competitors and customers in the retail segment.

When it comes to fixed broadband service, the total number of subscription for fixed broadband reached 14.2 million at the end of 2019. Out of this number, nearly 9.6 million connected through xDSL, 2.9 million through fiber Internet, one million through cable Internet, and the rest through other infrastructures. Thus, the ratios of

---

[7] Public switched telephone network.

[8] Voice over Internet protocol.

subscriptions have become 67.6% for xDSL, 20.4% for fiber, 7% for cable, and 4.9% for the others (ICTA, 2019a).

The ratio of broadband subscriptions per 100 inhabitants for Turkey is relatively low compared to OECD countries. According to key ICT Indicators of OECD, Turkey has the third lowest ratio with 16.9%, trailing only Colombia and Mexico (as of June 2019). The overall OECD average of 31.9% is almost twice that of Tukey.[9]

During the last decade, the number of xDSL subscribers has increased by 45%. On the other hand, the number of subscribers has increased nearly 19-fold for fiber and three-and-a-half-fold for cable. XDSL, and accordingly Türk Telekom has been losing ground against rival infrastructures, especially fiber.

Thanks to the boom in the number of fiber subscriptions that directly relates to investment in infrastructure, the share of fiber in total has started approaching the OECD average of 26.8%. When it comes to cable, its weight remains far below the OECD average of 33.5% as of June 2019.[10] This shows that cable's potential has not been utilized enough. As the share of cable Internet approximates to the OECD average, the dominance of xDSL, and accordingly Türk Telekom, will vanish.

Türk Telekom is maintaining its domination in the retail segment even though its market share has recorded a significant loss—nearly 26 points—over the last ten years. As the sole owner of the xDSL network, Türk Telekom has a 61.8% market share, of which 51.8% comes from xDSL and 10% from fiber, as of June 2019. Türk Telekom's main rivals are Türkcell Superonline, with a 15.6% market share, Türksat Cable Internet (6.7%), Vodafon Net (6.5%), Demirören Digital (3.5%), Türknet (2.2%), Millenicom (1.6%), and others (2.2%) (ICTA, 2019b).

Although Türk Telekom has lost 26 points over the last ten years, its market share is still far higher than the EU average. As of July 2019, incumbents controlled 40% of subscriptions in the EU. There are just six member countries where the incumbents' market shares exceeded 50% and only one country has a ratio above 60% (EU, 2020).

On the other hand, while Türk Telekom's market share has incurred a loss of 26 points over the ten years, EU incumbents have lost eight points over the same period. These comparative trends indicate Turkey has been approximating to the EU in respect of retail competition level.

Two indicators should also be noted in regulatory analysis: the price and speed of the Internet. According to the latest OECD statistics (from 2017), Turkey's Internet speed is nearly half the OECD average of 15.25 megabits per second.[11] This is the third worst among OECD countries, after Colombia and Mexico.

As for the price comparison, Turkey's price for low users[12] is the sixth cheapest, at 12.42 USD as of 2016. For high users,[13] the landscape is similar; Turkey is the eighth cheapest country in the OECD, at 24.47 USD for the same year.[14]

---

[9] https://www.oecd.org/sti/broadband/oecdkeyictindicators.htm (last visited on July 15, 2020).

[10] https://www.oecd.org/sti/broadband/broadband-statistics/ (last visited on July 15, 2020).

[11] See Footnote 10.

[12] 20 GB/month, 0.250 Mbps and above.

[13] 200 GB/month, 25 Mbps and above.

[14] See Footnote 10.

From this point of view, it can be argued that Internet users in Turkey have been rendered low-quality broadband service at a relatively cheap price. However, even the relatively low price does not seem to proportional to the current low-quality service.

Briefly, Türk Telekom has been exposed to expanding competitive pressure by its rivals at both markets. In the fixed voice line, which has had a downward trend in terms of its portion in total voice traffic, Türk Telekom has been losing ground due to growing VoIP usage and increasing market shares of TCPs in the various retail sub-segments.

As for fixed broadband lines, Türk Telekom's rivals have been gaining considerable ground in the retail market thanks to the growing share of fiber infrastructure. However, the cable, which is another alternative for infrastructure competition, has not been able to keep up with fiber. Thus, unlike OECD, cable infrastructure could not afford to put sufficient competitive pressure on Türk Telekom.

3 Legal Framework for Regulation of Fixed Telecom Services

In line with the EU, determination of regulatory obligations is based on market analysis studies conducted as per the "Ordinance on Market Analysis"[15] prepared based on the Articles 6 and 7 of the ECL. The Ordinance determines the procedures and principles for market analyses conducted to ensure establishment, protection, and improvement of competition, and the regulatory obligations imposed on the operators with significant market power.

The ICTA should carry out market analysis studies every three years on its own initiative or upon justifiable request at least from one operator. Market analysis studies include defining the relevant market, examining the need for regulation in the relevant market, analyzing competition level in relevant markets, identifying the operator or operators with significant market power, and determining the regulatory obligations for the operators in question.

As per the Ordinance, the ICTA conducts the market analysis study in a participatory manner. Prominent shareholders such as the Competition Authority, NGOs, and principle operators deliver their views on the draft study. These views might be posted on the ICTA website.

A relevant market is defined by considering its product/service and geographical dimensions. Throughout the definition of relevant product/service market, certain factors should be taken into account, where applicable: the purpose in terms of functionality, substitutability of products or services in terms of price, switching costs, behaviors of users and operators, mandatory use of bundled or ancillary product/services, comparative tariff or price analysis, potential competition, and entry barriers.

---

[15] Ordinance on Market Analysis. *Official Gazette*, dated November 27, 2012, and numbered 28480 (Turkish version).

Definition of geographical market should include the area in which operators are authorized to provide services, the coverage of the operator's network and the possibility of providing services in other areas, similarities in supply conditions of products or services with respect to geographic coverage, and users' preferences in local, regional, or national contexts.

A decision that regulation is necessary in the specified relevant market should be made when the following conditions are cumulatively met: high and non-transitory barriers to entry, market structure that does not tend toward effective competition within the relevant time horizon, insufficiency of competition law alone to adequately address the market failure, and the effects of wholesale remedies on competition at the retail level.

The Ordinance deems no effective competition in the relevant market in case one or more operators have significant market power individually or an operator has significant market power jointly with others.

Besides the market share, various parameters need to be considered when determining relevant market if an operator individually or a group of operators jointly obtains significant market power. These parameters include control of infrastructure that is not easily duplicated, technological superiority, lack of countervailing buying power, economies of scale or scope, vertical integration, barriers to entry and lack of potential competition.

The Ordinance stipulates that remedies to be imposed on operators with significant market power in the relevant market are supposed to be appropriate, proportionate, and oriented to the source of the problem. The ICTA is authorized to impose one, some, or all of the remedies of provision of access and/or interconnection, tariff control, publication of reference access or interconnection, facility sharing and co-location, nondiscrimination, transparency, and cost accounting and accounting separation on operators that have significant market power.

The ICTA has discretion to impose further remedies on operators that significant market power in the relevant market to take an existing or potential competition problem out without conducting a new market analysis in case the justifications are based on objective reasons.

Existing remedies may be amended as a result of the market analysis depending on the structure of the relevant market. In case no operators have significant market power in the relevant market any longer, the abovementioned remedies cannot be imposed, so the existing ones are to be withdrawn. Modification or withdrawal of the remedies can be carried out without renewing market analysis only if the objective justifications are put forward.

The basic principles of access and interconnection obligation are articulated in Article 15 of ECL. "Ordinance of Access and Interconnection"[16] elaborates on the basic principles and brings detailed regulations for clarifying procedures.

If an operator somehow avoids access obligation, the ICTA is entitled to force this operator to fulfill its obligation. An operator obliged to provide access might only

---

[16] Ordinance on Access and Interconnection. *Official Gazette* dated September 8, 2009, and numbered 27343 (Turkish version).

restrict access upon the ICTA's approval through setting out objective criteria that the security of network operations or network integrity or data protection cannot be assured or interoperability of networks is not possible.

The ICTA might order operators with significant market power to prepare reference access offers including interconnection. In such cases, operators should prepare and submit their reference offer to the ICTA within a maximum of three months following the imposition date of the obligation. The ICTA has discretion to demand some amendments in the access offer. The approved access offers are obliged to be published on the Internet.

The ICTA is entitled to bring responsibilities on operators that are put under access obligation, such as equality, nondiscrimination, accounting separation or cost accounting, co-location, facility sharing, transparency and clarity, to be based on cost and reasonable profit and to render access services with fair conditions.

As per the ECL, operators performing in the market should meet the negotiation demand for interconnection. In case the parties could not afford to deal, the ICTA is authorized to rule that the operators in question meet the interconnection request. Interconnection is to be provided at any interconnection point that can be proved to be technically and economically feasible. In case the operator requires interconnection from any other point, it incurs the cost of the point that will be provided.

The ECL stipulates that tariffs might be in the form of subscription fees, fixed charges, call charges, line rental, and similar price items. Operators are free to determine the tariffs by considering regulations in force. The ICTA has authority to approve monitor and supervise the tariffs as per the related provisions of ECL and "Tariff Ordinance,"[17] in case an operator or operators are determined to have significant market power. The ICTA is also held responsible for blocking anti-competitive tariffs such as price squeezing and predatory pricing.

In the approval process, the ICTA should consider that tariffs are to be based on the costs of efficient service provision and should not contain excessive charges, discriminatory practice, and discounts aiming at distorting competition. Throughout tariff approval, ICTA uses either the method based on the costs of efficient service provision falling to the individual service or the price cap method, or both of them. As per the Tariff Ordinance, operators with significant market power are required to deliver detailed documents for verification and monitoring the tariffs.

In accordance with the ECL, the ICTA might hold operators with significant market power responsible for preparing and publicizing reference access offers including interconnection. Reference access offers should be prepared according to related regulations of the ICTA that cover at least the items specified in the Annex-1 of Ordinance on Access and Interconnection.

Operators demanding access should not be forced to pay for the services that are not directly related to the requested services. In order to ensure this, services are to be unbundled and priced. Furthermore, the timeframes and conditions of reference access offer should be determined precisely. Reference offers are principally

[17] Tariff Ordinance. *Official Gazette* dated November 12, 2009, and numbered 27404 (Turkish Version).

renewed annually. On the other hand, operators are given the right to come up with amendments in their offers in circumstances where required by changing market conditions and technological developments. The ICTA approves the offers that it considers eligible. The operators are obliged to publish their approved reference access offers and to provide with access under conditions specified in their offers.

In circumstances where an operator can construct its facilities on or under the public domain or third parties' properties, or where it is authorized to use such properties or granted the right to expropriation, the ICTA, taking social interest and efficiency into account, might bring obligations of sharing the facilities or properties with the other operators in return for a reasonable charge. The ICTA might also impose obligations on operators to provide any kind of co-location including physical co-location for the equipment of other operators within their own facilities in return for a cost-based charge.

Detailed principles and procedures are regulated by "Communique on Procedures and Principles of Co-location and Facility Sharing."[18] As per the Communique, facility sharing refers to the use or sharing of related facilities and devices, including pipes, channels, towers, and towers used in electronic communication service by other operators. Co-location is defined as physical partner placement of the operator's system or devices required for access, in common areas located in the premises of the joint settlement obligation and suitable for the use of the relevant operators, including the joint settlement obligation.

As mentioned above, the obligation of nondiscrimination is directly addressed by the ECL and the Ordinance on Access and Interconnection. The basic principle of nondiscrimination is that the operators with significant market power should provide services to third parties on a nondiscriminatory basis and upon the same conditions and qualities that apply to their subsidiaries or partners or partnerships.

Like the nondiscrimination obligation, transparency is accepted as an important complement of access and interconnection obligation by the ECL. As per the Ordinance on Access and Interconnection, the ICTA might oblige operators with significant market power to make available the technical specifications, network characteristics, terms and conditions of supply and usage, fees, and other specified information.

The ICTA has discretion to impose accounting separation or cost accounting obligations on operators with significant market power in the relevant market. Operators under accounting separation should keep separate accounts for their activities. The ICTA might directly audit such an operator or have it audited if the operator fulfills its obligations in compliance with principles and procedures on accounting separation and cost accounting. Operators might be held responsible for publishing documents and information within the scope of obligations. The ICTA can publish such documents and information on its own initiative.

[18] Communique on Procedures and Principles of Co-location and Facility Sharing. *Official Gazette* dated December 2, 2010 and numbered 27773 (Turkish version).

4 Regulation of Fixed Voice Services

The first market analysis study on fixed voice services at the retail level was published in the first quarter of 2006. The second round of the study was conducted in 2009. The last one, which deregulated the retail market, was put into effect in 2014. Since retail competition has accelerated over the last ten years, the market analysis studies of 2009 and 2014 will be assessed.

Deregulation of the market was based on the fact that there are no entry barriers; mobile voice service put competitive pressure on the fixed voice services that lead competition to exist on its own under the market conditions; the regulations for wholesale services correct the market failures; the obligations of wholesale line rental and price squeezing at the voice originating market will facilitate the competition in the market; and competition law does not remain insufficient to cope with market imperfections.

The methodology of market analysis studies is observed to remain identical for all. The ICTA has conducted detailed analyses on the conditions of demand and supply substitution in the market definitions.

The 2009 Market Analysis Study determined three different relevant service markets: the local telephone call market, offered in a fixed location; the intercity telephone call market, offered in a fixed location; and the international telephone call market, offered in a fixed location. The geographical market was defined to be Turkey (ICTA, 2009).

The 2014 Market Analysis Study has defined two relevant service markets. The first one is local telephone calls, intercity telephone calls, and calls toward the mobile direction market provided from a fixed location for individual and corporate users, including non-geographic numbers and voice on broadband. The second one is international telephone call market provided from a fixed location for individual and corporate users, including non-geographic numbers and voice on broadband. The geographical market was determined as Turkey (ICTA, 2014).

Both market studies considered the criteria of existence of high and non-transitory barriers to entry, which are the existence of a market with the potential to be competitive on its own dynamic and the sufficiency of ex-post competition law correcting market failure to decide if regulation.

Based on assessments of the abovementioned criteria, the 2009 Market Analysis Study concluded that the relevant market should be regulated. Türk Telekom was determined to have efficient market power through forward-looking assessment on market share, control of infrastructure that was not easily duplicated, vertical integration, economies of scale and scope, distribution network, and buying power. Finally, it was proposed that Türk Telekom be put under the obligations of provision of access, carrier selection and carrier preselection, nondiscrimination, transparency, tariff control, and invoice provision. The ICTB approved the Market Analysis Study by its Decision.[19]

---

[19] ICTB Decision dated December 8, 2009, and numbered 2009/DK-10/625.

However, the 2014 Market Analysis Study ruled that the market could be deregulated because there were no entry barriers, competition law could correct market failures, the regulatory obligations on the wholesale market were effective, and the market had the potential to become competitive on its own dynamics thanks to strong mobile substitution. Based on this assessment, efficient market power analysis was not conducted. In conclusion, it was proposed that the regulatory obligations imposed on Türk Telekom be removed (ICTA, 2014); these proposals were approved by ICTB Decision.[20]

Contrary to previous studies, two factors can be asserted to play an important role in the deregulation decision of the ICTA. The first is the ICTA's assessment on supply substitution, in which the mobile call service is thought to be substitutable with a fixed call service (ICTA, 2014). There is no doubt that mobile substitution can put competitive pressure on Türk Telekom, an operator with significant market power.

The second factor relates to regulations in wholesale markets. In respect of fixed voice services, Türk Telekom has had various regulatory obligations imposed on it, both in calls originating at fixed network market and call termination at fixed network market. The ICTA (2014) has specifically addressed the remedies of wholesale line rental and price squeezing at calls originating at fixed network market that were to enhance competition in the market.

At the time the retail market was deregulated, the regulatory obligations proposed by the 2012 Market Analysis Study on Call Originating at Fixed Network Market were in effect. Türk Telekom was imposed to the remedies of access and interconnection, nondiscrimination, transparency, tariff control, accounting separation and cost accounting, carrier selection and carrier preselection, co-location, and facility sharing (ICTA, 2012a). These proposals were put into force by the ICTB's Decision.[21]

In addition, as per the Decision of ICTB,[22] Türk Telekom was obliged to abide by the remedies of price squeezing and wholesale line rental obligations in the call originating at the fixed network market.

The next Market Analysis Study on Call Originating at Fixed Network Market was conducted in May 2017. The ICTA (2017) proposed the same regulatory obligations as the previous study for Türk Telekom, which were approved by the Decision of the ICTB.[23] The said obligations are still in force.

As for call termination at fixed network, when the retail market was deregulated, the regulatory obligations proposed by the 2012 Market Analysis Study on Call Termination at Fixed Network Market were in force. Türk Telekom was imposed to the remedies of access and interconnection, nondiscrimination, transparency, tariff control, accounting separation and cost accounting, and publication of reference access offers (ICTA, 2012b).

[20] ICTB Decision dated March 26, 2014, and numbered 2014/DK-SRD/171.

[21] ICTB Decision dated January 3, 2013, and numbered 2013/DK – SRD/6.

[22] ICTB Decision dated 26.03.2014 and numbered 2014/DK-SRD/172.

[23] ICTB Decision dated 08.05.2017 and numbered 2017/DK-SRD/152.

These obligations were put into force by the ICTB's Decision.[24] Furthermore, the obligations of interconnection, transparency, and nondiscrimination were brought for all authorized operators as per the same Decision.

The following Market Analysis Study on Call Termination at Fixed Network Market was conducted in December 2016. The ICTA (2016) proposed the same regulatory obligations as the previous study for Türk Telekom and authorized operators, which were approved by the Decision of the ICTB.[25] The said obligations are still in force.

As touched upon earlier, TCPs have presented great advancement in the fixed voice market after deregulation so that their market shares in intercity calls, international calls, and calls to mobile segments exceed those of Türk Telekom, as of 2019.

5 Regulation of Fixed Broadband Services

The first market analysis study on fixed broadband services was conducted in 2006. Following this study, three further ones were presented, in 2010, 2012, and 2019.

In each study, Türk Telekom was determined to have significant market power and be exposed to regulatory obligations. All of the studies considered the EU's related regulations in their technical assessments. Hence, the methodology and analyses of market definition, determination of significant market power, and regulatory obligations are deemed to be in conformity with EU's technical assessment approach.

Since retail competition is observed to have been enhancing during the last ten years, the last three market analysis studies will be considered and assessed here. The methodology of market analysis studies seems to remain almost the same.

In all three studies, the ICTA has conducted in-depth analysis on retail demand substitution, wholesale demand substitution, and wholesale supply substitution while defining relevant product/service market.

As for definition of the relevant geographical market, the deep-dive analyses could not be observed. In other words, the studies did not consider the criteria ordered by the Ordinance on Market Analysis. However, the latest market analysis study provides exhaustive geographical market assessment that is compatible with the said Ordinance.

The relevant market definitions were identical in the 2010 and 2012 studies. The relevant product/service market was determined as the fixed wholesale broadband Internet services (xDSL, ATM, FR, ME Internet), including bit stream access at the IP level. The relevant geographical market was defined to be Turkey (ICTA, 2010b; ICTA, 2012c).

In parallel with technological developments, the relevant market definition has changed and two separate relevant markets have been defined as the wholesale local access services, including all the separated access types, and the wholesale broadband

[24] ICTB Decision dated 03.01.2013 and numbered 2013/DK-SRD/7.

[25] ICTB Decision dated 21.12.2016 and numbered 2016/İK-SRD/5042.

access services (fiber, xDSL, ME Internet), including bit stream access and resale services at the IP and Ethernet levels in the study (ICTA, 2019b).

The relevant geographical market was again determined to be Turkey. However, although the study determined the geographical market as Turkey, the ICTB defined two regional geographical markets by its Decision dated December 31, 2019, numbered 2019/DK-SRD/338. This issue is important with respect to deregulation and will be discussed later on.

What needs to be highlighted here is that, up until 2019 market study, fiber had not been accepted as a substitute broadband service to xDSL. Thus, broadband services over the fiber network had not been subjected to regulatory obligations, which meant potential entry barriers for the TCPs and restrictive competitive pressure on the incumbent. This drawback was eliminated by the latest market analysis study. From now on, fiber operators with significant market power are obliged to fulfill regulatory obligations. On the other hand, the cable network is still being construed not to be a substitute service with xDSL and fiber, so it falls outside the regulatory obligations.

All of the market studies have considered the same three criteria to find out whether there needs to be regulatory requirement in the relevant market. These criteria are existence of high and nontransitory barriers to entry, a market having potential to be a competitive market on its own dynamic, and sufficiency of ex-post competition law enforcement to cope with market failure. Finally, all of the studies concluded that the relevant markets necessitate regulation (ICTA, 2010b; ICTA, 2012c; ICTA, 2019b).

The 2010 Market Analysis Study is thought to have provided superficial analysis with respect to assessment on three criteria. On the contrary detailed analysis on the necessity of the regulation of relevant markets were conducted in the 2012 and 2019 Studies.

By reference to European Regulators Group (2008), both market analysis studies have taken into account the indicators of existence of sunk costs, control of infrastructure not easily duplicated, technological advantages or superiority, easy or privileged access to capital or financial resources, economies of scale and scope, vertical integration, barriers to develop distribution and sales network, and products or service diversification (ICTA, 2012c; ICTA, 2019b).

As seen in the assessment of regulatory requirement, all of the market studies in question have resorted to the same criteria by considering forward-looking assessment throughout the analysis of competition level in relevant markets, which are: market share, control of infrastructure not easily duplicated, economies of scale and scope, potential competition, and stabilizer buying power. At the end of the analyses, Türk Telekom was determined to own significant market power, thereby having been imposed to various regulatory remedies (ICTA, 2010b; ICTA, 2012c; ICTA, 2019b).

As expected, the regulatory obligations on Türk Telekom, especially the ones related to market access and interconnection, have become more detailed over time, mostly due to technological developments.

The ICTA (2010b) proposed the below remedies on Türk Telekom, all of which were approved and put into force by the Decision of the ICTB.[26]

- Providing wholesale broadband access, including xDSL resale, xDSL bit stream access, ATM, FR, and Metro Ethernet resale.
- Providing bit stream access at the level of IP.
- Nondiscrimination.
- Transparency.
- Publication of reference access offers.
- Tariff control.
- Cost accounting and accounting separation.
- Co-location and facility sharing.

The below remedies proposed by ICTA (2012c) were fully put into practice by the Decision of the ICTB.[27]

- Providing wholesale broadband access, including naked xDSL.
- Publication of reference access offers, including service level commitments and sanctions in case of failure to comply with those commitments.
- Nondiscrimination including announcement of technical information and conditions of new products or services and technical changes in currently delivered services at least two months before putting in practice or changes in tariffs or campaigns at least one month before putting in practice and delivering performance indicators to ICTA on service level commitments for each operators utilizing the services in three-month periods.
- Transparency, including publishing average values of performance indicators for each operator benefitting from the services in a three-month period.
- Tariff control and cost-based tariff.
- Cost accounting and accounting separation.
- Co-location and facility sharing.

The ICTA 2019 Market Analysis Study diverged from the previous ones in two aspects. Firstly, the regulatory obligations have been relatively elaborate; secondly, the proposed regulatory obligations have not been fully approved by the Decision of the ICTB.[28]

As stated above, the ICTA (2019b) has defined two separate markets. Thus, peculiar remedies have been determined and detailed for each market. The obligations below have been proposed to be imposed on Türk Telekom for the wholesale local access services market.

- Providing access for copper networks that cover full-unbundled access to the local loop, shared access to the local loop, and unbundled access to the sub-local loop.

---

[26] ICTB Decision dated January 7, 2010, and numbered 2010/DK – 10/20.

[27] ICTB Decision dated 11.01.2013 and numbered 2013/DK-SRD/29.

[28] ICTB Decision dated December 31, 2019, and numbered 2019/DK-SRD/338.

- Providing access to fiber networks through virtual unbundled local access; this obligation must include delivery of the traffic at the first local station without limiting or squeezing, enabling alternative operators to diversify their services, identification of virtual local area networks for the services such as voice and broadband and multicast/unicast traffic, delivery of traffic through Ethernet, provision of a Digital Subscriber Line Access Multiplexer port management opportunity to alternative operators, enabling alternative operators to question line failures to control final users' devices.
- Providing access to physical infrastructure elements including SSG Internet service.
- Co-location and facility sharing.
- Tariff control, including cost-based tariff and price squeezing.
- Preparation and publication of reference access offers, including service level commitments and sanctions in case of failure to comply with those commitments.
- Nondiscrimination, including the principle of equivalence of input and delivery of performance indicators to ICTA on service level commitments for each operator utilizing the services in a three-month periods.
- Transparency, including publishing the average values of performance indicators for each operators benefitting from the services in a three-month period.
- Cost accounting and accounting separation.

The remedies to be proposed on Türk Telekom for wholesale broadband access services market are as follows (ICTA, 2019b).

- Providing access at the level of Ethernet Layer 2 for xDSL and fiber bit stream, including provision of Digital Subscriber Line Access Multiplexer port management opportunity to alternative operators, enabling alternative operators to question line failures, to control over final users' devices, identification of virtual local area networks for the services such as voice and broadband and multicast/unicast traffic, delivery of traffic through Ethernet, and enabling alternative operators to diversify their services.
- Co-location and facility sharing.
- Tariff control, including cost-based tariff and price squeezing.
- Preparation and publication of reference access offers, including service level commitments and sanctions in case of failure to comply with those commitments.
- Nondiscrimination, including the principle of equivalence of input and technical information and conditions of new products or services and technical changes in currently delivered services at least two months before putting into practice, changes in tariffs or campaigns at least one month before putting into practice, and delivering performance indicators to ICTA on service level commitments for each operator utilizing the services in a three-month period.
- Transparency, including publishing average values of performance indicators for each operator benefitting from the services in a three-month period.
- Cost accounting and accounting separation.

As highlighted before, the abovementioned regulatory obligations have been approved by ICTB with some exceptions and revisions. According to the ICTB Decision,[29] virtual unbundled local access obligations imposed on Türk Telekom for both markets were postponed until July 1, 2020. Furthermore, ICTB has exempted Türk Telekom from access obligations and tariff control for three districts—Çukurova, Balçova, and Karşıyaka—where competitive market environments exist.

This Decision is of great importance because, for the first time, ICTB has defined geographical markets locally and deregulated them. This development can be expected to represent a new era of local deregulation process that is most likely to stimulate retail competition. It needs to be stressed that Türk Telekom has only been immunized from access and tariff obligations and therefore remains to be exposed to the rest of the regulatory remedies that imply ICTB's cautious approach toward deregulation.

In general, the 2019 Market Analysis Study has two distinctive features. First, it determines two different markets and two sets of detailed regulatory obligations for each market. This will increase the choices of accessing the networks with lower cost and higher quality for the competitors of Türk Telekom. Second, the fiber network is included in the market definition together with xDSL. Thus, the fiber infrastructure has started to be regulated. This will probably accelerate the already upward trend of fiber's share in total fixed broadband service that can be expected to erode Türk Telekom's market power.

From an historical perspective, market analysis studies on which regulations have been based for the operators with significant market power have evolved in line with the EU's approach. They have been getting involved in more detailed regulatory obligations for expanding and diversifying market definitions that pave the way for TCPs to have more options for rendering services to the final users. This will most likely enhance competition in retail markets.

On the other hand, two prominent critiques can be raised for regulation policy toward fixed broadband. Firstly, the latest market analysis study was conducted six years after the previous one was published, even though it should be repeated by the ICTA within a maximum of three years, as per the Ordinance on Market Analysis. If the market analysis study had been made on time, namely in 2016, TCPs would have obtained the abovementioned advantageous rendered by the 2019 Market Analysis Study earlier, which would have led to a more competitive market structure.

Secondly, unlike fiber, the cable network is still kept outside the regulatory obligations on the grounds that broadband services over cable network are being delivered in a limited geographical part of Turkey. As mentioned earlier, the share of cable infrastructure in total fixed broadband service has been quite low, around 7%, in Turkey compared to OECD countries. This implies an opportunity to enhance structural and retail competition by means of weighting to the cable network.

[29] ICTB Decision dated December 31, 2019, and numbered 2019/DK-SRD/338.

6 Conclusion

Turkey's regulatory approach to the telecommunications market has been evolving in the same direction as that of the EU. Enhancing retail competition through effective wholesale regulations has been at the core of this approach. Accurate market definitions, in-depth market analysis, and correctly demarcated regulatory obligations on the operators having significant market power are vitally important to reach this goal.

After nearly two decades of liberalization in the Turkish telecommunications market, both retail and infrastructure competition have recorded significant improvement in the fixed voice market. The market power of the incumbent monopoly, Türk Telekom, has been eroding by means of mobile and VoIP substitution of PSTN, on the one hand, and strengthening TCPs in retail market, especially in the sub-segments of intercity calls, calls to mobiles, and international calls, on the other.

Fixed voice traffic shifting rapidly to mobile, growing up of VoIP usage, deregulation of retail market, and robust regulations in wholesale market are the prominent factors for enhancing competition in the fixed voice market. From this point of view, one can assert that Türk Telekom no longer has sufficient power to determine economic parameters, such as price by acting independently of its competitors and customers in the retail market.

The Turkish broadband market can be described as having a low subscription ratio, lower retail market competition, and low-quality service provision with a cheaper price than the OECD and the EU average. Still, the rivals of network owners, namely Türk Telekom, have gained considerable ground in the retail market thanks to a growing share of fiber infrastructure in total traffic and effective regulations. However, cable has not been able to keep up with fiber in Turkey, unlike in other OECD countries. Therefore, cable infrastructure has not generated competitive pressure on Türk Telekom to the extent necessary.

The latest regulation that was put into force with delay is to enhance retail competition since it has imposed two sets of detailed regulatory obligations for two separate markets covering fiber, which pave the way for diversified options for getting access to network with lower cost and higher quality on the condition of efficient enforcement and effective monitoring. On the other hand, strengthening cable infrastructure by making the required investment and expanding market definition in such a way that it includes cable service, thereby imposing regulatory remedies on the cable infrastructure owner, will accelerate the course of retail competition as well.

References

EU. (2020). *Broadband market developments in EU*. Digital economy and society index report 2020.

European Regulators Group. (2008). *Report on guidance on the application of the three criteria test*. ERG (08) 21 Report on three criteria test final 080604 (June 2008). Available at: http://www.ictregulationtoolkit.org/document?document_id=3661

ICTA. (2009, December). *Market analysis study on call services over fixed network* (Turkish version).

ICTA. (2010a). *Quarterly market data report*. The fourth quarter report of 2010 (Turkish version).

ICTA. (2010b, January). *Market analysis study on wholesale broadband access market including data stream access* (Turkish version).

ICTA. (2012a, December). *Market analysis study on call originating at fixed network market* (Turkish version).

ICTA. (2012b, December). *Market analysis study on call termination at fixed network market* (Turkish Version).

ICTA. (2012c, December). *Market analysis study on wholesale broadband access market including data stream access* (Turkish version).

ICTA. (2014, March). *Market analysis study on call services over fixed network* (Turkish version).

ICTA. (2015). *Quarterly market data report*. The Fourth Quarter Report of 2015 (Turkish version).

ICTA. (2016, December). *Market analysis study on call termination at fixed network market* (Turkish version).

ICTA. (2017, May). *Market analysis study on call originating at fixed network market* (Turkish version).

ICTA. (2019a). *Quarterly market data report*. The Fourth Quarter Report of 2019 (Turkish version.)

ICTA. (2019b, December). *Market analysis study on wholesale local access and wholesale broadband access markets* (Turkish version).

TELKODER. (2020). *Assessments on sectoral regulations in 2019* (Turkish version). Available at: https://telkoder.org.tr/wp-content/uploads/2020/06/TELKODER-2019-Degerlendirme.pdf

Türk Telekom. (2019). 2019 annual report of Türk Telekom. Available at: http://www.ttyatirimciiliskileri.com.tr/FaaliyetRaporlariEN/2019-Annual-Report.pdf

Bülent Gökdemir is a senior competition expert working in the area of digital economy, network/infrastructure industries, manufacturing sectors and service sectors at Turkish Competition Authority and a part time lecturer at Middle East Technical University. He lectures on Competition and Regulation Economics and Digital Economy. He holds LLM from University of Dundee, M.Sc from Hacettepe University and Ph.D. from Hacettepe University. He has several publications in the field of competition law and economics, regulation of network industries, universal service obligation and productivity.

Regulation of Fiber and the Internet

Emin Köksal

Abstract This chapter examines the regulatory issues in the Turkish broadband market, focusing on fiber. Following a historical background that emphasizes the evolving regulatory and business environment in the broadband market, I describe the current situation and the current regulatory issues related to fiber deployment. Thereafter, targets that were included in the National Broadband Plan are examined and compared with the current state of the market. The main conclusion of the chapter refers to a lack of vision in both demand-side and supply-side regulatory policies.

Keywords Internet · Fiber · Regulation · Competition · Broadband

1 Introduction

Broadband internet has been considered as an indispensable infrastructure for the development of nations. In the last decade, fiber internet has become a prominent technology for broadband access. Countries have set targets, developed strategies, and organized action plans for the deployment of fiber internet and its penetration in society.

Albeit relatively late, Turkey prepared its national broadband plan in 2017 and defined ambitious targets for the 100th anniversary of the republic, 2023. However, the current state of the broadband market indicates a situation that is far from reaching those targets. Most of the problems in the market refer regulatory issues. This chapter examines those issues in the Turkish broadband market focusing on fiber.

The chapter is organized as follows. The next section summarizes the historical background from dial-up to fiber, emphasizing the developing business and regulatory environment. The third section outlines the current state of the market and current regulatory issues. The fourth section focuses on the targets and their comparison with the current state. The fifth section provides a discussion and some recommendations. The final section concludes.

E. Köksal (✉)
Bahçeşehir University, Istanbul, Turkey
e-mail: emin.koksal@eas.bau.edu.tr

M. Eroğlu and M. Finger (eds.), *The Regulation of Turkish Network Industries*,
https://doi.org/10.1007/978-3-030-81720-6_19

2 Historical Background from Dial-Up to Fiber in a Developing Regulatory Environment

To understand the current situation and the foundation of the regulatory issues in the Turkish broadband market, some background information is required. Therefore, in this section, the transforming legal and business environment in the Turkish broadband market is briefly summarized along a historical path.

2.1 Before the Regulatory Reform

Initial internet service providing attempts in Turkey appeared at some academic institutions in early 1990s. The first connection to a global Internet backbone from Turkey was established by Middle East Technical University (METU) and the Scientific and Technological Research Council of Turkey (TUBİTAK) in 1993 under TR-NET, which was an informal organization to introduce the Internet.

Meanwhile, some important legal developments had been realized. In 1994, by changing the main telecommunications act,[1] monopoly rights of providing basic telecommunication services were given to a newly established state-owned enterprise, namely Turk Telekom (TT). According to the new legislation, although basic telecommunication services would be provided by TT itself, other value-added services—including internet services—could be supplied through some hybrid methods such as revenue-sharing agreements. TT also had the authorization to allow the facilities of private and public initiatives on data transfer (OECD, 2002). While revenue-sharing agreements made it possible to attract private investors who would build and operate the infrastructure, the authorization regime kept everything under the management of TT. Such an arrangement could increase the value of the company, which would seek to be privatized in later years.

Through increasing demand in a short period of time, particularly from academia, the access supplied by TR-NET began to be insufficient. The increasing interest motivated METU and TUBİTAK to approach possible funders from the government. As a response to this approach, in 1995 TT announced a tender to find private investors who would build and operate an internet backbone under a revenue-sharing agreement. A consortium of GlobalOne, Satko, and METU became the highest bidder by offering 70% of its revenue to TT. The new network, which consisted of a triangle with corners in Turkey's three largest cities—İstanbul, Ankara, and İzmir—was called TURNET (Akgül, 1997).

Deployment of TURNET allowed the business model of internet service providing in Turkey. Since TT and its revenue partners preferred to keep their activities at the infrastructural level, retail access services were supplied through simple contracts with private internet service providers (ISPs). Many Turkish corporations showed

[1] Law no. 406, Postal and Telegram Act.

interest in this business model. A competitive environment in the retail market increased the number of ISPs in parallel to the dramatic increase in internet penetration. Between 1996 and 1999, the number of Internet users grew by approximately eight-fold and reached over 1% of the population. The number of ISPs increased to almost 80 by 1999 (see Wolcott & Çağiltay, 2001).

However, like TR-NET, through the increasing number of connections TURNET commenced to experience capacity problems by 1997. The government initiated a process to establish a new country-wide backbone called TTNET. It became operational in 1999, with the purpose of providing infrastructural services to carry the traffic from/to different ISPs. Basically, it supports access through the public switching telephone network (PSTN) and the cable TV network, both of which are owned by TT. Contrary to the limited access capacity of TURNET, TTNET network provided 140 points-of-presence (POP) for access, which serves every city in Turkey (Özyiğit, 2003). TT followed the same business strategy of positioning itself as a wholesaler and allowed ISPs as retailers.

2.2 The Regulatory Reform and TT's Entrance to the Retail Market

Meanwhile, the telecommunications industry in Turkey quickly started to be liberalized, a process that was accelerated by the economic crises the country experienced in 1999 and then again in 2001. The urgent financial liquidity need of the government was supplied via loans from international institutions—the International Monetary Fund and the World Bank—on the condition that structural reforms would be made in state-owned industries, including telecommunications.

Among these reforms were the establishment of an independent regulatory agency and allowing private third parties to access the telecommunications market. Moreover, the basis of the European Union (EU) regulatory framework, which Turkey has been obliged to streamline, required similar reforms. In 2000, with a new act,[2] the Telecommunications Authority (TA) was established as a national regulatory authority. This act also set a deadline to terminate the legal monopoly of TT in order to allow third parties to enter the telecommunications markets.

Shortly after the regulatory reform, TT changed its business strategy and decided to become active in retail internet services. It introduced a new brand called TTNET (the same name as the backbone) and started to launch aggressive tariffs to increase its penetration in the retail market. Moreover, along with dial-up services, it began to offer asymmetric digital subscriber line (ADSL) services by 2000. With this new access technology, subscribers were served through copper telephone lines, but with relatively high speed and quality. However, the retail access of ADSL had been exclusively provided by TT, and there was no regulation to force TT to provide wholesale access to the private ISPs.

---

[2] Act no. 4506, Telegram and Telephone Act.

Following complaints from the ISPs, in 2001 the Turkish Competition Authority (TCA) initiated an investigation concerning the predatory conduct of TT in dial-up services and its exclusionary conduct in ADSL services. Although the claim of exclusionary conduct was critical for the survival of ISPs, the TCA thought that it could be remedied through an ex-ante regulation by the TA, rather than ex-post intervention.[3] However, as of 2005[4] no attempt of an ex-ante regulation had come from the TA. According to Köksal and Ardiyok (2015), the main reason for this reluctance was related to the TA's lack of capacity at that time to make such a regulation, and the privatization process of TT, which was receiving considerable affection from the government. During this time, TT's share in the retail internet access market increased dramatically, from 2% in 2000 to 76% in 2005.[5]

2.3 *Privatization, Restructuring of TT, and the Rising DSL Technology*

Starting from 1993, a series of privatization efforts were made to privatize fixed telecommunication services in an environment without any regulatory framework (Atiyas & Dogan, 2009). Therefore, they were all canceled by the Constitutional Court. Finally, in 2005, TT was privatized to Oger Telecom with a value of 6.55 billion USD for 55% of its share.

In its merger review of privatization, the TCA played an important role in order to restructure the electronic communication market in Turkey. It set two conditions for its approval to the transaction: (1) divestiture of cable TV network from TT, and (2) restructuring the retail internet service business (TTNET) as a separate legal entity. While the divesture of cable TV network from TT indicated the initiation of a facility-based competition, establishing TTNET as a legal entity aimed to create transparency and to hold TT only on the wholesale level (see Köksal & Ardiyok, 2015).

The restructuring and the privatization of TT was accompanied by the rising DSL technology in retail market. In August 2005, after a long struggle between TT and ISPs, bitstream access tariffs were released by the TA and ISPs became capable of providing retail ADSL services by getting wholesale access that was 29–35% lower than the retail prices. However, after the privatization, the new management of TT and TTNET focused on increasing their penetration in ADSL services. TTNET's aggressive retail tariffs in order to attract large customer numbers had neutralized the discount for ISPs and created exclusionary effects for them. While TTNET increased

---

[3] TCA's decision dated October 2, 2002, and numbered 02-60/755-305.

[4] In 2005, the TA imposed an obligation on TT to provide resale bitstream access with the Board decisions dated July 25, 2005, and numbered 2005/475 and dated December 15, 2005, numbered 2005/876.

[5] Author's own calculation based on the compiled reports of the ICTA.

its subscribers in ADSL services from 1.5 million in 2005 to 2.7 million in 2006, alternative ISPs could only supply a number of subscribers.[6]

Following complaints by the ISPs, the TCA again arrived on the scene. In 2007 it initiated a detailed investigation of both TT and TTNET, which in 2008 led to TT being cleared of any wrongdoing in the wholesale market since it is under the jurisdiction of the TA. However, TTNET was investigated for its alleged activities. The TCA found that TTNET (with TT) applied price squeeze in the broadband market.[7] This called for another ex-ante regulatory intervention to prevent the anti-competitive pricing behavior of the incumbent operator, TT.

2.4 New Electronic Communication Act and a Partial Success in Service-Based Competition

While alternative ISPs were dealing with the exclusionary conducts of TT, a new electronic telecommunication act[8] was enacted in September 2008. The aim of this act, which had been worked on since 2005, was to give a new form to old and ineffective legislation. The act harmonized the regulations on telecommunications with the 2003 regulatory framework of the EU.

According to Köksal and Ardiyok (2015), the new act was lacking in certain key areas. First, while the new legislation introduced some key elements to specify the regulatory authority's independence, the relevant ministry continues to hold power in key matters. Second, despite fastening the fragmented provisions of the telecommunications law, the entire legislation regarding electronic communications continues to have a complex hierarchy, with some key matters addressed only by secondary legislation.

However, despite its deficiencies, it is important to emphasize that the new act is an output of eight years of efforts and experiences of the players in the market and the regulatory authority. With the new act, the name of the regulatory authority was also changed to the Information and Communications Technologies Authority (ICTA). In a few years, the new act—coinciding with some key accomplishments in practice, such as the commercial availability of bitstream access in 2008, implementation local-loop unbundling (LLU) access in 2009, and rules for price squeeze tests—initiated competitive pressure on TT. Furthermore, the total market share of alternative ISPs in DSL services increased gradually, from 3.5% in 2007 to 7.6% in 2010.[9]

---

[6] ICTA's quarterly market data reports.

[7] TCA's decision dated November 19, 2008, and numbered 08-65/1055-411.

[8] Law No. 5809, Electronic Telecommunication Act.

[9] Author's own calculation based on the compiled reports of the ICTA.

2.5 *Emergence of Fiber and the Regulatory Holiday*

The introduction of mobile broadband services in 2009, growing penetration of internet access through cable TV networks, and competition with alternative ISPs in DSL services, forced TT to invest in next-generation access networks (NGANs), namely fiber. In addition to its existing country-wide fiber backbone network, it started to invest in its fiber access network in metro areas. The length of the access network increased four-fold between 2009 to 2013, from 17,000 km to 59,000 km.[10] The number of fiber-to-the-building/home (FTTB/H) subscribers reached 624,000, which constituted 7% of the total broadband subscribers in 2013.

However, TT is not alone in its fiber deployment. Turkcell Superonline, a subsidiary of the largest mobile operator, Turkcell, appeared as a serious challenger against TT. Since 2010, Turkcell Superonline has provided retail fiber access. It reached 2 million fiber home passes in 12 major cities (ICTA, 2012b). By 2013 it had reached 570,000 FTTB/H subscribers. This indicated a 50/50 share in the fiber access market with the incumbent operator, TT. Behind this emerging competition in the market, there was an aggressive price competition. To increase their FTTB/H subscriber bases, both operators charged competitive prices, similar to their DSL access (ICTA, 2012a).

At the regulatory side, in order to support the deployment of NGANs, ICTA announced a regulatory holiday for FTTB/H investments.[11] On 3 October 2011, it was announced that FTTH/B access would not be included in the market analysis over the next five years or until the share of FTTH/B access reaches 25% of the fixed broadband. In that sense, the regulatory holiday seems an indication that ICTA has an intent to induce a facility-based competition.

In addition to the regulatory holiday, secondary legislation concerning the rights of way was enacted by the Ministry of Transportation (hereafter, "the Ministry") at the end of 2012.[12] This legislation sought to decrease the transaction costs in front of the fiber investments through putting access obligations to the existing infrastructure owners.

3 Current Situation in the Market and Current Regulatory Issues

This section explores the current situation of the Turkish broadband market and the regulatory concerns, emphasizing fiber deployment.

[10] ICTA's quarterly market data reports.

[11] ICTA's decision dated September 27, 2011, and numbered 2011/DK-10/504.

[12] Official Gazette dated December 27, 2012, and numbered 28510.

3.1 Market Situation

The regulatory holiday, which excludes the FTTH/B access from the market analysis, temporarily exempts operators from sharing their access network. In fact, the regulatory holiday has ambiguous effects on fiber deployment. At first sight, increasing fiber backbone and access networks, as well as the growing penetration of FTTH/B subscribers, indicate a positive effect. However, no scientific or sectoral publications have yet analyzed the effects of the regulation. The regulatory holiday officially terminated at the end of 2017, but the ICTA added fiber access to its market analysis at the end of 2019.[13]

Table 1 shows the fiber network lengths of the incumbent and alternative operators between 2009 and 2019. As expected, the fiber networks of the operators have increased during the last 10 years. However, neither TT's nor the alternative operators' deployment indicates stability over time.

Considering the investment efforts of the operators during the last 10 years, one could argue that there has been a facility-based competition between the incumbent and the alternative operators. This is true to a certain extent. While backbone investments of both incumbent and alternatives have increased, the alternatives' access network has enlarged far more than that of the incumbent.[14] This is expected since TT has already an existing network. However, alternative operators had to build their own access network because TT has no obligation to share it.

Table 1 Fiber network lengths between 2009 and 2019

Year	TT's network in thousand km	Change (%)	Alternatives' networks in thousand km	Change (%)
2009	128		17	
2010	131	2.3	38	123.5
2011	150	14.5	39	2.6
2012	168	12.0	42	7.7
2013	182	8.3	45	7.1
2014	193	6.0	52	15.6
2015	212	9.8	57	9.6
2016	228	7.5	63	10.5
2017	256	12.3	68	7.9
2018	282	10.2	73	7.4
2019	304	7.8	87	19.2

Source TELKODER (2020), ICTA's quarterly market data reports

[13] ICTA's decision dated December 31, 2019, and numbered 2019/DK-SRD/338.

[14] In its quarterly market data reports, the ICTA has published both the backbone and the access network lengths of the operators.

Another thing that must be mentioned in relation to this issue is the mobile services of the alternative operators. Turkcell and Vodafone, which have fixed broadband units, are mainly dealing with mobile services. By 2016 the mobile operators had started to provide 4G-LTE services and carry their land traffic over their fiber networks. Thus, the fiber investments of each operator do not aim to be used solely for getting fiber subscribers.

On the other hand, during the last 10 years, parallel to the access investments of the operators, the number of FTTH/B subscribers has increased, as shown in Table 2. As in the fiber network lengths, the increase in the number of subscribers does not indicate stability over time. While in the early phase of the regulatory holiday the number of FTTH/B subscribers had increased dramatically, it began to cease in later years. Currently, the share of FTTH/B subscribers is 23% of total fixed broadband subscribers (see Table 3).

Based on Table 2, one could argue that there has been a substitution from DSL to FTTH/B subscribers. Although this may be true to a certain extent, it is impossible to prove based on the ICTA's quarterly market data. However, it is widely known that, as the incumbent operator (TT) upgraded its copper network to fiber, subscribers have been incentivized to upgrade their subscription from ADSL to FTTB.

Broadband over cable networks has also increased but in an unstable manner. The advertised speed of connection has increased over time and converged to the speeds offered in FTTB/H connection. Along with its budget tariffs bundled with cable TV, its subscription base has increased (see Table 3). However, its effect would be limited

Table 2 Increase in the number of fixed broadband subscribers by technology (2011–2019)

Year	xDSL (%)	Cable (%)	FTTB/H (%)
2011	2.0	68.1	73.4
2012	−2.0	8.7	141.5
2013	0.0	−2.8	85.0
2014	2.3	14.8	22.1
2015	5.3	12.6	14.8
2016	8.5	17.1	15.2
2017	11.5	12.2	21.2
2018	9.7	12.7	19.9
2019	1.8	16.3	14.7

Source TELKODER (2020), ICTA's quarterly market data reports

Table 3 Fixed broadband subscribers by technology in 2019

	xDSL	Cable	FTTB/H
Number of subscribers	9,662,248	1,084,446	3,213,298
Share in total fixed broadband (%)	69.2	7.8	23.0

Source ICTA (2019)

in the competition between the facilities because of its limited network coverage (see Köksal & Ardiyok, 2015).

Finally, the absolute comparison in time emphasizing fiber may also be advanced by international comparison. As Table 4 represents, the OECD average in fixed broadband penetration rates are 10.6% in DSL, 10.7% in cable, and 8.9% in fiber. The penetration rates in Turkey are 11.6% in DSL, 1.3% in cable, and 3.9% in fiber.

Table 4 clearly shows that the broadband penetration in Turkey currently lags behind most of the OECD countries. Despite the network investments and increasing

Table 4 Broadband penetration rates (percent) by technology in OECD countries (December 2019)

Country	xDSL	Cable	Fiber	Other	Total
Switzerland	23.7	12.6	9.4	0.8	46.5
France	26.4	6.5	10.5	0.7	44.1
Denmark	12.9	14.6	15.3	0.8	43.6
Netherlands	15.0	20.3	8.0	0.0	43.3
Norway	7.1	11.2	22.9	1.1	42.3
South Korea	1.1	6.1	35.0	0.0	42.3
Germany	30.4	10.0	1.7	0.1	42.2
United Kingdom	10.8	20.8	7.0	2.2	40.9
Canada	31.0	7.9	1.2	0.0	40.1
Belgium	18.5	21.0	0.4	0.1	40.0
Sweden	4.7	6.5	28.0	0.1	39.3
Greece	38.5	0.0	0.1	0.1	38.7
Iceland	14.2	0.3	24.0	0.1	38.6
Portugal	4.7	11.6	19.5	2.7	38.6
Luxembourg	17.6	3.6	15.6	0.2	37.0
New Zealand	11.0	1.1	18.7	3.9	34.7
Czech Republic	8.6	5.7	6.1	14.3	34.7
United States	5.9	22.0	5.5	1.2	34.7
Australia	18.3	7.5	6.9	1.7	34.5
Hungary	6.9	16.0	8.7	1.1	32.6
Spain	5.5	4.8	21.7	0.6	32.6
Estonia	8.6	7.2	13.9	2.9	32.6
Finland	6.3	8.3	17.7	0.3	32.5
Japan	1.2	5.4	26.0	0.0	32.5
OECD	*10.6*	*10.7*	*8.9*	*1.5*	*31.8*
Slovenia	9..0	8.4	12.0	0.4	29.8
Ireland	17.7	7.5	3.3	1.1	29.7
Slovak Republic	8.7	3.2	9.5	7.9	29.3

(continued)

Table 4 (continued)

Country	xDSL	Cable	Fiber	Other	Total
Italy	11.9	0.0	2.0	15.0	28.9
Austria	17.2	10.0	0.8	0.3	28.4
Lithuania	4.5	0.7	21.3	1.8	28.4
Israel	17.4	8.5	1.5	0.0	27.4
Latvia	5.7	0.9	18.4	1.5	26.6
Poland	4.7	7.4	4.9	3.2	20.3
Chile	2.1	9.6	5.0	1.3	18.0
TURKEY	*11.7*	*1.3*	*3.9*	*0.3*	*17.3*
Mexico	5.4	6.0	3.8	0.2	15.4
Colombia	3.1	8.5	2.0	0.7	14.2

Source OECD (2020)

number of subscribers in recent years, the international comparison signifies a relatively poor performance. The next subsection elaborates the current regulatory problems that might be the cause of such an unsatisfactory performance.

3.2 Current Regulatory Problems Date Back to the Past

Following the regulatory reform in early 2000s, the ICTA adopted a ladder of investment (LOI) approach to provide a competitive environment, which was also compatible with the EU regulations. Basically, the LOI approach indicates a transitory policy from service-based competition to facility-based competition. Initially, through some regulatory assistance, alternative operators were incentivized to serve over incumbent's infrastructure. Then, as their subscriber bases grow, they were encouraged to climb the investment ladder through investing in network elements (see Cave, 2006).

The ICTA's LOI approach is based on promoting service-based competition through offering a gradual level of access to TT's facility. Starting in 2004, the ICTA had asked TT to open its network to alternative ISPs that allow them to provide gradually increasing value-added services from resale to bitstream access and then to LLU. However, the first step of LOI, resale arrangement, could only be available in 2007. For the implementations of bitstream access and the LLU, the alternative operators had to wait until 2008 and 2009; respectively. According to Atiyas and Renda (2007), the ICTA could not manage to launch appropriate secondary legislation to promote service-based competition because it had initially worked without a strong legal foundation and it also lacked human resources with expertise on regulation. More specifically, as Köksal and Ardiyok (2015) mentioned the ICTA's strategy to open the incumbent's infrastructure was based on a pure pricing strategy. Details about access conditions, level of service quality, and process of dispute resolution were not mentioned comprehensively. Therefore, these deficiencies allowed TT to conduct non-price tactics to prevent entry of alternative ISPs for a long time.

Hence, almost a decade has passed with attempts to open the incumbent's network in an inefficient way. Once the resale, bitstream, and LLU applications became commercially available, fiber access appeared as an attractive access method to replace DSL services both for the consumers and the suppliers. However, similar problems, such as a lack of secondary legislation and deficiency in implementation have been valid for the fiber deployment of the alternative operators.

Primarily, the deployment of NGANs and related passive infrastructural elements mostly require sunk cost investments that may only be returned in the long run. Therefore, instead of duplicating networks, sharing existing infrastructural elements with minimal upgrades may be seen as a feasible way in almost all parts of the world. In Turkey, rights of way regulation[15] was enacted in 2012 as a milestone that forces the infrastructure owners to share their existing infrastructure. However, the implementation of the regulation indicates some major problems in practice.

Firstly, the regulation does not explicitly compromise some public utilities that have country-wide infrastructure. For instance, according to the regulation, the General Directorate for Highways (KGM), the Turkish Electricity Transmission Cooperation (TEİAŞ), Turkish State Railways (TCDD), the Petroleum Pipeline Corporation (BOTAŞ), and even municipalities do not have an explicit obligation to share their infrastructure (Ulaştırma Denizcilik ve Haberleşme Bakanlığı, 2017). In fact, such a deficiency creates impediments in front of the fiber deployment at both intercity and inner-city levels. Since infrastructure owners such as KGM, TEİAŞ, TCDD, and BOTAŞ only open their networks voluntarily, competition at intercity level become less intense. Such a situation gives the entrant operators two choices: build their own network (which is costly) or be dependent on the incumbent telco's fiber network. On the other hand, passive infrastructure of municipalities has great importance for fiber deployment in metro areas. In some parts of the world, municipalities have played a crucial role under different business models, ranging from building their own and renting them to the operators, or building and operating methods (see Mölleryd, 2015). In Turkey, however, alternative operators have been discontented regarding the municipalities' conduct. It is widely claimed that their request for the deployment of fiber network was rarely permitted by the municipalities (TELKODER, 2020).

Another issue related to the implementation of right-of-way regulation indicates exclusionary and exploitative conducts of the right-of-way owners (TELKODER, 2020; Ulaştırma Denizcilik ve Haberleşme Bakanlığı, 2017). A widely expressed exploitative problem is the amounts demanded by the right owners, which are far above the mentioned fees in the regulation. Some right-of-way owners also require the operators to build extra elements. The fees demanded usually contain extra elements that are not mentioned in the regulation. On the other hand, certain behaviors of the right-of-way owners may be exclusionary because they do not respond to the demands of the operators or delay them. Since the regulation obliged to use existing

[15] Official Gazette dated December 27, 2012, and numbered 28510.

infrastructure, such delaying or non-responding behaviors become exclusionary for the operators.

Similar to the right-of-way regulation, facility sharing regulation[16] has indicated inefficiencies and implementation problems. In addition to the coordination between the existing network owners not being achieved effectively, some public infrastructure owners have had difficulty sharing facilities at their network. The latter case has become concrete through the conduct of municipalities. While some municipalities have denied sharing their own facilities, some others have prevented the sharing of facilities built by the operators on their infrastructure (TELKODER, 2020). Neither the Ministry nor the ICTA has solved those problems effectively.

However, as happened earlier, alternative operators carried the issue to the TCA. In their complaint against the subsidiary of the İstanbul Metropolitan Municipality ISTTELKOM, alternative operators claimed that their requests for the deployment of fiber networks were rarely permitted. If the necessary permission is given, some unfair conditions are applied. More specifically, it is claimed that ISTTELKOM gets the propriety of the network built by the alternative operators and does not allow the operators to rent it to third parties. Through a detailed investigation, the TCA decided that the behavior of ISTTELKOM is abusive since it tries to monopolize the infrastructure in the city of Istanbul. Therefore, it ordered the cancelation of the articles of the facility sharing contract, which prevented the operators from renting their built network to third parties.[17] Again, such a development shows that the absence of detailed regulation leaves the issue to the competition authority. From this perspective, as Ardıyok and Oğuz (2010) mentioned for the earlier interventions, competition rules have offered an alternative way to implement competition in the industry.

For efficient facility sharing, effective coordination between the operators is required under supervision. In May 2018 operators that have their own infrastructure (TT, Turkcell, Vodafone, Türksat, and the association of telco operators) signed a protocol under the supervision of the Ministry. The aim of the protocol was to mutually open the fiber networks of the operators. Thus, infrastructure of the operators would be used in an effective manner and fiber deployment investments directed to the most efficient areas. However, the protocol has not yet been implemented.

Finally, related to the coordination of facility sharing and also for the supervision of the rights of way regulation, an electronic database that displays available facilities and network elements should be organized. Such an initiative started in 2016 under the name of Electronic Communication Infrastructure Information System (EHABS) and rules and procedures enacted by the ICTA in 2018.[18] The system aims to provide a digital map of cable types, spare capacities, and all current infrastructure that may be used for fiber deployment. Such a database would provide valuable information and facilitate facility sharing via transparency and supervision. However, like the protocol between the operators, the system has not been activated yet (see TELKODER, 2020).

---

[16] ICTA's decision dated March 12, 2013, and numbered 2013/DK-ETD/187.

[17] TCA's decision dated April 11, 2019, and numbered 19-15/214-94.

[18] ICTA's decision dated July 23, 2018, and numbered 2018/DK-TED/236.

The regulatory deficiencies and implementation problems mentioned in this subsection are not exhaustive. However, the mentioned problems are the most important issues advancing the fiber deployment at the supply side.

4 Targets and Comparison with the Current State

This section focuses on the targets and their comparison with the current state.

4.1 Target for Digital Society Via Fiber

Empirical findings in the literature suggest that penetration and the speed of the internet are correlated with the growth and development level of the countries (see Kelly & Rossotto, 2012; Minges, 2016). Although some countries have defined targets for their digital development, others have concentrated on their specific needs without a concrete target or a plan. Most of the plans of countries for their digital development indicate distinct targets are based on penetration rates or speed of the broadband internet. More recently, those targets have consisted of a combination of speed and penetration rate in the society (see European Commission, 2010, 2016).

The Digital Agenda for Europe (European Commission, 2010) defined the targets for the deployment and penetration of high speed broadband access. The most significant targets in this agenda are to provide 100 Mbit/s internet to 50% of the EU population and to provide minimum 30 Mbit/s broadband speed to all EU citizens. Regular updates for these targets have been discussed and amended along the technological and economic development. Thus, the European Commission (2016) adopted a new plan, under the name of European Gigabit Society, that defines more ambitious targets for 2025. In that document, the targets were updated to provide minimum 100 Mbit/s broadband speed to all EU citizens.

The European Commission's recommendations apply only to the EU Member States. Turkey is not obliged to adopt the targets and recommendations. However, starting in the 2000s in some strategic and political documents of governments, some targets started to be mentioned. In those documents[19] the targets were expressed in penetration rates without specifying any concrete targets regarding connection speed and its availability for households. Later, both in government programs and the Ministry's strategic plans, targets began to be expressed in terms of the number

[19] For example, the Information Society Strategy and Action Plan prepared by the government in 2006.

of internet subscribers,[20] although still no definite targets were mentioned in terms of speed and its availability for households.

However, in recent years more qualified targets for the digital society can be seen in the political parties' and governments' programs, as well as ministries' action plans. The 2015–2018 Information Society Strategy and Action Plan of the Ministry of Development is one of them (Kalkınma Bakanlığı, 2015). This plan had two fundamental targets to be achieved by 2018: increasing the broadband penetration rate to 70% (including mobile) and raising the number of FTTH/B subscribers to 4 million. Later, in the National Broadband Strategy and Action Plan (2017–2020) these targets were updated and extended for 2020 and 2023, the 100th anniversary of the Turkish Republic (Ulaştırma Denizcilik ve Haberleşme Bakanlığı, 2017). The next subsection will elaborate these targets and their current achievement level.

4.2 Comparison With the Current State

The National Broadband Strategy and Action Plan (hereafter, the National Broadband Plan) was prepared with the participation of various stakeholders under the coordination of the Ministry. It was designed as a follow-up to the 2015–2018 Information Society Strategy and Action Plan of the Ministry of Development (Kalkınma Bakanlığı, 2015). The motto of the National Broadband Plan is "broadband from everywhere to everyone". The key principles in the plan are as follows:

- Developing broadband infrastructure across the country
- Developing fiber infrastructure across the country
- Increasing the capacity and speed of broadband access
- Accomplishing competitive and market-based sectoral achievements
- Increasing demand for broadband services.

In that context, the National Broadband Plan consists of three strategic goals:

- Provision of the supply of broadband
- Provision of the demand for broadband
- Provision of both the supply and the demand of broadband.

Under these strategic goals are 25 action plans. On the supply side, most of the actions are related to facilitating the deployment of fiber infrastructure, including the need of developing new secondary regulations. There are also some actions for increasing the penetration of broadband, both over satellite and cable networks.

The actions related to the demand for fixed broadband indicate decreasing the taxes and other fiscal burdens on broadband services. The National Broadband Plan also

[20] In the strategic plan of the relevant ministry prepared in 2008, the target was 11 million subscribers by 2013. In the government program prepared in 2011, the target was 15 million broadband subscribers by 2015.

Table 5 Comparison of current situation and targets for 2020 and 2023

	2020 Q1	2020 target	2023 target
Fixed broadband penetration rate (%)	17.5	20	30
Number of fiber subscribers (million)	3.3	5	10
Availability 100 Mbit/s connection for households (%)	0.1	50	100
Availability 1 Gbit/s connection for households (%)		–	20

Source ICTA (2020), Ulaştırma Denizcilik ve Haberleşme Bakanlığı (2017)

calls for the gap to be filled with digital divides. Specifically, it suggests facilitating the access of handicapped and socially disadvantaged groups.

Table 5 provides the targets in the National Broadband Plan in terms of fixed broadband penetration, number of fiber subscribers, and availability of 100 Mbit/s and 1 Gbit/s speed for households, both for 2020 and 2023.

Table 5 also compares the current (as of first quarter of 2020) situation and the targets for 2020. While it appears that the goal for fixed broadband penetration for 2020 has been partially achieved, the number of fiber subscribers remains at just 3.3 million, which is still less than one-third of the goal for 2020. As the COVID-19 crisis increases the demand for new internet subscriptions, fixed broadband penetration goal for 2020 may not be difficult to achieve. However, the goal for the number of fiber subscribers seems likely to be accomplished in the coming years.

For the connection speeds, the targets indicate the capacity of the home pass infrastructure that can provide at least 100 Mbit/s and 1 Gbit/s speeds. The ICTA has not yet published home pass statistics of the operators. However, as of the first quarter of 2020, the number of fixed broadband subscribers with 100 Mbit/s connection constituted only 0.1% of the total number of subscribers.

5 Discussion and Recommendations

Considering the regulatory issues in broadband internet, the main challenge in consolidating competition in the market and increasing penetration in the society is the lack of a coherent vision for the broadband internet. Neither a service-based nor a facility-based competition approach has been explicitly made to increase the competition in the market. Similarly, in order to increase the penetration in the society, there has not been a concrete vision that takes into account the socio-economic facts of the country. These criticisms are also valid for the deployment and penetration of NGANs.

The lack of a coherent vision is mostly related to the absence of willpower of the relevant institutions and, mostly, politicians. This can be seen quite clearly by looking at the political decisions since the 1990s. The privatization attempts of TT, without a regulatory framework, was historically the primary sign of this lack of vision. In addition, it has to be remembered that the liberalization movement and the regulatory framework were initiated by external factors, such as financial need after economic crises and obligations related to the EU accession process.

After the regulatory reform, almost a decade passed with ineffective attempts to open the incumbent operator's infrastructure to the alternative operators. Such a service-based competition experience has failed mostly because of implementation problems. Moreover, without proper application of the LOI approach, the facility-based competition was delayed. Meanwhile, the state-owned cable network was not put to use as an element of effective facility-based competition. Neither a privatization prospect nor a sound strategic investment plan has been put forward. Despite a brand-new act and accompanied secondary legislation, the absence of a coherent approach has meant only minimal achievement.

A similar story has been observed for the deployment of fiber. The regulatory holiday might be a good way to increase the facility-based competition in fiber access. However, without effective passive infrastructural sharing at the backbone level, this has not created a satisfying outcome. The right-of-way and facility sharing regulations require consistent updates considering the implementation problems in practice.

The chaos detailed in Sect. 3.2 between the public infrastructure owners and the operators may be effectively solved by the central government creating a cooperative climate between them. In addition, proper updates in the regulations may contribute to this climate by transparently and explicitly defining the responsibilities of the infrastructure owners and the duties of the operators. The execution and the management of the operations may be administered centrally by a newly founded center in order to facilitate the coordination.

The lack of vision can also be seen to attract foreign investment. Electronic communication is subject to regulation in almost all countries. As long as the regulations are compatible with those of other countries, it may be easy to attract foreign investors. Moreover, returns on investments are important for operators that will make infrastructure investments. Therefore, a predictable regulatory environment is critical. In fact, Turkey is fortunate to be an EU candidate, in order to follow a concrete model along the EU directives. However, one could argue that Turkey has not adopted the directives appropriately. For a long time in the electronic communication sector, including internet service providing, Turkey has attracted little foreign investment, and their numbers have declined during the last decade.

The lack of a coherent vision has also been experienced on the demand side to increase the penetration of broadband in the country. Two facts should be mentioned on that issue. First, Turkey's socio-economic and demographic facts have been ignored while developing regulations and related policies. However, as reported in a large-scale study by the Berkman Center for Internet & Society (Benkler et al., 2010), supply-side policies are quite limited to increase the penetration and the usage of the internet in societies. In fact, demographic and social factors are the main determinants of both access and usage of broadband internet. As Anıl and Köksal (2015, 2016) showed, demographic, social, and geographic factors determine households' adoption of internet and the usage of individuals in Turkey. Therefore, socio-economic and demographic factors should be considered when developing penetration strategies. Moreover, financial mechanisms, such as universal service funds should be directed towards disadvantaged geographical areas in order to increase access and internet literacy.

Secondly, other demand-side measures for the general population can be designed that could relatively decrease the cost of access and increase penetration. In that context, decreasing taxes for broadband access country-wide and elimination of all taxes in low penetrated regions should be considered. Moreover, value-added and special consumption taxes for information and communication end-user devices should also be decreased or completely abolished, along with a coherent penetration strategy.

6 Conclusion

This chapter examines the regulation of the broadband internet in Turkey. After a historical background that emphasizes the evolving regulatory and business environment in the broadband market, I explained the current situation and the current regulatory issues related to fiber deployment, then examined targets in the National Broadband Plan and compared them with the current state of the market. The following conclusions can be drawn.

On the supply side of the broadband market, neither a service-based nor a facility-based competition approach has been effectively employed by the ICTA and the Ministry to ensure competition. This has persisted for fiber deployment in the last decade. Despite the regulatory holiday in fiber access, the lack of effective passive infrastructural sharing regulations at the backbone level prevented a satisfactory outcome from being achieved. Alternative operators have still faced impediments, both regarding service-based and facility-based competition. Therefore, the right-of-way and facility sharing regulations require consistent updates to eliminate the implementation problems in practice.

On the other hand, on the demand side, socio-economic, demographic, and geographic factors have been ignored while developing regulatory policies to increase the penetration of broadband in society. However, as the empirical findings show, demographic and social factors are the main determinants of both access and usage of broadband internet. Therefore, those factors should be taken into account while developing penetration strategies, especially to increase access in disadvantaged geographical areas and to cover vulnerable social groups. Moreover, other measures such as abolishing taxes for broadband access and decreasing special consumption taxes for information and communication end-user devices should also be considered to decrease the cost of access and increase the penetration.

All of these issues require a coherent vision for the development of broadband internet at both supply side and demand side. Without such a coherent vision, it appears that the targets set out for the 100th anniversary of the republic will not be met.

References

Akgül, M. (1997). *İnternette TURNET'i aşmak!* Internet World-TR. http://www.bilkent.edu.tr/~akgul/otesi.html

Anıl, B., & Köksal, E. (2016). Türkiye'de İnterneti Kimler Ne için Kullanıyor? *Marmara Üniversitesi İktisadi ve İdari Bilimler Dergisi, 38*(1), 1–13. https://doi.org/10.14780/iibd.61602

Ardıyok, Ş, & Oğuz, F. (2010). Competition law and regulation in the Turkish telecommunications industry: Friends or foes? *Telecommunications Policy, 34*(4), 233–243. https://doi.org/10.1016/j.telpol.2009.10.002

Atiyas, I., & Dogan, P. (2009). The political economy of liberalization of fixed line telecommunications in Turkey. *Mossavar-Rahmani Center for Business and Government*. http://www.hks.harvard.edu/m-rcbg/rpp/Workingpapers/RPP-2009-01_Atiyas_Dogan.pdf

Atiyas, I., & Renda, A. (2007). Telecommunications. In S. Ülgen (Ed.), *Second generation structural reforms: De-regulation and competition in infrastructure industries: The evolution of the Turkish telecommunications, energy and transport sectors in light of EU* (pp. 25–75). Centre for Economics and Foreign Policy Studies (EDAM) and Centre for European Policy Studies (CEPS).

Benkler, Y., Faris, R., & Gasser, U. (2010). Next generation connectivity: A review of broadband Internet transitions and policy from around the world. In *Berkman centre for internet & society* (Issue February). http://cyber.law.harvard.edu/sites/cyber.law.harvard.edu/files/Berkman_Center_Broadband_Final_Report_15Feb2010.pdf

Cave, M. (2006). Encouraging infrastructure competition via the ladder of investment. *Telecommunications Policy, 30*(3), 223–237. https://doi.org/10.1016/j.telpol.2005.09.001

European Commission. (2010). *A digital agenda for Europe.*

European Commission. (2016). *Connectivity for a competitive digital single market—Towards a European gigabit society.*

ICTA. (2012a). *Pazar Analizi Çalışmaları: Toptan Fiziksel Altyapısına Erişim Pazarı.* https://www.btk.gov.tr/uploads/pages/onceki-pazar-analizleri/fsae-pa-kgaid-2012.pdf

ICTA. (2012b). *Yerlatı Tesislerinin Paylaşımına İlişkin Erişim Yükümlülüğü Düzenlemeleri: Ülke Uygulamaları ve Türkiye İçin Öneriler.* http://btk.gov.tr/kutuphane_ve_veribankasi/raporlar/arastirma_raporlari/dosyalar/Yeralti_Tesislerinin_Paylasimi.pdf

ICTA. (2019). *Üç Aylık Pazar Verileri Raporu (2019 4. Çeyrek).*

ICTA. (2020). *Üç Aylık Pazar Verileri Raporu (2020 1. Çeyrek).*

Kalkınma Bakanlığı. (2015). *2015–2018 Bilgi Toplumu Stratejisi ve Eylem Planı.* http://www.bilgitoplumu.gov.tr/en/wp-content/uploads/2016/03/Information_Society_Strategy_and_Action_Plan_2015-2018.pdf

Kelly, T., & Rossotto, C. M. (2012). Broadband strategies handbook. *The World Bank.* https://doi.org/10.1596/978-0-8213-8945-4

Köksal, E., & Anil, B. (2015). Determinants of broadband access and usage in Turkey: Do regions matter? *Topics in Middle Eastern and African Economies, 17*(1), 113–132. https://doi.org/10.2139/ssrn.2362376

Köksal, E., & Ardiyok, Ş. (2015). Reviewing regulatory policy for broadband in Turkey. *Competition and Regulation in Network Industries, 16*(4), 354–377. https://doi.org/10.1177/178359171501600403

Minges, M. (2016). Exploring the relationship between broadband and economic growth. In *World development report* (No. 102955; Background Papers–Digital Dividends). http://documents1.worldbank.org/curated/en/178701467988875888/pdf/102955-WP-Box394845B-PUBLIC-WDR16-BP-Exploring-the-Relationship-between-Broadband-and-Economic-Growth-Minges.pdf

Mölleryd, B. (2015). *Development of high-speed networks and the role of municipal networks* (No. 26; OECD Science, Technology and Industry Policy Papers). https://doi.org/10.1787/5jrqdl7rvns3-en

OECD. (2002). *Regulatory reform in the telecommunications industry.* http://www.oecd.org/regreform/1840797.pdf

OECD. (2020). *Broadband portal*. http://www.oecd.org/sti/broadband/broadband-statistics/
Özyiğit, A. (2003). *Türkiye'de İnternet*. Internet 10 Yaşında Etkinliği. http://www.internetarsivi.metu.edu.tr/10yil_konusma.php
TELKODER. (2020). *2019 yılı Sektör Düzenlemelerine Yönelik Değerlendirmeler*. https://telkoder.org.tr/wp-content/uploads/2020/06/TELKODER-2019-Degerlendirme.pdf
Ulaştırma Denizcilik ve Haberleşme Bakanlığı. (2017). *Ulusal Genişbant Stratejisi ve Eylem Planı (2017–2020)*. http://www.sp.gov.tr/upload/xSPTemelBelge/files/lxlRY+Ulusal_Genisbant_Stratejisi_ve_Eylem_Plani_2017-2020.pdf
Wolcott, P., & Çağiltay, K. (2001). Telecommunications, liberalization, and the growth of the internet in Turkey. *Information Society, 17*(2), 133–141. https://doi.org/10.1080/019722401750175685

Emin Köksal is an associate professor of economics at Bahçeşehir University. He teaches Managerial Economics at MBA programs and Industrial Organization at the undergraduate program. He also teaches courses such as Platform Economics and Competition Policy in Digital Markets. He publishes extensively on competition and regulatory issues in telecommunications, energy, and platform industries. He has experience in platform business models, network neutrality regulations, and internet usage. He is the associate editor-in-chief of the journal *Competition and Regulation in Network Industries*. As an economic consultant, he has led economic analyses for antitrust cases and regulatory assessments.

Regulation of Mobile Communications Markets in Turkey

Barış Yüksel

Abstract This chapter explores the emergence and development of the mobile communications markets in Turkey and focuses on the effects that the legal framework has had on the overall market structure and its performance. Historical background concerning the allocation of radio frequency spectrums to mobile network operators is examined and the impacts of different methodologies adopted during resource allocation processes are scrutinized. I conclude that the design of a Global System for Mobile Communications and IMT-2000/Universal Mobile Telecommunications System (3G) auctions led to permanent damage in the market structure and that these damages were partially remedied by the relative success of the IMT-Advanced auction. The reasons for justifying state intervention to mobile communications markets are analyzed and the relevant legal framework envisaged for such intervention is laid down. Regulations imposed by the Information Communications and Technology Authority on all operators and those that are only imposed on operators with significant market power are identified, the practical use of such regulations in the mobile communications markets are set out, and the net effects of these regulations on society are assessed. I argue that the current rules regarding the formation of subscription agreements lead to net welfare losses and offer suggestions for how these losses may be negated are presented. I also discuss regulations that specifically aimed to create a more competitive mobile communications market by way of curbing the market power of Turkcell and facilitating the entry of mobile virtual network operators. The chapter shows that these regulations were successful in increasing competition between mobile network operators, but that the attempts to introduce mobile virtual network operators have failed. I further explain that while the ICTA followed the EU in regulating the conventional mobile communications services, it failed to do so with respect to net neutrality rules and that this may create problems in markets for over-the-top services and lead to consumer harm.

Keywords Mobile communication · Regulation · Competition · Liberalization

B. Yüksel (✉)
BASEAK, Istanbul, Turkey
e-mail: byuksel@baseak.com
URL: https://www.baseak.com

M. Eroğlu and M. Finger (eds.), *The Regulation of Turkish Network Industries*,
https://doi.org/10.1007/978-3-030-81720-6_20

1 Introduction

Mobile communications services are provided wirelessly by exploiting specific radio frequency spectrums (RFS), which are scarce resources that should be utilized and managed in the best interests of society (Walden, 2018, p. 16). RFSs that are used for the provision of mobile communications services are owned by the state and may not be subject to private property. RFSs are allocated by the state to a limited number of mobile network operators (MNOs) for a pre-defined period. While the scarcity of frequency spectrum acts as a natural limit with respect to the number of MNOs that may operate in the upstream level that is, operating a mobile network) of the mobile communications market, it may be possible for new players to enter the downstream level (that is, providing retail mobile communications services to end users) by utilizing the networks (especially RFSs) of existing MNOs. Such players, which do not have their own RFSs or networks and use those of MNOs are called mobile virtual network operators (MVNOs). However, mainly due to tax-related complications, there are no MVNOs in Turkey (ICTA, 2017b)[1] and the downstream level of the Turkish mobile communications market is a tight oligopoly that is solely comprised of MNOs. The limited competition between MNOs is also threatened by structural failures in the market such as network externalities that create advantages for first movers. In addition, as per Universal Services Act No. 5369, mobile telecommunications services are considered universal services that must be available to all citizens in Turkey, in a non-discriminatory manner, with affordable prices and in accordance with predetermined minimum quality standards, and existing MNOs must ensure that these requirements are complied with.

The above-mentioned idiosyncrasies of the mobile telecommunications market (use of publicly owned scarce resources, oligopolistic market structure, network externalities, and universal service requirements) necessitate state intervention and the characteristics of this market are important in designing the relevant legal framework for state intervention. Mobile communications markets are shaped by rapid technological developments, which confer significant benefits to consumers by enabling them to access a wide variety of services with affordable prices, while simultaneously leading to frequent changes in market structure. Therefore, state intervention in this market should consist of up-to-date rules that can be revised (and removed, if needed) promptly. The state should refrain from over-intervention so as to avoid stifling the dynamism. The New Regulatory Framework (NRF) in the European Union (EU), which consists of the Framework Directive, EU Authorization Directive, EU Access and Interconnection Directive, EU Universal Services and User's Right Directive and EU Communications Privacy Directive, may be considered an example in that respect (European Parliament and the Council, 2002a; 2002b; 2002c; 2002d: 2002e).

[1] Some firms in Turkey enter into revenue-sharing agreements with MNOs. However, as these firms only contribute to their brand names and do not provide mobile communications services themselves or get to determine the terms and conditions of sales, they do not exert meaningful competitive pressure on MNOs and instead serve a complementary purpose.

Electronic communications markets in Turkey, including mobile communications markets, are currently regulated in accordance with the Electronic Communications Act No. 5809 (ECA), which amended and repealed the majority of the legacy Telegram and Telephone Act No. 406. The ECA is implemented by the Information Communications and Technology Authority (ICTA), an independent regulatory authority (Irmak, 2013) that takes into consideration the general policy strategies adopted by the Ministry of Transportation and Infrastructure (Ministry) while regulating the market. Although the ECA is modeled after the NRF, previous legislation still has some effects on the regulation of mobile communications market in Turkey as MNOs provide Global System for Mobile Communications (GSM) and IMT-2000/Universal Mobile Telecommunications System (3G) services in accordance with concession agreements[2] that were drafted before the entry into force of the ECA.

In Turkey, GSM services were first provided by Telsim Mobil Telekomünikasyon A.Ş. (Telsim) and Turkcell İletişim Hizmetleri A.Ş. (Turkcell) in 1994 under the terms of revenue-sharing agreements signed between these operators and Türk Telekomünikasyon A.Ş. (Türk Telekom), which was a vertically integrated state-owned entity that was responsible for the provision of all telecommunication services.[3] In 1998, Telsim and Turkcell signed concession agreements with the Ministry and were granted 25-year licenses for the provision of GSM services. In 2000 İŞ-TIM Telekomünikasyon Hizmetleri A.Ş.[4] (İŞ-TIM) and Türk Telekom were granted the right to provide GSM services. İŞ-TİM signed a concession agreement with the Ministry on October 27, 2000 and began operating in March of 2001 under the *Aria* brand name. Türk Telekom signed a concession agreement on January 11,2001 and began offering GSM services on December 14, 2001, through its brand *Aycell*. In 2004, Aycell and Aria merged their activities under the name of TT-TIM İletişim Hizmetleri A.Ş. and continued operating in the GSM market through the *Avea* brand. In 2015 Türk Telekom acquired sole control of *Avea,*[5] which was later rebranded as *TT Mobil*. In 2006, Telsim was acquired by Vodafone. As of 2020, there are three MNOs in the Turkish mobile communications market, namely, Turkcell, Vodafone and TT Mobil. These MNOs provide GSM, 3G and IMT-Advanced (4.5G) services. Per the 2019 Q4 Market Report of ICTA, Turkcell is the market leader with a 40.5 percent customer-based market share, followed by Vodafone (31.1 percent) and TT Mobil (28.4 percent) (ICTA, 2020, p.68).

The first section of this chapter lays down the general legal framework that shaped the mobile communications market in Turkey, focusing on the spectrum management policies, design of RFS auctions and *ex ante* regulations of the ICTA. The second

[2] In Turkey, the concession agreements in the telecommunications markets are defined as the agreements between the Authority and the operators related to the telecommunications services to be provided and/or the telecommunications networks to be constructed and operated nationwide by a limited number of operators.

[3] The auction for the privatization of Turk Telekom was finalized on June 24, 2005.

[4] İŞ-TIM was formed by Telecom Italia Mobile and Türkiye İş Bankası.

[5] Turkish Competition Board (TCB) Decision dated July 28, 2015, and numbered 15–32/454–138.

section focuses on market competition and examines the strategies of ICTA that aim to ensure a level playing field between existing MNOs and to facilitate entry of MVNOs with a view to add service-based competition (that is, competition between firms that use the same infrastructure in the upstream level while acting independently in the downstream level) in addition to the existing infrastructure-based competition. The last section briefly assesses the regulations in the mobile communications markets and lays down expectations regarding the future. This chapter concludes with remarks on the impacts of the regulatory environment on the mobile communications markets in Turkey so far.

2 Legal Framework Shaping the Mobile Communications Markets in Turkey

The legal framework in Turkey had crucial impacts on the way in which the mobile communications markets emerged and developed. The most critical issues that shaped the mobile communications markets were (i) the way in which RFSs to be used for the provision of mobile communications services were allocated to MNOs and (ii) *ex ante* regulations in the market.

2.1 Historical Review of the Allocation of RFS Bands

The authorization regime under the ECA and the Authorization Directive of the ICTA is modeled after the EU Authorization Directive. Any operator that provides electronic communication services must be authorized by the ICTA. The default authorization in the ECA is through notification. Operators that do not need to be granted a right to use a resource (e.g., RFS, mobile number, and satellite position) are merely required to notify the ICTA before providing electronic communications services and they may start their operations as soon as the ICTA notifies them that they fulfill the conditions set forth in the Authorization Directive. As far as the mobile communications markets are considered, MVNOs that do not require the ICTA to allocate mobile network codes (MNC) and mobile numbers for their use may be authorized via notification. Per Article 1.7. of the "Definition, Scope and Term of Electronic Communications Services, Networks and Infrastructures," which is a secondary legislation issued by the ICTA to provide clear definitions and explanations regarding all electronic communications services provided in Turkey, MVNOs that are authorized via notification shall apply to hosting MNOs in case they require MNC or mobile numbers and these should be allocated to MVNOs in accordance with the agreements to be signed between MNOs and MVNOs.

In case operators need to obtain a right to use a resource for providing electronic communications services, they must apply to the ICTA. The ICTA may decide to restrict the number of operators that may be granted with a right to use scarce resources. In the mobile communications market, RFSs are the most valuable scarce resources that may only be allocated to a limited number of MNOs. The number of MVNOs that may obtain a right to use MNCs and mobile numbers is not limited.

The way in which RFSs are allocated has a significant impact on the structure of the market and the performance of MNOs. Therefore, RFS allocation and pricing policies should not only aim to maximize the revenues that the state would generate, but also to incentivize the efficient use of RFSs and focus on general economic benefits of the public (GSMA, 2019).

In Turkey, the right to use RFS bands (900 MHz[6]) for mobile communications services was first granted to Telsim and Turkcell in 1998. These two operators were already providing GSM services in accordance with the terms of their revenue-sharing agreements with Türk Telekom and each paid USD 500 million for the right to use RFSs (Yayla, 2009, p. 80). No auction was held and the spectrum prices were determined unilaterally by the state rather than as a result of a competitive process. During the period when the mobile communications market remained a duopoly, Turkcell's market share constantly exceeded that of Telsim and the gap widened gradually. One of the main reasons why Turkcell consistently held a higher market share was that Telsim's activities were suspended between November 1995 and July 1996 due to its breach of the revenue-sharing agreement, and this negatively affected Telsim's reputation (Yayla, 2009, p. 81).

In 1999, the government decided to grant licenses to three more MNOs, which would provide GSM services over 1800 MHz RFS band. Two licenses were to be granted through an RFS auction and the third license would be purchased by Türk Telekom for the same price paid in the auction. The auction, which was held in 2000, was designed as a sealed bid auction and it was decided that two licenses would be auctioned sequentially. The purchase price of the first RFS block would constitute the price floor for the second RFS block. The result of the auction proved that the design was highly problematic as it allowed İŞ-TIM to exploit the loophole in the auction design and raise the cost of entry[7] significantly (Binmore & Klemperer, 2002, p. 4) by offering an extremely high license fee for the first block (USD 2.526 billion), thereby preventing the sales of the second RFI block (Yayla, 2009, p. 81). Türk Telekom paid the same fee for the third license. The combined effects of late

[6] The main difference between 900 MHz RFS band and 1800 MHz RFS band is that the former provides better coverage and indoor penetration. This means that in order to achieve the same coverage, MNOs that use the 1800 MHz frequency band must build a larger number of cell towers compared to SPs that use the 900 MHz frequency band. Also, regardless of how many cell towers are built, the 1800 MHz frequency band signals are unable to penetrate through thick walls, so subscribers who reside or work in buildings with thick walls may not receive cell reception.

[7] For a contrasting view suggesting that İŞ-TIM did not intentionally offer such a high price and the result of the auction was an example of a "winner's curse" rather than an exploitation of a loophole, see Atiyas & Doğan, 2007.

entry,[8] high license fees paid for the licenses, comprehensive coverage obligations imposed by the concession agreements,[9] a lack of national roaming regulations,[10] and the inferiority of 1800 MHz RFS band vis-à-vis 900 MHz RFS band[11] made it very difficult for the new entrants to challenge the first comers (even after the consolidation) and Turkcell managed to retain a high market share[12] in the GSM market until the entry into force of the ECA.

The 3G auction, which took place in 2008 and saw four licenses with asymmetrical bandwidths (all comprising of blocks within the 2100 MHz RFS band) auctioned off, had a different design. To prevent "anti-competitive spectrum hoarding"[13] each existing MNO was allowed to obtain only one license and sequential auctions were held for each asymmetrical block. Turkcell, Vodafone, and Avea purchased three of the licenses, but the fourth license was not purchased due to lack of demand by a new entrant (Yayla, 2009, p. 112). This lack of demand may have stemmed from the fact that some critical requirements in the tender specifications (such as minimum coverage obligations and expected network expansion rates) were applied symmetrically to incumbent MNOs and potential new entrants. This placed potential new entrants at a competitive disadvantage as they lacked an established subscriber base and did not own any network elements (Erşen, 2009, p. 101). Introduction of the 3G technology and the *ex ante* regulations of the ICTA, which are explained in detail in the following sections, changed the competitive dynamics of the market to a certain degree and Turkcell lost approximately 10 percent market share between 2008 and 2015 (when the 4.5G auction took place).[14]

The largest RFS auction (in terms of total bandwidth allocated for the provision of mobile communications services) in Turkey was held in 2015 and a total of 390.4 MHz of bandwidth, comprising blocks of 800 MHz, 900 MHz, 1800 MHz,

---

[8] Network externalities are beneficial for the incumbent operator as they help it to preserve its strong position and provide a first-mover advantage when the market is opened to competition (Köktürk, 2012, p. 9).

[9] Both İŞ-TIM and Türk Telekom were required to cover 50 percent and 90 percent of the administrative units with a population over 10,000 within two and five years, respectively.

[10] For a detailed analysis of the disputes between İŞ-TIM, Telsim and Turkcell concerning national roaming, see TCB Decision dated 09.06.2003 and numbered 03–40/432–186. Although the TCB initially held that Turkcell and Telsim's constructive refusal to provide national roaming services to İŞ-TIM amounted to abuse of collective dominance, this decision was later overturned in administrative review process due to procedural irregularities and the TCB's decision did not have any effects in practice.

[11] See footnote 12.

[12] According to ICTA's 2009 Q1 Market Report, Turkcell, Vodafone and Avea had subscriber-based market shares of 56 percent, 27 percent and 17 percent respectively as of Q1 2008 (ICTA, 2009, p. 28.).

[13] Anticompetitive spectrum hoarding could be described as market players individually or jointly acquiring or retaining spectrum quantities greater than their foreseeable technical needs, with the effect of distorting competition (ERG-RSPG, 2019, p. 14).

[14] According to ICTA's 2016 Q1 Market Report, Turkcell, Vodafone and Avea had subscriber-based market shares of 47.6 percent, 29.3 percent and 23.1 percent, respectively as of Q1 2015 (ICTA, 2016a, p. 56).

2100 MHz, and 2600 MHz RFS bands, were offered as 20 separate packages,[15] and 365.4 MHz of bandwidth, corresponding to 18 separate packages, was sold. Two packages comprising of blocks of 2600 MHz RFS band were not sold as they were reserved for a fourth MNO and there was no demand for new entry. Total bandwidth allocated to mobile communications services in Turkey tripled as a result of the auction and jumped from 183.8 MHz to 549.2 MHz (ICTA, 2016b, p. 23).

The 4.5G auction was designed in a way that would address the peculiarities of the Turkish mobile communications market. While the same model in the 3G auction was implemented for the allocation of packages comprising of blocks of 800 MHz and 2600 MHz RFS bands, different models were created for the allocation of packages comprising of blocks of 900 MHz, 1800 MHz, and 2100 MHz RFS bands. Turkcell and Vodafone, which already held almost all available blocks of 900 MHz RFS band, were not allowed to bid for the largest block of 900 MHz RFS band and Avea was the sole bidder. Avea, which already held a significant portion of all available blocks of 1800 MHz RFS band, was not allowed to bid for the largest block of 1800 MHz RFS band. No spectrum caps were implemented for blocks of 2100 MHz RFS bands and all three packages were acquired by Turkcell after rounds of competitive auctions. It was further decided that MNOs would be allowed to use the RFS bands allocated to them in a technology-neutral manner, on the condition that specific obligations stemming from the concession/right of use agreements were complied with. The 4.5G auction assumed the role of a technological leveler in the Turkish mobile communications market by harmonizing the composition of the RFS band portfolios of all three MNOs. Between 2015 and 2020, the competition became more intense and Turkcell lost an additional 7.5 percent market share. As of 2020, the gap between the market shares of MNOs is narrower than ever (Fig. 1 and Table 1).

2.2 Ex Ante Regulations Under the ECA

Article 1 of the ECA specifies that the aim of the ECA is "*to create effective competition, to ensure the protection of consumer rights, to promote the deployment of services throughout the country, to ensure efficient and effective use of resources, to promote new investments and technological developments in communications infrastructure, network and services through regulations and inspections in electronic communications sector.*" The ECA utilizes two types of *ex ante* regulations to reach these goals: (i) general obligations that may be imposed on all operators, and (ii) special obligations that may only be imposed on operators that possess a significant market power (SMP) in a relevant market susceptible to *ex ante* regulation. The latter aim to identify operators with market power that operate in markets characterized by structural imperfections and subject these operators to special responsibilities,

[15] Three packages each for 800 MHz, 900 MHz, 1800 MHz, and 2100 MHz RFS bands and eight packages for the 2600 MHz RFS band.

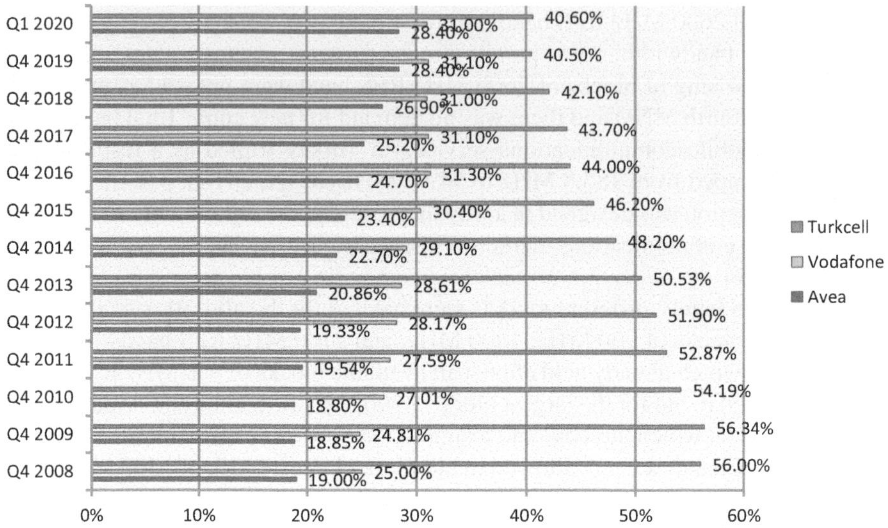

Fig. 1 MNOs Market Shares over time based on Number of Subscribers (2008–2020[16]). *Source* ICTA

Table 1 Evolution of the Mobile communications market in Turkey

Year	Development
1994	Turkcell and Telsim entered the market through revenue-sharing agreements
1998	Turkcell and Telsim signed concession agreements to provide GSM services
2000	İŞ-TİM (Aria) signed a concession agreement to provide GSM services
2001	Türk Telekom (Aycell) signed a concession agreement to provide GSM services
2004	Aria and Aycell merged to form Avea
2006	Telsim acquired by Vodafone
2008	Turkcell, Vodafone and Avea signed concession agreements to provide 3G services
2015	Türk Telekom acquired the sole control of Avea
2015	Turkcell, Vodafone and Avea signed concession agreements to provide 4.5G services

whereas the former aim to ensure that all electronic communications markets are functioning properly.

While these obligations serve important purposes, they also have costs. As a general consensus, a regulation is only considered to be socially beneficial if its net effect is positive (OECD, 1997). While benefits of regulations tend to be more visible, identifying their costs requires a more thorough assessment. Costs of regulations may be categorized under three sub-groups: (i) direct costs, (ii) indirect costs, and (iii) enforcement costs (Renda et al., 2013). Direct costs consist of administrative burdens

[16] Market shares of MNOs before 2008 were not regularly monitored and are not accessible.

and compliance costs borne by businesses, citizens, civil society organizations, and public authorities. Some of these costs are one-off costs (e.g., technical infrastructure investments); whereas others are recurrent costs (employee costs, training costs, regular reporting costs, etc.) (Renda et al., 2013, p. 24–26). The most common examples of indirect costs of regulations are price increases that stem from additional costs borne by firms and negative effects on competition through strengthening of entry barriers, altering market structure in ways that would facilitate anti-competitive cooperation and benefiting firms with market power through asymmetrical effects.[17] Enforcement costs are costs that must be borne to monitor compliance with the regulations and to effectively and rapidly impose sanctions when an incompatibility is detected (Renda et al., 2013, p. 30).

While examining general and SMP-based regulations in Turkish mobile communications market, comments on the net effects of these regulations are also presented throughout the following sub-sections.

2.2.1 General Regulations in Mobile Communications Markets

Obligations that are imposed on all operators in mobile communications markets are as follows:

- Providing interconnection (ECA Article 16/2 and the By-Law on Access and Interconnection)
- Facility sharing and collocation (ECA Article 17 and the By-Law Regarding the Sharing of Cellular System Antenna Facilities and Wireless Access Networks)
- Number portability (ECA Article 32 and the By-Law on Number Portability)
- Respecting consumer and end user rights (ECA Articles 47–52 and the By-Law Regarding Consumer Rights in Electronic Communications Sector).

Operators must abide by these regulations, regardless of the market power they hold, because the regulations are deemed necessary for the healthy functioning of the markets, efficient and effective use of resources, and maximization of consumer welfare. For example, all operators are required to provide interconnection in order to ensure that the number of contacts with whom consumers may communicate is not limited by strategic behaviors. Although the logic behind imposing interconnection obligations in the absence of market power may be discussed, this would be futile in practice as the ICTA is of the opinion that all MNOs hold a monopoly position (and therefore SMP) for terminating calls over their own networks and an obligation to provide interconnection at regulated tariffs and on a non-discriminatory basis is

[17] For a discussion regarding the asymmetrical effects of the General Data Protection Regulation that led to the strengthening of incumbent platforms' market power, see Damien, 2020. Similar effects could arise in the mobile communications market due to regulations with high direct and indirect costs when these are applied symmetrically to all operators, without taking their scales into consideration.

imposed as an SMP-based obligation in any case (ICTA, 2017a). Similarly, facility sharing and colocation obligations prevent waste of resources and minimize harm to the environment, while rules concerning number portability ensure that operators would not be able to raise switching costs by locking in their subscribers. Moreover, obligations regarding consumer rights ensure that operators do not exploit the information asymmetry between themselves and consumers and that the quality of service provided to consumers do not fall below a certain threshold.

The net benefits of general regulations concerning interconnection, facility sharing and colocation, and number portability are quite clear, while their costs (especially administrative costs as they do not require constant monitoring) are relatively low. On the other hand, the effects of regulations that aim to protect consumers may be more ambiguous. Under a broad interpretation of the term "consumer protection," respective regulations in Turkish mobile communications market include those related to equal access to mobile communications services (ECA Article 47), transparency and access to information (ECA Article 49), content and formation of subscription agreements (ECA Article 50), personal data protection (ECA Article 51), and minimum quality of service (QoS) (ECA Article 52). Most of these regulations seem to be objectively necessary due to the idiosyncrasies of the mobile communications market. For example, as mobile communications services are crucial in today's world, equal access must be ensured. Furthermore, factors such as the complicated nature of the services provided, high information advantages of operators, and low negotiation capabilities of end users justify the need for the rest of these requirements. As long as it is clear that there is a need for regulation, the emphasis should be on minimizing its costs, especially when these costs are not negligible.

Two of the above-mentioned regulations stand out due to their high costs[18]: minimum QoS requirements and rules regarding the formation of subscription agreements. With respect to the former, the regulations in Turkey are in accordance with the international best practices and there is no evidence suggesting that they lead to overregulation or social welfare loss. However, rules in the latter category lead to high costs without creating any visible benefits.

With respect to the formation of subscription agreements, Article 50/1 of the ECA sets out that these agreements shall either be made in writing or electronically. As signing written agreements with all subscribers leads to significant wastage of resources and considerable transaction costs, forming such agreements electronically could generate large efficiencies. The ICTA has the discretion to determine the methods and procedures concerning the formation of subscription agreements electronically in a way that would ensure that the identity of consumers can be verified with sufficient accuracy. Currently, Article 7 of the By-Law Regarding Consumer Rights in Electronic Communications Sector requires that subscription agreements

[18] Personal data protection regulations may also have high direct and indirect costs, but the requirements in the ECA do not far exceed the general requirements set forth by the Personal Data Protection Act No. 6698.

may only be formed via secure electronic signature.[19] However, consumers must acquire a certificate from a qualified electronic certificate service provider (by way of physically visiting the premises of such a provider) and use a secure electronic signature creation device (Kamu Sertifikasyon Merkezi, 2020) to sign a subscription agreement using a secure electronic signature. Due to these complicated procedures, secure electronic signature is rarely used and, in practice, almost all subscription agreements are still made in writing. Negative impacts of this inefficient regulation may be negated if the ICTA amends the relevant By-Law, identifies an alternative electronic signature technology (e.g., a form of electronic signature, such as biometric signature, that fulfills the criteria for advanced electronic signature in eIDAS[20]) to sign subscription agreements electronically and defines reliable criteria for authenticating the identity of the consumers (e.g., definition of strong customer authentication in Article 4 of the PSD2 (European Parliament and the Council, 2015b) may be imported in the By-Law by the ICTA).

In addition to the foregoing, ICTA should refrain from relying on the provisions of the GSM and 3G concession agreements, which were drafted prior to the entry into force of the ECA and grant the ICTA with certain regulatory powers that are alien to the modern economic rationale of the ECA, such as the right to interfere with the tariffs in the absence of market power. The 13th Circuit of the Council of State (CoS) held that the ICTA's intervention to Avea's tariffs based on the concession agreement was against the law.[21] A similar reasoning should be adopted for all concession agreement based *ex ante* regulations that openly contradict with the *ratio legis* of the ECA.

2.2.2 SMP-Based Regulations in Mobile Communications Markets

Article 7/3 of the ECA sets out that the ICTA "*may identify the operators with SMP in the relevant markets by conducting market analyses. The ICTA may also impose obligations on operators with SMP with the aim of ensuring and promoting an effective competition environment.*" Obligations that may only be imposed on SMP operators are:

- Tariff regulations (ECA Article 13(2)(b))
- Accounting separation and cost accounting obligations (ECA Article 21)
- Access obligations (ECA Article 16 and the By-Law on Access and Interconnection).

[19] Secure electronic signature is defined in Article 4 of the Electronic Signature Act No. 5070 and corresponds to "qualified electronic signature" in the eIDAS (European Parliament and the Council, 2014).

[20] Although the Electronic Signature Act only defines secure electronic signature and remains silent about advanced electronic signature, this may not prevent ICTA from adopting the definition in the eIDAS as the ECA grants the ICTA with sole discretion to determine the procedures regarding the formation of subscription agreements electronically.

[21] CoS Decision dated 18.06.2014 and numbered E.2010/2441-K.2014/2526.

The procedure for market analysis and SMP designation, which is set out under the By-Law on Market Analysis, comprises the following steps:

- Definition of relevant market
- Assessing whether there is a need for regulation in the relevant market
- Analyzing the competitiveness of the market
- Designation of operators having SMP
- Imposition of remedies on operators with SMP.

The by-law adopts competition law principles when defining the market. Although there is no explicit reference, criteria that shall be taken into account are derived from these principles.

While identifying whether a need exists for *ex ante* regulation in a relevant market, the by-law adopts the three-criteria test recommended by the European Commission (European Commission, 2014). According to this three-criteria test, a market may be subjected to SMP-based *ex ante* regulations only if the following three criteria are cumulatively satisfied:

- High entry barriers are present
- The market is not expected to become competitive within a reasonable period, and
- Competition law is not sufficient to cure market failures alone.

As of 2020, only one mobile communications market satisfies the above criteria, namely the market for mobile call termination. In its analysis of mobile call termination market, the ICTA defined separate relevant markets for terminating mobile calls and SMS/MMS services over the 2G, 3G and 4.5G networks of each MNO. This approach was also in accordance with the recommendation of the European Commission.[22] The ICTA held that all MNOs should be designated as operators with SMP due to the absence of actual and potential competition for call termination services. To remedy potential problems in the relevant markets *ex ante*, the ICTA decided to impose the following seven obligations on Turkcell, Vodafone, and Avea[23]:

- Obligation to provide interconnection (imposed on all operators including MVNOs)
- Non-discrimination obligation (imposed on all operators including MVNOs)
- Transparency obligation (imposed on all operators including MVNOs)
- Obligation to prepare and publicize reference interconnection offers
- Tariff controls
- Accounting separation and cost accounting obligations
- Collocation obligation.

[22] Wholesale voice call termination on individual mobile networks (Market 2) is listed among the markets that should be susceptible to *ex ante* regulation within the EU Commission's Recommendation (European Commission, 2014).

[23] ICTA Decision dated April 19, 2017 and numbered 2017/DK-SRD/131.

It should be noted that the mobile termination rates (MTR) are determined by the ICTA asymmetrically due to the differences between the mobile traffic, revenues, number of subscribers, RFS portfolios, and market entry dates of MNOs. In that vein, Avea, which is the latest entrant and the MNO with the lowest market shares (in terms of mobile traffic, revenue and number of subscribers) is allowed to charge a higher MTR (2.96 kr/dk) than Vodafone (2.58 kr/dk) and Turkcell (2.50 kr/dk) (ICTA, 2017a, p. 53). The European Commission also recognizes using asymmetrical MTRs as a tool to promote competition in the mobile communications market as a valid approach under certain circumstances (European Commission, 2009b, p. 9).

3 Regulations Aiming to Create a Competitive Mobile Communications Market

As explained in Section I.A, Turkcell held a large market share and considerable market power in the retail mobile communications services market due to such factors as its first-mover advantage, favorable RFS portfolio, and certain strategic practices that are laid down under Section II.A below. When the ECA entered into force, restoring competition in the retail mobile communications services market became one of the priorities of the ICTA and the ICTA took important steps with that goal in mind. Mobile portability regulations and asymmetrical interconnection fees explained in Section I.B are examples of regulations that aim to restore and promote competition in the retail level. The ICTA also tried more unconventional methods to curb Turkcell's market power and introduce more competition in the market via imposing a ban on Turkcell's exclusionary on-net/off-net price differentials and designating Turkcell as an operator with SMP in the market for call origination and access to mobile networks, and requiring it to provide access to all MVNO's over-regulated tariffs and under pre-defined conditions. With the benefit of hindsight, it can be seen that the former had been a success, whereas the latter was an utter failure.

3.1 Ban on Exclusionary On-Net/off-Net Price Differentials

More than 55 percent of all mobile subscribers in Turkey were using Turkcell in Q1 2008 and accordingly, a large majority of the calls were being terminated on Turkcell's network. Vodafone and Avea were required to pay regulated MTRs to Turkcell for calls that were originated on their networks, but terminated on that of Turkcell. Therefore, the MTR's charged by Turkcell constituted a natural price floor for the retail off-net fees (RF_{OFF}) of other MNOs. In order to gain a competitive advantage vis-à-vis its competitors, Turkcell had adopted a strategy to consistently keep its on-net fees (RF_{ON}) below its MTRs. As a result of this strategy, Turkcell guaranteed that other MNOs could never offer a cheaper alternative to communicate with Turkcell's subscribers and this allowed Turkcell to exploit the network externalities stemming from its first-mover advantage. Turkcell's strategy would have been unsustainable in theory if MTRs were calculated by using the "pure LRIC[24]" approach, as suggested by the European Commission (European Commission, 2009a, para 19), as keeping RF_{ON} below MTRs would have led to losses under such a scenario. However, because the ICTA chose to include certain joint and common costs to MTRs (thereby preferring the LRIC + approach), Turkcell was able to profitably maintain this strategy.

Following complaints raised by other MNOs, the ICTA made Decision No.149[25] on March 25, 2009, which required Turkcell keep its RF_{ON} above its MTRs. As Turkcell was not designated as an SMP operator on the retail level, the legal bases for this decision were questionable (it was unclear whether the relevant obligation was based on the provisions of the concession agreement or Turkcell's SMP in the market for call origination and access to mobile networks). However, the Plenary Session of Administrative Law Chambers held, in its decision dated October 23, 2017, and numbered E.2015/3397-K.2017/3222, that such obligations could be imposed based on Turkcell's SMP in the market for call origination and access to mobile networks and that the Decision No. 149 was in accordance with the law.

The scope of Decision No. 149 was subjected to amendments during the course of time.[26] The Decision No. 142, which required Turkcell to charge RF_{ON} above its cost of on-net call origination (which was 1.71 times higher than its MTRs) was the most interventionist form of the said decision. The Decision No. 149 and all the amending decisions were finally voluntarily revoked by the ICTA on August 16, 2016, with Decision No. 370.[27] The ICTA removed these obligations because over-the-top (OTT) services became much more significant and pricing strategies regarding voice and SMS services became a less relevant parameter of competition between MNOs.

Decision No. 149 and the following amendments had crucial impacts on the retail mobile communications services market and constituted an important part of the ICTA's overall strategy to curb Turkcell's market power. The most quantitively visible effect of these decisions was the decrease in the proportion of on-net calls within total calls. While on-net calls constituted 88.67 percent of all mobile traffic in Q1 2009 (just before the introduction of the Decision No. 149), this dropped to 73.59 percent within a few months in Q3 2009 (ICTA, 2011, p. 49). The gap between on-net and off-net calls continued to narrow, and as of Q3 2016 (just before the revocation

[24] According to the definition adopted by the European Commission, a pure LRIC approach implies the exclusion of costs that would not have been avoidable if the wholesale termination service had been discontinued.

[25] ICTA Decision dated March 25, 2009, and numbered 2009/DK-07/149.

[26] ICTA decisions; dated April 25, 2012, and numbered 2012/DK-07/181, dated March 13, 2013, and numbered 2013/DK-ETD/142, dated January 6, 2014, and numbered 2014/DK-ETD/21.

[27] ICTA Decision dated August 16, 2016, and numbered 2016/DK-ETD/370.

of the Decision No. 149), on-net calls amounted to 51 percent of the total traffic (ICTA, 2016a, p. 51). During the period in which the Decision No. 149 remained in force (Q1 2009–Q3 2016), Turkcell's market share based on mobile traffic dropped by almost 20 percent from 45.22 percent to 37.2 percent. Although it is difficult to argue that the Decision No. 149 was the sole factor for the foregoing changes in the market dynamics, it had been an effective tool for the ICTA.

3.2 Regulations Aiming to Promote Service-Based Competition

To remedy problems associated with the oligopolistic structure of the retail mobile communications services market by way of promoting new entry, the ICTA tried to facilitate the MVNO business model via *ex ante* regulations. In 2009, the ICTA held that Turkcell possessed SMP in the market for call origination and access to mobile networks as a result of its first analysis of the relevant market.[28] Turkcell was required to grant MVNOs access to its network on transparent and non-discriminatory terms over regulated access tariffs and in accordance with the provisions set forth under a reference access offer to be approved by the ICTA.

The ICTA renewed this market analysis in 2013 and maintained its position regarding the designation of Turkcell as an operator with SMP and decided to impose the same *ex ante* obligations on Turkcell. However, the Turkish Competition Authority (TCA) did not agree with the findings of the ICTA and raised its concerns regarding the inaccuracy of the ICTA's conclusions concerning the three-criteria test and the market power held by Turkcell (ICTA, 2013). The TCA referred to the European Commission's statement that the market for call origination and access to mobile networks should not be susceptible to *ex ante* regulations as it had the potential to become more competitive in the near future (European Commission, 2007). The TCA also pointed out that all three MNOs had similar networks and controlled infrastructure that may not be easily replicated by MVNOs. The TCA added that the MNOs had technological advantages, benefited from economies of scale and vertical integration, and controlled large-scale distribution and sales networks when compared to MVNOs. In light of the foregoing, the TCA argued that the market realities did not support the ICTA's conclusions that the market should be susceptible to *ex ante* regulation or that Turkcell unilaterally possessed SMP.

The ICTA did not agree with the TCA's concerns and kept its initial assessments intact, designating Turkcell as the sole MNO with SMP. Yet, as a result of the action for annulment brought by Turkcell, the relevant market analysis conducted in 2013 was overturned by the 13th Circuit of the CoS.[29] In the said decision, CoS repeated some of the concerns that the TCA had raised in response to the public consultation of the ICTA. While the CoS did not agree with the TCA's opinion that the conditions of

---

[28] ICTA Decision dated December 8, 2009, and numbered 2009/DK-10/626.

[29] CoS Decision dated March 5, 2015, and numbered E.2014/1750-K.2015/895.

the three-criteria test were not satisfied (thereby siding with the ICTA and confirming that the market should be susceptible to *ex ante* regulation), it decided that the data at hand did not support the ICTA's conclusion that Turkcell is the sole operator with SMP. The CoS referred to the analyses conducted by the Slovenian and the Spanish telecommunication authorities and reminded that the European Commission had vetoed the Slovenian authority's decision to designate a single operator with SMP in a market where the market shares of the first three MNOs were 52.9 percent, 29.5 percent, and 9.1 percent respectively (a larger asymmetry in favor of the leading operator compared to the Turkish mobile communications market) while approving the Spanish authority's decision to designate all operators as operators with SMP in a similar setting. Keeping the Commission's approach in mind and taking into account the fact that there is no considerable difference between the number active base stations owned by the MNOs in Turkey, the CoS decided that the ICTA's failure to assess whether more than one MNO holds SMP collectively was not in accordance with the law. ICTA reconsidered its market analysis in accordance with the relevant CoS decision and conducted a detailed collective SMP analysis. However, ICTA did not change its conclusion and decided to designate Turkcell as the sole operator with SMP.[30]

Despite these ex ante obligations imposed on Turkcell, no MVNO managed to successfully enter the retail mobile communications market in Turkey. This was mainly due to a double-taxation problem stemming from the rules concerning Treasury Share payments of operators. Per Additional Article 37 of the Act No. 406, both the MNOs and operators that provide services over the networks of MNOs (that is, MVNOs) must pay a Treasury Share amounting to 15 percent of their gross sales. MVNOs purchase services from MNOs at the wholesale level. Although the price paid to MNOs at the wholesale level includes the Treasury Share, MVNOs are required to pay an additional Treasury Share over their gross sales to end users at the retail level. This creates an important asymmetrical cost disadvantage for MVNOs vis-à-vis MNOs and it removes all incentives for market entry.[31]

Finally, when renewing its market analysis in 2017, the ICTA changed its position and decided that the market for call origination and access to mobile networks should not be subjected to ex ante regulations because the conditions set forth under the three-criteria test had not been satisfied.[32] The ICTA pointed out that the reason why there were no MVNOs in the market was due to the double-taxation problem rather than structural imperfections stemming from SMP of Turkcell or other MNOs. With respect to indicators of market power in the retail and wholesale levels, the ICTA opined that the asymmetries between MNOs in the retail level were withering away and that MNOs held very close positions in the wholesale level as the composition of their RFS portfolios became much closer after the 4.5G auction and the sizes of their networks were similar. Eventually, the ICTA noted that the dynamics of the retail

[30] ICTA Decision dated March 22, 2018, and numbered 2018/DK-SRD/93.

[31] For a detailed discussion of the double-taxation problem and a proposal to remedy this problem, see Yüksel & Demirkan, 2017.

[32] ICTA's Decision dated April 12, 2017, and numbered 2017/DK-SRD/124.

market did not justify further intervention as the data at hand was indicative of intense competition between existing MNOs. The ICTA added that any problem that may arise in such a market could be addressed via competition laws in any case (hence, the third criterion of the three-criteria test is not satisfied). Therefore, call origination and access to mobile networks market in Turkey was deregulated in 2017.[33]

4 Assessment of the Regulations and Expectations for the Future

Until the entry into force of the ECA and the formation of the ICTA, the regulations concerning the mobile communications market in Turkey were neither coherent nor effective as there were no clear policy objectives upon which they were based. Following the principles enshrined in the ECA, which were generally in line with those in the EU's NRF, the ICTA managed to deal with some of the most apparent structural problems in the market, contributing to competitiveness, ensuring effective resource management and preventing MNOs from attempting to abuse or misdirect consumers. Despite their general positive impacts, some of the ICTA's regulations proved ineffective (such as those aiming to facilitate the entry of the MVNOs) and even led to negative effects (such as those concerning the formation of agreements), either due to internal defects or other external factors. However, as the ICTA closely monitored the developments in the EU and decided not to diverge from these best practices to a significant extent, this ensured that the general policies pursued by the ICTA were the right ones and its mistakes were isolated to specific cases.

Having said that, it seems that the ICTA has not kept up with the latest developments in the mobile communications markets as it has failed to address the emerging structural concerns that may justify additional intervention. In 2008, when the ECA entered into force, concerns regarding the mobile communications market were mostly confined to those that were discussed previously within this chapter. Most importantly, there was a clear distinction between conventional mobile communications services (that is, voice and SMS services) and OTT services and the regulations entirely focused on the former. However, the lines dividing the two type of services disappeared over time as the quality, quantity, and functionality of OTT services increased dramatically and many applications that can easily substitute, if not supersede, conventional services were introduced.

These technological changes led to the emergence of brand-new market dynamics and unprecedented problems. The most apparent structural risk stemmed from MNOs' ability to control the Internet via traffic management practices that would allow them to prioritize some applications or to slow down and even block some others. This was deemed as a threat to the openness of the Internet in general and the competitiveness of the mobile communications market in particular as most of

[33] Although the decision to deregulate the market was taken in 2017, this decision became effective on April 12, 2019, per the ICTA Decision dated April 3, 2019, and numbered 2019/DK-SRD/097.

the MNOs became vertically integrated in the OTT service level and could have the incentives to favor their own applications or to hinder the activities of competing applications. In the EU the TSM Regulation was introduced to protect the open Internet (European Parliament and the Council, 2015b). The TSM Regulation laid down the rules concerning net neutrality and prohibited any traffic management practices that go beyond reasonable traffic management measures, by blocking, slowing down, altering, restricting, interfering with, degrading, or discriminating between specific content, applications or services and in the absence of objective justifications.

Unlike the EU, the ICTA did not prioritize net neutrality and did not issue any rules to control the traffic management practices of the MNOs. The MNOs in Turkey are currently free to engage in all kinds of traffic management practices as long as those practices do not violate other legislations such as Act No. 4054 on the Protection of Competition. As of 2020, some MNOs in Turkey use traffic management practices both to favor specific major third-party applications such as Instagram, Twitter, WhatsApp, and Youtube[34] and their own OTT services.[35] While such practices may confer significant benefits to consumers, they may also render market entry more difficult and distort the playing field. Hence, it is important for the ICTA to realize the threats and incorporate certain regulations that would enable it to distinguish between harmful and beneficial traffic management practices. Some of these practices may have already exacerbated the problems in various OTT services markets and created certain problems that may prove considerably difficult to remedy in the near future. In order to avoid further and possibly permanent damages, the ICTA should design new rules that would safeguard the openness of the Internet, recognize and protect consumers' rights concerning the choice of content, and prevent MNOs from stifling competition by engaging in unreasonable traffic management, while carefully refraining from over-regulation that would prohibit efficient and welfare enhancing practices.

5 Conclusion

The legal framework in the early years of the mobile communications markets in Turkey was generally shaped in accordance with the emerging needs of the operators that desired to engage in mobile communications services, and the priority was to enable the provision of these services rather than designing an appropriate market structure. While these early policies did incentivize first movers (especially Turkcell) to invest heavily in mobile communications services, the lack of regulatory foresight allowed them to take advantage of the structural market failures and to establish

---

[34] Vodafone offers certain packages to its subscribers whereby the traffic generated by specific applications do not reduce the overall Internet quote and are effectively free (https://www.vodafone.com.tr/Red/red-pass.php).

[35] Turkcell's own music streaming, messaging, cloud and video streaming applications do not reduce the overall Internet quote of Turkcell subscribers and are effectively free for them (https://www.turkcell.com.tr/tr/internetinden-yemeyen-uygulamalar).

considerable entry barriers at a very early stage. Consequently, a duopoly with a strong market leader emerged. This situation was exacerbated due to the flaws in the design of the initial RFS auction, as the market ended up with four players rather than the initially planned five. Combined effects of the first-mover advantages enjoyed by Turkcell and Telsim, the high costs imposed on İŞ-TIM and Türk Telekom due to the flaws in the auction design, excessive coverage obligations in the concession agreements of the new entrants, and the lack of a sound regulatory framework that would facilitate entry led to further consolidation in the market and only three MNOs remained.

After the entry force of the ECA and the modernization of the regulatory framework in 2008, the ICTA worked hard to rectify the mistakes of the past and to create a more competitive and efficient mobile communications marketplace, as envisaged in the ECA. The ICTA did well to remedy the asymmetries between existing MNOs via minimizing expansion barriers and switching costs and preventing Turkcell from using its market power to further exploit structural failures. However, the ICTA's *ex ante* regulations were not able to increase service-based competition in the downstream level through the introduction of the MVNOs, mainly because the disincentives created by double taxation could not be negated.

The ICTA also relied on *ex ante* regulations to fulfill other goals pursued by the ECA, such as protecting consumers, ensuring effective use of resources, and promoting deployment of services throughout the country. While most of the regulations imposed by ICTA seemingly created net social benefits, the ICTA's strict approach concerning the formation of subscription agreements in the mobile communications market had been highly inefficient and led to high and unnecessary transaction costs as well as waste of resources.

While following the best practices in the EU for regulating the conventional mobile communications services provided by the MNOs, the ICTA failed to recognize the changing market dynamics and increasing significance of OTT services, which led to new structural problems and a need to update the former regulatory policies.

As to RFS allocation policies, the 4.5G auction was designed carefully, in accordance with the delicacies of the Turkish market, and it significantly reduced technological asymmetries stemming from differences in the compositions of RFS portfolios of MNOs. However, despite ICTA's success in using the 4.5G auction as a technological leveler in the Turkish mobile communications markets, it was still not able to increase the number of MNOs in the market. While this may be due to a lack of demand, it may also be the case that right incentives for new entrants (such as asymmetrical and light investment and coverage obligations) were not created.

All in all, the modern legal framework established by the ECA and the effective use of *ex ante* regulations by the ICTA covered up some of the wounds created by the early policy mistakes, but the structure of the mobile communications market, which comprises only three players and an ever-present market leader, is far from being optimal for maximizing the welfare of the society. Furthermore, ICTA's failure to keep up with the EU and to prioritize OTT services may already have created problems in the markets for OTT services and the ICTA should revise the current regulatory framework in line with the EU's best practices as soon as possible.

References

Atiyas, İ, & Doğan, P. (2007). When good intentions are not enough: Sequential entry and competition in the Turkish mobile industry. *Telecommunications Policy, 31*(8–9), 502–523. https://doi.org/10.1016/j.telpol.2007.05.012

Binmore, K., & Klemperer, P.D. (2002) The biggest auction ever: the sale of the British 3G telecom licenses. *Economic Journal 112: 74–96*. http://www.paulklemperer.org

ERG & RSPG (2019, June). *ERG-RSPG report on the management of radio spectrum in order to avoid anticompetitive hoarding*. https://rspg-spectrum.eu/wp-content/uploads/2013/05/rspg09_278_erg_rspg_report_on_radio_spectrum_competition_issues_0906041.pdf

Erşen, C. (2009, March). *Telekomünikasyon Hizmetlerinin Yetkilendirmesinde İhale Yöntemleri:Ülke Uygulamaları ve Türkiye İçin Öneriler*. Bilgi Teknolojileri ve İletişim Kurumu Uzmanlık Tezi. https://www.btk.gov.tr/uploads/thesis/telekomunikasyon-hizmetlerinin-yetkilendirmesinde-ihale-yontemleri-ulke-uygulamalari-ve-turkiye-icin-oneriler.PDF

European Commission (2007, December). *Commission Recommendation of 17 December 2007 on relevant product and service markets within the electronic communications sector susceptible to ex ante regulation in accordance with Directive 2002/21/EC of the European Parliament and of the Council on a common regulatory framework for electronic communications networks and services* (2007/879/EC). https://eur-lex.europa.eu/legal-content/EN/TXT/PDF/?uri=CELEX:32007H0879&from=EN

European Commission (2009a, May). *Commission Recommendation of 7 May 2009 on the Regulatory Treatment of Fixed and Mobile Termination Rates in the EU* (2009/396/EC). https://eur-lex.europa.eu/legal-content/EN/TXT/PDF/?uri=CELEX:32009H0396&from=EN

European Commission (2009b, May). *Commission Staff Working Document accompanying the Commission Recommendation on the Regulatory Treatment of Fixed and Mobile Termination Rates in the EU - Explanatory Note*. https://ec.europa.eu/smart-regulation/impact/ia_carried_out/docs/ia_2009/sec_2009_0600_en.pdf

European Commission (2014, October). *Commission Recommendation of 9 October 2014 on relevant product and service markets within the electronic communications sector susceptible to ex ante regulation in accordance with Directive 2002/21/EC of the European Parliament and of the Council on a common regulatory framework for electronic communications networks and services* (2014/710/EU). https://eur-lex.europa.eu/legal-content/EN/TXT/PDF/?uri=CELEX:32014H0710&from=EN

GSMA (2019, May). *Auction Best Practice GSMA Public Policy Position*. https://www.gsma.com/spectrum/wp-content/uploads/2019/05/Auction-Best-Practice.pdf

ICTA (2009, July). *Türkiye Elektronik Haberleşme Sektörü Üç Aylık Pazar Verileri Raporu – 2009 1. Çeyrek*. https://www.btk.gov.tr/uploads/pages/slug/2009-q1.pdf

ICTA (2011, February). *Türkiye Elektronik Haberleşme Sektörü Üç Aylık Pazar Verileri Raporu – 2010 4. Çeyrek*. https://www.btk.gov.tr/uploads/pages/slug/2010-q4.pdf

ICTA (2013, January). *Mobil Şebekelere Erişim ve Çağrı Başlatma Pazarı – Kamuoyu Görüşlerini İçeren Döküman*. https://www.btk.gov.tr/uploads/pages/slug/mobilerisimcagribaslatma-pa-kgid-2012.pdf

ICTA (2016a, December). *Türkiye Elektronik Haberleşme Sektörü Üç Aylık Pazar Verileri Raporu – 2016 3. Çeyrek*. https://www.btk.gov.tr/uploads/pages/slug/2016-q3.pdf

ICTA (2016b). *Mobil İletişimde Yeni Teknoloji 4.5G*. https://www.btk.gov.tr/uploads/pages/slug/45g-brosur.pdf

ICTA (2017a, April). *Mobil Çağrı Sonladırma Pazarı Pazar Analizi*. https://www.btk.gov.tr/uploads/pages/elektronik-haberlesme-sektorunde-pazar-analizleri/ek-3-mcspa-kgid-2017.pdf

ICTA (2017b, April). *Mobil Şebekelere Erişim ve Çağrı Başlatma Pazarı Pazar Analizi.* https://www.btk.gov.tr/uploads/pages/onceki-pazar-analizleri/ek-1-msecbpa-nd-2017.pdf

ICTA (2020, February). *Türkiye Elektronik Haberleşme Sektörü Üç Aylık Pazar Verileri Raporu – 2019 4. Çeyrek.* https://www.btk.gov.tr/uploads/pages/pazar-verileri/4-ceyrekraporu-2019-final.pdf

Irmak, A. İ. (2013, June). *Türk Kamu Yönetiminde Bağımsız İdari Kurulların Yeri ve Bilgi Teknolojileri ve İletişim Kurumu.* Bilgi Teknolojileri ve İletişim Kurumu Uzmanlık Tezi. https://www.btk.gov.tr/uploads/thesis/turk-kamu-yonetiminde-bagimsiz-idari-kurullarin-yeri-ve-bilgi-teknolojileri-ve-iletisim-kurumu.PDF

Kamu Sertifikasyon Merkezi (2020). *Biyometrik İmza ve Nitelikli Elektronik İmza Karşılaştırması.* http://kamusm.bilgem.tubitak.gov.tr/dosyalar/beyazbulten/Biyometrik_Imza_ve_Nitelikli_Elektronik_Imza_Karsilastirmasi_Whitepaper.pdf

Köktürk, N. S. (2012). *Telekomünikasyon Sektöründe Dikey Bütünleşik Firma Davranışlarının Düzenlenmesi: Sektörel Düzenleme ve Rekabet Hukuku* (Uzmanlık Tezleri Serisi No: 120). Rekabet Kurumu. https://www.rekabet.gov.tr/Dosya/uzmanlik-tezleri/120-pdf

OECD (1997). *Regulatory Impact Analysis – Best Practice in OECD Countries.* http://www.oecd.org/regreform/regulatory-policy/35258828.pdf

Renda, A., Schrefer, L., Luchetta, G., & Zavatta, R. (2013). *Assessing the Costs and Benefits of Regulation.* Centre for European Policy Studies. http://www.economistiassociati.com/files/cba_study_sg_final_0.pdf

Walden, I. (2018). *Telecommunications Law and Regulation* (5th ed.). Oxford University Press.

Yayla, F. (2009, April). *Frekans İhalelerinin İhale Teorisi Kapsamında Değerlendirilmesi: Dünya Uygulamaları ve Türkiye İçin Model Önerisi.* Bilgi Teknolojileri ve İletişim Kurumu Uzmanlık Tezi. https://www.btk.gov.tr/uploads/thesis/frekans-ihalelerinin-ihale-teorisi-kapsaminda-degerlendirilmesi-dunya-uygulamalari-ve-turkiye-icin-model-onerisi.PDF

Yüksel, B., & Demirkan, H. (2017). Elektronik Haberleşme Sektöründe Çifte Vergilendirme Problemi ve Mobil Şebeke İşletmecileri Üzerindeki Asimetrik Etkileri. *Mali Hukuk Dergisi, 13*(145).

Decisions

CoS 13th Circuit Decision dated 18.06.2014 and numbered E.2010/2441-K.2014/2526.

CoS 13th Circuit Decision darted 05.03.2015 and numbered E.2014/1750-K.2015/895.

ICTA Decision dated 25.03.2009 and numbered 2009/DK-07/149.

ICTA Decision dated 08.12.2009 and numbered 2009/DK-10/626.

ICTA Decision dated 25.04.2012 and numbered 2012/DK-07/181.

ICTA Decision dated 13.03.2013 and numbered 2013/DK-ETD/142.

ICTA Decision dated 06.01.2014 and numbered 2014/DK-ETD/21.

ICTA Decision dated 16.08.2016 and numbered 2016/DK-ETD/370.

ICTA Decision dated 12.04.2017 and numbered 2017/DK-SRD/124.

ICTA Decision dated 19.04.2017 and numbered 2017/DK-SRD/131.

ICTA Decision dated 22.03.2018 and numbered 2018/DK-SRD/93.

ICTA Decision dated 03.04.2019 and numbered 2019/DK-SRD/097

TCB Decision dated 09.06.2003 and numbered 03–40/432–186.

TCB Decision dated 28.07.2015 and numbered 15–32/454–138.

Legislation

EU

Directive (EU) 2015/2366 of the European Parliament and of the Council of 25 November 2015 on payment services in the internal market, amending Directives 2002/65/EC, 2009/110/EC and 2013/36/EU and Regulation (EU) No. 1093/2010, and repealing Directive 2007/64/EC (PSD2) (2015b). *Official Journal,* L 337/58, 23 December, pp. 35–128.

Directive (EU) 2002/19/EC of the European Parliament and of the Council of 7 March 2002 on access to, and interconnection of, electronic communications networks and associated facilities (2002a). *Official Journal,* L 108/7, 24 April, pp. 7–20.

Directive (EU) 2002/20/EC of the European Parliament and of the Council of 7 March 2002 on the authorization of electronic communications networks and services (2002b). *Official Journal,* L 108/21, 24 April, pp. 21–32.

Directive (EU) 2002/21/ EC of the European Parliament and of the Council of 7 March 2002 on a common regulatory framework for electronic communications networks and services (2002c). *Official Journal,* L 108/33, 24 April, pp. 33–50.

Directive (EU) 2002/22/EC of the European Parliament and of the Council of 7 March 2002 on universal service and users' rights relating to electronic communications networks and services (2002d). *Official Journal,* L 108/51, 24 April, pp. 51–77.

Directive (EU) 2002/58/EC of the European Parliament and of the Council of 12 July 2002 concerning the processing of personal data and the protection of privacy in the electronic communications sector (2002e). *Official Journal,* L 201/37, 31 July, pp. 37–47.

Regulation (EU) No. 2015/2120 of the European Parliament and of the Council of 25 November 2015 laying down measures concerning open Internet access and amending Directive 2002/22/EC on universal service and users' rights relating to electronic communications networks and services and Regulation (EU) No. 531/2012 on roaming on public mobile communications networks within the Union (TSM Regulation) (2015a). *Official Journal,* L 310, pp. 1–18

Regulation (EU) No. 910/2014 of the European Parliament and of the Council of 23 July 2014 on electronic identification and trust services for electronic transactions in the internal market and repealing Directive 1999/93/EC (eIDAS Regulation) (2014). *Official Journal,* L 257/57, 28 August, pp. 73–115.

Turkey

Act No. 4054 on the Protection of Competition (1994). *Official Journal,* 22140, 13 December.

By-Law on Access and Interconnection (2009). *Official Journal,* 27343, 8 September.

By-Law on Market Analysis (2012). *Official Journal,* 28480, 27 November.

By-Law on Number Portability (2009). *Official Journal,* 27276, 2 July.

By-Law Regarding Consumer Rights in Electronic Communications Sector (2017). *Official Journal,* 30224, 28 November.

By-Law Regarding the Sharing of Cellular System Antenna Facilities and Wireless Access Networks (2016). *Official Journal,* 29910, 6 December.

Electronic Communications Act No. 5809 (2008). *Official Journal,* 27050 (recurring), 5 November.

Electronic Signature Act No. 5070 (2004). *Official Journal,* 25355, 23 January.

Telegram and Telephone Act No. 406 (1924). *Official Journal,* 59, 21 February.

Universal Services Act No. 5369 (2005). *Official Journal,* 25856, 25 June.

Barış Yüksel graduated from Bilkent University Faculty of Law in 2010 and completed his LL.M. in Bilkent University Law and Economics Program. Before joining BASEAK in 2021 as of counsel, he worked in the competition law and regulation departments of respected law firms. Barış is highly experienced in competition law and served numerous domestic and international clients during his career, providing legal advice and representing his clients before the

Turkish Competition Authority in a wide array of cases. Barış is specialized in telecommunications and energy regulations and he consulted a number of regulated firms as well as the Information Communications and Technologies Authority of the Republic of Turkey. He is a member of the Ankara Bar Association.

Privatization of Turkey's Network Industries

Remzi Özge Aritürk

Abstract Privatizing network industries has always been a cumbersome and challenging task, all over the world. Successful privatization of network industries depends heavily on restructuring and regulation, as well as relevant policies implemented before, during, or after privatization and vice versa. Moreover, network industries' political commodity feature, institutional endowment, and customs of the society and the government are important country-specific factors in privatization, which may differ and affect privatization cases across countries. In this context, this chapter addresses the history of privatization of network industries in Turkey's electricity and telecommunications industries, which have made significant headway to date. On the other hand, the fact that considerable time has passed since both privatization cases gives us the opportunity to make post-privatization reviews. At this point, our responsibility is to discern the contribution of regulation, technological development, and other relevant governmental policies—besides the anticompetitive conducts of privatized incumbents—from that of privatization policies to the liberalization of network industries, which is the final goal. Therefore, the other purpose of this chapter is to learn lessons from what has been found in previous experiences and benefit from their guidance to avoid undesirable outcomes.

Keywords Privatization · Network industries · Energy · Electricity · Telecommunication

1 Introduction

Privatization is an economic transition process by which the ownership, property, or operating rights of a government-owned business are transferred to the private sector. The granting of licenses, public service concessions, property, ownership, or operating rights may be subject to privatization. The scope and extent of privatization differ depending on the economic development, ownership of production means and

R. Ö. Aritürk (✉)
Turkish Competition Authority, Ankara, Turkey
e-mail: roariturk@rekabet.gov.tr

M. Eroğlu and M. Finger (eds.), *The Regulation of Turkish Network Industries*,
https://doi.org/10.1007/978-3-030-81720-6_21

the identity, existence and lineage of the dominant class(es) previously in the country. In the early years of the Turkish Republic, there was no noteworthy industrial bourgeoisie, since a precapitalist system of production used to prevail in Anatolia during and before the Turkish War of Independence and a revolutionary national bourgeoisie capable of developing capitalism was missing (Timur, 1997). Consequently, etatism was adopted and implemented for the purpose of rapid industrialization of the country and promotion of private sector and public engagement in the economic activity when necessary (Ergan, 1998). In accordance with etatism, state economic enterprises were established and operated in order to produce from fast-moving consumer goods in the early stages to heavy industries and then to network industries. Therefore, privatization in Turkey involves a wide range of industries, such as textile or sugar plants, electricity grid, telecoms, and oil companies. However, there is an important difference between the grounds of privatization of network industries and competitive industries and they are not equally agreeable. As competitive industries are free of significant market failures and market mechanisms are able to tame undertakings for the sake of consumers, private ownership outperforms public ownership in terms of efficiency in such markets (Armstrong et al., 1994), particularly taking into account noneconomic motivations of public ownership.

2 Network Industries, Their Dilemma with Failures, and Privatization in the Corner

Since the invention of the wheel, presumably, network industries have increasingly been a large part of the world economy, including bazaars, roads, railroads, credit/debit cards, telecommunications, electricity grid, gas network, software, search engines, etc. They are fundamentally comprised of a fixed durable network that requires high sunk capital to operate. Thanks to the nodes and links, consisting of the whole network, network externalities have arisen as a consequence of this complementarity between subunits of network (Economides, 2004), which has eliminated a massive amount of transaction costs for both users and providers, and amplified any benefit or harm brought to the system via quasi-multiplier effect of network effect. Notwithstanding their significant unique benefits, network industries display features of markets with either a natural monopoly or dominant player since barriers to entry are very high, due to wasteful duplication of infrastructure and strong network effects. Hence, network economies face insignificant effective competition, whereby they do not have sufficient motivation to operate more efficiently or charge lower prices to consumers. This historical market failure calls for regulation or public ownership depending on the country's economic conditions and policy agenda (Newberry, 2001). On the other hand, regulatory failure stemming from inefficient performances of regulatory authorities, which are not free of information asymmetries, is the other edge of double-edged sword of utility policy. Therefore, in light of this dilemma, the choice of regulation versus public ownership is principally about the choice of trust in markets through competition versus state control, both of which are accompanied by discretionary regulation.

Policy response to the network utility issue depends largely on institutional endowment and the political reflexes and customs of the society in the country. The political choice for ownership, structure, and regulation might undergo significant alterations subsequent to an ideological shift within public administration and varying cyclical conditions such as technology, international competitive pressure, and financing needs (Newberry, 2001). Nonetheless, sustainable operational efficiency associated with adaptation capacity to innovation and external changes is the indispensable focus of utility policy, and herein, competition outweighs regulation in many respects, particularly in terms of costs, prices, and technological development. Taking regulatory failure into consideration, it is in public's best interest to confine regulation to the natural monopoly part of network and open remaining services provided over network to competition, irrespective of public or private ownership of network utility. Yet, it has been more difficult to develop competition with public-owned utilities because they are suitable for being political instruments, even though encouragement of privatized monopolies with respect to efficient operation and lower prices is another challenge for regulation. In addition, public ownership, by and large, causes underinvestment due to a low rate of return as a result of politically undervalued tariffs giving rise to inefficiency in allocation and production. Thus, privatization is prerequisite for attaining benefits of competition as long as it is preceded by restructuring and succeeded by effective regulation, since a genuine liberalization of a network industry is comprised of meticulous design and realization of restructuring and privatization besides vigilant fulfillment of regulation. Although the main point of privatization is not about privatization income, the financing problem as a burden on Treasury will be relaxed as well as efficiency gains. In many cases, network utilities were sold below their market value for fear that the buyer would not cover costs and gain sufficient profit for reinvestment considering contingency analysis with respect to credibility problems of regulation and policy making (Newberry, 2001). Still, neither income level nor urgency of privatization should be a top priority for privatization agenda since timely and proper restructuring and regulation will have long-lasting effects on total welfare through a smoothly running liberalization process.

3 Retrospect of Privatization in Turkey

Turkey had implemented an import substitution industrialization (ISI) policy since the 1930s on account of limited export income and increasing import goods prices (Boratav, 2003), respectively, in agricultural production, fast-moving goods, consumer durables, heavy industries, intermediate and investment goods, and infrastructure such as electricity and telecoms. Due to limited export income (as opposed to increasing import expenditures), a global oil crisis, and populist domestic policies inconsistent with development plans, a severe balance-of-payments crisis broke out in the end of the 1970s (Bahar, 2005). Pursuant to economic stability measures, a neoliberal structural adjustment program was put into effect in 1980 and Turkey ended its ISI policy, initiated the transition to a free-market economy, and adopted export-led development and industrialization, along with the aim of bringing foreign investment into the country. In line with these developments, a privatization program was commenced to increase efficiency of public enterprises, then promote the development of the capital markets, find adequate financial resources in order to finance investment requirements of infrastructure sectors due to growing demand, and fund the budget deficits (Özata, 2013).

Considering the accomplishments of privatization and liberalization processes with respect to network industries compared to the ones in other sectors, this chapter focuses on electricity and telecoms privatization cases in Turkey. On the other hand, as has been clearly and repeatedly witnessed in numerous cases all around the world, privatization is not only privatization unless it is merely change of ownership and the primary goal of it is to add some significant revenue for once to the Treasury of the country. Instead, it is a temporary transition process, closely associated and having a cause-and-effect relationship with restructuring and liberalization. Restructuring is an important preparation phase prior to privatization, not only because it would be more cumbersome to implement it later on, but also because it provides potential buyers with necessary transparency and predictability. On that account, restructuring adequately before privatization provides potential buyers with necessary observation and testing opportunities with respect to market prices, regulatory competence, and policy credibility so that they could value the business accurately (Newberry, 2001). In addition, what happens in privatization does not stay in privatization, but long lasts thereafter in liberalization. Economics, technological development, and political factors have paved the way for liberalization, and in many cases, it is related to ownership policies as well (Armstrong et al., 1994). In fact, liberalization is a consecutive and eventual phase that restructuring and privatization prepare as the final playing field for undertakings. Apart from efficiency gains, competition as a desired fruit of liberalization can alleviate asymmetric information problem and its unavoidable result of regulatory failure besides regulatory capture and policy credibility problems (Armstrong et al., 1994). After all, liberalization may either be able to put competition into effect and competition exerts pressure on the privatized incumbent utility, or call for more effective and tighter regulation if it does not turn out to be fruitful and competition is too weak to function properly. Thus, it is integral to cover both restructuring and liberalization processes to some extent when addressing privatization cases.

4 Privatization of Electricity Utility in Turkey

Before dealing with privatization cases, it would be helpful to provide a brief overview of the development and structural changes of related institutions, as well as the legal framework of Turkish electricity industry along the road to comprehensive reforms.

The nascent Turkish electricity industry started to emerge with the establishment of a 15 MW capacity thermal power plant (TPP) in İstanbul Silahtarağa in 1913 (Erol, 2007). Early on, electricity services were carried out through Etibank, the Mineral Research and Exploration Institute (MTA), and the Electrical Power Resources Survey and Development Administration (EİEİ) in Turkey. Subsequently, the General Directorate of State Hydraulic Works (DSİ), the General Directorate of Provincial Bank (İller Bankası), and local administrations got involved. As electricity services had become prevalent because of increased generation and extended distribution, an institutional structure became indispensable and these services were incorporated within Turkish Electricity Authority (TEK), which was founded in 1970 by Law No. 1312 in order to integrate and coordinate all electricity supply activities under a single institution. To this end, power plants and transmission facilities were taken over by TEK, yet distribution networks remained within local administrations until 1982, when they were transferred to TEK by Law No. 2705 as well; thereby, TEK became a vertically integrated, publicly owned organization that was in charge of all electricity operations in Turkey. However, in 1984 Law No. 3096 came into effect, which allowed private sector to generate, distribute, and trade electricity. The scope and method of private sector participation in the electricity industry changed in 1994 and 1997. In 1993, TEK, by the Council of Ministers Decision No. 93/4789, was divided into two public economic enterprises: the Turkish Electricity Generation and Transmission Corporation (TEAŞ) and the Turkish Electricity Distribution Corporation (TEDAŞ). Thereafter, in 2000, by Council of Ministers' Decision No. 2000/1312, TEAŞ was separated into three public economic enterprises: the Turkish Electricity Generation Corporation (EÜAŞ), the Turkish Electricity Transmission Corporation (TEİAŞ), and the Turkish Electricity Trade and Contracting Corporation (TETAŞ).

4.1 *Periodical Privatization Cases and Attempts in Turkish Electricity Industry Until Comprehensive Reform*

The earliest private sector participation in the electricity sector in Turkey actually took place just three years after the proclamation of Turkish Republic, in 1926 by Council of Ministers' Decision, through which concession regarding generation, transmission, distribution, and trade of electricity in Kayseri and its periphery was given to Kayseri ve Civarı Elektrik A.Ş. (Kayseri A.Ş.) for 56 years. By the expiration date, in 1982, all facilities in the region had been transferred to TEK at no cost, yet as a result of application within the scope of Law No. 3096, Kayseri A.Ş. was again assigned to be the incumbent company in 1988, operating rights of electricity facilities in the region were transferred to the company, and it currently operates in the 18th distribution region, in accordance with its license and contract renewed in 2009.

Apart from the early exceptional example of Kayseri A.Ş., there were additional discrete cases. By the Council of Ministers' Decisions, Çukurova Elektrik A.Ş. (ÇEAŞ) regarding Seyhan hydroelectric power plant (HPP) in 1953 and Kepez Elektrik Ticaret A.Ş. (Kepez A.Ş.) regarding Kepez HPP in 1956 were given concessions for operating power plants and sale of generated electricity energy to consumers whose installed capacity was over 500 kW for 49 and 60 years, respectively. After the enactment of Law No. 3096, in 1988 each company was assigned to be the incumbent company responsible for generation, transmission, distribution, and trade of electricity for 70 years upon their application, whereas concession agreements of both companies were terminated in 2003 by Ministry on account of noncompliance of their transactions and businesses with concession agreements and legislation. Similarly, in 1989 Aktaş Elektrik Ticaret A.Ş., by Council of Ministers' Decision, was given the concession for generation, transmission, distribution, and trade of electricity in the 14th distribution region for 30 years, although the concession agreement was terminated in 2002 for breaching the concession agreement and legislation.

From a legislative perspective, Law No. 3096 initially provided a legal foundation and enabled the private sector to engage in electricity industry in relation to existing publicly owned facilities. The method adopted under the law was transfer of operating rights (TOR), by which the operating rights of existing facilities were assigned to a private company for a particular period of time, as exemplified by abovementioned cases. This contract duration has been long enough that an authorized private company is responsible for not only operation and maintenance of existing facilities, but also a necessary investment for the publicly owned infrastructure. The second method adopted under the law was the build–operate–transfer (BOT) model, firstly developed by World Bank in the 1980s, for newly built generation facilities. While previously infrastructure investments used to be financed through government debt, this became increasingly difficult due to high public deficits, low financial credibility, currency risk, and reluctance of financial institutions to fund such investment projects (Özer, 2012). As an ipso facto alternative to undercapitalization and public debt for financing infrastructure investments, BOT agreements were launched first in the 1980s. With regard to the BOT model, a private company was given a concession with respect to operation of a generation plant for a significant period of time—initially up to 99 years and later 49 years—in return for building it and the plant is transferred to public ownership at no cost by the expiration date of the concession agreement. Because of currency risk, economic instability and predictability problems, as well as expected high profit margins of investors vis-à-vis country risk, these BOT contracts were designed to include guarantee of purchase for generated electricity as well as tax exemptions in 1994 through Law No. 3996 and Council of Ministry Decision No. 2011/1807. Also, in 1997, a third method as build–own–operate (BOO) model was adopted within Law No. 4283. According to this model, a generation plant was built and operated under private ownership and it was possible to provide guarantee of purchase by Treasury under law.

In the face of increasing demand for electricity energy and insufficient installed capacity and generation, private sector participation into generation segment of the industry started with the BOT and BOO models in Turkey. Taking into account the

Table 1 Expiration dates of BOO power plants

Power plant	Installed capacity (MW)	Fuel type	Expiration date
Gebze CCGT	1540	Natural gas	2019
Adapazari CCGT	770	Natural gas	2019
İzmir CCGT	1520	Natural gas	2019
Ankara CCGT	770	Natural gas	2019
İskenderun import coal FP	1210	Import coal	2019
Total	5810		

Source EÜAŞ annual reports and wholesale and retail electricity sector inquiry report of TCA

abovementioned constraints with respect to alternative investment opportunities, the governments might have then resorted to these models in order to ensure security of supply. However, the power purchase agreements (PPA) with Treasury grants entailed purchase commitments for a certain amount of electricity energy at predetermined price formulas over long periods (15–30 years). For instance, the power plants under BOO schemes largely started to operate in the early 2000s, for around 16 years. As seen below, most of BOO and BOT contracts with purchase guarantees have recently expired. Unfortunately, some of these contracts did not undergo a competitive procedure and their price schemes were designed in such a way that the return on these investments took place faster than free-market conditions. The main benefits of these contracts could have been realized through competition for the market approach and included performance criteria to decrease costs (Başaran & Bağdadioğlu, 2010), considering that the electricity generated and purchased within the scope of these contracts did not take place within the competitive part of the market (Tables 1, 2, and 3).

As seen in the graph below, even though the electricity demand fluctuated annually in Turkey between 2003 and 2018, the generation of electricity within the scope of BOO, BOT, and TOR contracts remained much more stable and their ratio decreased gradually based on the expiration of some of these contracts. Indeed, the most dramatic decrease of their ratio in generation became apparent in 2019 after these contracts expired. This shows that their operation continued outside the competitive market structure. Furthermore, after expiration dates, some of these important power plants with a significant capacity were shut down when EÜAŞ opted not to sign a new PPA with them. Although these contracts were utilized as an instrument to ensure security of supply, they were not designed and concluded meticulously enough that power purchase guarantees at high prices for a significant period for these plants caused a huge amount of stranded cost for all stakeholders in electricity industry. Considering that the main goal of electricity industry reforms is to form a market structure that ensures the transfer of investment risks from consumers to generators and market participants, these contracts have had a retardant effect on

Table 2 Expiration dates of BOT power plants

Power plant	Installed capacity (MW)	Fuel type	Expiration date
Trakya CCGT	478	Natural gas	2019
Uni-Mar CCGT	478	Natural gas	2019
Esenyurt CCGT	180	Natural gas	2019
Ova CCGT	253	Natural gas	2017
Birecik HPP	672	Hydro	2016
Others	325	–	–
Total	2386		

Source EÜAŞ annual reports and wholesale and retail electricity sector inquiry report of TCA

Table 3 Expiration dates of TOR power plants

Power plant	Installed capacity (MW)	Fuel type	Expiration date
Çayırhan TPP	620	Lignite	2020
Hazar I-II HPP	29,8	Hydro	2022
Total	649,8		

Source EÜAŞ annual reports and wholesale and retail electricity sector inquiry report of TCA

liberalization process of electricity industry in Turkey, placing the burden of commercial risks on the state and, therefore, consumers. Unfortunately, competition could not take place in these competition-for-the-market-type contracts. However, these contracts are considered as a starting point for the liberalization process in generation. Taking lessons from these non-competitive methods, a more fundamental and ultimate restructuring of electricity industry became necessary (Graph 1).

In addition to the abovementioned private sector participation cases in the electricity industry, there were conspicuous privatization attempts through transfer of ownership or TOR method. Firstly, in 1991, by Council of Ministers' Decisions No. 91/2325, concessions regarding generation, transmission, distribution, and trade of electricity were given for 30 years in the fourth, seventh, 11th, and 18th distribution regions, respectively. Nevertheless, these assignments were abolished by Council of Ministers' Decisions No. 96/8811 in 1996, although the 10th Department of Council of State later annulled this abolishment in order to comply with related assignments with Law No. 3096. On the other hand, the privatization attempt of TEK through sale of ownership by the Law No. 3974 was annulled by the Constitutional Court in 1994. The fact that electricity generation, transmission, and distribution activities

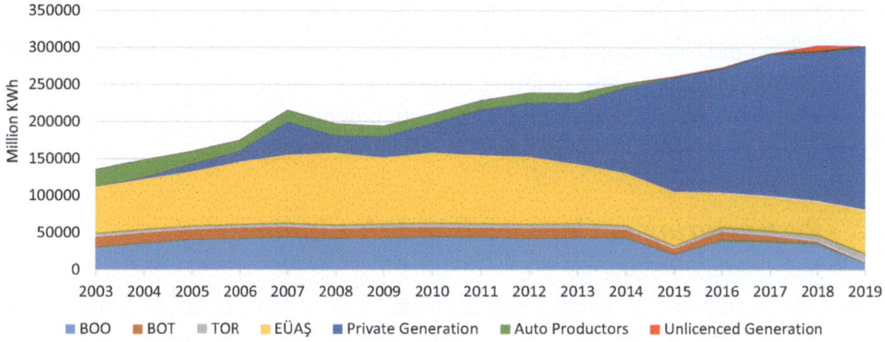

Graph 1 Distribution of electricity generation in Turkey by generation institutions (2003–2019). *Source* EÜAŞ annual reports and wholesale and retail electricity sector inquiry report of TCA

were considered as strategic public services closely related to country security, that no measures against monopolization were taken and ownership transfer of generation facilities making use of natural resources were found unconstitutional. The Constitutional Court also ruled that provisions regarding private sector involvement into electricity generation, transmission, and distribution activities were subject to public law rather than private law and had to be designed as concessions under the judicial review of the Council of State. Consequently, by Law No. 4046, High Council for Privatization, Directorate of Privatization Administration, and Privatization Fund were founded and "privatization" found place within constitution.

In 1996, the number of distribution regions increased to 29 as a result of changes to their territory. In addition to the four privatized distribution regions that had existed until then, it was declared that 25 additional distribution regions would be privatized for 30 years and related tenders were received until April 1997. Tenders received with respect to five regions were canceled, as they were found to be insufficient, and tenders with respect to other five regions were canceled due to disputes about assignment issues. The remaining 15 agreements were sent to the Council of State for approval; the Council nullified four of these agreements and approved the other 11. However, the privatization process regarding these agreements could not be concluded due to various disputes. By the enactment of Electricity Market Law No. 4628 (EML) in 2001, these agreements were labeled as "*existing agreements*" and it was planned to modify them until a certain date that would be determined by Energy Markets Regulation Authority[1] (EMRA). Last but not least, it would be useful to briefly mention the 1998-dated decision of the Turkish Competition Board (TCB), a conditional approval decision regarding distribution privatization attempts. The first condition of TCB was the removal of exclusivity clause within concession agreements, whereby distribution and retail operations would be separated and retail activities would be open to competition. Important conditions that the TCB laid down were that fixed

[1] EMRA was initially established as Electricity Market Regulatory Authority, but renamed after the enactment of Natural Gas Market Law No. 4646.

price application for power purchase agreements between TEDAŞ and distribution companies would be replaced with a free pricing mechanism within a predetermined upper and lower range, and the non-discriminatory treatment obligation to purchasers with equal status would be included within agreements.

4.2 Restructuring at Crossroads to Privatization

Despite its segmented structure, each segment of electricity industry constitutes a sufficiently large economic activity and thus a requirement for a holistic and coordinated legal and political framework arose because of the fact that each segment completes and effects each other significantly.

4.2.1 Market Structure and EML

After numerous diverse privatization cases and attempts, the enactment of the EML in 2001 and successive regulations and documents had been a milestone in the liberalization of electricity industry in Turkey. Discontentment with BOT and BOO models and their unprosperous effects had given rise to a search for more competitive model indeed (Atiyas et al., 2012). The EML brought an electricity market, the liberalization level of which was planned to increase within a certain program and timeline, particularly based on bilateral contracts and a balancing mechanism for real-time equality between demand and supply. One of the most important novelties that the EML brought was the establishment of regulatory authority, EMRA, with wide powers supported by financial and administrative autonomy. As mentioned above, public assets were legally unbundled into EÜAŞ, TEİAŞ, and TETAŞ. The EML envisaged that generation activities would be implemented by EÜAŞ and its affiliate companies, private generation companies, auto producers, and auto producer groups; transmission as a natural monopoly would be implemented by TEİAŞ under public ownership, subject to regulation. Also, EÜAŞ would sell all the electricity it generated to TETAŞ initially. Distribution activities would be implemented by TEDAŞ, its affiliate companies, and private distribution companies in regions specified within their licenses, whereas TETAŞ would take over all of TEDAŞ's and TEAŞ's energy sale-and-purchase agreements. In other words, TETAŞ also undertook energy purchase agreements within the scope of BOT, BOO, and TOR agreements. Among market activities, wholesale activities would be implemented by TETAŞ and private wholesale companies, and retail sale services would be assumed by distribution companies owning retail licenses besides retail companies. In order for an undertaking to conduct any of these activities in electricity industry, it had to obtain a license from EMRA and meet the required conditions in the legislation. According to the EML, the market model would be comprised of bilateral contracts and complementary balancing mechanism for real-time equilibrium between supply and demand, which would be operated by National Load Dispatch Center within TEİAŞ.

Another critical change brought by EML was gradual liberalization of the demand side through introducing "*eligible consumer*" notion; thereby, a market based on bilateral contracts was designed via opening both supply and demand side to competition. According to the EML, EMRA would determine and announce an eligible consumer threshold—that is, an annual consumption amount of electricity energy—every year no later than the end of January, and consumers exceeding that level would have the right to choose their suppliers. On this subject, the most striking provision in the EML was that non-eligible consumers were allowed to choose their supplier as well as eligible consumers. The supplier concept hereby referred to not only a distribution company with a retail license, but also EÜAŞ and its affiliate companies, private generators, auto producer or their groups, wholesale, or retail companies.

Eventually, electricity generation and distribution facilities would be privatized in accordance with the judicial review of the 1st Department of Council of State. With regard to the related review, it was not possible to privatize HPPs through transfer of ownership as these facilities cannot be dissociated from natural resources, TPPs could be privatized separate from natural resources they made use of through transfer of ownership, and it was not possible to privatize distribution facilities and concessions to implement distribution services through transfer of ownership—but rather—through TOR, as regulated within EML, for up to 49 years. The EML also constrained foreign ownership with respect to privatization in electricity market activities to such an extent that it was not allowed to have a market share that would provide control power in electricity generation, transmission, and distribution sectors.

4.2.2 Privatization Strategy Document and Road to Privatization

Despite its privatization provisions, EML did not involve a timetable for related privatizations. In 2004, in order to determine the implementation of privatization and liberalization process of the electricity sector, the High Planning Council endorsed "*Electricity Energy Sector Reform and Privatization Strategy Document*" (PSD) through Decision No. 2004/3. Compared to EML, PSD appeared to shift its focus away from privatization of distribution regions and toward expediting and facilitating this process (Turkish Competition Authority, 2015); thereby, PSD diverged from those principals regarding liberalization within the scope of EML in some aspects. Firstly, PSD envisaged that distribution companies are entitled to supply electricity exclusively to non-eligible consumers. Although this change decelerated demand-side liberalization, it was not withdrawal of eligible consumer concept, a gradual transition to liberalization instead, depending on an eligible consumer threshold. Considering the public good feature of electricity services and bounded rationality of consumers, a gradual liberalization in demand side of electricity sounds more promising. Second, the eligible consumer threshold would be pegged at 7.8 GWh until 2009 and 2011 were set as a target year by which the market would be completely open to competition. However, this plan was abandoned by a faster annual decrease that extended over a longer period, such as 480 MWh in 2009, 30 MWh in 2011, and 1400 KWh in 2020. As gradual liberalization refers to a constant yet extending over a longer period of decrease in eligible consumer threshold, this timeline seemed more coherent. Third, PSD also aimed to form "*Transition Period Contracts*" (TPCs), which included contracts signed between distribution companies and portfolio generation companies,[2] HPPs held by TETAŞ and EÜAŞ as well as TETAŞ, for supplying a minimum of 85% of the forecasted power load for non-eligible consumers. TPCs contained regulated prices and included a five-year transition period succeeded by bilateral contracts based on market prices. Thereby, a smooth transition to free market was aimed (Turkish Competition Authority, 2005). Fourth, contrary to EML, which envisaged regional tariffs and accordingly direct income support, the PSD stated that a scheme would be developed to equalize tariffs across all distribution regions. Notwithstanding that the EML aimed to design regional tariffs that reflected regional costs, thereby eliminating cross-subsidies between regions, the PSD put regional tariff implementation aside and focused on uniform regulated tariffs until the end of transition period. However, regional tariff implementation has not yet been put into practice. Last but not least, subsequent to the PSD, an amendment in the EML, through Law No. 5398 in 2005, rendered it possible for distribution companies to establish generation facilities without any restriction on the condition that they hold generation license and separate accounts of each company. Initially, the EML restricted the engagement of distribution companies in generation activities to the extent that annual electricity generated by the facilities owned by distribution undertaking could not exceed 20% of total electricity consumption in the relevant region. Unfortunately, this amendment implied a regress as it diverged significantly from a liberalization model based principally on unbundling of a vertically integrated structure, which started with the split of TEK into TEAŞ and TEDAŞ. Thus, PSD gave precedence to privatization rather than liberalization in the restructuring of the electricity industry.

Although efficiency gains of introducing effective competition in the long run were much higher vis-à-vis privatization, potential high revenues of privatization prevailed in both the restructuring and privatization processes. The PSD put the privatization of distribution facilities before that of generation facilities in its schedule on the grounds that it would provide generation and wholesale companies with reliable purchasers. This argument sounds reasonable considering that it would be more difficult to deal with loss/theft problem (Atiyas et al., 2012), to take commercial risks for operational efficiency, and accordingly to sign contracts with private generation companies under public ownership. However, alleged risks regarding management of distribution activities under public ownership would be avoided through efficient management in accordance with a strict course of action (Atiyas et al., 2012). After all, competition potential is the highest, inter alia, in generation activities, while it is limited in retail activities (Atiyas et al., 2012).

[2] These companies refer to generation plants that were planned to be privatized among all publicly owned plants transferred to EÜAŞ.

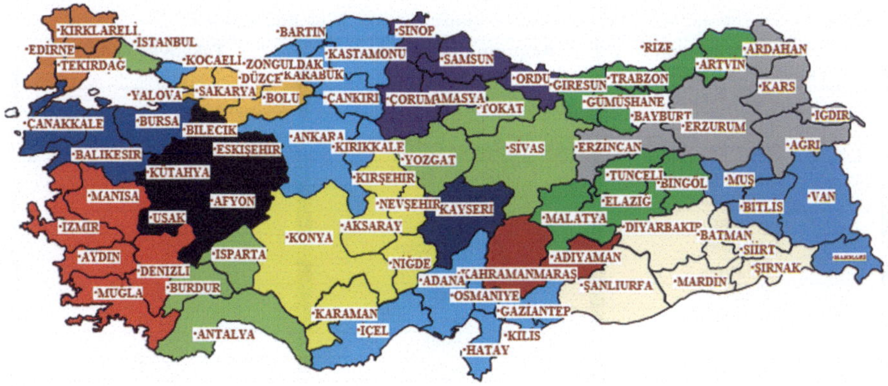

Fig. 1 Dispersion of distribution regions and related ownerships

In light of the Constitutional Court's decision and the judicial review of Council of State, the Directorate of Privatization Administration (ÖİB) planned to implement privatization of TEDAŞ through the TOR method. To this end, Turkey's electricity distribution system held by TEDAŞ was split into 21 regions, and as Kayseri A.Ş. (apart from Menderes and Göksu) was granted operating rights previously, the remaining 18 distribution regions would be privatized. Subsequently, corporatization was completed by March 2005 for every distribution region and TEDAŞ signed TOR contracts with each of these companies, followed by the granting of distribution and retail licenses by EMRA to each company. Then, distribution companies signed energy supply contracts with TETAŞ, which would be concluded by the sale of distribution companies' shares to investors. Even though the privatization of all distribution companies was intended to be completed by 2006, the process started in 2008 and it was not until 2012 that the whole process was finalized, with a significant time lag as can be seen in detail from Fig. 1 and Table 4.

4.2.3 An Apposite Touch from Turkish Competition Authority

It is necessary to mention the TCB's 2005-dated opinion[3] regarding privatization of distribution facilities. In its opinion, the TCB stated that a vertically integrated structure allowed within EML was likely to lead distribution companies to discriminate in favor of its own generation and retail companies and against rival companies besides cross-subsidization between its companies in vertical segments; that

[3] In Turkey, the TCB takes part in privatization processes in two phases. Firstly, in the pre-notification phase before the tender is announced, the TCB provides its opinion regarding the privatization and may stipulate some conditions in case it is likely to create a market structure that would significantly impede effective competition. Second, after the conclusion of tender process, the TCB's authorization is obligatory for the transaction to become lawful and the TCB, in this phase, may again stipulate some conditions to meet competitive concerns.

Table 4 Privatization dates of distribution regions and acquiring undertakings

Distribution Region	Acquiring Undertaking	Takeover Year
Kayseri ve Civarı	Kayseri A.Ş.	1990
Aydem	Bereket Grubu	2008
Başkent	Enerjisa	2009
Sakarya	Akcez OGG	2009
Meram	Alarko-Cengiz	2009
Osmangazi	Yıldızlar SSS Holding A.Ş.	2010
Uludağ	Limak	2010
Çamlıbel	Cengiz-Kolin	2010
Yeşilırmak	Çalık Enerji	2010
Çoruh	Aksa	2010
Fırat	Aksa	2010
Trakya	IC Holding	2011
Akedaş	Akedaş	2011
Boğaziçi	Cengiz-Kolin	2012
Akdeniz	Cengiz-Kolin	2012
Gediz	Bereket Grubu	2012
Aras	Kiler-Çalık Enerji	2012
Dicle	İşkaya-Doğu	2012
Vangölü	Türkerler	2012
İstanbul Anadolu Yakası	Enerjisa	2012
Toroslar	Enerjisa	2012

vertical integration would increase barriers to entry, particularly for non-vertically integrated suppliers, and exclude independent suppliers out of the market, which would deteriorate because of the increased market power of incumbent suppliers strengthened by supplier switching costs for consumers; and that the industry would be dominantly composed of a limited number of vertically integrated undertakings characterized with an inadequate competition. Even though the TCB expressed its reservations about vertical integration of distribution activities, not only with retail but also generation activities, it did not stipulate any condition with respect to unbundling of generation and distribution since it was already allowed within EML and the subject of pre-notification was privatization of distribution and retail activities. The TCB also noted that the course of action regarding distribution privatizations was shaped substantially in line with the aim of security of supply and encouragement of foreign investment, while adequate structural measures were not taken into consideration regarding the establishment of competition and protection of consumers. Consequently, the pre-notified privatization method was characterized as restraining competition in retail market, which was essential for lower prices. Moreover, the TCB drew attention to the unbundling of distribution and retail activities prior to privatization, so bidders could value related assets accurately and legal risks would be minimized, or, at least, the unbundling of distribution and retail activities was critical before the transition period ended. In conclusion, TCB identified ownership separation between distribution and retail activities as the most ideal and unproblematic method as it minimized barriers to entry into the market and stipulated legal separation as a second-best and minimum condition to be implemented preferably before tenders. If not, it would be implemented through imposing obligations on the acquiring company that it be completed by the end of transition period, that is, by the end of 2012. The opinion of the TCB had such an impact that the EML was amended in 2008 and it was envisaged that legal separation of distribution, generation, and retail activities would be accomplished by January 2013. Unfortunately, this amendment was not comprehensive and detailed enough to address legal separation issues effectively, which was verified by investigations by both EMRA and Turkish Competition Authority (TCA), resulting in fines and additional measures. Moreover, in its License Regulation EMRA introduced some exemptions until January 2016 with respect to legal separation, although the fact that distribution companies are subject to legal separation is clearly stated within the up-to-date Electricity Market Law No. 6446 that replaced the EML in 2013. The fact that distribution privatizations extended over a long time led to an asymmetry among distribution companies in terms of proceedings with legal separation. It is evident that lagging structural and behavioral measures inhibited competition and preserved market power in markets. The TCA conducted three preliminary investigations regarding market foreclosure in retail and distribution markets in 2014, but did not rush to open investigation and took some time, despite substantial and various solid evidences of anticompetitive conduct; it sent its opinion along with a long and detailed list of "don'ts" that were supposed to be terminated or avoided immediately. Nonetheless, after receiving numerous complaints, the TCA conducted three investigations regarding six large distribution regions and fined relevant undertakings for abuse of dominance, in both distribution and retail markets, and accordingly anticompetitive market foreclosure and prohibited a great number of conducts investigated.

Regarding distribution privatizations, TCB in its authorization decisions introduced a specific dominance threshold in order to prevent dominance all over Turkey based on the assessment that all of the distribution companies would be potential rivals to each other. In this respect, TCB prohibited those privatizations that would lead to the market share of a distribution company and its affiliates, all of which belong to the same undertaking, over 30% of total distribution sales. TCB stated that the ground for such a threshold was that it was necessary to ensure adequate number of potential rivals with competitive power in order to form a competitive retail market. Consequently, TCB approved most of the privatizations unconditionally and some of them that would lead to coordinated risk in terms of competitive conduct among rivals conditionally, and prohibited some of them for exceeding the dominance threshold.

On the other hand, the 2009-dated Electricity Energy Market and Supply Security Strategy Document (EMSSD), taking into consideration all developments since PSD apart from the blackout that occurred in 13 cities of Western Anatolia, stated that a transition period was necessary to form a competitive free electricity market and it aimed to carry out necessary amendments within legislation and take measures in order this period to avoid posing any risk to supply security. Thus, the EMSSD ushered in a new period in which competition in the wholesale market was to be developed (Turkish Competition Authority, 2015). A brief process of liberalization and privatization of electricity market is seen below (Fig. 2).

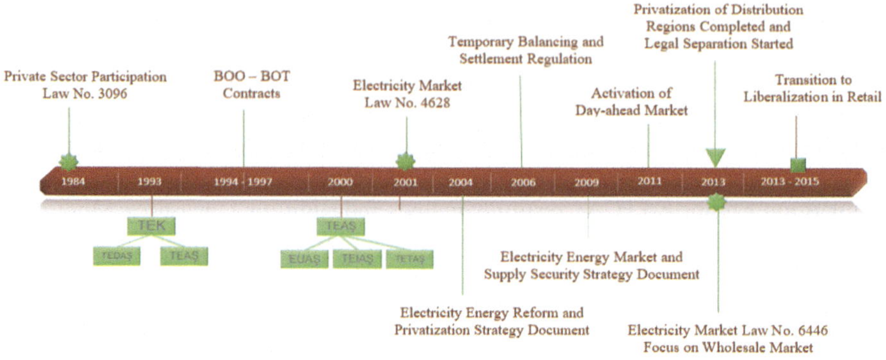

Fig. 2 Liberalization process of electricity market in Turkey. *Source* wholesale and retail electricity sector inquiry report of TCA

4.2.4 Developments in Generation Activities

The liberalization and privatization processes of generation activities took a prolonged path due to commitments from the past—particularly long-term BOO and BOT contracts—as well as the persistent significant role of public engagement and gradual investment of private sector depending on incentives and regulations based on energy policy. Firstly, PSD envisaged that publicly owned generation assets, 18 TPPs and 27 HPPs, would be grouped into portfolio generation companies (PGC) under EÜAŞ, except for some HPPs. It was intended that TPCs signed between PGC and distribution companies would prevent the creation of market power in the markets, ensure a gradual transition of public generation companies into the free market, and make them more attractive to investors in this period. After the expiration of TPCs, PGCs would be privatized. As mentioned above, since competition potential is highest in generation and limited in retail on the condition that it is shaped by an accurate restructuring and kept track by regulation, it is better to start privatization from generation and then proceed to distribution in order to make the liberalization process most fruitful. Nonetheless, Turkey opted for the reverse path, due to the reasons mentioned above. The fact that privatization of distribution companies took place in a way that allowed the companies to vertically integrate to both generation and retail has a restrictive effect on competition in these segments, as they contain the incumbent company that used to operate exclusively prior to privatization. However, this has a higher restrictive effect on retail markets due to switching costs and behavioral factors for consumers compared to generation. Unless a block of generation facilities, the sum of which amounts to a market power is transferred to the same private undertaking through privatization, the structure of generation segment will be fragmented and abuse of unilateral market power will hardly be a competition concern aside from anticompetitive collusive practices. As the graphs below show, the proportion of private generation has increased significantly since the beginning of restructuring and started to dominate generation activities after 2012, completion of distribution privatization. The main factors for this outcome are an increase in the new investments of private sector, individual privatizations of publicly owned generation facilities, as well as expiration of BOO and BOT contracts. The remarkable change of public versus private engagement in electricity generation is visible in the following graphs. While the proportions of private generation versus EÜAŞ were 8.62% and 48.17%, respectively, in 2006, they were 72.7% and almost 20%, respectively, in 2019. The proportion of private generation grew almost ninefold, while that of EÜAŞ more than halved. Additionally, the proportion of EÜAŞ's installed capacity decreased from 58.3% to 21.6% vis-à-vis and that of private generation increased from 6.87% to 67.56%. This implies a gradual yet constant increase in the participation of the private sector in generation activities. Another important point is that BOO and BOT contracts have recently expired, as clearly visible in the graphs below. This development has been very welcome since it means more room for competition has finally been available. This is because both BOO and BOT contracts were granted guarantee of purchase, thereby out of competition. This is clearly visible in the graphs below, as the proportion of BOO and BOT in electricity generation has been considerably higher than that in installed capacity. Privatization of generation facilities within the scope of BOT contracts will proceed along with PGCs so that they will be added up to private generation (Graph 2).

On the other hand, the relative position and function of EÜAŞ is another significant issue, despite its constantly decreasing share in generation. Even though the proportion of EÜAŞ in generation and installed capacity has steadily decreased, EÜAŞ continues to be the largest single generation entity. As the graphs below show, EÜAŞ was by far the largest generation entity among top 10 generation groups by 2019 and the rest of the market is fragmented. Since the market power or dominance concept in electricity industry, particularly in generation and wholesale segments, is distinct from any other goods or services markets due to its peculiar features, the market power/dominance threshold is likely to be much lower in this industry,

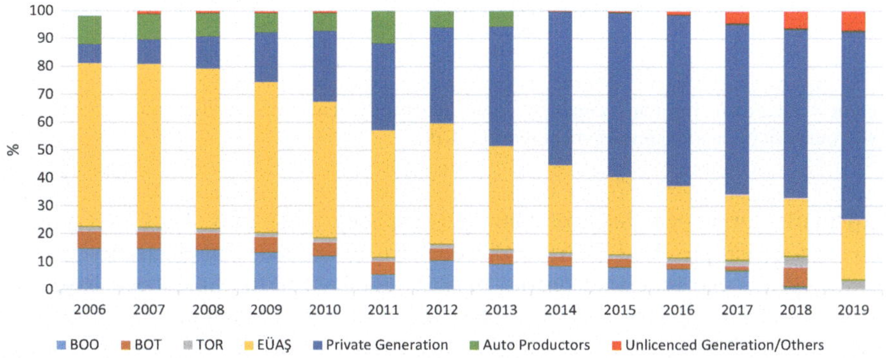

Graph 2 Proportion of generation institutions in total installed capacity in Turkey (2006–2019). *Source* EÜAŞ annual reports

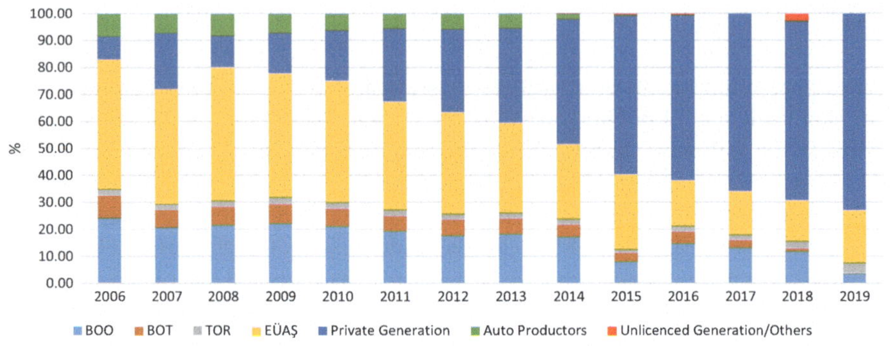

Graph 3 Proportion of generation institutions in total electricity generation in Turkey (2006–2019). *Source* EÜAŞ annual reports

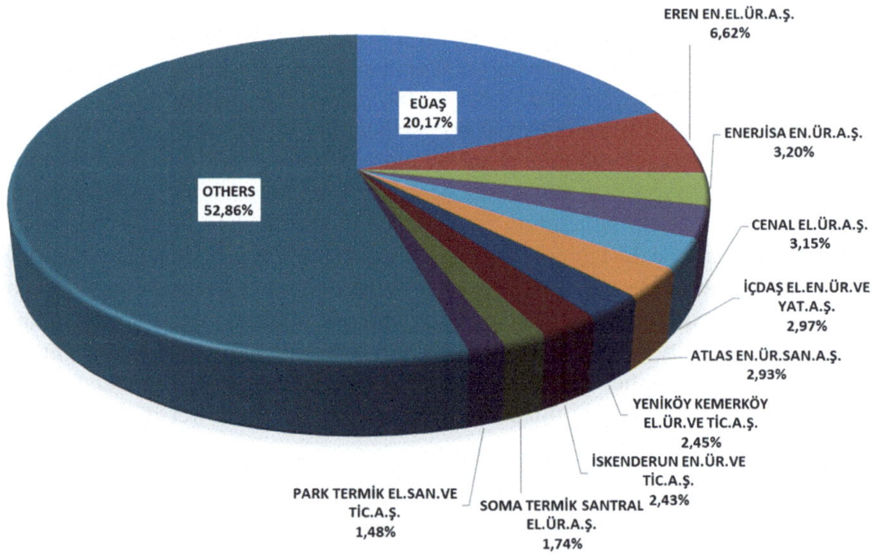

Graph 4 Shares of top ten companies within electricity generation in 2019 (percent). *Source* EMRA electricity market development report (2019)

particularly during peak loads or contingent capacity constraints. Thus, this generation capacity has provided EÜAŞ with a decisive effect on cost of electricity, market clearing price, and regulated tariffs. In fact, regulated tariffs have been based on the cost of electricity provided from TETAŞ and EÜAŞ; the high price of electricity purchased from BOO and BOT contracts meant that the cost of electricity from BOO and BOT contracts was to be blended with the cost of electricity provided from PGCs in order to balance the end cost of electricity. Furthermore, it can be inferred that EÜAŞ has made use of PGCs to avert deviations with respect to market clearing price and to help keep it within a plausible range in the wholesale market. This assigned "benevolent" task of EÜAŞ can be discerned from the lower proportion of EÜAŞ in generation vis-à-vis the higher proportion of EÜAŞ in installed capacity. In other words, contrary to BOO–BOT generation versus installed capacity asymmetry, which is in favor of generation, EÜAŞ did not seem to resort to engaging in generation as much as possible, but rather when "necessary" depending on predetermined goals. Considering that BOO and BOT contracts have recently expired, there will be less need for PGCs, so it is expected that more of remaining generation facilities will be privatized. However, at least some significant HPPs that are not among PGCs for the security of supply grounds will stay within EÜAŞ, and EÜAŞ may make most of these critical capacities as an emergency valve in case of contingencies to avert price spikes in the wholesale market.

The last noteworthy effect of privatization with respect to market structure is that the market parameters of two identical markets, in terms of the number of operating undertakings with the same capacity, would be different depending on whether private or public ownership is prevalent (Wolak, 2003). In this respect, the impact assessment on market structure of a privatization of a generation facility is likely to be misleading if the relevant market is used to host considerable publicly owned generation facilities, the operational goals of which are to ensure supply and system security, as well as interfering with price deviations and so forth rather than profit maximization (Turkish Competition Authority, 2015). In other words, market power issues that were lacking previously in the relevant market dominated by publicly owned facilities are likely to arise, particularly in those markets characterized by inadequate competition, after privatization. Therefore, although publicly owned facilities have been privatized individually in Turkey instead of as a large portfolio—which is less risky in terms of market power issues—competitive assessment of privatization of public generation facilities should henceforth be implemented thoroughly and concentration analyses of privatizations need to be evaluated based on more sensitive criteria rather than traditional, static ones to avoid abuse of market power issues arising.

4.3 Conclusion for Electricity

Turkey somehow started to pave the way for private sector participation into the electricity industry in the 1980s due to a financial impasse of public finance and fundamental economic policy change in the face of growing demand and related reinvestment needs. However, earlier models such as BOO and TOR contacts were not successful as they were not designed and put into effect meticulously along with the country risk, even though they could have been suitable competition for the market tools then. Since 2001, a fundamental electricity reform has been taking place within Turkey's policy agenda, and although these reforms have brought critical institutions and legal framework for the sake of healthy market structure, privatization has been a decisive motivation for restructuring and similarly revenue generation has been a main focal point for privatization. Although this reform meant a shift toward competition in the market, which necessitated a competitive market structure ex-ante, the order of privatization program and the subsequent strategy diverged from this goal. Privatization started with distribution-retail segment, which promised relatively limited potential for competition, and generation segment having highest potential for competition followed thereafter. Furthermore, the revenue maximization motivation of privatization, instead of growing competition, may not have attracted the most efficient investors, but did attract the most influential ones, particularly regarding distribution facilities (Atiyas et al., 2012). Nevertheless, short-term revenue maximization motivation is likely to cause high electricity prices for end-consumers in the medium/long term to cover privatization bids. This means that operation and investment risks remain within consumers instead of investors, as happened with BOO and BOT contracts. On the other hand, privatization is expected to decrease the loss/theft cost if controlled by effective regulation. Lastly, delays in privatization in either distribution or generation segment, regardless of the underlying reason, have postponed and even diminished efficiency gains in the industry.

5 Privatization of Telecommunications Industry in Turkey

In most countries, telephone services started being run by private companies on a local scale, without regulation and subject to market rules in mid-nineteenth century (Arıöz, 2005), whereas telegraph and postal services were under the control of states to, respectively, meet military-bureaucratic needs and to implement overseas communications as a result of development of global trade (Aygün, 2008). Although telegraph and postal services are part of the telecommunications industry, they are beyond the scope of this chapter because they are still under the control of state in Turkey and have already lost their importance thanks to superseding innovations in other segments of the industry.

5.1 Historical Background and Respective Changes

Before delving into the main privatization case, a concise historical background of the development and structural changes about related institutions besides legal framework of telecommunications industry in Turkey would provide a catch-up work regarding the industry situation.

The first telephone system in Turkey was established in İstanbul in 1911 by American Western Electric (through its affiliate company Dersaadet Telefon A.Ş.) according to a 30-year concession. Though it was confiscated by the state during WWI, it continued to provide telephone services until 1936 for İstanbul and periphery and was nationalized in 1936 in return for £800,000 (İçöz, 2003). After the foundation of Turkish Republic, Telegraph and Telephone Law No. 406 (TTL) came into effect in 1924 in order to regulate telegraph and telephone services operated by private companies until then; the Postal, Telegraph, Telephone Administration (PTT) was founded in 1936 and given exclusive rights to provide telecommunication services all over the country. The PTT was brought under the control of the Ministry of Transportation by Law No. 3613 in 1939 and restructured as public economic enterprise in 1953 by Law No. 6145. After abandonment of ISI policy and adaptation of neoliberal structural adjustment program, the government that came to power in 1983 concentrated investments on infrastructure while restraining public investments in manufacturing. Although Turkey had an underdeveloped telephone network with a narrow coverage area[4] by 1982, it was ranked first among European ITU members in terms of trunk line increases between 1982 and 1992; and also, internal financing was used for extending and developing telecommunications network during this period (İçöz, 2003).

5.2 Market Performance, Restructuring, and Legal Framework

The first segment of the telecommunications industry to be opened to competition in Turkey was the sale of telecommunications devices through the privatization of affiliate companies of PTT. However, the first fundamental step for restructuring was unbundling of telecommunication activities from PTT and corporatization of these activities through establishment of Türk Telekomünikasyon A.Ş. (TTAŞ) by Law No. 4000 in 1994. The same law authorized the Ministry of Transportation to grant licenses with respect to supply of pager, data network, smart network, Cable-TV, payphone, satellite systems, phone book, and similar value-added telecommunication services. Since increasing demand for various telecommunication services could not be adequately met, TTAŞ transferred a significant amount of these value-added services to private companies based on a revenue-sharing model (Ardıyok, 2001). Also in 1994, TTAŞ initiated GSM services (Pan-European Digital Mobile System—GSM 900) with two different private undertakings—Turkcell İletişim Hizmetleri A.Ş. (Turkcell) and Telsim Mobil Telekomünikasyon Hizmetleri A.Ş. (Telsim[5])—based on revenue-sharing agreements, although Turkcell and Telsim were individually granted license for GSM activities in 1998. On the other hand, attempts to privatize telecommunications activities started along with Law No. 4000 and took place several times due to annulment decisions of Constitutional Court. In fact, four attempts were made to form the legal basis for privatization of telecommunication activities, including Decree Law No. 509 in 1993. Constitutional Court annulled those provisions addressing privatization in related laws (Decree Law No. 509, Law

[4] The number of subscribers was 1.67 million, awaiting lines to be connected were 1.83 million, average waiting period was seven years, and more than 70% of rural area was not covered by the network then (Ardıyok, 2004).

[5] Currently owned and operated by Vodafone Group through its affiliate in Turkey Vodafone Telekomünikasyon A.Ş.

No. 4000 in 1994, Law No. 4107 in 1995, and Law No. 4161 in 1996), either totally or partially.

In accordance with TTL, which was subsequently amended by Law No. 4000, Law No. 4107, Law No. 4161, and related Constitutional Court decisions, shares of TTAŞ, up to 49%, would be privatized in such a way that 10% would be transferred, free of charge, to the Directorate-General of Postal Services, 5% would be transferred to company employees and small savers, and the remaining 34% would be sold to natural or legal persons. Evaluation of TTAŞ shares and implementation of privatization transactions, by operation of Law No. 4046, were assigned to Directorate of Privatization Administration (ÖİB), and approval of block sale was assigned to Council of Ministers. In 1998, the Council of Ministers decided to implement privatization through a block sale of 20% of TTAŞ shares to a strategic partner and public offerings for 14% of shares domestically and internationally. In 2000, an attempt to privatize 20% of TTAŞ shares to a strategic partner through block sale was unsuccessful as there was no bid received in the related tender. Thereafter, another tender took place with an increased proportion (33.5% of TTAŞ shares), for block sale, but the tender again could not be completed (İçöz, 2003).

The second fundamental reform for restructuring and liberalization in the telecommunications industry was the enactment of Law No. 4502 in January 2000, whereby TTAŞ was converted from a public economic enterprise to corporation. This law represented a turning point as, before the privatization process started, it established a sector-specific regulatory authority (the Telecommunications Authority (TA)[6]), then transferred regulatory powers to the TA, brought additional provisions regarding privatization of TTAŞ and universal service obligation, and envisaged that the legal monopoly of TTAŞ would be abolished in voice communication and infrastructure by 2004. Also in 2001, Law No. 4673 was enacted, which transferred power to grant licenses from Ministry of Transportation to TA, removed the restriction for the proportion of shares of TTAŞ to be privatized except "golden share,"[7] and accordingly rendered it possible to abolish all legal monopoly rights of TTAŞ even before December 31, 2003, on the condition that public shares of TTAŞ decreased below 50%, even though the legal monopoly rights of TTAŞ were actually abolished by January 1, 2004, and opened to competition.

Regarding GSM services markets, in 2000 İş-Tim Telekomünikasyon Hizmetleri A.Ş. (Aria—under the joint control of Türkiye İş Bankası A.Ş. and Tim International N.V.) and Aycell Haberleşme ve Pazarlama Hizmetleri A.Ş. (Aycell—affiliate company of TTAŞ) were individually granted third and fourth licenses as GSM operators after related tenders and entered into GSM services market six years after the entry of the first two operators. Turkey differs from other countries in this respect as the incumbent operator entered into GSM services market so much later that the first two entrants already gained highly significant customer base until then.

[6] The name of the authority changed to Information and Communication Technologies Authority (ICTA) in 2008 by the operation of Electronic Communications Law No. 5809.

[7] This refers to veto powers regarding national security, state of emergency issues, and some investment decisions above a certain amount.

Indeed, in 2003 Aria and Aycell merged into Avea İletişim Hizmetleri A.Ş. (Avea), which was under the joint control of Türkiye İş Bankası A.Ş. and TTAŞ, whose market shares in terms of subscriber numbers and turnover, respectively, were 13% and 4%.[8] Finally, in 2015, TTAŞ acquired sole control of Avea. In fact, the GSM services market, the growth potential of which had been initially underestimated, achieved rapid growth after 1998, when the revenue-sharing model was abandoned and the operating undertakings were granted concessions.

Lastly, the Cable-TV network (Cable-TV), which was the best alternative infrastructure to public switched telephone network (PSTN) at the time, was initiated as a pilot project in Ankara in 1989 and then tendered to private companies (participating companies) based on a revenue-sharing model (approximately 70% of revenue was allotted for TTAŞ and 30% for private companies that were also responsible for network investment and maintenance) in 1997 for nine big cities for 10 years. These agreements included an option to extend the contract duration by the end of the seventh year of the contract period. By 2004, there were approximately one million Cable-TV subscribers in Turkey, with a capacity of two and a half million subscriber, along with six operators, although the design of revenue-sharing model did not provide operators with adequate income for network investment. Although it was initially developed merely in order to provide uninterrupted and smooth television broadcasting, it became a true alternative for users in terms of broadband data transfer besides other value-added services thanks to technological advances (Arıöz, 2005) and made it possible to provide consumers with all services more cheaply and cost-effectively. As a potentially true alternative to PSTN after liberalization, as in many countries, Cable-TV was inadequate compared to PSTN then in Turkey. The underlying reason for this underdevelopment was, in both Turkey and other countries, the fact that the incumbent operator possessed Cable-TV as well as PSTN, thus did not attach necessary importance to Cable-TV, and did not allow the revenue-sharing model to function in such a way that encouraged participating companies to extend investment and coverage of network. The undervaluation of Cable-TV is actually an ipso facto result of the avoidance of product-service cannibalization by TTAŞ, which possessed two alternative infrastructures. However, Cable-TV then entailed 11 out of Turkey's 13 largest cities in terms of income share and covered one-third of the country's entire population. On the other hand, the legal monopoly did not include establishment and operation of Cable-TV since the 1994-dated amendment. Nevertheless, TTAŞ had a de facto monopoly on Cable-TV because neither the Ministry of Transportation before 2000 nor TA later granted any license to any other undertaking.

5.3 *Privatization of TTAŞ*

After setting the necessary legal framework, the Council of Ministries' Decision No. 2004/793 deemed that 55% of TTAŞ shares would be privatized through block sale

[8] TCB Decision No: 03–81/970–399.

method, although both public offering and block sale were options initially. Despite the 1998-dated valuation of TTAŞ shares, a new valuation was carried out due to the changing structure of TTAŞ and economic conditions. Finally, 55% of TTAŞ shares were transferred to Oger Telecom Joint Venture Group, which actually belonged to Saudi Oger Ltd. and Oger Telecoms LLC, in return for US$6.5 billion in 2004, and in 2008, 15% of TTAŞ shares that belonged to the Treasury went public (Malkoç, 2009).

5.3.1 Another Appropriate Intervention of Turkish Competition Authority

Depending on the legal framework, competition authorities play not only a significant role within the liberalization process in the post-privatization period, but also as a gatekeeper of market structure during privatization of network industries. As in electricity industry privatization, the TCB's critical involvement in pre-notification process had a decisive influence on material aspects of TTAŞ privatization and proved to be a good example of the fact that TCB pursued even minimal competition opportunities in the relevant markets such as tax issues which was totally another state policy subject (Temel, 2012).

The TCB remarked that privatization of TTAŞ as a whole would continue double monopoly, which was comprised of both PSTN and Cable-TV. The TCB pointed out that such privatization was contrary to the goal of establishment of competition between two alternative infrastructures and would lead to a market structure in which prices would not go down, product variety would be restricted, costs of rivals would be determined by the incumbent, and motivation for growing Cable-TV would be eliminated. The TCB highlighted that privatization should not be revenue-oriented, but instead be implemented in order to form a market structure that lent itself to liberalization. The TCB clearly stated that legal separation of Cable-TV and PSTN would not be sufficient to build a competitive market structure; even though liberalizing access to the core network was essential for building competition in retail activities, it was not adequate to reach targeted competition level in the market as rivals would still be dependent on infrastructure that was under the control of incumbent operator and its monopolistic conditions. Therefore, TCB enunciated that ownership unbundling of Cable-TV from TTAŞ was indisputable and should be transferred to an undertaking different from acquirer of TTAŞ in order to prove that liberalization was desired strategically, to form competition in access services to network, to forestall monopoly so far preserved, and to materialize the solid effects of liberalization.

Secondly, the TCB underlined the requisite for legal separation of wholesale and retail activities regarding Internet services on PSTN no later than six months after the transfer of TTAŞ in order to avoid cross-subsidization between wholesale and competitive retail activities and abuse of dominance through exclusionary conducts.

Lastly, taking into consideration of partial substitutability and imminent convergence between GSM and PSTN services, the TCB noted that in case the dominant undertaking in GSM services (Turkcell) won the tender, TCB would not approve that acquisition. The justification for this decision was that it would have otherwise led to double dominance in both PSTN and mobile services, strengthening dominance in GSM services market, and to substantial concentration in voice services and access to wholesale activities in GSM and PSTN markets.

As a result of CB's opinion, Cable-TV and related ownership and operation rights were excluded from TTAŞ privatization and Cable-TV was transferred to Türksat A.Ş. (Türksat), which had actually been established for the provision of satellite services by Law No. 5335 in 2005. Regarding legal separation of wholesale and retail Internet services, TTNET A.Ş. was established in 2006 as the Internet service provider of TTAŞ in retail Internet activities. Lastly, the owners of the first two winner bids did not belong to the undertaking controlling Turkcell.

5.3.2 Post-privatization Process and Assessment

Even though the liberalization process in telecommunications was initiated in 1994, noteworthy and credible efforts actually started in 2000, although the process was subsequently prolonged for more than four years. Privatizations can represent important opportunities to make up the leeway and may be used as a springboard to enable competition flourish in relevant sectors (Turkish Competition Authority, 2004). As a one-shot yet market-scale, fundamental, direct, ex-ante restructuring intervention with long-term effects, privatization is too valuable to sacrifice its substance and method for the sake of revenue maximization, particularly considering regulatory failure, regulatory capture, inaccurate policy, and incumbents' anticompetitive conduct risks thereafter.

Unfortunately, regarding the market performance of the sector, the development of Internet access services in Turkey fell far behind, and by 2004, it still could not catch up with the transformation that had started abroad at the end of the 1990s. Since necessary measures for service variety, innovation, low price, and high quality through vigorous competition between infrastructures had not been taken on time, a super-dominant undertaking with a 95% market share in retail Internet services was created, besides existing monopoly in the local loop. This was obvious from Cable-TV's high idle capacity vis-à-vis TTNET's super-dominance. Despite having a capacity of almost two and half million subscribers, Cable-TV provided only 40,000 customers with Internet services by 2004. The highly time-sensitive nature of telecommunications markets had already been proved when the last two entrants had struggled to penetrate into GSM services market and, for that very reason, had to merge. The importance of timely restructuring and regulation can also be observed in the fact that TTAŞ did not start to provide broadband Internet access services (ADSL) until 2005 and continued providing narrowband Internet services to maximize its revenue.

In fact, privatization of TTAŞ was implemented largely in accordance with TCB's opinion. Nonetheless, not all possible risks with respect to liberalization consisted of those reservations put forward by TCB. As clearly pointed out in TCB's opinion, although it had been legally possible to grant licenses for establishing and operating an alternative data network since 1994, it was not until 2006 that TK finally granted licenses to participating companies of Cable-TV[9] and later other undertakings. More importantly, Cable-TV has not been privatized yet and has been operated by Türksat since 2006. However, it seemed that TCB assumed privatization of Cable-TV soon after that of TTAŞ. This policy was not consistent with either liberalization policy or grounds for TTAŞ privatization, or the 2004 government program stating that competition in telecommunications would be developed and a market structure hosting alternative service providers for users would be ensured. Aside from the persisting inertia of Cable-TV privatization, Cable-TV has never posed a threat to TTAŞ, particularly for fixed broadband Internet services since then. Türksat has preserved the network, but no significant investment and aggressive operation have been made to create a true rival against TTAŞ and it has been continuously underutilized. From 2006 to 2019, Cable-TV (for television broadcasting) subscription remained almost the same (at around 1.25 million), while Cable-TV Internet subscriptions increased from 40,000 to almost one million (ICTA, 2019; Telkoder, 2016). As Graph 5 shows, GSM services and mobile Internet turned out to be a significant segment of the industry due to convergence and great progress in information technologies, and it outnumbered fixed Internet services significantly in terms of subscriber numbers. The telecommunications industry stands out among other network industries in terms of technological progress, innovation potential, variety of value-added services, as well as the substantiality of network effects and development of competition (Laffont & Tirole, 2001). Compared to other capital-intensive network industries, telecommunications, particularly for the last three decades, have made groundbreaking progress, the impacts of which on market performances may have been confounded with outcomes of related policies such as privatization, liberalization, or regulation. This applies mostly when the implementation of such policies coincides with the emergence of breakthroughs in the relevant industry, just as it happened in Turkey. In this context, one should not confound competitive pressure on the incumbent operator stemming from GSM services or other alternative infrastructures with the impacts of privatization, regulation, and liberalization policies in Turkey. In other words, if there had not been any leapfrogging in the telecommunications industry, the competitive picture of telecommunications would have been worse than it is today and in favor of the incumbent operator.

On the other hand, by 2019 the monthly average usage per capita for fixed broadband was 113.6 GB, while it was only 5.9 GB for mobile Internet (ICTA, 2019). The importance of data services is clearly seen within mobile services as well. While in 2008 only 1.3% of total revenue came from data services for mobile operators, this proportion rose to 73.8% in 2019 (ICTA, 2020, 2019). Considering the tremendous growth in data services vis-à-vis other telecommunication services and the

[9] However, a disagreement arose between Türksat and participating companies regarding the ownership of Cable-TV and each side asserted that it had ownership, as Article 25 (which related to the issue of revenue sharing agreements) was controversial about what would happen to the ownership of Cable-TV facilities when participating companies were granted licenses.

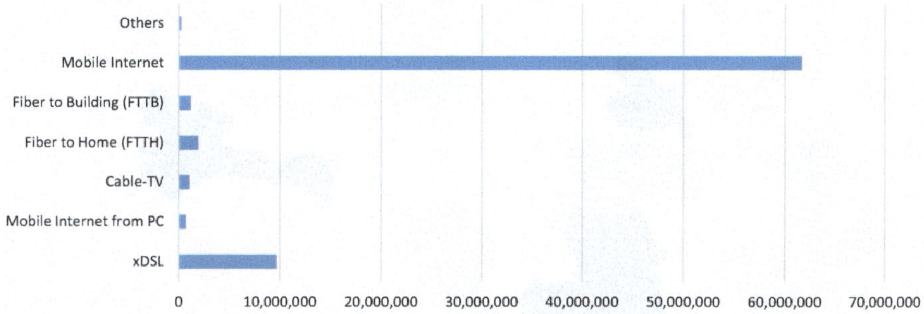

Graph 5 Market shares of Internet service providers in terms of subscriber numbers by 2019/4. *Source* ICTA Turkey electronic communication sector quarterly market data report (2019/4)

still comparatively limited capacity of mobile Internet services, fixed broadband Internet services represent a massive segment of the industry in terms of data usage. As Graph 6 shows, TTAŞ has been preserving its leading position in fixed broadband Internet services while Cable-TV kept its relatively insignificant share in the market. This was not the competition between two network infrastructures that TCB anticipated, as it had predicted immediate Cable-TV privatization.

Currently, three undertakings possess telecommunications infrastructures and supply wholesale access services, as well as retail ones in Turkey: TTAŞ (ISDN, xDSL, and fiber optic), Turkcell Superonline (fiber optic), and Türksat (Cable-TV). In Fig. 3, only Turkcell Superonline operates in the yellow-painted cities, only Türksat operates in blue-painted cities, and both operate in green-painted cities, while TTAŞ operates in every city. Also, as shown in Graph 6, despite developing a rival fiber-optic network besides Cable-TV and resale companies, the incumbent operator still dominates the fixed broadband Internet services market.

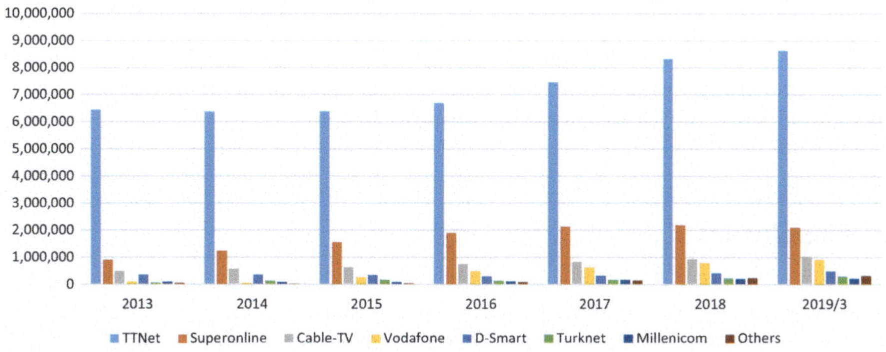

Graph 6 Distribution of Internet subscribers among Internet service providers (2013–2019). *Source* ICTA Turkey electronic communication sector quarterly market data report (2019/4) and https://btdunyasi.net/turkiyenin-internet-gercegi/

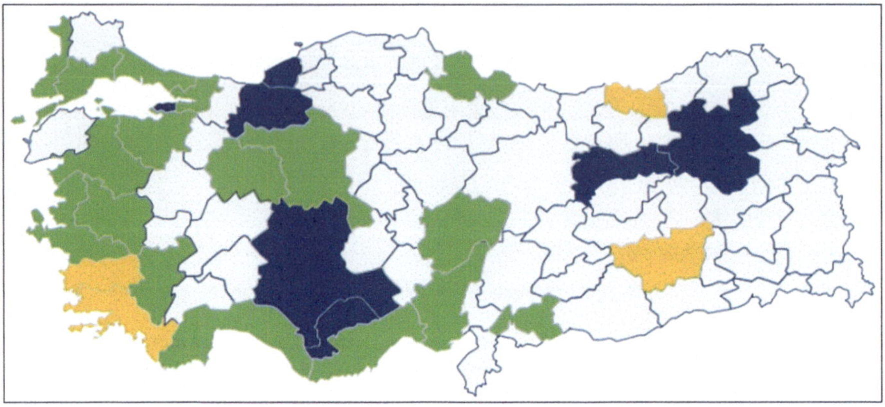

Fig. 3 Distribution of Turkcell and Cable-TV in Turkey (2020). *Source* https://btdunyasi.net/turkiyenin-internet-gercegi/

5.4 Conclusion for Telecommunications Industry

The TTAŞ privatization case proves how a proper privatization could be interrupted through follow-up regulatory failures as well as inconsistent policy choices and anti-competitive abusive conducts of the incumbent operator. Aside from the de facto time lag in relation to any policy put into effect, delayed license grants from regulatory authority in various segments of telecommunications industry after necessary legal framework set inhibited liberalization, despite its expected supporting role. Just as competition law is not about protecting competitors, but competition itself, privatization is not about bestowing (direct/indirect/unspoken) privileges upon privatized incumbent operator, but about developing and encouraging liberalization in the market. Competition potential is highest when there are alternative networks and lower when it is about access to the only network. The Cable-TV example, besides delays about other license grants with respect to infrastructure management and related legal framework, all provide unfair advantages in favor of the incumbent operator, albeit at the ultimate cost of consumer benefit.

6 Conclusion

In an economic environment that carries risks of market failure, regulatory failure, regulatory capture, mistaken or unpunctual policies, as well as anticompetitive conduct, privatization is an irreversible, one-shot restructuring opportunity. However, it still needs to be preceded/accompanied/succeeded by accurate regulation and timely economic policies that continuously pay attention to and respond immediately to what each stakeholder in the industry points out and requests. Liberalization is like raising a child: Doing a few things wrong may very well offset the impacts of doing many things right; privatization, if it is implemented as it should be, might undo some past mistakes or neglects of either regulation or policy choices on condition that all lessons are taken from bitter experiences.

Among all of the grounds for privatization of network industries, some are not necessarily totally escapable simply by implementing privatization. The political commodity feature of goods and services provided over network industries, accompanied by a lack of predictable long-term government policies, keeps some serious risks of public ownership over network industries alive by means of regulatory failure and inaccurate or tardy regulations, as well as biased government policies. Considering liberalization as the ultimate aim, relevant authorities and policy makers are responsible for hearing each stakeholder of the industry equally or even fairly for the sake of building competition whenever possible and designing regulation of the core network, which is not open to competition, as a leverage to develop competition in competitive segments. Today, hearing and responding stakeholders other than the incumbent operator, such as new entrants, consumer groups, and competition authorities, is likely to mitigate restrictive effects of countries' institutional endowments with respect to privatization and liberalization. On the other hand, country-specific situations are non-negligible facts that need to be addressed realistically. For instance, regarding the electricity industry, Turkey had idiosyncratic constraints like stranded costs arising from past commitments with long-term effects and they have been plausible justification of significant public sector engagement within industry until recently. However, such extensions of temporary public engagements should not become habits (even if well-meaning or paternalistic) in the form of market performances of remaining public enterprises or biased regulations. Therefore, on the one hand, implementing privatization of important public enterprise/s in a segment of the industry and, on the other hand, keeping significant public engagement that have decisive effects on market outcomes, even after expiration of certain constraints, contradict and counteract each other in the context of liberalization.

Continuing from the electricity case, it has been shown in many countries that ownership unbundling of the core network from competitive retail activities could have provided more permanent, inexpensive, and efficient results compared to legal separation as envisaged by competition authorities. Nevertheless, the Cable-TV case in Turkey as an unbundling of ownership example in telecommunications has also proved that mere unbundling of ownership is necessary but not sufficient and, additionally, must be privatized as a significantly competitive value to a viable third party in a way that best helps to serve liberalization. This third party needs to be as equally or more efficient than the incumbent operator. Both examples have provided some important takeaways. Firstly, in highly time-sensitive markets like network industries, never put off until tomorrow what you can do today. In other words, as restructuring, privatization, and regulation complete and interact strongly with each other, they cause long-lasting externalities that ultimately land on the shoulders of consumers. Therefore, it is in the public's best interest not to pass on privatization's responsibility to regulation or expected public policies and vice versa. Second, privatization, as a crucial restructuring tool, is a whole process and should not be left incomplete.

Liberalization requires a holistic approach and should be concluded sector-wide with minimal constraints. In fact, restructuring is best accomplished before privatization so that not only bidders but also every other stakeholder in the industry can clearly take into account market facts and legal framework, thereby calculate the true risks of the market, and accordingly take suitable action. In this respect, despite its benevolent intention, public engagement as a significant market player in the industry after privatization takes place is contradictory to liberalization, since the motivations of public enterprises are different from those of market-oriented ones. Thus, their performances and effects on market outcome are unpredictable, which makes privatization policy questionable. Similarly, regulation subsequent to privatization needs to serve only liberalization and is not be used as an indirect public engagement in markets. Fourth, network industries are comprised of various segments, some of which are vertically related and others are substitutable to each other. No matter what the relation in between is, the order and starting point of privatization process may have considerable effects on benefits of liberalization in such industries. For instance, whether to start privatization from generation activities or distribution retail activities in electricity industry is likely to affect the proportion of vertically integrated undertakings in the market, as well as the cost of electricity and level of regulated tariffs. Privatization needs to start from where potential efficiency gains are the highest and continue to the one with lower potential. Therefore, a decisive privatization policy should break the persisting reluctance that stems from institutional endowment, economic background, political habits, and the tendency for political capitalization of network industries in the relevant country, to yield network industries to liberalization. Privatization is meant to be the retirement of public engagement as a market player that needs to be succeeded by trust in markets firstly and then monitoring, regulating, incentivizing, and deterrent mechanisms, all of which contribute to liberalization. Eventually, it is difficult to make a healthy evaluation regarding unfinished privatization cases, as in Turkey. Just like leapfrog innovations and alternative investments of rivals, both privatization and regulation have a mission to leave the field clear for the market, and policy makers are responsible for not rendering privatization as the weak link in the liberalization chain.

References

Ardıyok (2001). Türk telekomünikasyon A.Ş.'nin özelleştirilmesi: doğum sancilari. Rekabet Dergisi, https://www.rekabet.gov.tr/Dosya/dergi/5-pdf Accessed November 9, 2020.

Ardıyok (2004). Yerel telekomünikasyon hizmetlerinde rekabet. Rekabet kurumu, Ankara https://www.rekabet.gov.tr/Dosya/uzmanlik-tezleri/48-pdf Accessed November 4, 2020.

Arıöz A. (2005). "Telekomünikasyon sektöründe serbestleşme süreci", rekabet kurumu, Ankara https://www.rekabet.gov.tr/Dosya/uzmanlik-tezleri/69-pdf Accessed November 3, 2020.

Armstrong, M., Cowan, S., & Vickers, J. (1994). *Regulatory reform: Economic analysis and British experience*. The Massachusetts Institute of Technology Press.

Atiyas, İ, Çetin, T., & Gülen, G. (2012). *Reforming Turkish energy markets political economy, regulation and competition in the search for energy policy*. Springer.

Aygün, E. (2008). Posta hizmetlerinde serbestleşme süreci ve rekabet hukuku uygulamaları. Rekabet Kurumu, Ankara https://www.rekabet.gov.tr/Dosya/uzmanlik-tezleri/48-pdf Accessed November 2, 2020.

Bahar, O. (2005). Türkiye ekonomisinde 1980 öncesi uygulanan sanayileşme politikaları. *Balıkesir Üniversitesi Bandırma İktisadi ve İdari Bilimler Fakültesi Akademik Fener Dergisi, 2*(4), 64–90. https://dergipark.org.tr/tr/download/article-file/203556 Accessed September 20, 2020.

Başaran, A. A., & Bağdadioğlu, N. (2010). *Türkiye Elektrik Sektöründe Reform Süreci, Regülasyon ve Rekabet Politikası.* Sosyoekonomi/2010-EN, https://dergipark.org.tr/en/download/article-file/197673 Accessed October 20, 2020.

Boratav, K. (2003). *Türkiye İktisat Tarihi 1908–2002.* İmge Kitabevi, 7. Baskı, Ankara.

Economides, N. (2004). Competition policy in network industries: An introduction. https://papers.ssrn.com/sol3/papers.cfm?abstract_id=386626, Accessed October 2, 2020.

Ergan, N. G. (1998). Etatism in Turkey. *Hacettepe Üniversitesi Edebiyat Fakültesi Dergisi 15*(2), 49–57. https://dergipark.org.tr/en/download/article-file/610929 Accessed September 5, 2020.

Erol, E. (2007). *Türkiye'de Elektrik Enerjisinin Tarihi Gelişimi: 1902–2000.* Doctoral Dissertation, İstanbul Üniversitesi Sosyal Bilimler Enstitüsü, İstanbul, http://nek.istanbul.edu.tr:4444/ekos/TEZ/42928.pdf Accessed September 12, 2019.

İçöz, Ö. (2003). *Telekomunikayon Sektöründe Regülasyon ve Rekabet.* Rekabet Kurumu, Ankara https://www.rekabet.gov.tr/Dosya/uzmanlik-tezleri/16-pdf Accessed November 12, 2019.

Information and Communication Technologies Authority. (2019). *Türkiye Elektronik Haberleşme Sektörü Üç Aylık Pazar Verileri Raporu.* Ankara, https://www.btk.gov.tr/uploads/pages/pazar-verileri/4-ceyrekraporu-2019.pdf#page=51&zoom=100,0,0 Accessed December 10, 2019.

Information and Communication Technologies Authority. (2020). *Türkiye Elektronik Haberleşme Sektörü Üç Aylık Pazar Verileri Raporu.* Ankara https://www.btk.gov.tr/uploads/pages/pazar-verileri/turkiye-haberlesme-raporu-002.pdf Accessed December 10, 2019.

Laffont, J-J., & Tirole, J. (2001). *Competition in telecommunications.* Munich lectures in economics. The Massachusetts Institute of Technology Press.

Malkoç, E. (2009). *Özelleştirme ve Türk Telekom Örneği.* Doctoral Dissertation, Kadir Has Üniversitesi Sosyal Bilimler Enstitüsü, İstanbul. http://academicrepository.khas.edu.tr/bitstream/handle/20.500.12469/1865/0050068ErdalMalkoc.pdf?sequence=1&isAllowed=y Accessed November 20, 2019.

Newberry, D. M. (2001). *Privatization, restructuring and regulation of network utilities.* The Massachusetts Institute of Technology Press.

Özata. (2013). Macroeconomic impacts of privatization: The case of Turkey. *Journal of Business, Economics and Finance* ISSN: 2146–7943, *2*(4), https://dergipark.org.tr/en/download/article-file/374582 Accessed October 28, 2019.

Özer. (2012). *Yap-İşlet-Devret Modeli ve Türkiye Uygulamaları.* Ziraat Mühendisliği Ocak-Haziran 2012 Sayı: 358, Ankara. https://dergipark.org.tr/en/download/article-file/946587 Accessed October 20, 2019.

Telkoder. (2016). Türkiye'nin İletişimde İkinci Baharı—Kablo TV Şebekesi. https://telkoder.org.tr/wp-content/uploads/2017/06/T%C3%BCrkiye%E2%80%99nin-%C4%B0leti%C5%9Fimde-%C4%B0kinci-Bahar%C4%B1.pdf Accessed December 15, 2020.

Temel, E. (2012). *Özelleştirme Uygulamaları ve Rekabet Politikası.* Rekabet Kurumu, Ankara https://www.rekabet.gov.tr/Dosya/uzmanlik-tezleri/134-pdf Accessed November 12, 2019.

Timur, T. (1997). *Türk Devrimi ve Sonrası.* İmge Kitabevi, 4. Baskı, Ankara.

Turkish Competition Authority. (2004). *Türk Telekomünikasyon A.Ş.'nin Özelleştirilmesine İlişkin Rekabet Kurulu Görüşü.* Ankara.

Turkish Competition Authority. (2005). *Türkiye Elektrik Dağıtım A.Ş.'nin Özelleştirilmesine İlişkin Rekabet Kurulu Görüşü.* Ankara.

Turkish Competition Authority. (2015). *Elektrik Toptan Satış ve Perakende Satış Sektör Araştırması.* Ankara.

Wolak, F. A. (2003). *Designing competitive wholesale markets for Latin American Countries,* CNE, Spain.

Remzi Özge Aritürk is a senior competition expert at the Turkish Competition Authority. He holds a BBA from Ankara University and an MPA from Columbia University in the city of New York concentrated in Global Energy Management and Policy. He also completed an Annual Training on Regulation in Energy Utilities in Florence School of Regulation in European University Institute. He has been working as a case handler for antitrust cases and conducting merger analyses in energy, telecommunications, IT, media, entertainment, tourism, professional activities, cement, construction, metals, glass, and health services, and contributing to the preparation of secondary legislation of competition law since 2005. His recent studies mainly focus on energy and climate change policies, behavioral economics, and zero-price markets.

Public–Private Partnerships in Turkey

Uğur Emek

Abstract For three decades now, governments across the world have enabled the private sector to deliver public infrastructure services through public–private partnerships (PPPs). A PPP is a long-term contract between a contracting agency and a private party for the delivery of public services, in which the private party undertakes considerable risk by bundling construction and operation contracts into a single PPP contract. Remuneration is linked to performance measures prespecified in the contract. This chapter investigates PPP enabling rules, procedures and institutions, and practices in Turkey. The Turkish Government has aggressively appealed to PPPs, especially in the wake of the Global Financial Crisis of 2008, because central banks in advanced economies printed large amount of money to help firms in difficulty, and this money flowed to developing countries such as Turkey as inexpensive and easy finance. While the renovation of aged facilities and the making of new investments were postponed due to unavailability of public funds through PPPs, authorities have largely neglected general standards of good governance, which require the establishment of enabling institutions, rules, and procedures that surround PPPs. This chapter suggests that PPPs may meet the country's growing needs for *improved infrastructure.* However, Turkey's experience suggests that unless convenient enabling rules, procedures, and methodologies exist, PPPs may result in unexpected outcomes harming social welfare.

Keywords Public-private partnership · BOTs · Contracts · Privatization

1 Introduction

For several years now, decision-makers in Turkey have promoted public–private partnerships (PPP) for the delivery of infrastructure services. Infrastructure services such as energy and transportation contribute to the reduction of production costs, the cheaper delivery of the products produced to the target markets, and the increased mobility of employees, especially due to urban transportation services. Education

U. Emek (✉)
Başkent University, Ankara, Turkey

M. Eroğlu and M. Finger (eds.), *The Regulation of Turkish Network Industries*,
https://doi.org/10.1007/978-3-030-81720-6_22

and health services help improve human capital. Improving human capital, lowering production costs, and providing workers with easy access to workplaces will increase the competitiveness of an economy and social welfare. According to projections made by international organizations (e.g., Abadie, 2014: 2; McKinsey & Company, 2013: 3), there is a need for trillions of dollars of infrastructure investment worldwide. Global Infrastructure Hub,[1] a G-20 initiative, provides data on the infrastructure that 49 countries need in order to maintain their potential growth rates by taking into account the size of past infrastructure investments and the quality of existing investments. According to this dataset, Turkey requires infrastructure investment with a value of US$975 billion (based on 2015 prices and exchange rates). Under current trends, governments may finance investment expenditure with a value of $569 billion and still needs additional finance for the remaining $405 billion.

Due to this considerable financing gap, Turkey—like most developing countries—refers to PPPs for the provision of the infrastructure investments it requires. PPPs differ among themselves according to risk sharing mechanisms, financing methods, and fee payment mechanisms. They are based on long-term contracts signed between the contracting agency responsible for providing infrastructure services and a private company or consortium. Within the scope of these contracts, new investments (greenfield investments) can be made, and the operating rights of existing public facilities may be transferred to the private party (brownfield investments). The private sector also enters contracts with financial institutions for project finance, subcontractors for the delivery of certain services, and insurance companies in order to ensure the provision of the relevant service. Taxpayers and/or service users pay a service fee. Within this framework, PPP contracts appear to have three stakeholders: (a) contracting agencies, responsible for providing infrastructure services; (b) private sector (investors and financial institutions); and (c) final beneficiaries (service users).

Many governments across the world use PPP procedures for practical reasons. They attempt to rapidly deliver necessary investments, which are difficult to finance through available public funds, transfer significant part of the project risks to the private sector, and benefit from the technology and capabilities of the private party (Grimsey & Lewis, 2004). According to the generally accepted principles by OECD countries, a government should appeal to PPP methods to the extent that they provide value for public money (OECD, 2008). The life-cycle cost of a PPP project should be lower than traditional public procurement[2] within the framework of this approach. Oliver Hart (2003), who received the Nobel Prize for Economics in the field of incomplete contract theory in 2016, stated that a PPP project can perform better when service delivery standards can be identified more easily than the construction standards. In such projects, the private party will design and build the relevant facility by considering all expenses that may arise in the life cycle of the project. By

[1] Retrieved June 18, 2020 from https://www.gihub.org/countries/turkey/.

[2] The traditional procurement route involves separating design from construction. Contractors are invited to submit tenders for the construction of the project. Contracting agencies would then design, finance, and operate the facility built by the contractor.

linking the payments to service delivery performance through reward and penalty mechanisms, the value for money (VfM) may increase in these projects.

Although many governments have embraced PPPs, they have also attracted considerable criticism regarding the efficiency and sufficiency of PPPs in the service delivery. Successful implementation of PPP projects is not easy. Such projects are attractive for governments from the perspective of political economy, because authorities may deliver infrastructure investments by avoiding the control of gatekeepers in the budgetary process and by putting the liabilities arising from these projects off-balance sheet. Nevertheless, end users who are not involved in drafting of a PPP contract pay the cost of a relevant project. A note prepared for the World Bank (de la Torre and Rudolph, 2015) identified the various risks arising from PPP practices. Unless they are used carefully, PPP contracts can result in a greater capital expenditure than traditional public procurement. Moreover, regardless of whether PPP or traditional procurement is chosen, large-scale mega investment projects may create numerous risks in terms of technical, financial, managerial, environmental, social, administrative, and legal aspects (Flyvbjerg, 2014).

In order for PPP projects to contribute to social welfare, well-designed rules and procedures should be in place. In this framework, a detailed road map should be identified, a comprehensive regulation should be put in force, a central unit consisting of competent experts should be established, and the framework and content of PPP project cycle management should be well defined. A well-designed PPP contract must be compatible with general standards of good governance, which require the enabling institutions, rules, and procedures surrounding PPPs to be put in place. Such a set of policies would ensure efficiency in the service delivery and reduce the government's willingness to provide overly generous guarantees. Thus, the interests of taxpayers and service users, who are the final payers of services delivered, will be pursued.

This chapter starts by explaining PPPs and their (dis)advantages and then reviews PPP contracts in Turkey according to sectors, types, and years. I subsequently evaluate institutional and legal structure in the country. Finally, I conclude that enabling institutions, rules, and procedures surrounding PPPs in Turkey remain immature despite the delivery of significant number of PPP contracts. Therefore, government and parliament still have a lot of work to do, such as writing and using PPP contracts effectively and efficiently.

2 Conceptual Framework of PPPs

This section first attempts to describe what a PPP means and subsequently identifies the advantages and risks involved in PPP contracts in light of available literature and international best practices. It is expected that these observations would serve as a guide for decision makers and practitioners in the field.

2.1 Definition of PPPs

In international practices, private participation in infrastructure in the form of build–operate–transfer (BOT), design–build–operate–transfer (DBOT), build–lease–transfer (BLT), transfer of operating rights (ToOR), lease, concession, and joint venture are usually named as PPPs (Grimsey & Lewis, 2004, pp. 52–55; Yescombe, 2007, pp. 1–14). These procedures differ in terms of the risks and returns undertaken by the parties identified by contracts. There is no generally accepted definition on PPPs. The PPP Knowledge Lab (2018) defined a PPP as "a long-term contract between a private party and a government entity, for providing a public asset or service, in which the private party bears significant risk and management responsibility, and remuneration is linked to performance." PPPs typically do not include service contracts or turnkey construction contracts, which are categorized as public procurement projects or the privatization of utilities where there is a limited ongoing role for the public sector.

From this definition, it is still likely to specify the following ingredients that are generally included in PPP contracts:

- Governments enable the private sector to deliver public assets or services.
- The contract period is long (e.g., 25–30 years).
- The private party elected sets up a new company (special-purpose vehicle, or SPV) to provide the public service in question.
- Governments regulate and supervise the delivery of service in terms of standard, quality, and quantity.
- The SPV carries out fixed capital investment and the operation of the facility together by bundling construction and management contracts (such as BOT or BLT).
- The private party may also take over and renovate existing public facilities (e.g., ToOR).
- At the end of the contract period, the SPV may transfer the ownership of the facility to government entity in return for a predetermined fee (such as BOT) or no charge.
- The government may pay the fee for the service itself, as the end user, or enable the SPV to collect the fee from users, or government and service users may pay jointly.
- The government may grant specific guarantees to the private party and retain certain risks arising from a PPP contract in order to improve total value of the contract.

2.2 *Benefit*[3]

- The VoM may increase in the long run by transferring certain risks to the SPV. Within the risk transfer scheme, the whole-life cost of project, including the design, investment, operation, and maintenance, is taken into account due to the bundling.
- The VoM is likely to benefit from the technological know-how and other innovative solutions of the private sector, such as efficiently supplying a public asset or service.
- Private financing is mobilized in addition to the public funds that fail to meet growing demand in infrastructure services.
- The private party may complete the necessary infrastructure investments on time and budget without sacrificing quality.
- Business opportunities may be emerged for small entrepreneurs as subcontractors in areas such as construction works, electrical works, security services, cleaning services, and maintenance–repair services.
- The capabilities and skills of domestic firms, especially small- and medium-sized enterprises (SMEs) that participate in large PPP projects, may be developed by partnering or subcontracting. Thus, SMEs may gain experience and have the capacity to bid on infrastructure projects abroad.
- The competitiveness of the country may increase and the economy can be stimulated through advanced infrastructure investments.

2.3 *Risk*

- Costs of developing, procuring, and operating PPP projects may be greater than traditional public procurement in certain cases.
- Private financing comes at a price. Large-scale PPP contracts need project finance that is at least 70–80% of the upfront capital expenditure. Financial institutions should believe that cash flow of relevant project is sufficient to meet all the financing, construction, and operating expenses, including the equity return. To the extent that a PPP contract involves omissions and errors, the SPV would either fail to get a project finance or access to finance expensively, which is likely to lead to failure for a relevant PPP contract. PPP contracts are incomplete by nature; therefore, inefficiencies arise because it is hard to foresee and contract about the uncertain future (Hart, 2003)
- PPP contracts that generate revenue in local currency and borrow in foreign currency would become more open to developments/shocks in international financial markets that may not be controlled and managed by a SPV.
- Risks may not be transferred completely to the private entity. The SPV and its lenders would be unwilling to accept risks outside their control (such as risks

[3] I benefited from World Bank (2016) for the subsections of benefit and risk.

arising from exchange rates and inflation). In addition, the SPV would prefer stability in public policies and regulations, and therefore be reluctant to undertake political and regulatory risks. Otherwise, even if a PPP contract takes place, its cost would be greater than it should be.

- As long as a private party has a significant degree of supervisory power over service delivery, commercial interests would be superior to public interests. This may then imperil the availability and quality of public service. To prevent this, service standards should be clearly identified in the PPP contracts.
- The government has the ultimate responsibility for the availability and quality of essential public services. When contracting authorities fail to understand the content of PPP contracts and fail to effectively monitor service standards, expected benefits from PPP contracts are unlikely to emerge.
- Contractors would obtain information about PPP projects and data sooner than government entities. Due to this information asymmetry, they may report incorrect and incomplete information with respect to service standards.
- Due to the long-term and complex nature of PPP contracts, it is unlikely that all possibilities and events that may emerge in construction and operation periods will be forecast. Therefore, contracts may be renegotiated after they come into force. Nevertheless, value for public money emerged in competitive tenders may disappear to the extent that the terms and conditions of a PPP contract differ from its initial structure due to revisions.

2.4 Lessons

International best practices indicate that rules and procedures should be clearly specified to engage in PPPs. For this, specific methodologies should be developed in order to appraise the benefits and risks that may arise during the whole life cycle of PPP projects. The cost of a proposed PPP should then be compared with the cost of traditional procurement. In this regard, a contracting agency should check whether a proposed PPP contract would contribute to the value for public money at the outset. In international practices, this comparison is called a public sector comparator (PSC) (Grimsey & Lewis, 2004 and OECD, 2008).

On the other hand, affordability of availability payments[4] and minimum revenue guarantees[5] granted by government under PPP contracts should be evaluated within the framework of long-term budget constraints (OECD, 2008; Posner et al., 2009). In this respect, authorities should consider budget deficits and borrowing requirements in the coming years. Contracting agencies should then ensure that an undue burden is not put on taxpayers, service users, and future generations. Therefore,

[4] An availability payment is a periodic payment by the contracting agency to the contractor for providing infrastructure services that conform to the quality and/or criteria specified in the PPP agreement.

[5] When real project revenue is lower than that specified at a PPP contract, the difference is paid by the government under a revenue guarantee scheme.

the affordability of commitments arising from long-term contracts should also be taken into consideration in the project development phase, in addition to the relative effectiveness of PPP projects against traditional procurement. Finally, the rights and obligations of government specified in PPP contracts must be reported transparently in public financial statements.

3 PPPs in Turkey

The background of private participation in infrastructure in Turkey goes back to the Ottoman Empire. The public service concession, which emerged as a way of delivering public services such as rail, electricity, and gas, was an important means of private participation in infrastructure throughout the nineteenth century. In accordance with international developments, governments attached importance to state-owned enterprises in order to deliver infrastructure services in the first half of the twentieth century. However, since the 1980s, especially due to the unavailability of public funds, PPPs in various fields have been used to increase private financing in infrastructure.

Today, there is no generic PPP-enabling legislation that comprehensively regulates a PPP project cycle management from project development to tendering, from selecting the winning bid to writing and implementing contracts, so on. Instead, various legislations exist in such sectors as energy, transport, health, and education, which regulate different PPP procedures such as BOT, BOO, BLT, and ToOR.[6] Although a central PPP unit was recently established, its mandate and capacity is debatable in terms of coordination and oversight in the PPP process.

In BOT and BOO procedures, the SPV finances and builds a new facility and operates it over the contract term. In return, the government provides a revenue guarantee scheme and then pays the difference when the actual revenue collected from service users falls short of the guaranteed one (contingent liability). At the end of the contract period, the SPV transfers the relevant facility to the government free of charge in working conditions. Conversely, the BOO method, which was only implemented for five thermal power plants in the 1990s, enables the SPV to maintain ownership of the facility even after the end of contract.

The BLT method works similar to the BOT method, with a few exceptions. In the BOT procedure, primary consumers are service users, whereas in the BLT method the government is itself a basic consumer and pays availability payment to the SPV in exchange for service delivery (direct liability).

In the ToOR method, the existing public facilities are transferred to the private party, which renovates and operates the facility during the duration of the contract. The private entity pays a fee for the operating rights of the facility and, at the end

---

[6] See Emek (2015) for how the existence of various legislations instead of a generic regulation may lead to the coordination failure.

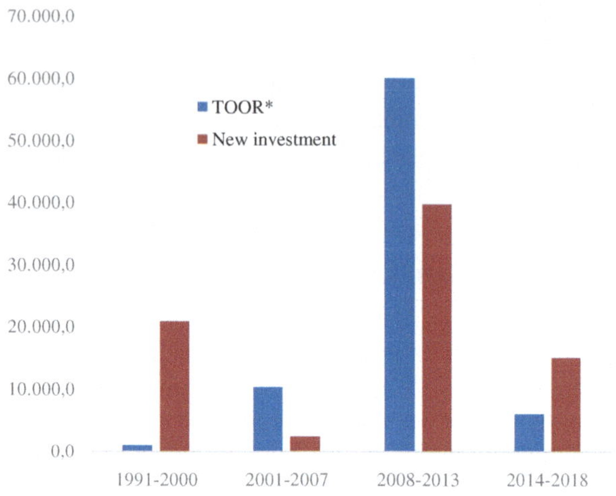

Fig. 1 Total value of PPP contracts (millions of US dollars) * Transfer of operating rights. *Source* Presidency of Strategy and Budget

of the contract period, again transfers the relevant facility to the government under working conditions as well.

Figure 1 depicts developments with respect to PPP contracts having been delivered in Turkey. As mentioned above, in the ToOR method the private company is charged for the operating right of facility. The monetary values indicated in Fig. 1 for a ToOR procedure include these transfer fees. New investment, on the other hand, represents upfront investment expenditures made in BOT, BOO, and BLT procedures. In the 1990s, new power plants were established in the electricity sector thorough BOT and BOO procedures, and operating rights of power generation plants were transferred to the private sector. The total value of new investment and fees of operating rights are $21 billion and $1 billion, respectively, for the period from 1991 to 2000. After Turkey's 2001 Financial Crisis, new investments with the PPP method were suspended and operating rights of existing facilities were frequently transferred in order to generate revenue for the government. The new investment and values were approximately $2.4 billion and 10.4 billion, respectively, between 2001 and 2007. The Turkish Government encountered a large amount of private finance after the Global Financial Crisis (GFC) emerged in 2008. As is well known, central banks in the developed economies printed a large amount of money to overcome the destructive effects of GFC. A significant part of this money flowed to developing countries, including Turkey. Thus, Turkey actively entered PPP contracts during 2008–2014 due to generous and inexpensive private finance. Over this period, the total amount of ToOR and new investment contracts was $40 billion and $60 billion, respectively.

Among these contracts, the most striking is the İstanbul Airport project, both in terms of the contract size and the preferred procedure. In this contract, the private party has committed to make a new airport investment and then to pay an annual lease fee to the government over contract period. Therefore, the İstanbul Airport contract itself includes two methods: BOT and ToOR. The investment value of the airport is $11.7 billion. According to the ToOR contract, the operator will make a total payment

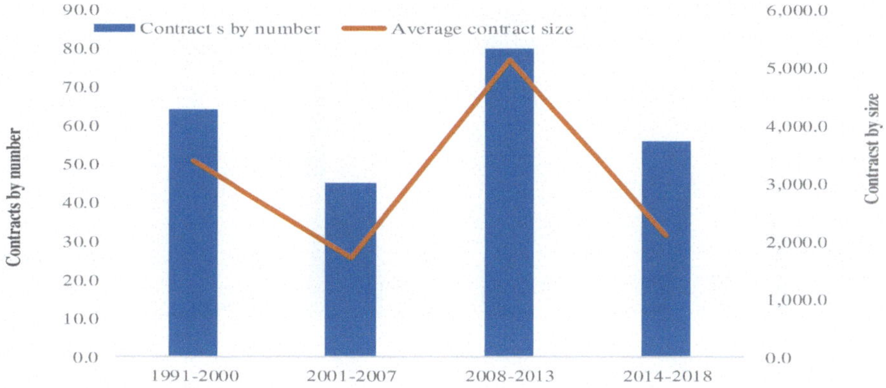

Fig. 2 Average contract size (million US dollars). *Source* Presidency of Strategy and Budget

of $25 billion to the Ministry of Transport and Infrastructure (MoTI) over 25 years. When central banks in developed countries began to tighten their monetary policies in 2012–2013, international funds started to return from developing countries to home. As a result, the private sector had difficulty finding financing for new PPP contracts. Therefore, the number and value of contracts signed decreased after 2014. The investment value in the contracts signed during this period was $15 billion and the total fee of operating rights was $6 billion.

Figure 2 involves the number of contracts and average contract size for the same periods. Average contract size is calculated by dividing the total value of the contracts by the number of contracts. After 2008, both the number of contracts and their sizes started to grow rapidly. Contracts worth billions of dollars entered into force, especially in the energy, healthcare, and transportation sectors. As mentioned above, it is not easy to write long-term and complex PPP contracts prepared about an uncertain future. For this reason, the monetary values of the projects should be kept as low as possible and these methods should not be used in areas where technology changes rapidly (Hart, 2003). The Turkish Government failed to follow this principle, then increased the value of the contracts, but also used PPP procedures in the health sector, where technology changes quickly. As a result, contracts had difficulty reaching project finance and, as time elapsed, unit costs projected in the feasibility studies increased. In addition, the parties frequently revised contracts upon request of financial institutions. As a result, investments were made over budget and over time. Delays and cost overruns led to the deterioration of the effectiveness of projects (Emek, 2017b).

Figure 3 contains breakdown of contracts by sector. The energy sector leads in terms of the number of contracts. Airports have the highest contract value, followed by the energy and road sectors. In recent years, new investments have been made in transport sectors such as airport and road. In the energy sector, especially operating rights of electricity generation and distribution facilities were transferred. Figure 4 denotes the dispersion of contracts by type. The most commonly used procedures

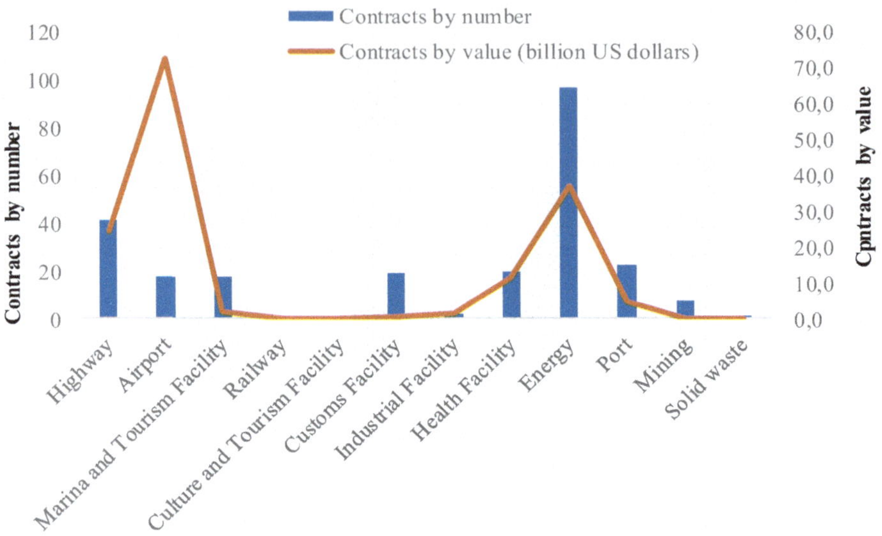

Fig. 3 Contracts by sector. *Source* Presidency of Strategy and Budget

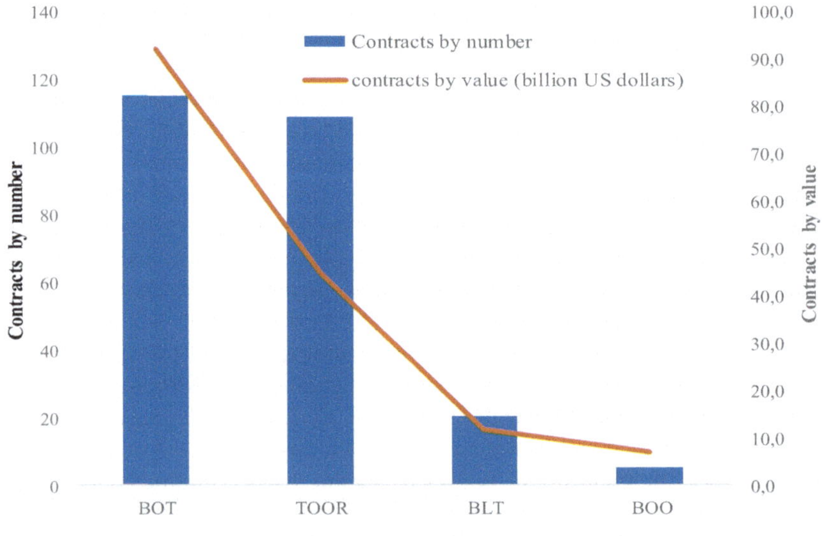

Fig. 4 Contracts by method. *Source* Presidency of Strategy and Budget

in terms of number and value of contracts are the BOT, where new investments are made, and the ToOR, where existing facilities are transferred.

4 Legal and Institutional Structure

Turkey has had accession negotiations with the European Union (EU). In this context, it needs to undertake regulations of the EU on PPP/concessions within the framework of the public procurement chapter. As an opening benchmark, the EU requests that Turkey align the legislation for public contracts and concessions with its procurement directives, at the same time ensuring the coherence of any further legislative initiative on public–private partnerships with the transposed acquis. As a response to this benchmark, Turkey stated that it aims to make the necessary regulations on PPPs, which have gradually become a more common practice all around the world in Turkey's Program for Alignment with the Acquis (2007–2013). SIGMA, a joint initiative of the OECD and the EU, examined Turkey's PPP legislation and concluded that it is fragmented and unclear (SIGMA, 2008). Various definitions and rules in different legal texts cause inconsistencies in practice. The inconsistency and legal uncertainty caused by the multi-part legislative structure adversely affect the incentives of (foreign) private enterprises and, therefore, the healthy functioning of the PPP market in Turkey. Sigma recommended that Turkey should develop a framework legislation regulating feasibility studies, tender processes, risk sharing, and accounting. Moreover, a central PPP unit responsible for the implementation of the legislation would increase the efficiency in the operation of the system.

Figure 5 represents the project development and monitoring processes in BOT and BLT procedures. The line ministries have a central role in the development,

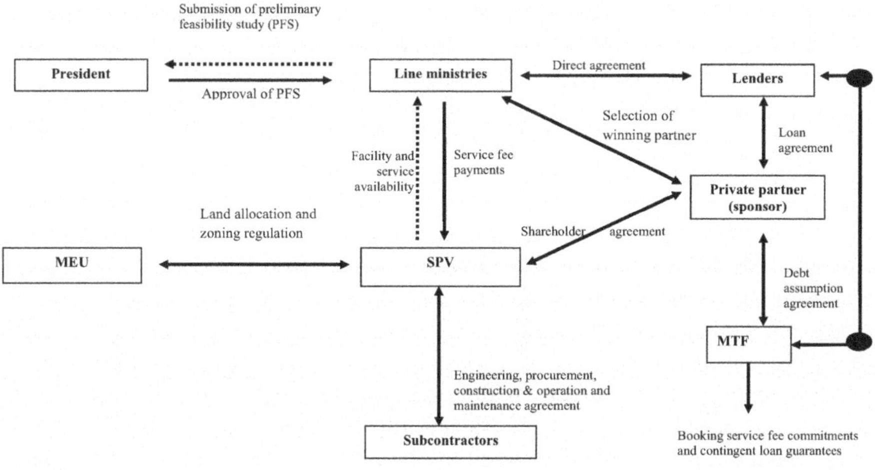

Note: Dashed arrows indicate offers, straight arrows indicate responses. MEU: Ministry of Environment and Urbanization, MTF: Ministry of Treasury and Finance; SPV: Special Purpose Vehicle

Source: Adapted from Emek, (2017b).

Fig. 5 Institutional framework

procurement, and execution of PPP contracts. The relevant ministry needs to receive an authorization decision from the president of Turkey before initiating tendering process. The Presidency of Strategy and Budget (PoSB) administers the authorization process as the secretariat of the president. The authorization process begins when the relevant ministry submits a prefeasibility report to the president. While evaluating this report, the PoSB considers the economic, financial, and social added value of the proposed project and its compatibility with sectoral, regional, and national policies. In addition, it receives opinion from the Ministry of Treasury and Finance (MoTF) on the coordination, budgeting, and monitoring of liabilities arising from PPP contracts. It then prepares its final opinion and presents it to the president. If these institutions find the technical preparations insufficient, they may request that the relevant ministry make changes, additions, and deletions in the project. If the authorization decision is positive, the line ministry initiates the tender process and signs a contract with the winning bidder. The relevant ministry can also make direct agreements with financial institutions to recognize step-in rights. Direct agreements specify under what circumstances and how lenders step in to the project company's position in the contract, when the private party fails to maintain its obligations under the PPP contract.

The MoTF, along with the PoSB, is responsible for the allocation of appropriations to the annual budgets for liabilities arising from PPP contracts and booking the rights and obligations arising from these contracts. In addition, the MoTF enters debt assumption contracts with financial institutions. As a general condition, a PPP contract includes provisions regarding termination of contract before the end of its end and takeover of facilities by contracting agencies. If a PPP contract is terminated before its term for the reasons specified in the contract, the Ministry assumes debt used for the construction of facility and contracting agencies take over the relevant facility under the debt assumption agreement.

On the other hand, upon request by the relevant ministry, the Ministry of Environment and Urbanization (MoEU) establishes the right of construction to the SPV on lands belonging to the Treasury in order to realize the investment. Moreover, the zoning plans of the places where PPP projects would be constructed are made by the MoEU, if requested by the relevant ministry.

Finally, the successful bidder (a company or a consortium) would create a SPV before signing a PPP contract. The SPV would finance, design, invest, and operate the relevant facility. The SPV is prohibited from engaging in other projects because its only objective is the delivery of works and services specified in the PPP contract. The SPV can raise capital directly without carrying the debt or other liabilities of the parent organization, even if it serves benefit of the parent organization (Yescombe, 2007). Thus, the project finance raised by the SPV does not influence the balance sheet of parent companies. Second, the SPV protects government entities and all financial institutions involved in the PPP contract from default risk. In this respect, the project revenue is allocated primarily to the maintenance and repair expenses necessary for the continuation of the project, and then to the payment of commercial and financial debts, structural loans, and finally equity holder(s) shares.

5 Analysis: Project Selection and Accounting

Projects need to be financially attractive for the private party to engage in a PPP contract. PPP projects contain risks as well as opportunities. The main priority of the contracting agency is to optimally transfer project risks to the private party within the framework of contract rules. Efficient allocation of risk is of prime importance to achieve VfM through PPPs. Risks should be allocated to the party that is best able to manage them at the lowest cost (Yescombe, 2007). For this reason, it is necessary to evaluate the project risks in detail and to investigate the effects of risk sharing on the contracting parties and the total cost of the project. As a result, in the project development process, contracting agencies should evaluate project priorities, project risks, procedures (traditional public procurement or PPP contract), and tender rules within the framework of objective criteria.

On the other hand, the actual performance of PPP contracts may deviate in practice from what is expected at the outset. Therefore, the performance of PPPs should be monitored regularly over the contract period. OECD (2008, pp. 54–56) evaluated the practices of Australia, Brazil, France, Hungary, and UK and stated that the performances of PPP projects in these countries are monitored through a basket of indicators, which involve:

- Efficiency measures defined in terms of inputs and outputs (such as the provision of a bridge service at a certain level of vehicle traffic in return for service fee)
- Effectiveness measures for outcomes (such as quantity, level of coverage of area or population)
- Service quality measures
- Financial performance measures
- Process and activity measures.

Further, the International Public Sector Accounting Standards Board (IPSASB, 2015) issued a standard (IPSAS 32—Service Concession Arrangements: Grantor) to set out the accounting requirements of the contracting agency in 2011. The Standard requires that liabilities (direct or contingent) arising from PPP contract be recorded in financial statements under the accrual basis of accounting principles when the contracts are signed.

5.1 Appraisal Value for Public Money

Regulations of BOT and BLT procedures request relevant ministries to first prepare a prefeasibility report. The report should analyze proposed projects in terms of technical, financial, economic, environmental, social, and legal feasibility. Risks including service fees and revenue guarantees should be analyzed and an optimal risk sharing scheme should be developed. As a result, the rationale for delivering relevant public service through the PPP procedure instead of traditional procurement should be presented comparatively. Relevant ministries should receive authorization from the president based on this report. As mentioned above, the PoSB, as the president's secretary, should examine these reports within the framework of economic, regional, and national priorities.

Despite explicit regulations, contracting agencies fail in practice to fulfill these requirements while preparing feasibility reports. These reports are not made available to the public on the grounds that they contain commercial confidentiality. For this reason, it is not easy to make a detailed and objective evaluation of the methodologies, findings, and results of the reports. Fortunately, other studies have evaluated the preparation process of PPPs rather than individual feasibility reports. For instance, World Bank (2018: p. 208) surveyed the legal and institutional structure, preparation, procurement, and contract management of PPPs in 135 economies, including Turkey's. As mentioned above, the report indicates PPPs' prioritization requirement in legislation consistent with public investment prioritization and central approval before tendering is available. Nevertheless, no specific methodologies have been developed for economic analysis, fiscal affordability assessment, risk identification, comparative assessment (VfM analysis), financial viability, or bankability assessment. Further, there is not even a legal requirement for preparation of standardized PPP transaction documents, market-sounding assessment, and disclosure of results of preparation assessments. Market assessment methodology is a procedure that assesses potential interest from financiers and contractors, providing insights into the likely level of market interest and providing the procuring authority with an opportunity to adjust the project scope if necessary to ensure the private sector participation and improve competition. When preparation assessments results are disclosed to potential bidders through the tender documentation, the informed bidders can submit better-prepared bids (World Bank, 2018).

Due to the lack of specific methodologies for the preparation of projects, practitioners may act carelessly in the processes of identifying needs and procedures, preparing feasibility reports, and writing contracts. In this context, those who participated in the studies of the Tenth Development Plan (2014–2018) Public–Private Partnership Special Expertise Commission stated, in most projects, the contracting agencies do not prepare a well-designed feasibility study. According to the participants, feasibility reports contain deficiencies and flaws. Companies that believe that the contracts would be revised after winning the tender attempt to bid almost in every tender. However, systemic problems that may arise after the tender both delay projects and cause transaction costs (Bakanlığı, 2014).

Those who participated in the Eleventh Development Plan (2019–2023) Public–Private Partnership Special Expertise Commission, published four years after the previous report, mainly said the same thing. Participants stated that the procuring agencies consider the feasibility study as an attached document required to complete the tender dossier. They also reported that the quality of the preparation processes of PPPs should be increased by attaching maximum importance to feasibility studies (Bakanlığı, 2018).

5.2 Accounting

As mentioned above, a number of PPP contracts have been signed with direct (shadow tolls and/or service fees) and contingent (such as minimum revenue or loan guarantees) liabilities in the transport sector as well as healthcare. However, the long-term affordability of PPPs with regard to the country's budgetary capacity and priority has barely been considered while the government promoted or implemented PPPs. Further, these liabilities have been booked off-balance sheet. The government and proponents of PPPs considered the off-budget feature of the PPP contracts as an advantage. For instance, the then undersecretary of the Ministry of Health (Tosun, 2014) once stated that the liabilities from the contracts would not appear in the government's debt portfolio. Thus, the adverse economic effects of the understatement of the actual level of the government's liabilities have been entirely neglected. However, unless long-term affordability is assessed beforehand, it is still possible that the off-budgetary treatment of PPPs may prompt a government to fund more projects than it can afford under intertemporal budget constraints (Posner et al., 2009). Fortunately, in addressing this issue, the MoTF published a Communication in July 2015 that sets out the accounting requirements for contracting agencies in a PPP contract in line with the aforementioned IPSAS 32—Service Concession Arrangements. If this communication is effectively enforced, it would threaten the off-budget status of the future projects and result in a sharp decrease in the number of the PPP contracts to be promoted in coming years (Emek, 2017a).

The MoTF started to book assets and liabilities at the end of 2015. As of 2017, it was monitoring the assets and liabilities of 58 PPP contracts of central agencies (Bakanlığı, 2018). Twenty-one of these belong to the BLT procedure, which delivers contracts of hospitals and other contracts fall under the scope of BOT method. Twelve of the 58 contracts are customs facilities belonging to the Ministry of Customs and Trade, 18 are tunnels, ports and marinas of the MoTI, and seven are bridge and highway projects referring the General Directorate of Highways. Twenty-three PPP projects of local administrations are monitored in their accounting records. In this framework, the financial status of 81 PPP projects within the scope of general government is monitored.

However, the MoTF does not make this information monitored available to the public in the financial statements of relevant agencies such as balance sheets and income statements. Further, the Turkish Court of Accounts stated in its audit report (Sayıştay, 2019) that the Ministry of Health does not put the liabilities and assets of PPP contracts in financial statements, even more than four years after the aforementioned Communication entered into force. Moreover, the MoTF does not recognize both the contingent and direct liabilities of PPPs in its public debt reports, which contradicts the philosophy of IPSAS-32.

6 Conclusion

Due to growing need, the aging of existing facilities, and the rapid increase in population, the need for infrastructure investments has been growing worldwide. In this regard, PPPs have been used for the provision of infrastructure investments and services since the 1990s. Within the framework of these procedures, existing facilities are transferred to the private sector through the ToOR method, and also new investments are made. The expectation was that the private finance would promptly realize the necessary infrastructure investments to ensure efficiency in service provision.

Turkey has a fragmented legal and institutional structure for the development of PPP contracts. It also hardly has a clear PPP program. Different line ministries in different sectors have simultaneously launched PPP contracts without fully understanding their fiscal impact. Indeed, contracting agencies have entered most PPP contracts due to political demands. However, international experience with PPPs suggests that a strategic roadmap is essential to introduce well-designed enabling rules, procedures, and institutions in order to secure VfM in the service delivery. In this context, the Tenth Development Plan of Turkey (MoD, 2014) suggested "[a] strategy paper be prepared as a future road map for the PPP model implementations and a framework law be prepared in order to compile the scattered PPP legislation in force" (par, 594). The MoTF has been authorized to introduce a comprehensive strategy for public procurement within the framework of accession negotiations with the EU. As an opening benchmark, the EU has requested, "this strategy would include all reforms necessary in terms of legislative alignment and institutional capacity-building in order to comply with the procurement *acquis* with time-schedules and milestones."[7] The strategy paper would have covered PPPs as well as other public contracts. As of the time of writing, there is no such a document in place.

Flyvbjerg (2014) stated that technology and designs are often non-standard in mega projects, which are large-scale, complex ventures. This leads to "uniqueness bias" among decision-makers in the different ministries, who tend to see their projects as singular. Decision-makers think that their projects are different from other projects and are therefore reluctant to learn from the experiences of others. This is exactly what has happened in Turkey. Legal structure does not automatically ensure cooperation. As a result, coordination has failed. Contracting agencies have been concerned about losing their influence on PPPs in their sectors and responsibilities. Furthermore, they were unwilling to even partially share their authority with others (Emek, 2015). For instance, hospital projects have not benefited from the experience of the MoTI, or the MoTI has not appealed regarding highways, bridges, and tunnels to the practices of Ministry of Energy and Natural resources on power plants. Legally, the PoSB appears to be a central unit for coordinating the PPP project development process. In fact, after pre-authorization of a project, it has no role in the finalizing feasibility studies, preparing tender procedures, and writing contracts. In this context, the PoSB should be empowered even more in terms of legal authority and human resources to overcome this disorganization.

[7] Retrieved July 8, 2020 from https://www.ab.gov.tr/5-public-procurement_70_en.html.

The contracting agencies pay little attention to the preparation of feasibility studies. Although regulations require the procuring agencies to prepare well-designed feasibilities, in practice they are unable to do this, because there are few specific methodologies for the appraisal of financial, economic, environmental, and fiscal implications. Further, they rarely consider the final impact of risks arising from PPP contracts. In this context, the Tenth Development Plan (MoD, 2014 states "[t]he PPP process has a structure that may charge the public sector with longterm fiscal liabilities, and it has to be designed and managed appropriately." It then suggests that institutional capacity based on project planning and management expertise should be improved (par 582). The Eleventh Development Plan (MoD, 2019) again emphasized the importance of feasibility studies and risk analysis and suggested that risk sharing with the private sector should be balanced and budgetary obligations should be taken into account (par 204). At time of writing, few initiatives on the issue have appeared in public.

References

Abadie, R. (2014). *Capital project and infrastructure spending: Outlook to 2025.* PwC Public Sector Research Centre publication.

Bakanlığı, K. (2014). *Onuncu Kalkınma Planı (2014–2018), Kamu Özel İşbirlikleri Özel İhtisas Komisyonu Raporu.* Kalkınma Bakanlığı Yayınları.

Bakanlığı, K. (2018). *On Birinci Kalkınma Planı (2019–2023), Kamu Özel İşbirlikleri Özel İhtisas Komisyonu Raporu.* Kalkınma Bakanlığı Yayınları.

de la de la Torre. A. & Rudolph, H. (2015). The seven sins of flawed public-private partnerships. Working paper 102050. World Bank. Retrieved July 8, 2020 from https://openknowledge.worldbank.org/handle/10986/23595

Emek, U. (2015). Turkish Experience with Public-private partnerships in infrastructure: Opportunities and challenges. *Utilities Policy, 37*, 120–129.

Emek, U. (2017a). Public-private partnerships in the turkish healthcare sector: Policy, procedure and practice. In N. Kakabadse & N. Mouraviev (Eds.), *Public–private partnerships in transitional nations: Policy, governance and praxis* (pp. 86–108). Scholars Publishing.

Emek U. (2017b). Sağlık Sektöründe Kamu-Özel İşbirliği Sözleşmeleri: Beklenti ve Gerçekleşme. *Hacettepe Üniversitesi Hukuk Fakültesi Dergisi, 7*(1), 141–168.

Flyvbjerg, B. (2014). What you should know about megaprojects and why: An overview. *Project Management Journal, 45*(2), 6–19.

Grimsey, D., & Lewis, M. K. (2004). *Public private partnerships: The worldwide revolution in infrastructure provision and project finance.* Edward Elgar.

Hart, O. (2003). Incomplete contracts and public ownership: Remarks and an application to public-private partnerships. *The Economic Journal, 113*(486), 69–76.

IPSASB (International Public Sector Accounting Standards Board). (2015). IPSAS 32—service concession arrangements: grantor, international public sector accounting standards board.

Mckinsey & Company. (2013). *Infrastructure productivity: How to save $1 trillion a year.* McKinsey Global Institute.

MoD (Ministry of Development). (2014). Tenth Development Plan (2014–2018). Retrieved July 8, 2020. https://sbb.gov.tr/wp-content/uploads/2018/11/The_Tenth_Development_Plan_2014-2018.pdf

MoD (Ministry of Development). (2019). Eleventh Development Plan (2019–2023). Retrieved July 8, 2020. https://www.sbb.gov.tr/wp-content/uploads/2020/03/On_BirinciPLan_ingilizce_SonBaski.pdf
OECD (Organisation for Economic Co-operation and Development). (2008). *Public–private partnerships: In pursuit of risk sharing and value for money*. OECD.
PPP Knowledge Lab. (2018). What are public private partnerships? Retrieved July 8, 2020. https://ppp.worldbank.org/public-private-partnership/overview/what-are-public-private-partnerships
Posner, P., Ryu, S. K., & Tkachenko, A. (2009). Public–private partnerships: The relevance of budgeting. *OECD Journal on Budgeting*, Volume 2009/1, s. 1–26
Sayıştay. (2019). *Sağlık Bakanlığı 2018 Yılı Sayıştay Denetim Raporu*. Retrieved July 8, 2020 from https://www.sayistay.gov.tr/tr/Upload/62643830/files/raporlar/kid/2018/Genel_B%C3%BCt%C3%A7e_Kapsam%C4%B1ndaki_%20Kamu_%C4%B0dareleri/SA%C4%9ELIK%20BAKANLI%C4%9EI.pdf
SIGMA (Support for Improvement in Governance and Management). (2008). *Peer assistance in public procurement and concessions/PPPs. Final report*. October, Paris.
Tosun, N. (2014). Eski hastaneler, modern şehir hastanelerinde birleştirilecek. *Sağlık Düşüncesi ve Tıp Kültürü Dergisi*, 30: 18–23. Retrieved June 18, 2020 from http://www.sdplatform.com/Soylesiler/51/Eski-hastaneler-modern-sehir-hastanelerinde-birlestirilecek.aspx.
World Bank (2016). Government objectives: Benefits and risks of PPPs. Retrieved June 18, 2020 from https://ppp.worldbank.org/public-private-partnership/overview/ppp-objectives
World Bank (2018). *Procuring infrastructure PPPs 2018*. Retrieved September 11, 2020 from http://pubdocs.worldbank.org/en/256451522692645967/PIP3-2018.pdf. https://ppp.worldbank.org/public-private-partnership/library/benchmarking-ppp-procurement-2017
Yescombe, E. R. (2007). *Public-private partnerships: Principle of policy and finance* (1st ed.). Elsevier.

Uğur Emek is currently a Professor of Economics and serves as a Chair of Department at Başkent University, Ankara, Turkey. Before joining academia, he served as a planning expert responsible for evaluation of PPPs in the State Planning Organization (now, Presidency of Strategy and Budget). His main research interests include public procurement & public-private partnerships in infrastructure; competition & regulation in network industries; anti-corruption; and Turkish economy. He has numerous publications in the aforementioned fields in Turkish and English.

Regulatory Authorities in Turkey

Engin Saygin

> *[The administrative or regulatory state is] a massive and enormously complex system of [...] regulations that governs literally every aspect of our lives -from the way we get medical care or even the food we eat, from the design of our cars to where and how we can build anything, whether it be a pond or a new power plant.*
> William Kovacs
> Former Senior Vice President, Environment, Technology & Regulatory Affairs of the US Chamber of Commerce

Abstract "Regulatory authorities," which are also known as "independent regulatory agencies," "regulatory and supervisory agencies," "autonomous authorities," or "supreme councils," play crucial roles and perform different functions within the administrative structure of Turkey, as they do in other countries. The list of these authorities in Turkey is provided by the Public Financial Management and Control Law No. 5018. There are 11 regulatory authorities on this list: (i) the Radio and Television Supreme Council, (ii) the Information Technologies and Communication Agency, (iii) the Capital Markets Board, (iv) the Banking Regulation and Supervision Agency, (v) the Energy Market Regulation Authority, (vi) the Public Procurement Agency, (vii) the Turkish Competition Authority, (viii) the Public Oversight, Accounting, and Auditing Standards Authority, (ix) the Personal Data Protection Authority, (x) the Nuclear Regulatory Agency, and (xi) the Insurance and Private Pension Regulatory and Supervision Agency. All have been established as public legal institutions. Due to the limited scope of this paper, it is impossible to analyze all authorities that have regulatory power in Turkey. Therefore, we will focus on the authorities that have power and duties in Turkish network industries, especially the Energy Market Regulation Authority and the Information Technologies and Communication Agency.

E. Saygin (✉)
Faculty of Law, Department of Administrative Law, Ankara Yıldırım Beyazıt University, Ankara, Turkey
e-mail: esaygin@ybu.edu.tr

M. Eroğlu and M. Finger (eds.), *The Regulation of Turkish Network Industries*,
https://doi.org/10.1007/978-3-030-81720-6_23

Keywords Regulatory authorities · Independence · Administration · Judicial review · Turkey

1 Introduction

The development of regulatory authorities is not a new phenomenon, either around the world or in Europe, including Turkey. It is well known that such authorities emerged in the late nineteenth century. The first independent federal commission was the Interstate Commerce Commission of the USA, which was formed in 1887 as a result of mounting public indignation in the 1880s against railroad malpractices, and was terminated in 1995 (Federal Register, 2020). The emergence of regulatory agencies in the USA is intimately related to two elements (Carrapico & Farrand, 2016: 7). The first element is that the State apparatus was seen as too politically unstable to function as a strong foundation for economic growth, as it was considered as too reliant on election outcomes and shifting political interests. To have a more consistent and consistent approach on which the markets could rely, it was agreed that efficiency could only be accomplished by competent regulators in an apolitical sense. The second element is that as a result of the privatization process, the public has demanded that the private sector be regulated and its ability to deliver goods and services to society be regulated. Therefore, regulatory bodies have emerged as the best operational option for regulating liberalization while remaining independent of the political system. Their primary responsibilities include gathering and processing data, as well as developing effective solutions to practical regulatory issues.

Following the USA, a number of regulatory authorities have been established in Europe. Notably, the emergence of regulatory authorities in the 1980s and 1990s across European states and the 1880s–1910s in the USA seems to be cases of similar devices adopted for different motives (Lodge, 2008: 286). Many reasons can be given for the emergence of European regulatory authorities (OECD, 2014: 131. For the reasons mentioned by Turkish academics, see Çağlayan, 2019: 227–228; Ulusoy, 2019: 249; Atay, 2006: 261).

Expertise:Regulatory authorities are closer to the regulated sector than ordinary bureaucracy and can more easily gather information on this sector. Experts are thus more likely to work for Regulatory authorities than traditional bureaucracies because of their more flexible organizational structure.

Flexibility: Since regulatory agencies have more autonomy, they are better able to adapt legislation to changing circumstances. They can also quickly adapt to new technologies or technological expertise.

Credible commitments: Regulatory agencies are shielded from day-to-day political pressures and electoral constraints, have a longer time span than politicians, and can thus boost the legitimacy of governments' pro-market and fair-regulatory commitments.

Stability: Regulatory agencies have a stable and consistent regulatory framework for similar reasons.

Efficacy and Efficiency: Regulatory authorities produce higher regulatory results as a result of stability, which ensures that rules are less likely to change suddenly and unexpectedly.

Public Participation and Transparency: Regulatory authorities' decision-making processes are more accessible and transparent than those of ministerial agencies, making them more responsive to diffused interests like those of customers.

Decision-Making Costs: Delegation to regulatory authorities lowers decision-making costs because, in the face of policy conflict, majorities are more easily found to "let some else decide," particularly if the political benefits and losses from policies are unclear.

Blame-shifting: When there are regulatory errors or unpopular decisions, regulatory authorities allow politicians to avoid taking responsibility.

Political Uncertainty: When politicians create a policy, they are aware that it may be changed in the future if a new party wins elections. To avoid this, strategy should be kept apart from politics. Thus, regulatory authorities are a means for politicians to fix policies so that they will last beyond their term of office.

The growing number of regulatory authorities is one of the most significant developments in the administrative structure of Turkey since the 1980s. As in other countries, they play crucial roles and perform different functions within the administrative structure. The list of these authorities in Turkey is provided by the Public Financial Management and Control Law No. 5018 and comprises 11 regulatory authorities: (i) the Radio and Television Supreme Council, (ii) the Information Technologies and Communication Agency, (iii) the Capital Markets Board, (iv) the Banking Regulation and Supervision Agency, (v) the Energy Market Regulation Authority, (vi) Public Procurement Agency, (vii) the Turkish Competition Authority, (viii) the Public Oversight, Accounting and Auditing Standards Authority, (ix) the Personal Data Protection Authority, (x) the Nuclear Regulatory Agency, and (xi) the Insurance and Private Pension Regulatory and Supervision Agency, all of which have been established as public legal institutions. Due to the limited scope of this paper, it is impossible to analyze all authorities that have regulatory power, so we will focus on the authorities that have power and duties in Turkish network industries, especially the Energy Market Regulatory Authorities (hereinafter, the EMRA) and the Information Technologies and Communication Agency (hereinafter, the ITCA). This paper consists of five parts. The first part provides terminology and the classification of regulatory authorities. The second part explains the constitutional and legal status of regulatory authorities. The third part clarifies the independence and structure of the EMRA and the ICTA in the light of the latest legislative amendments. The fourth part analyzes the powers and duties of the EMRA and the ICTA. The final part emphasizes the fundamental issues of the judicial review of regulatory decisions in Turkey.

2 Terminology and the Classification of Regulatory Authorities

A variety of terms are used to describe **Regulatory Authorities** in the Turkish literature, including "regulatory and supervisory authorities (***düzenleyici ve denetleyici kurumlar***)," "independent regulatory agencies (***bağımsız idari otoriteler***)," **Regulatory Authorities** (***düzenleyici kurumlar***)," "supreme authorities (***üst kurullar***)," and autonomous authorities (***özerk kurumlar***)" (Ulusoy, 2000: 253; Ulusoy, 2003; Atay, 2006: 261–262). Turkish administrative law scholars have long agreed to qualify these authorities as "independent regulatory agencies (***bağımsız idari otoriteler***)" (Atay, 2000: 60) in their academic studies, as many of their European counterparts use the French term, "*autorité administrative indépendante*". However, the Grand National Assembly of Turkey (Turkish Parliament) has chosen the term of "supreme council (***üst kurul***)" since 1994 with the establishment of the Radio and Television Supreme Council, an administratively and financially autonomous and impartial public legal authority for the regulation and supervision of radio, television and on-demand media services.

As with the title of regulatory authorities, there are different views on the origin of regulatory authority in Turkey. Sever (2015: 213) argued that the Central Bank of the Republic of Turkey (*Türkiye Cumhuriyet Merkez Bankası*) can be seen as the first regulatory agency in Turkey because, according to the Law on the Central Bank of the Republic of Turkey No. 1715, the primary objective of the Central Bank is to "*regulate the money market and the circulation of money*". Duran claimed that the Insurance Supervisory Agency (*Sigorta Murakabe Kurumu*), established in 1959, is the first regulatory agency (Azrak, 2001: 22. See also Sayhan, 2003: 85) because the Insurance Supervision Law No. 7397 envisaged the issuance of regulations governing insurance-related bodies and their activities. Nevertheless, we accept that the Capital Markets Board of Turkey is the first regulatory and supervisory authority in the context of international standards because the Capital Markets Law was enacted in 1981 and the Board has been authorized to make detailed regulations for organizing the markets and developing capital market instruments/institutions.

The European literature contains two broad groups of regulatory authorities. Thatcher argued that the first group of authorities are sometimes called "***economic regulators***" regulating competition including utility regulators, general competition authorities, and financial bodies. They have power over such matters as investigating and preventing "unfair competition," issuing enforcing and altering licenses, determining interconnection among networks, and determining price caps (Thatcher, 2001: 4). The second group of regulatory authorities are called "***social regulators***" because they are responsible for promoting "public interest" goals other than competition including agencies for the environment, safety (at work or food for instance), and racial and gender equality. They were given powers that included setting standards, issuing licenses, prohibiting supply, enforcement of legal requirements, and provision of information (Thatcher, 2001: 5). However, the Turkish literature comprises three broad groups of regulatory authorities. Ulusoy argued that the categorization of regulatory authorities is based on ***the structure of the decision-making body***, ***the characteristic of the regulation***, and ***the nature of the activity area***. Based on the structure of the decision-making body, regulatory authorities are classified into two groups: *authorities with committee-type decision-making bodies* and *authorities with non-committee-type decision-making bodies*. In Turkey, all regulatory authorities have committee-type decision-making bodies. In accordance with the characteristic of regulation, regulatory authorities are divided into two groups: *those with general powers* and *those with specific powers*. In this sense, the Public Procurement Agency and the Turkish Competition Authority have general jurisdiction in their activity areas, whereas the Capital Markets Board, the Banking regulation and Supervision Agency, and the EMRA have regulatory powers in their specific activity areas. Finally, in accordance with the nature of activity area, there are regulatory authorities *in the field of fundamental rights and freedoms and in the field of economics*. The Radio and Television Supreme Council can be seen as the first regulatory authority in the field of fundamental rights and freedoms, whereas the Capital Markets Board, the Banking Regulation and Supervision Agency, and the EMRA can be identified as regulatory authorities performing in the field of economics (Ulusoy, 2003: 11–13; See also Duran, 1997: 3; Tan, 2005: 13).

3 Constitutional and Legal Status of Regulatory Authorities

While there is not a general regulation or a specific article for each regulatory authority in the Turkish Constitution, it is argued that there is a constitutional base for these authorities in the Constitution. The Constitutional Court of Turkey (see Docket No: 2002/32, Judgment No: 2003/130, Date: 20.22.2003) and scholars argue (Ulusoy, 2011: 18) that Article 167 of the Constitution of Turkey can be accepted as a constitutional ground for regulatory authority because, according to Article 167, "***the State shall take measures to ensure and promote the sound and orderly functioning of the markets for money, credit, capital, goods and services; and shall prevent the formation of monopolies and cartels in the markets, emerged in practice or by agreement.***" Similarly, it is argued that Article 172 of the Constitution can be seen a legal base because, according to the article, "***the State shall take measures to protect and inform consumers; shall encourage their initiatives to protect themselves.***"

According to Ulusoy, the constitutional obligation of regulation with regard to the public and private economic activities does not mean that regulation should be effectuated by regulatory authorities (Ulusoy, 2011: 18) because Parliament has discretionary power to implement regulation via regulatory authority or conventional administrative agencies (Ulusoy, 2003: 140–141). Thus, the Constitution does not reveal any preference concerning the "origin" of regulation.

Among 11 regulatory authority, only the Radio and Television Supreme Council has a constitutional foundation: "*The Radio and Television Supreme Council, established for the purpose of* regulation *and supervision of radio and television activities, is composed of nine members. The members are elected, on the basis of number of members allocated to each political party group, by the Plenary of the Grand National Assembly of* Turkey *from among the candidates, twice the number of which is nominated by political party groups in proportion to their number of members. The formation, duties and powers of the Radio and Television Supreme Council, and qualifications, election procedures and term of office of its members shall be regulated by law.* (Article 133 of the Constitution)".

Other regulatory agencies have been established by specific laws. Relevant laws lay down the general rules for the management and scrutiny of a sector and, and the same time, establish an independent regulatory authority with the requisite powers over that sector (Sezen, 2007: 322). The names, relevant legislation, establishment dates, and board member numbers of all these regulatory authority in Turkey are given in Table 1.

Table 1 Regulatory authority in Turkey as of August 2020

Authority and official website	Relevant legislation	Establishment year	Number of board members	Term of office of members*
Capital markets board-https://www.cmb.gov.tr/	Capital market law numbered 6362	1982	7	–
Radio and television supreme Council-https://www.rtuk.gov.tr/en	Article 133 of the constitution and the law on the establishment of radio and television enterprises and their media services numbered 6112	1994	9	6 years
Turkish competition authority-https://www.rekabet.gov.tr/en	The law on the protection of competition numbered 4054	1994	7	–
Banking **Regulation** and supervision agency-https://www.bddk.org.tr/Home	Banking law numbered 5411	1999	7	–

(continued)

Table 1 (continued)

Authority and official website	Relevant legislation	Establishment year	Number of board members	Term of office of members*
Information technologies and communication agency-https://www.btk.gov.tr/	The law on the establishment of information technology and communication institution numbered 2813	2000	7	–
Energy market **Regulation**-https://www.epdk.gov.tr/Home/En	Electricity market law numbered 6446 and the law on the utilization of renewable energy resources for the propose of generating electrical energy numbered 5346	2001	7	–
Public procurement agency-https://www.ihale.gov.tr/	Public procurement law numbered 4734 and the law on public procurement contracts numbered 4735	2002	10	–
Public oversight, accounting and auditing standards authority-https://www.kgk.gov.tr/Home	Public oversight, accounting and auditing standards authority's organization and responsibilities decree law numbered 660	2011	9	–
Personal data protection authority-https://www.kvkk.gov.tr/en/	The law on the protection of personal data numbered 6698	2016	9	–
Nuclear regulatory agency-https://ndk.gov.tr/tr/	The decree law numbered 702 concerning the organization and duties of the nuclearregulatory authority and amendments of certain laws	2018	5	–

(continued)

Table 1 (continued)

Authority and official website	Relevant legislation	Establishment year	Number of board members	Term of office of members*
Insurance and private pension regulatory and supervision agency	Presidential decree numbered 47 on the organization and duties of the insurance and private pension regulatory and supervision agency	2019	5	–

*Pursuant to Provisional Article 34 of Decree No. 703 dated July 2, 2018, the term of office of the chairmen and members of regulatory authoritiesexcept the Radio and Television Supreme Council have been amended. According to the amendment, (a) the term of office of those who have completed their fourth year in office ends without any further action on the date of publication of this Article; and (b) the term of office of those who have not completed their fourth year in office ends when they complete their fourth year in office without any further action
Article 4 of the Presidential Decree No. 3 on Appointment of High Ranking Public Executives in Public Institutions of 10 July 2018 regulates the term of office of the chairmen and members of regulatory authorities except the members of the Personal Data Protection Authority that are chosen by Parliament cannot exceed the term of office of the president

4 Independence and Structure of Regulatory Authorities in Network Industries

The characteristic of independent regulatory authorities is shown by their names. "***Independent***" implies that regulatory authorities are organically and functionally from the executive branch; "***administrative***" implies that they are administrative organizations so they possess public legal personality; "***authority***" implies that they are not consultative organs, but they are regulatory and specific final decision-making organs in their areas (Tan, 2005: 15).

The concept of independence is closely related to the structure of the organization and has several dimensions. For example, the independence of regulatory authorities consists of organic and functional independence (Çağlayan, 2019: 228; see also Sezen, 2007 for the discussions on the independence of regulatory agencies). Organic independence implies that the members of the decision-making organ of authority must be endowed with certain statutory securities. Functional independence implies that there must not be any kind control authority of any units within the executive branch on the acts and actions of authority that may result in the invalidity of the acts and actions of authority. Thus, one may accept that there are five crucial elements for independence of regulatory authorities (Irion et al., 2017: 9–10):

Status and powers: Regulatory authorities must be fully independent based on their legal status and powers. If any institution or individual outside the court can overturn the judgment and/or issue instructions, then autonomy will be limited and they should be authorized to make binding judgments beyond the recommendations.

Financial autonomy: Regulatory authorities must have sufficient financial resources. With sufficient financial resources, there is always a risk of independence and effective operation.

Autonomy of decision-makers: Both of nomination and appointment processes should be designed in such a way that significant structural distortions in the decision-making process are avoided.

Knowledge: Human resource is one of the most important element for institutions' good performance. Therefore, regulatory authorities should also be equipped with sufficient human resources and adequate expertise to perform their duties.

Transparency and accountability: Regulatory authorities must meet a minimum level of transparency and accountability for their decisions in order to maintain a balance of relative austonomy.

When the status of regulatory authorities in Turkey is examined, one could observe that there is no standard model regarding the status and power of authorities. They are covered by different authorities, organizational properties, and different degrees of autonomy according to their sectors. Therefore, we will take a close look at the organizational structure of the EMRA and ICTA in order to understand their independence.

4.1 Structure of the EMRA

Turkey's Electricity Market Law No. 4628 (hereinafter Law No. 4628), which was ratified on February 20, 2001 and enacted on March 3, 2001, called for the establishment of the Electricity Market Regulatory Authority. As a result, Turkey's first independent electricity regulatory agency in the energy market became operational in 2001. A subsequent Natural Gas Market Law No. 4646, adopted on April 18, 2001, empowered the Authority with regulatory oversight over the gas market. Later, the Authority expanded its regulatory oversight role to also include the petroleum and liquefied natural gas sectors through the Petroleum Market Law No: 5015 in 2003 and the Liquefied Petroleum Gases Market Law No: 5307 in 2005 (Activity Report, 2018: 1). As a result of the growing jurisdictional power, the name of the Electricity Market Regulatory Authority has been changed as the Energy Market Regulatory Authority with the 30th Article of Electricity Market Law numbered 6446.

The objective of Law No. 4628 is "*to ensure the development of a financially sound and transparent electricity market operating in a competitive environment under provisions of civil law and the delivery of sufficient, good quality, low cost and environment-friendly electricity to consumers and to ensure the autonomous* regulation *and supervision of this market*" (Article 1).

According to Law No. 4628, the EMRA is a public entity with administrative and financial autonomy. The EMRA is headquartered in Ankara and related to the Ministry of Energy and Natural Resources. As the EMRA can establish representative offices in distribution regions in order to carry out customer relations (Article 4/1 and 4/4), a liaison office in İstanbul has been opened.

The EMRA consists of The Energy Market Regulatory Board (hereinafter the Board), Presidency, and the service units (Article 5 of the law No. 4628). It uses its powers via the Board when performing its duties. The representative and decision-making body of the EMRA is the Board (Articles 4/5 and 4/6). The first Board assumed duty on November 19, 2001 by swearing an oath before the First Chairmanship of the Supreme Court of Appeals.

The Board consists of seven members, including the chair (chairman/chairwoman) and the vice chair. The president of the Republic appoints the chair and the second chair of the EMRA. The chair of the board is also the president of the EMRA (Article 5/1 and 5/3). The organization and coordination among the board members and the service units are performed by the president. The president is responsible for the enforcement and the execution of the Board decisions and representation of the board. These responsibilities also include the communication to the public of the works of the board. All decisions of the EMRA, except for the powers entitled to the staff by the president, are made by the board. The board assigns tasks to or authorizes the president in subjects it deems appropriate. The second chair assumes all duties and powers of the president in the president's absence.

The board members are appointed by the president of the Republic among persons holding a law, political sciences, administrative sciences, public administration, economics, engineering, management, or finance degree from a four-year university and having at least 10 years of experience in public institutions and organizations or the private sector (Article 5/2).

The board members cannot be removed from office before the end of their terms of office. However, if the president or board members are convicted of breaching the prohibitions set forth in Article 6 of the Law No: 4628 or crimes offended in relation with their assigned duties, they are considered to have lost the qualifications required to be a civil servant as per the Civil Servants Law No. 657. Also, if they are unable to work for more than three months because of an illness or accident or because their health conditions are determined to be unsuitable for working by a committee of doctors, then without necessarily waiting for three months, they shall be removed from office by the president of Turkey (Article 6/5).

In case the presidency and board memberships become vacant before the end of terms of office by any reason, appointments are made to the vacant posts within one month (Article 5/5). The board members, whose terms of office end, can be re-elected.

The board meets whenever it deems necessary, but at least once a week. The meetings are chaired by the president or, in the absence of the president, by the second chair. The agenda of each meeting is prepared by the president or, in the absence of the president, by the second president and notified to the board members (Article 8/1-a).

The board meets with an absolute majority and the decisions are made with the majority of the participants (Article 8/1-a).

The service units of the EMRA and the duties and responsibilities of the same, the titles and numbers of the cadres are set forth by the Law on Organization and Duties of the Energy Market Regulatory Authority No. 4628 and Regulation on the Organizational Structure of EMRA and the Working Principles and Procedures Applicable to its Employees, published in the Official Gazette number 29004 dated May 18, 2014. Hereunder are the service units as defined in Law No. 4628;

(a) Electricity Market Department
(b) Natural Gas Market Department
(c) Petroleum Market Department
(d) Liquefied Petroleum Gases Market Department
(e) Tariffs Department
(f) Audit Department
(g) Expropriation Department
(h) Legal Department
(i) Strategy Development Department
(j) Human Resources and Support Services Department
(k) Office of Press and Public Relations Counselor
(l) Special Bureau for the Board
(m) Special Bureau for the President.

4.2 *Structure of the ICTA*

Policymaking, regulation, and operation functions are separated by Law No. 4502 dated January 27 2000, amending Laws No. 406 and 2813, which are the basic laws of the telecommunications sector. While policymaking is the responsibility of the Ministry of Transportation, Maritime Affairs and Communications, regulatory and supervisory functions were given to the Telecommunications Authority (CEERWG, 2020).

To remove the legislative untidiness, create competition in the sector, lessen the uncertainties for operators, and allocate resources to research and development, the Electronic Communications Law came into force on 10 November 2008 and the name of the Authority changed to the ICTA (CEERWG, 2020).

According to Article 5 of the Law on the Establishment of the Information and Communication Technologies Authority No. 2813, "*With this law hereby, Information and Communication Technologies Authority is founded, as a public legal entity with public administrative and financial autonomy, in order to fulfill the duties and exercise the powers assigned to it by laws. The Authority is composed of the Information and Communication Technologies Board and Presidency Organization.*

The Authority shall act independently while performing its duties. No organ, authority, institution or person can give orders or instructions to the Authority. The Authority is associated with Ministry of Transport."

The Information and Communication Technologies Board (hereinafter the ICTB) is the decision-making body of the ICTA. The ICTB is composed of seven members in total, including one chair and six members. The chair of the board is also the president of the ICTA (Article 5/5).

The chair of the ICTB is the highest-ranking executive officer of the ICTA. The competence to manage and represent the ICTA belongs to the chair. When needed, the chair may transfer his/her competence of representation in writing. The board, based on a recommendation by the chair, elects one of the members as vice chair. The vice chair acts as a proxy holder on behalf of the chair if he/she is on vacation, sick, on a home-abroad mission, deposition, or not present for other reasons (Article 5/6).

Service units of the ICTA include the main service body (headquarters), composed of legal consultancy, departments, and directorates; advisory and support service units; and local body units organized under the name of regional directorates. Service units are determined in accordance with the by-law proposed by the ICTA and put into force upon the approval of the president of Republic in accordance with the activity field, tasks and functions of the ICTA as defined in Law No. 2813 (Article 5/4).

Thus, in order to ensure EMRA's and ICTA's operational autonomy, the relevant laws mandate that the members of these regulatory authorities cannot be dismissed before the expiry of their terms of office. The laws also provide for EMRA's and ICTA's financial autonomy from the government by empowering it to fund its activities through fees charged to the energy industry. However, although these arrangements are international best practice, there is a concern that appointments and EMRA's and ICTA's autonomy have occasionally been compromised in practice. It has been suggested that the president of the Republic may address the issue by appointing highly experienced energy and financial sector professionals who have wide-spread recognition and market confidence (Dilli and Kari, 2015: 49).

5 Powers and Duties of Regulatory Authorities in Network Industries

Powers and duties are given to ensure that regulatory authorities operate in an effective and coordinated way, with enhanced accountability and clear operational independence.

5.1 Powers and Duties of the EMRA

The EMRA performs the duties and using the powers it has been entrusted by the:

(a) Electricity Market Law No. 4628
(b) Natural Gas Market Law No. 4646
(c) Petroleum Market Law No. 5015
(d) Liquefied Petroleum Gases (LPG) Market Law No. 5307
(e) Electricity Market Law No. 6446.

As addressed above, the objective of the aforementioned laws is to establish a financially viable, stable, and transparent energy market that functions as per the provisions of private law and within a competitive environment to ensure the independent regulation and supervision of the market in order to provide sufficient electricity, natural gas, petroleum, and LPG of good quality to consumers, at a low cost, in a reliable and environment-friendly manner. In order to reach this objective, the Authority is responsible for and entitled to (Activity Report, 2018: 1–2). ***In the electricity market***: Grant and renew licenses defining the rights and liabilities of the legal persons pertaining to their authorized activities; prepare the existing contracts within the scope of the transfer of operational rights as per Law No. 4628; establish performance standards by monitoring the market performances; prepare, improve, and execute the secondary legislation, audit the license holding legal persons; prepare, modify, and execute regulated tariffs; and ensure that the market activities are in compliance with the Electricity Market Law. ***In the natural gas market***: Grant licenses and certificates defining the rights and liabilities of the real and legal persons dealing with the import, transportation, distribution, storage, trade, and export of natural gas; examine the market and system operation; prepare, amend, execute, and supervise the distribution and customer services regulations; examine the cost-reflective prices; prepare, modify, and execute the regulated tariffs; and ensure that the market activities are in compliance with the Natural Gas Market Law. ***In the petroleum market***: Grant licenses for establishing and operating the facilities to perform refining, processing, lube oil production, storage, transportation, eligible consumer and bunker activities, distribution and transport of liquid fuel and vendor activities; determine the special conditions for the scope of the licensed activities and procedures, rights, and liabilities stemming from the licenses, limiting the activities within the scope of the licenses, carrying out the activities and make the necessary arrangements. ***In the LPG market***: Grant, update, temporarily stop, or cancel licenses for establishing and operating the facilities for the purposes of distribution, transport, vendor activities, storage, production of LPG tubes, filling, and maintenance; determine the special conditions for the scope of the licensed activities and procedures, rights, and liabilities stemming from the licenses, limiting the activities within the scope of the licenses, carrying out the activities and making the necessary arrangements.

Thus, the EMRA is an autonomous authority responsible for:

- Regulating and supervising the electricity market
- Monitoring all energy market activities and has the power to:
  - Issue licenses for generation, transmission, distribution, and supply
  - Prepare, enforce, amend, and implement electricity market legislation in order to create performance standards
  - Establish and supervise the tariff pricing mechanism for consumers that legislation prevents from choosing their own retail supplier
  - Impose sanctions for improper activities (e.g., monetary penalties or license suspension); and resolve disputes between license holders.

5.2 Powers and Duties of the ICTA

According to Article 5 of the Law on the Establishment of the Information and Communication Technologies Authority No. 2813, the Board shall conduct the following services and exercise the following powers in addition to the functions assigned in this Law and other legislation:

- The board shall make the secondary regulations regarding the electronic communications sector, taking into consideration the international principles and practices, and make decisions following discussions on the subject matters concerning the missions of the Authority.
- Assign the vice presidents, first legal consultant, heads of departments, and regional directors.
- Determine the strategic plan, performance criteria, objectives and targets, quality of service standards of the Authority, and establish the human resources and working policies to determine the service units and their missions of the Authority.
- Discuss and decide on the budget proposal prepared in line with the strategic plan, objectives, and targets of the Authority.
- Fulfill other duties assigned by laws.

Law No. 7253 Amending Law No. 5651 on the Regulation of Internet Broadcasts and the Prevention of Crimes Committed through Such Broadcasts (Published in the Official Gazette, Date: 31/07/2020 and No: 31202) has introduced significant amendments regarding social media and extending the power of ICTA. The amendments have aimed to regulate social media platforms by introducing "social network providers" into the Internet Law and their obligations in terms of operations. Therefore, the amending law has provided Law No. 5651 with a new Supplementary Article 4, requiring that foreign-based social network providers (SNPs) whose platforms are accessed from within Turkey more than 1 million times a day (qualified SNPs) appoint a real or legal person representative that is authorized to receive service of administrative and court orders and respond to other notifications and requests made by service providers under the law. It also requires that qualified SNPs file a report with the ICTA every six months, reporting information regarding compliance with blocking and removal orders and statistical information on personal blocking and removal requests made by private persons.

Qualified foreign-based SNPs that do not appoint a representative and notify the ICTA of the appointment are subject to a sanctions regime with five steps, with the severity of each step increasing in the face of continued noncompliance:

i. The ICTA is to instruct qualified SNPs that do not have a representative in Turkey to appoint one. SNPs that do not comply within 30 days of the ICTA's notification will face an administrative fine of 10 million Turkish lira.
ii. SNPs that do not appoint a representative within 30 days of their first fine will be fined an additional 30 million Turkish lira.
iii. If the SNP insists on not complying, the ICTA is to forbid the SNP from providing advertising services to real and legal persons whose tax residences

are in Turkey and are thus subject to Turkish income or corporate taxes, as well as bar the conclusion of new advertising contracts and money transfers related to these.

iv. If the SNP still refuses to comply and does not appoint a representative within three months of the issuance of the administrative decision forbidding advertising services, the ICTA may request that a court order the reduction of the bandwidth assigned to the SNP's services by 50%.

v. If the SNP continues to not comply, the ICTA may ask the court to order a 90% reduction in bandwidth. Should the court decide to order a lesser reduction to the bandwidth, this second reduction cannot be less than 50%. On complying at any point of the sanctions regime, only one-fourth of the fines will be payable and the additional sanctions will be lifted.

Thus, it must be addressed that the amending law has considerably increased the administrative and judicial fines for foreign SNPs and obliges them to open communication channels in Turkey. It is argued that the sanctions provided for foreign SNPs are too harsh at the final levels, as 90% bandwidth throttling will effectively block access to such networks.

6 Judicial Review of Regulatory Decisions

Under the lenses of the principal-agent theoretical framework,judicial review is considered as one of the predominant ex-post control mechanisms to overcome the "democratic and accountability deficit" that results from the delegation of regulatory powers from elected politicians to independent bureaucrats (Mejia, 2020: 2). According to Bignami, the allegations of unlawful behavior that plaintiffs bring to courts for judicial review can be theoretically distinguished considering three types of violations to substantive principles of administrative decisions (Cited in Mejia, 2020: 3–4). These principles are fully applicable for Turkish legal system.

- ***Rule of law***: Judges are responsible for enforcing the limits of public administration and ensuring that bureaucracies follow the legislative attributions granted by legal acts, constitutions, or presidential decrees. In this context, a violation to the principle of rule of law is linked with a failure to comply with the purposes and limits set down in laws passed by political principals. Arguments of a breach against this principle in the text of judicial appeals in the world, including Turkey indicate aspects like jurisdictional incompetence, purpose of the decision forbidden by law, violation of the law, error of law, inconsistency with applicable statutes, in excess of statutory jurisdiction and illegality, among other claims.
- ***Fundamental rights***: This principle deals with the protection of fundamental rights and freedoms established by constitutions against public institutions. Examples are very similar for all countries, and these include the lack of proportionality and equality of an administrative act, unfairness and abuse of power, attempts

against the freedom of expression and association, the right to privacy and human dignity, personal liberties, and the right to engage in trades.

- ***Policy rationality***: When the existing discretionary rules and the limits of public administration leave decision making to bureaucratic discretion, the courts assess the legality of an administrative measure against criteria related to sound policy formulation.

As stated above, Turkish regulatory authorities have been established as public legal institutions; therefore, they are all public legal entities, unlike the French counterparts. As the Turkish Constitution requires administrative control over any type of public entity including public economic entities and activities, the mandatory administrative control over public economic entities and activities results from the principle of the unity of the public administration having the constitutional authority (Ulusoy, 2011: 17). The principle of the unity of the public administration is formulated by Article 123/1-2 of the Constitution, which states: "*The* administration *forms a whole with regard to its structure and functions ... The organization and functions of the* administration *are based on the principles of centralization and local* administration." In Turkish administrative law, the principle of decentralization means implementing some services to be offered to the society by statutory bodies established outside the State public entity. The decentralized administration that functions under the administrative tutelage of the central administration is divided into two parts: local administration and functionally decentralized organizations. In other words, the principle of decentralization is performed in terms of location or in terms of services. Implementation of some services and activities requiring technical knowledge and expertise by public corporate entities independently of the central administration is called decentralization in terms of services. Public corporate bodies established in accordance with this system are either public institutions or professional organizations of a public nature. As noted above, regulatory authorities have legal entities. Therefore, they have been accepted as a part of the functionally decentralized organization. Thus, administrative tutelage regulated Article 127/4 of the Constitution as follows over regulatory bodies in the framework of principles and procedures set for by law exists (Günday, 2013: 85).

> The central administration has the power of administrative tutelage over the local administrations in the framework of principles and procedures set forth by law with the objective of ensuring the functioning of local services in conformity with the principle of the integrity of the administration, securing uniform public service, safeguarding the public interest and meeting local needs properly.

In addition, Article 125 of the Constitution rules that all acts and actions shall be subject to judicial review. Due to this clear provision, the acts and actions of economic sectoral regulatory authorities adopted as a part of administration shall be subject to judicial review. There are no decisions of authorities outside the scope of judicial review. Sezen summarized the fundamental issues about judicial reviewas follows (Sezen, 2007: 325–326):

> While some of the board members complain about the insufficiency of sanctions themselves, some others stress that delays in legal proceedings make sanctions ineffective. Another

> problem in judicial review is that sectors are not yet familiar with the idea of regulation by [regulatory authorities], and they have a strong tendency to challenge almost all of [their] resolutions at court; [they] are usually unable to defend their decisions properly. Multinational companies who can start legal proceedings have the advantage of employing very powerful and experienced lawyers and can influence the court, while the [regulatory authorities] have to put their case with a small number of less experienced lawyers. Hence, in the legal proceedings, [they] may stage a weaker defense against the opposing parties. Furthermore, the [regulatory authorities] have a mistrust of courts. They believe they lack the expertise required to judge their collective decisions from expert knowledge of the sector.

7 Conclusion

Regulatory authorities were first established in USA and then in Europe and Turkey following the adoption of privatization and liberalization policies. There are 11 regulatory authorities in Turkey and all, except the Radio and Television Supreme Council, have been established by law and have public legal personality and enjoy administrative and financial autonomy.

Even though the relevant laws of regulatory authorities clearly regulate their administrative and financial autonomy, the possible need to strengthen "regulatory governance" in Turkey has been discussed for a long time. For instance, the OECD Study on Regulatory Reform in Turkey of 2002 (OECD, 2002) underlined the possible need to strengthen "regulatory governance" in Turkey and referred to "*rules for the standards, procedures and principles for the connection to and use of transmission and distribution networks, customer service, and the Balancing and Settlement Code (which establishes the detailed procedures and principles related to real-time balancing and financial reconciliation of the system), which were prepared by the related public industry participants and approved by* EMRA." The same report also stressed that care was needed in these arrangements if the EMRA could not initiate change and could only approve change coming up from industry, as industry participants could be expected to resist socially desirable change that may have been contrary to their own financial interests. In 2020, the same criticism can apply, more or less the same, to the EMRA.

We must also note that recent amending legislation, such as Law No. 5651 on the Regulation of Broadcasts via Internet and Prevention of Crimes Committed through Such Broadcasts and the Regulation on Amending the Regulation on the ICTA's administrative sanctions, have been proposed and came into force without any regulatory impact assessment(Gürkaynak et al., 2014: 1982).

References

Activity Report. (2018). *Energy market regulatory authority*. Available at https://www.epdk.gov.tr/Detay/Icerik/1-2335/emra-activity-reports

Atay, E. E. (2000). Fransız İdare Hukukunda Bağımsız İdari Otoriteler (BİO'lar): Rekabet Konseyi. *Perşembe Konferansları*, Rekabet Kurumu Yayınları, Ankara.
Atay, E. E. (2006). Bağımsız İdari Otoriteler ve Türkiye Uygulaması. *Gazi Üniversitesi Hukuk Fakültesi Dergisi X, 1–2*, 259–293.
Azrak, A. Ü. (2001). Dünyada ve Türkiye'de Bağımsız İdari Otoriteler ve İdarenin Yeniden Yapılanması Bağlamında Bunlara İlişkin Bazı Sorunlar. *Bağımsız İdari Otoriteler*, Rekabet Kurumu Yayını, Ankara.
Bignami F. (2016). Regulation and the courts: Judicial review in comparative perspective. In F. Bignami & D. Zaring (Eds.), *Comparative law and regulation: Understanding the global regulatory process*. Edward Elgar Publishing.
Carrapico, H., & Farrand, B. M. (2016). Dialogue, partnership and empowerment for network and information security: The changing role of the private sector from regulation adopters to regulation shapers. *Crime, Law and Social Change, 67*(3), 245–263.
CEERWG—The Central and Eastern European Working Group. (2020). Available at https://www.ceerwg.net/turkey
Çağlayan, R. (2019). İdare Hukuku Dersleri. Adalet Yayınevi, Ankara.
Dilli, B., & Nyman, J. (2015). Turkey's energy transition milestones and challenges, International Bank for Reconstruction and Development/The World Bank. Available at http://documents1.worldbank.org/curated/en/249831468189270397/pdf/ACS14951-REVISED-Box393232B-PUBLIC-EnergyVeryFinalEN.pdf
Duran, L. (1997). Türkiye'de Bağımsız İdari Otoriteler. *Amme İdaresi Dergisi, 30*, 1–10.
ERRA. (2015). *Independence of the national regulatory authorities*, G. Szörényi (Ed.). Available at https://erranet.org/wp-content/uploads/2016/03/ERRA-Survey-Regulatory-Independence_OPEN_final.pdf
Federal Register. (2020). *Interstate commerce commission*. Available at https://www.federalregister.gov/agencies/interstate-commerce-commission
Günday, M. (2013). *İdare Hukuku*, İmaj, Ankara.
Gürkaynak, G., Yılmaz, İ, & Taşkıran, N. P. (2014). New era for Turkish internet law: Will Turkey become another China or Iran? *Journal of Business and Economics, 5*(11), 1976–1982.
Irion, K., Ledger, M., Svensson, S., & Rsumovic, N. (2017). *The independence and functioning of the regulatory authority for electronic media in Serbia*. Study commissioned by the Council of Europe. Available at https://rm.coe.int/the-independence-and-functioning-of-the-regulatory-authority-for-elect/16808c9c75
Johannsen, K. S. (2003). *Regulatory independence in theory and practice—A survey of independent energy regulators in eight European countries*. AKF Forlaget.
Lodge, M. (2008). Regulation, the regulatory state and European politics. *West European Politics, 31*(1–2), 280–301. https://doi.org/10.1080/01402380701835074
Mejia, L. E. (2020). Judicial review of regulatory decisions: Decoding the contents of appeals against agencies in Spain and the United Kingdom. *Regulation & Governance*. Available at https://onlinelibrary.wiley.com/doi/epdf/https://doi.org/10.1111/rego.12302
OECD. (2002). *Regulatory reform in Turkey: Regulatory reform in electricity, gas and road freight transport*. OECD Publishing, Paris. Available at https://www.oecd.org/turkey/1840779.pdf
OECD. (2014). *Regulatory policy in Mexico: Towards a whole-of-government perspective to regulatory improvement*. OECD Reviews of Regulatory Reform, OECD Publishing, Paris. Available at https://doi.org/10.1787/9789264203389-en
Sayhan, Ş. (2003). Bağımsız İdari Otoritelerin Hukuki Niteliği. *İstanbul Üniversitesi Siyasal Bilgiler Fakültesi Dergisi, 29*, 91–100.
Sever, Ç. (2015). Türkiye'de Düzenleyici Kurumların Yapısı, İşlevi ve Dönüşümü. *Ankara Üniversitesi Hukuk Fakültesi Dergisi, 64*(1), 195–216.
Sezen, S. (2007). Independent regulatory agencies in Turkey: Are they really autonomous? *Public Administration and Development, 27*, 319–332.
Tan, T. (2005). *Ekonomi, Regülasyon ve Bağımsız Otoriteler.* Sempozyum-Danıştay ve İdari Yargı Günü 137. Yıl, Danıştay Yayın No: 72, Danıştay Matbaası, Ankara.

Thatcher, M. (2001). *Delegation to independent regulatory agencies in western Europe.* Paper presented at the 29th ECPR Joint Session of Workshop, Grenoble, France.
Ulusoy, A. D. (2000). *Regülasyon Kurumları Hakkında Genel Bir Değerlendirme.* 2000 Yılında İdari Yargı Sempozyumu, Ankara, Danıştay Yayını No. 59.
Ulusoy, A. D. (2003). Bağımsız İdari Otoriteler. Turhan Kitabevi, Ankara.
Ulusoy, A. D. (2011). The place of regulation within the public law in Turkey. In T. Çetin & F. Oğuz (Eds), *The political economy of regulation in Turkey.* London, Springer.
Ulusoy, A. D. (2019). *Yeni İdare Hukuku.* Yetkin, Ankara.

Websites of Regulatory Authorities in Turkey

Radio and Television Supreme Council, https://www.rtuk.gov.tr/en
Information Technologies and Communication Agency, https://www.btk.gov.tr/
Capital Markets Board, https://www.cmb.gov.tr/
Banking Regulation and Supervision Agency, https://www.bddk.org.tr/Home
Energy Market Regulation Authority, https://www.epdk.gov.tr/Home/En
Public Procurement Agency, https://www.ihale.gov.tr/
Turkish Competition Authority, https://www.rekabet.gov.tr/en
Public Oversight, Accounting and Auditing Standards Authority, https://www.kgk.gov.tr/Home
Personal Data Protection Authority, https://www.kvkk.gov.tr/en/
Nuclear Regulatory Agency, https://ndk.gov.tr/tr/
Insurance and Private Pension Regulatory and Supervision Agency (Not available as of August 2020).

Engin Saygin graduated from the Faculty of Law, Ankara University in 2000. He received his master degree (LL.M) from the University of Essex, the United Kingdom in 2003. He started his Ph.D. at Exeter University in 2004 and completed at Brunel University in 2007. His Ph.D. has been supervised by Prof. HélèneLambert and he passed his viva successfully with no revision with the title of "Improving Human Rights Compliance though National Human Rights Institutions: An Ombudsman for Turkey". After receiving his Ph.D., Mr. Saygın joined at Erzurum Atatürk University in 2009 and served as an assistant professor at the Faculty of Law until 2014. Currently, he is an associate professor at the Department of Administrative Law of the Faculty of Law, Ankara Yıldırım Beyazıt University. He teaches the courses of Introduction to Law, Constitutional Law, Administrative Law and the Law of Administrative Procedure at undergraduate level in English. Mr. Saygın also carried out researches as a visiting academic at the University of Groningen, the Netherlands in 2013 and the University of Manchester, the United Kingdom in 2019.

Relationship Between Competition Law and Sector-Specific Regulations

Şahin Ardıyok and Armanç Canbeyli

Abstract The relationship between national regulatory agencies (NRAs) and national competition agencies (NCAs) plays an important role in terms of having a harmonized enforcement and an effective free market economy. Accordingly, this chapter focuses on the interaction between NRAs and NCAs, while setting out the structural and practical distinctions between their respective regulatory functions. Along these lines, we also distinguish the technical regulations from the competition regulations and discuss the risk of double sanctioning, as well as exploitation of jurisdictional overlaps via forum shopping. In an effort to address these issues, we evaluate the efficiency of three essential methods: (i) legislative rules, (ii) bilateral cooperation protocols, and (iii) judicial review. Additionally, this chapter assesses the filtering method as an alternative for allocating the regulatory duties. In essence, this method provides that if the sector-specific regulation is capable of remedying the competition law failures in the market, the competition authorities may not engage in such failure. That said, if the behavior in question violates both the sectoral regulation and the competition regulation, but the sectoral regulation is not sufficient to remedy the antitrust concerns, then both authorities may engage with the issue based on their own legislations.

Keywords Competition · Regulation · Jurisdictional overlap · National competition agencies · National regulatory agencies · NCAs · NRAs · Technical regulations · Economic regulations · Forum shopping · Double sanctioning · Ne bis in idem · Ex ante regulations · Ex post regulations · Related market infringements · Filtering method · Turkey · Enforcement

Ş. Ardıyok (✉) · A. Canbeyli
Balcıoğlu Selçuk Ardıyok Keki Attorney Partnership, Istanbul, Turkey
e-mail: SArdiyok@baseak.com

A. Canbeyli
e-mail: ACanbeyli@baseak.com

M. Eroğlu and M. Finger (eds.), *The Regulation of Turkish Network Industries*,
https://doi.org/10.1007/978-3-030-81720-6_24

1 Introduction

Regulation has been one of the most popular topics of discussion among legal scholars in Turkey for decades. From technical regulations to economic regulations and sector-specific regulations, the question started as *why* regulation is needed and then evolved into *where* regulation is needed. Although the debates on these questions have covered reasonable ground, one question that still needs more discussion is *Who should be responsible for regulation?* Behind this question is the fundamental issue of the jurisdictional friction between the national regulatory agencies (NRAs) and the national competition agencies (NCAs), and the search for best practice requires us to differentiate between the strong suits of the NCAs and NRAs. Such a distinction will then be used to allocate responsibility for technical regulations, economic regulations, and competition law enforcement. In this study, we seek to diagnose the key issues that are important for a fair and effective distribution of authority among NCAs and NRAs in Turkey.

We start by assessing the root causes of the jurisdictional frictions between the NRAs and the NCAs. Thus, we examine the necessities and functions of the competition regulations, as well as their sector-neutral application. In the same vein, we make a distinction between technical regulations and competition regulations. We argue that technical regulations require an in-depth sectoral experience and know-how and would therefore ideally fall within the jurisdiction of the NRAs.

On the other hand, also argue that NCAs have a better understanding of the competition area and, therefore, should take charge of enforcing the competition rules. While we focus on the Turkish jurisdiction, we will also assess the United States (US) and European Union (EU) examples on the issue and elaborate on the methods for clarifying the jurisdictional overlap in Turkey. Lastly, we delve into a proposed solution for the jurisdictional friction by establishing a tentative structure (namely, the filtering method).

2 The Need to Protect Competition and the Basic Characteristics of Enforcement

In order understand the need to protect competition, the roots of this regulatory discipline can be traced back into the history of economics. Competition rules made a sensational entrance at the end of the 1800s. Before the Sherman Act of 2 July 1890, introduced by Ohio Senator John Sherman, the USA was going through rapid industrialization and one of the terms that defined the era was "consolidation." In a demonstration of what the free market economy looks like without the balancing impact of competition rules, so-called robber barons took pride in how they managed to consolidate the market and increase their profits. With John Davison (J.D.) Rockefeller controlling an estimated 90% of the oil industry as well as the increasing grip of John Pierpont (J.P.) Morgan over the railway industry, there was an overwhelming need to balance the scale via antitrust rules. Along the lines of the *three John's rumble*[a] came the rise of Theodore Roosevelt and his trustbuster policies.

Competition rules aim to safeguard the benefits arising from the concept of Competition, the technical and legal meaning of which includes both the state of being in Competition and the state of the market structure (Ateş, 2008). Hence, the natural habitat of enforcement for Competition rules is the consolidation of bilateral and multilateral commercial relationships that we have come to know as the modern economic system.

In an effort to maximize consumer welfare (and eventually the total welfare via innovation and creative destruction), the need to protect Competition led enforcers to impose a set of regulatory rules onto the markets for goods and services. The main goal of the legislators with Competition legislation was to ensure that the market players were not refraining from Competition to the advantage of profit maximization and the disadvantage of welfare maximization.

At this sharp intersection of law and economics, NCAs are expected to protect consumer welfare whilst refraining from distorting the natural economic order. Operating on a fine line, adaptability and flexibility have proven to be some of the core assets for this field of enforcement. In an effort to meet the challenges of this multidimensional task, competition law practice needed to introduce sector-neutral rules in order to draw a line in the sand and set the basic principles that were applicable regardless of the sectoral differences.

However, in the ever-developing world of commerce, new markets are constantly being created while pre-existing markets are being redefined by new innovations and technology. Hence, competition rules had to be adaptable to the different characteristics of different markets, as well as to changes in the characteristics of the same markets because of commercial and technological advances.

Therefore, one of the basic principles to be recognized in this field is the peculiarity of the sector neutrality requirement, which not only means that the same set of rules are applicable throughout all markets, but also requires those rules to be enforced on a case-specific basis with a tailored-to-adapt legislative backbone. In other words, antitrust enforcement needs to fully understand the legal and economic circumstances surrounding each case and respond with a bespoke enforcement action that recognizes the specific needs of the given case without deviating from the core principles underlying the need to protect competition.

Against this backdrop, protecting competition in the markets includes preventing the anti-competitive actions, along with mergers and acquisitions that may hinder competition. . Accordingly, NCAs are tasked with and equipped to trace the behavioral market failures caused by the actions of market players and penalize any violation. Such duties are mostly exercised on an ex post basis, except for the merger control regime, which operates as a preventive review mechanism.

However, behavioral failures traced by ex post enforcement are not the only type of market failures. Markets may also suffer from structural/permanent failures, which may require the firm presence ofex ante regulations designated for the sector or a sector-specific regulatory agency. Examples of such markets include railways, electricity, natural gas, and telecommunications, where there is a tendency for natural monopolies and network externalities. Sectors such as banking or securities may also require NRAs due to their complex structures and the information asymmetry (Ardıyok, 2019).

3 Allocation of Regulatory Duties

In accordance with the above explanations, we argue that various factors create the need for sectoral regulations and competition enforcement. Despite their sector-neutral characteristics, competition regulations are not fit to remedy every kind of market failure throughout all sectors.

One of the reasons for this is that some market failures, by their nature, exceed the bounds of competition law knowledge and require sectoral insights that an NCA may lack. Additionally, competition regulations are mostly designed on an ex post basis in order to intervene only after a market failure has occurred and been detected. However, this approach is often insufficient when we put it up against a permanent/structural market failure, which rather requires ex ante action.

In short, it can be argued that competition enforcement primarily aims to intervene in temporary or behavioral market failures. On the other hand, sectoral regulations are crafted to tackle the permanent and structural failures in the specified market.

Accordingly, allocating the regulatory duties in a structurally efficient and practically beneficial way is crucial for establishing a healthy and sustainable regulatory framework. There are two main questions that may navigate us through this shark tank:

(i) Which agency would enforce the relevant regulations more successfully?
(ii) Which agency would enforce the relevant regulations at a lower cost (in terms of time, resources, and efficiency)?

Expecting NRAs to take charge of sectoral regulations while NCAs enforce the competition rules is easier said than done. In other words, it is often challenging to make a clear distinction on when or to what extent an action falls within a given agency's jurisdiction. Different countries or even different sectors in a country may have different needs, and the answers to these questions may require different approaches. The new institutional school theories consider various elements, such as the size of the industry, the economies of scope, experience of economic regulation, and the regulation background of the relevant country. Therefore, we see different applications of these theories in different countries. By way of example, we see public utility commissions active in different sectors and sectoral regulators active at an interstate level (such as the FCC or FECR) in the federal state structure of the USA, where regulated markets on an individual state level are relatively small. On the other hand, Australia is an example of economy of scope and dedicates a department in the Australian Competition and Consumer Commission for the telecommunication sector instead of establishing a separate regulatory agency. Contrary to this approach,

Oftel in the UK used to regulate not only telecommunications issues, but also the competition rules in the telecommunication sector. Lastly, the same regulations may be enforced by different agencies, as exemplified by the USA, where the Competition rules are enforced by both the Antitrust Division of the Department of Justice and the Federal Trade Commission.

Accordingly, the pursuit of best practices in regulation should have a pragmatic perspective to better utilize the time and effort that will be put into regulatory enforcement by governmental institutions as well as market players. Therefore, we should assess managerial scope and transaction economics as well as all costs to determine the allocation of regulatory duties (Ardıyok and Oğuz, 2010).

3.1 Who Should Enforce Technical Regulations?

In light of the foregoing explanations, we see that competition law enforcement has a sector-neutral character that allows it to apply across all industries. While this is the natural course of enforcement for competition rules, it may also cause jurisdictional frictions between the NCAs and the NRAs. In an effort to avoid such friction, legislators and enforcers should distinguish between technical regulations, economic regulations, and competition rules..

Technical regulation relates to ensuring compatibility and standardization in the relevant market. Hence, technical regulation require an accumulation of immense expertise and industry-specific knowledge, which may require engineers or technicians to work alongside legal experts. An example of such efforts comes from the telecommunications industry, where the Turkish Information Technologies and Communication Authority (ITCA) operates a market surveillance laboratory to establish test and experiment procedures, follow standards, update procedures, and carry out tests for system compliance.

Accordingly, due to the required level of sectoral insights and technical expertise, we may conclude that NRAs are indeed the most viable candidates to carry out the technical regulations designated for a given market. In the same vein, we believe it is also beneficial to assess the allocation of duties for the economic regulations.

3.2 Who Should Enforce Economic Regulations?

Economic regulations require an in-depth understanding of the relevant industry, which is easier for the NRAs to access due to their sectoral capabilities. Regulating the economic parameters not only includes establishing the relevant rules, but also requires periodical follow-ups to assess whether the rules are applied properly.

Hence, NRAs constitute a better option to enforce the economic regulations than the NCAs. It is also worth mentioning that economic regulations are interlinked with the technical regulations in the sense that they require detailed cost assessments based on the sectoral dynamics. Therefore, delegating both the technical regulations and the economic regulations to the NRAs would potentially create a significant synergy.

That said, it is also important to assess the role of the NCAs in economic regulations. Normally, the NCAs would have a supportive or balancing function in economic regulations. By way of example, due to the circumstances of a given case, the NCAs may be better positioned to collect or process the market data required to verify whether the economic regulations are applied properly. There are even examples such as Australia, where the NCA takes charge of the economic regulations. On the other hand, undertakings may try to influence NRAs in an organized manner to shape the economic regulations in their favor. Accordingly, NCAs have a more coincidental relationship with the undertakings compared to the rather relationships of the NRAs. Hence, NCAs may also aid NRAs as a balancing factor against such institutional blindness that may decrease the efficiency and accuracy of the economic regulations.

3.3 Who Should Enforce Competition Rules?

On the flip side of the coin, and by extension, we also need to determine who will be responsible for enforcing the competition rules. The peculiarity of this answer emerges from the fact that competition rules are enforced throughout all sectors unlike technical regulations, which have a sector-specific enforcement.

Therefore, the question here boils down to choosing between (i) ***NCAs***, with vast competition law expertise, but limited sectoral insights, and (ii) ***NRAs***, which have economic regulation capabilities, but only limited understanding of competition rules. Accordingly, we need to reveal the differences between these two types of organizations' perspectives.

NCAs mostly tackle behavioral market failures and decide whether an action violates the competition rules, while the sectoral regulations focus on structural market failures and therefore are perceived to have a rather permanent function (Şahin, 2010).

Denis Carlton answered this ever-important question by saying that "Whoever can say no more easily protects competition better" (Ardıyok, 2019). Here, saying no stands for the authorities' ability to resist political pressures as well as pressure by interest groups. According to Carlton, the most successful ones at saying no are judges, followed by the NCAs (Ardıyok and Oğuz, p. 234) and thirdly the NRAs. Indeed, competition rules and NCAs are less affected by political pressures, which means that competition become less subject to change than other regulatory rules (Ulusoy, 2016, p. 9).

Accordingly, we may argue that NCAs, which have an accumulated expertise, experience, and institutional culture in competition, will perform better than the NRAs at protecting markets against non-competitive behaviors and reviewing mergers and acquisitions (Akçollu, 2003, p. 24).

4 Relationship Between Regulatory Agencies

Thus far we have touched upon the need for competition enforcement and the functions of sectoral enforcement. In the same vein, we have also discussed the peculiarities of the technical and economic regulations and the competition rules. From a structural point of view, all these issues revolve around the quintessential question of "*Who should be responsible for which* regulation*?*"

In an effort to answer this question, we consider it important to define the characteristics of the relationship between sector-specific regulations and competition law enforcement. Accordingly, sector-specific regulations and competition rules feature certain functional differences. As we emphasized in the foregoing sections, sector-specific regulations are mostly designed to tackle permanent/structural market failures. For this reason, the NRAs have more insights into the relevant market and are equipped with in-depth technical and economical knowledge/expertise, including information on costs and other dynamics used for price regulations.

On the other hand, competition rules are a better fit to combat behavioral market failures, with the advantage of sector-neutral enforcement. Competition rules also bring a more efficient mechanism for resource distribution because they both encourage companies to reduce their costs and pass those cost reductions on to consumers (TÜSİAD, 2014, p. 50).

Despite these functional differences, sectoral regulations and competition rules both pursue the same ultimate goal of remedying market failures. Considering that both regulatory disciplines aim to contribute to social welfare through better performing markets, we argue that the relationship between these regulatory disciplines should complement each other. In other words, the best way to optimize the respective functions of sector-specific regulations and competition rules is to establish a complementary relationship, where both regulators will enforce their expertise in harmony and efficiency.

In an effort to emphasize the macroeconomic importance of the harmony between competition regulations and the sector-specific regulations, the OECD stated: "As a result of the regulations maintained in line with the competition policies, it has been observed that the related economies have become more resilient to internal and external crises" (OECD, 2006).

In the absence of such a harmonious and complementary framework, regulators may start competing against each other to further and better regulate the market. Additionally, market players may also exploit this lack of harmony by engaging in so-called forum shopping. In such cases, authorities may produce different judgments on the same set of actions, which would deprive the markets of reliable jurisprudence and the protective cloak of legal foreseeability.

Hence, the NRAs and NCAs should refrain from simultaneously engaging in resolving a market failure. Double sanctioning is one of the greatest risks of not having a configured regulation framework. In other words, when the rules of engagement and the distribution of jurisdictional authority are not clear and harmonized between the regulatory agencies, this harms the complementary function of these agencies and violates the principle of "*ne bis in idem*". Accordingly, the section below discusses the risk of double sanctioning and the application of the *ne bis in idem* principle in regulatory affairs.

4.1 The Unlawful Act of Double Sanctioning a Single Infringement

Legislators and enforcers should keep in mind the old adage that "lightning never strikes twice" when dealing with jurisdictional overlaps between sectoral regulations and competition rules. Considering that both NRAs and NCAs are chasing the same ultimate goal of remedying market failures, the same action may be caught by both of the agencies.

Within the realm of economic regulation, an undertaking can receive different sanctions from different regulators for substantially the same set of actions. However, sanctioning the same undertaking via two different regulators would destroy the complementary harmony we drew attention to above.

The risk of double sanctioning is undoubtedly one of the most important points of caution in dealing with regulatory affairs. The main line of defense against such a risk is the ever-important principal of *ne bis in idem*. Originating from criminal law, this principle aims to ensure that an action is not prosecuted and sanctioned more than once.

Although the original intention of this principle was to protect people from double sanctioning for the same crime, the legislative rationale behind this notion exceeds the bounds of criminal law. When we consider the hefty fines and consequences of regulatory enforcement today, double sanctioning in regulatory enforcement emerges as a risk to be reckoned with.

Accordingly, we need to assess how this principle may be enforced in the field of regulatory enforcement. In so doing, we should begin with the landmark the Aalborg Portland[b] decision judgment from the European Court of Justice (ECJ), which is regarded as the primary ECJ decision that sets out the principles of the application of *ne bis in idem* in competition law (Karabel, 2015, p. 23). The Aalborg Portland decision provides that the event, the person, and the benefits protected by law should be the same in order to give raise to the application of *ne bis in idem* in competition law.

Ne bis in idem is also incorporated within Turkish law via Article 15 of the Torts Law, which states that, in the event that a single action results in more than one tort, the heaviest of the administrative fines defined for the tort violations will be imposed. Considering that the Torts Law is applicable to every violation that is deemed as a tort and subject to an administrative fine, we assess that the violations defined in the regulations and the competition legislation may also be assessed within this scope.

When we look at Turkish jurisprudential practice, we see that the Constitutional Court recognizes *ne bis in idem* as a primary principle of the state of law.[c] Similarly, the High Court of Appeals recognizes that a person cannot be punished twice for an action as long as (i) the person is the same and (ii) the action is the same.[d]

After touching upon the place of *ne bis in idem* in jurisprudence, it is worth examining this principle within the context of regulatory enforcement in Turkey. The Competition Board has recognized this principle in its Izmirgaz decision,[e] where the behaviors subject to investigation concerned both the competition rules and the rules set out in the Natural Gas Market Law No. 4646. Upon preliminary investigation, the Competition Board detected considerable signs of an abuse of dominance. However, the Competition Board reasoned that the relevant behaviors were also in violation of the natural gas legislation and the Energy MarketRegulatory Authority (EMRA) was already taking action to ensure that the violating behaviors were stopped. Accordingly, the Competition Board decided to close its preliminary investigation, considering that EMRA's enforcement would be sufficient to stop the violation and it would be better if the two agencies cooperated on the case. On the other hand, in the Enerjisa decision[1] the TCA investigated and penalized the abuse of dominance issues, despite the fact that the relevant behaviors were regulated by the *ex ante* regulations of EMRA.

In light of the foregoing, we find that double sanctioning the same person for an action that violates both the competition regulations and the sectoral regulations in markets with permanent failures and a lack of competition would go against the principles recognized by Article 15 of the Torts Law. In an effort to fully diagnose the double sanctioning risks, we should also focus on the type of markets that are structurally more inclined to experience such problems. Accordingly, we see that double sanctioning risks are often featured where there are related market infringements, where the market is open for competition, but at the same time has potential exposure to the economic regulations due to its ties with the upstream market. Considering that related market infringements constitute a natural interception point between competition rules and economic regulations, it is worth assessing the jurisdictional dynamic in these markets.

4.2 *Defining the Institutional Jurisdiction Over the Related Market Infringements*

We have argued above that the regulatory agencies in the markets with structural market deficiencies should carry out economic regulations. We also noted that *ex ante* regulations usually fall within the jurisdiction of NRAs, while *ex post* regulations to ensure competitiveness are better enforced by NCAs. Accordingly, the allocation of jurisdiction over ex ante and ex post application of regulations is one of the challenges of harmonizing regulatory practice.

In a market with an NRA, that regulator may invoke ex ante rules and oblige undertakings to act in certain ways. However, some of the actions allowed by the *ex*

[1] Refer to the TCA's decision dated August 8, 2018 and numbered 18-27/461-224.

ante rules may end up distorting competition in the market, which would require the NCA to take action. To that end, harmonizing this jurisdictional friction is essential for establishing healthy enforcement.

The first question to ask in such a case is whether the *ex ante* regulation of the NRA restricts the willful behaviors of the market players. If the investigated behavior is not ex ante regulated by the NRA or the ex ante regulations provide the market players with the opportunity to choose from a set of different actions (and the chosen action violates the competition rules), we can conclude that the NCAs should have the authority to investigate such actions. This approach is also an extension of the above-emphasized complementary relationship between the agencies.

Accordingly, a market player may distort the competition in the market by violating a positive obligation imposed by the NRA. In such cases, sectoral regulations have a special rule status and their application should be prioritized over the competition regulations (even if the action in question distorts competition). On the other hand, if the anti-competitive actions of a market player fall within a section of the regulatory framework, where it is free to act according to the market mechanism, than the NCAs should have jurisdiction over the matter, despite the fact that the relevant action was against the intended purposes of the economic regulations.

This may also be explained by the fact that economic regulations usually kick in when the market is not able to restore the expected level of competition due to certain structural/permanent market failures. competition rules, on the other hand, are naturally designed to filter and catch the anti-competitive behaviors within a market that is open to competition. Considering that the sector-specific regulations are imposed to create market outputs that are close to competitive markets, the test methods of the competition rules would not constitute an accurate tool. Accordingly, we can conclude that the economic regulations should be a substitute for the competitive pressure that disciplines the market (Dunne, 2015, p. 35).

By way of an example, assume that an NRA has the authority to set the prices in the relevant market. If it does so, disregarding the determined pricing would be a violation of the ex ante regulations. However, violating the price determination may also constitute abusive pricing and violate the competition regulation. In such a case, even if the pricing constitutes a competition law violation, the NRA would have authority over the violation, considering that the relevant market player has violated an ex ante regulation (that is, price determination).

However, if the NRA had imposed a price ceiling instead of setting a price, the firms in the relevant market could have determined their own prices within the given limits. In such a case, if the pricing of a relevant firm constitutes predatory pricing or amounts to a price squeeze, the NCA would have the authority to regulate such behavior despite the existence of the ex ante regulations. This is because the relevant market firm is not violating the ex ante price ceiling while imposing predatorily low prices. That said, such pricing is still distorting the competition in the market, and is therefore subject to the NCA's authority. An example of this approach can be traced in the 2008 TTNET decision of the TCA, where it determined that TTNET was engaged in a price squeeze in the retail broadband Internet services market by abusing its dominance in the wholesale broadband internet services market. Taking a similar approach with the Telefonica decision and in contrast to the Linkline decision in the EU, the TCA resolved that the relevant behaviors constitute a price squeeze.

In an effort to utilize the public sources effectively and ensure that each regulatory task will be allocated to the agency that is the most capable one for the task, we think it is beneficial to follow the foregoing pattern and define the jurisdiction on a case-specific basis.

4.3 Coordination of Regulatory Agencies in the USA and the EU

Competition law and sectoral regulations are, at their core, case-law-based practices. This is partly because of the sectoral flexibility and the need for adaptation that we discussed above. Hence, the jurisprudential guidance set forth within the enforcement history is important for the related parties when navigating their behaviors in a specific market or situation.

However, the reference to case law in this field of enforcement does not only mean the judgments of the relevant country. While domestic case law is one of the most important assets of enforcement, agencies should also keep track of the developments in other jurisdictions.

One reason for this is that we are looking at a rapidly globalizing world and the behaviors investigated by regulatory agencies begin to have more and more connection with other jurisdictions. Major players in digital markets, such as Google and Booking.com, are often investigated by regulators in different jurisdictions for similar reasons. Hence, it is important to utilize the best practices around the world for healthy enforcement.

On the other hand, since the establishment of the first modern NRA, the ICC, competition rules and regulatory practices have historic roots in the US market and the principles set therein have helped to shape the system of regulation we have come to know today. Across the Atlantic, we see that the Turkish competition law regime is closely modeled on the European system. Therefore, it also important to follow developments in Europe, which have a "North Star" kind of influence on enforcement. Accordingly, before looking more closely at the Turkish practice, it is worth delving into the key developments and case law that have set the basic concepts of the coordination among competition regulations and sectoral regulations.

When we assess US practice, we see that the foundation of the relationship between competition regulations and economic regulations was shaped by the Verizon-Trinko[f] and Linkline[g] decisions of the Supreme Court (Ardıyok, 2019).

At the time of the review, Verizon-Trinko had a monopoly over telecommunications infrastructure and was placed under a set of positive behavior obligations via regulation. In its decision, the Supreme Court assessed whether Verizon-Trinko had hindered its competitors' activities in the downstream market, where they needed to utilize Verizon-Trinko's infrastructure.

Against this backdrop, the Supreme Court assessed whether Verizon-Trinko's behaviors would be in violation of the competition rules if there have not been any economic regulation in place, and concluded that the only reason for Verizon-Trinko to grant access to its competitors is the obligation to do so imposed by the regulatory agencies. The Supreme Court also ruled that it was not possible to bring such a positive obligation solely based on the competition rules. Hence, the foregoing approach indicates that the competition rules will be enforced in regulated markets in the USA; however, it is crucial to separate actions that violate the competition rules from the actions that violate the economic regulation.

On the other side of the Atlantic, the benchmark decision to guide European practice also concerns the telecommunications sector, namely the Deutsche Telekom (DT)[h] decision. Similar to the US example, DT was in a monopoly position in some of the elements of the telecommunications infrastructure and subject to detailed regulations in both the upstream and downstream markets. However, unlike in the US example, the EU system exceptionally allows the NCAs to impose an obligation to supply. Hence, price squeezes could be caught by the competition rules if they are not caught by the sectoral regulations.

Assessing the allegations of a price squeeze by DT, the Commission first evaluated whether DT's infrastructure constituted an essential facility for the undertakings active in the downstream market. The Commission argued that the NRA had already obliged DT to provide access to its infrastructure, which showed that the infrastructure was indeed an essential facility. Accordingly, the Commission ruled that it was not necessary to issue an additional assessment based on the competition rules. In fact, this approach of the Commission is echoed in the Guidelines (Commission, 2009), which defines price squeezing as a type of refusal to deal, which required the input subject to violation to be essential for downstream activities (Commission, 2009, para. 80). The Guidelines also assert that the NCAs may not have to engage in this assessment if the NRA has already assessed this before imposing an obligation to supply (Commission, 2009, para. 83). On a related note, the necessity of essential facility status of the relevant input in price squeeze cases originates from the Commission's National Carbonising (NC),[i] British Sugar,[j] and IPS[k] decisions, which provide that the essential facility status is necessary to attribute responsibility (Hou, 2012). Accordingly, we can conclude that an action in violation of an economic regulation, the aim of which is to increase the competition in the market, would not automatically mean that the same action violates the competition rules.

4.4 Coordination of NCAs and NRAs in the Related Markets—The Turkish Practice

The previous sections have assessed the basic principles of how to allocate the regulatory duties, along with case law examples from the US and the EU. We also delved into the risks of jurisdictional overlap and double sanctioning.

Against this backdrop, we now assess the legal tools used in Turkey to establish and maintain cooperation and harmony among the regulatory agencies. Before assessing the details of the Turkish practice, we feel it is beneficial to touch upon the overall look of the regulatory environment in Turkey. In terms of the functioning of the institutional structure, certain legislative boundaries define the jurisdiction of the NCA and NRAs. These provide legal foreseeability in enforcement and serve as a primary source of reference to provide a framework for the jurisdictional overlaps.

However, when we assess the Turkish legislators' take on the issue, we see that it mostly provides the general principles and does not outline a detailed solution. Accordingly, much like the other jurisdictions, the legislative determinations are not entirely sufficient to sort out the jurisdictional issues between the NCA and the NRAs in Turkey. This is partly because of the inherent need for case-specific adjustments for this kind of problem and the strategic elbow room left by the relevant legislations to allow such adjustments. The dynamic and developing nature of Turkish enforcement also necessitates maintaining an active communication between the institutions.

Due to the foregoing peculiarities of the Turkish regulatory regime, we argue that it is important to support the legislative determinations with bilateral cooperation channels between the relevant institutions. This will give the Turkish institutions the opportunity to better understand their respective positions in any given situation and also allow them to harmonize different interpretations of the legislative boundaries.

Even though bilateral cooperation has proven to be an effective method of harmonization, the accumulated enforcement experience in Turkey demonstrates that judicial involvement remains critical in enforcement. In other words, while the legislative framework and bilateral cooperation mechanisms help to harness the potential for harmonization, we still need the judicial authorities to serve as a referee and provide jurisprudential guidance on how to implement harmonization.

In light of the foregoing peculiarities of the Turkish system, we break our analysis down into three topics: (i) enacting legislation to define boundaries, (ii) signing cooperation protocols among the regulators, and (iii) clarifying the jurisdictional issues via judicial review (Ardıyok, 2019).

4.4.1 Enacting Legislation to Define Boundaries

By virtue of Articles 1, 2, and Article 27 of the Competition Law, we see that the Turkish Competition Authority (TCA) and its decisional body the Competition Board are authorized to investigate and penalize (if necessary) the anti-competitive actions of all market players throughout the country. Additionally, the anti-competitive behaviors are defined in a broad manner in this legislation due to their general norm status (Gürzumar, 2006).

Although the determination of whether an action violates the Competition Law or not falls within the Competition Board's jurisdiction, the same action may also be in violation of a sectoral regulation. The initial examples of such jurisdictional friction occurred between the TCA and the ITCA. This is because the electronic communications market was opened for Competition before the other regulated markets, which made the accumulation of empirical experiences possible before the other examples.

Therefore, we may argue that the spearhead example of legislation that directly deals with the cooperation of regulatory agencies is the Electronic Communications Law No. 5809, where there are provisions concerning the relationship with the TCA and the ITCA. The European counterpart of this legislation, on which it is modeled, creates the so-called *three criteria test*. Accordingly, agencies may remedy a market failure by way of regulation only if (Commission, 2014): (i) the barriers to entry are high and permanent, (ii) it is not possible for the market to establish a competitive structure on its own in the future, and (iii) market failures cannot be solved solely by the application of Competition rules. This approach is based on the idea that free markets perform better than regulated markets and *ex ante* regulations regulations are always the second-best option.

Against this backdrop, we think it is beneficial to assess the relevant provisions in the Electronic Communication Law that deals with the jurisdictional boundaries. The Electronic Communications Law determines the ITCA's duties and regulative authorities in a broad manner. By virtue of Article 1, the ITCA has the duty to establish Competition in the relevant markets. However, this is not the same as the duty to maintain and protect competition. This difference is important for understanding the allocative differences between the responsibilities of the TCA and the ITCA. Accordingly, the ITCA is responsible for remedying the structural failures and trying to open the market to Competition, while the TCA is responsible for tackling the behavioral market failures caused by the anti-competitive actions of the market players.

Moreover, Article 6-1(b) of the Electronic Communications Law provides that the ITCA has the authority to investigate and penalize the anti-competitive actions that occur in violation of the Electronic Communications Law and its secondary legislation. This provision is important for distinguishing the lines between the ITCA and the TCA in the sense that it authorizes the ITCA to intervene in Competition violations that are against the regulations of the electronic communications legislation. Similarly, Article 7-3 gives the ITCA the authority to determine the firms that have significant market power and bring additional obligations on them to establish and protect a competitive structure.

Accordingly, once the ITCA brings an obligation based on Article 7-3, violations of such obligations may fall under the sectoral competition violations defined in Article 6-1(b) and caught by the ITCA. Article 7-1 also guides the ITCA in exercising its authority and provides that it may investigate and remedy anti-competitive behaviors in the electronic communications sector, with or without a complaint. The same provision also sets out that the ITCA is entitled to collect information on matters that fall within its duties. Needless to say, these provisions should not be interpreted as giving the ITCA the authority to determine whether or not a behavior violates the Competition Law.

Lastly, Article 6-1(b) also opens up a cooperation channel for the ITCA to take the TCA's opinion on the competition violations in the electronic communications sector. On the flip side of this coin, it is mandatory for the TCA to take ITCA's opinion on telecommunication-related investigations, due to the fact that the NRA is more knowledgeable and experienced on such issues.

Based on the foregoing explanations, it is worth mentioning that even if the relevant law has directly regulated the allocation of duties, the overall language is still open to interpretation on different levels. We think this is because of an intended legislative policy to provide flexibility to switch between ex ante and ex post regulations. Considering that the competitive markets generally perform better than the regulated ones, we think it may be beneficial to have a dynamic legislative infrastructure that sets out the main principles, instead of having detailed and rigid microeconomic regulations.

4.4.2 Signing Cooperation Protocols Among the Regulators

Bilateral protocols asserting the basic principles of cooperation are another tool for harmonizing the regulative interaction between the NCAs and the NRAs. Examples of such protocols are featured between the TCA and the ITCA, the EMRA, the Banking Regulation and Supervision Agency, and the Public Procurement Authority. Considering that the telecommunications sector has been experiencing these jurisdictional frictions longer than the other sectors, we will briefly assess the cooperation protocol in this sector to understand the general outlook of such protocols.[2]

The protocol between the TCA and the ITCA defines the main aim of the cooperation as preventing so-called "forum shopping", where undertakings apply to several agencies for the same problem. In terms of jurisdictional boundaries, the protocol mainly asserts that the ITCA will be engaged in the process only if the relevant action is an abuse of dominance and also is in violation of the *ex ante* regulations. However, this protocol is not legally binding and is instead designed as a gentleman's agreement between the two agencies. Hence, the administrative judicial authorities are also not bound by the protocol and may disregard it during an individual case.

In terms of cooperation, the protocol provides that the TCA must ask for accept the ITCA's opinion when investigating the electronic communications sector and take into account the economic regulations while rendering its decision. On the other hand, the protocol also provides that the ITCA may ask for the TCA's opinion while reviewing competition violations in its designated sector. However, the protocol notes that the ITCA shall have the TCA's opinion while making market assessments.

4.4.3 Defining Jurisdictional Boundaries by Judicial eview

The previous section touched upon the place and function of legislative provisions, as well as the bilateral cooperation protocols for harmonizing regulatory affairs. Accordingly, allocating jurisdictional duties and authority in the legislation provides

[2] For detailed information, please see https://www.btk.gov.tr/uploads/pages/elektronik-haberlesme-sektorunde-pazar-analizleri/btk-rk-protokol-22012015.pdf (last accessed August 9, 2020).

the interested parties with an increased level of foreseeability. However, we also understand that the legislators intentionally refrain from detailed provisions and rigid allocation rules to leave room for dynamic adaptation in enforcement.

On the other hand, we see that agencies establish a more detailed understanding via the cooperation agreements on a bilateral basis. Despite having a non-binding and somewhat vague presence in the regulatory framework, cooperation protocols are functional tools in the sense that they guide the regulators on how to interpret the legislative infrastructure and feature the signatories' understanding of the intended purposes of enforcement. They also allow the signatories to improve institutional communication, which enhances the cooperation on different levels. Agencies even organize regulatory workshops and expert training sessions to inform each other of the basic principles in their field of regulation. Such trainings and the enhanced level of communication allow the institutions to make educated decisions regarding when to reach out and cooperate on a case-specific basis.

The common point in the legislation and the cooperation protocols is that they try to allocate the regulatory duties before the agencies engage in a market failure. In other words, the assessment in these methods is a hypothetical one and the agencies are to follow this hypothetical assessment when determining who will engage in a specific case at hand.

However, legislation and protocols are not the only tools that define the jurisdictional boundaries. Because of their ex ante nature, legislation and protocols may not always fully cover and comply with the de facto realities of the market. This is where the judicial review option emerges. Under Turkish law, regulatory institutions are regarded as a part of the administrative structure and therefore subject to administrative judicial review(Ardıyok, 2002). Accordingly, parties may require a judicial review of a judgment rendered by a regulatory authority if they think that the regulatory duties were not allocated correctly. Hence, the administrative judicial authorities constitute the third option for allocating the regulatory duties. In fact, judicial reviewis even more determinative than the other options, considering that it has the judicial authority to determine how the legislation should be interpreted and to what extent the protocol should be taken into consideration.

When we assess the respective jurisprudence of the Council of State (**CoS**), we see that in four important decisions[l] concerning the telecommunications sector, the CoS assessed whether the TCA has the authority to investigate the abuse of dominance by the operators with effective market power in markets subject to the ITCA's ex ante regulations (Ardıyok, 2019).[3]

In its Telcoder decision, the TCA argued that the ITCA imposed price regulation in the market and therefore had no authority to investigate the price squeeze allegations against Türk Telekom.[m] However, the CoS canceled this decision by explaining that the TCA should have assessed whether the market players had the ability to make

[3] Under Turkish law, NCA and the NRAs are considered to be part of the governmental administration. Accordingly, the related parties are entitled to challenge such decisions before the Administrative Courts of First Instance (and subsequently the Regional Administrative Court and the Council of State). Hence, judicial review of administrative decisions is among the most prominent sources of guidance for how to interpret the rules and allocate the regulation duties.

decisions on their pricing (within the bounds of the ex ante regulation) in a way that would hinder the competition in the market. Accordingly, the CoS reasoned that as long as the undertaking in question has the behavioral freedom to engage in a competition law violation, the TCA has the authority and the duty to investigate such behavior. This approach is in line with our foregoing assessment and also with the Commission's ruling in the Deutsche Telecom[n] decision.

That said, the CoS also provides that violation of the ex ante regulations by the same action would not hinder the TCA's authority to investigate. However, we believe that, in such a case, the NRA would also investigate the matter and the TCA should wait for the NRA to render its decision.

In a similar ruling, namely the Kule A.Ş. decision, the TCA assessed Turkcell's imposition of anti-competitive exclusivity terms in its lease agreements and reasoned that the ITCA's mandatory facility-sharing regulations made it impossible to violate the competition regulations. In other words, the TCA ruled that compliance with the sector-specific (ex ante) regulations would be sufficient to remedy the market failure. However, with an accurate approach, the CoS stated that the potential of ex ante regulations to remedy the competition violations in the market is not sufficient and the TCA must investigate the de facto situation in the market to ensure that the ex ante regulations are remedying the relevant issues. The main principle we may infer from this judgment is that the TCA has a complementary function to the NRAs (Ulusoy, 2016, p. 8).

On the other hand, in its Bis Enerji A.Ş. decision, the TCA assessed allegations that the Turkish Electricity Distribution Company's (TEDAŞ) tariffs constituted price discrimination. However, the TCA reasoned that the relevant tariffs were subject to EMRA's regulations and would therefore fall outside the scope of its jurisdiction. Upon appeal, the CoS[o] ruled in favor of the TCA's reasoning and approved its decision. Hence, the CoS provides that if the ex ante regulations are taking the competitive functions' place in the market, the TCA should not engage in the issue.

However, even if the administrative judicial authorities' jurisprudence is mostly in line with the principles we have explained in the foregoing sections, the Borusan I and Borusan II decisions exceptionally provide an opposing reasoning. Accordingly, in the said decisions the CoS states that the ITCA and the TCA serve different purposes and are subject to different procedures.

The CoS also provides that the violation of ex ante regulations would not directly violate the competition rules. Moreover, the CoS further reasons that the even if the ex ante regulations remedy all of the anti-competitive harm due to a behavior, the TCA would still have the authority and duty to investigate such behavior.

Above, we have established that the NRAs and the NCAs share the same primary goal and function. Considering that both regulatory agencies aim to remedy market failures, we argue that it is factually inaccurate to conclude that the ITCA and TCA serve different purposes. For that, we find the Borusan decisions inaccurate and diverging from the previous decisional practice.

Additionally, we think it is beneficial to assess potential interpretations of the Boursan decisions' position regarding the TCA's duty to investigate anti-competitive behaviors, even if such behavior has been remedied by the ex ante regulations. We have emphasized that, in cases where the ex ante regulations have successfully remedied the relevant market failures, it is not necessary for the NCAs to engage in further intervention for the same issue. This would not only waste administrative resources but also pose a risk of double sanctioning.

Hence, we think the right way to interpret the position taken in the Borusan decisions is to provide that the TCA should only assess whether the ex ante regulations has successfully remedied the anti-competitive behavior in the market. If yes, then it is not necessary to continue investigating the issue. Under this approach, the TCA has fulfilled its duty to investigate the issue and also stay aligned with the previous decisional practice. That said, if we argue that the Borusan decision requires the TCA to investigate the issue even if the anti-competitive actions are remedied by the ex ante regulations, this approach would harm the complementary relationship and harmonized enforcement, and would also pose a risk of double sanctioning.

On the other hand, TCA also diverted from the decisional practice of the judicial authorities in its Enerjisa decision, where it investigated electricity distribution/retail companies for abuse of dominance. By way of background, the electricity distribution and retail operations used to be integrated under the control of the government. In an effort to open these services to competition, distribution and sales activities were privatized as a whole and on a regional basis. Consumers under a certain level of consumption would automatically fall within the portfolio of the retail company that was integrated with the distribution company ("Authorized Retailer"), while the consumers exceeding a certain consumption level would have the right to choose another retail company ("Independent Retailer") and pursue a better deal.

After the privatization, however, a legal separation was required between the distribution and retail sale activities through sectoral regulations upon the request of the NCA. Considering that there is only one distributor in every region, the Authorized Retailer's integration with the distributor raised concerns about whether this integration would hinder the Independent Retailers' ability to compete. Acting on these concerns, the TCA resolved that the Authorized Retailers are considered to be in dominant position in their region due to their close position with the distributor and being the Authorized Retailer for the below-the-limit consumers. Imposing a hefty administrative fine, the TCA assessed a number of behaviors that were considered to constitute abuse of dominance.

One of the behaviors scrutinized by the TCA was using the resources and manpower of the distribution company for delivering debt notices to households on behalf of the Authorized Retailer. In its judgment, the TCA reasoned that the distribution company was providing an unfair competitive advantage to the Authorized Retailer by supporting it from its resources. The TCA also provided that delivering debt notices was a considerable cost item for the Independent Retailers and they were in no position to procure such services, which restricted their ability to compete. Hence, the TCA concluded that such practices, which provided the Authorized Retailers with an advantage that would otherwise not be available to the Independent Retailers, would constitute abuse of dominance in the given market structure.

On the other hand, the same behavior that set the base for the fine was also regulated by sectoral legislations. Accordingly, Article 12 of the Consumer Regulation ("Previous Regulation"), which was applicable at the time of the decision, regulated the debt notice delivery services and made it possible for the Authorized Retailers to procure debt notice delivery services from the distributors in an implicit manner. Moreover, Article 9(2) of the Electricity Market Law No. 6446 also brought a non-discrimination obligation, including the distributor's relationship with the Independent Retailers. Further clarifying the issue, the Energy Markets Consumer Services Regulation ("New Regulation"), published on the *Official Gazette* dated 30 May 2018 and numbered 30,346, explicitly provided that the retailers could procure debt notice distribution services from the distribution company by virtue of Article 32(5).

Against this backdrop, we think that the TCA's Enerjisa decision deviates from the judicial applications in Turkey, as well as those in the US and the EU. Based on the current jurisprudence, it appears that the TCA steps into the NRAs' shoes and even audits their work. However, the more accurate approach against this issue would be to assess whether ex ante regulations are applied correctly and, if necessary, bring the NRA's decisions before the judicial review. If such a course of action is insufficient, then the appropriate approach would be to amend or replace the ex ante regulation.

Lastly, we believe it is also important to assess markets where there is a need for ex ante regulations, but the NRA have not imposed any. In such markets, the TCA would need to step in and remedy the relevant anti-competitive behaviors. In fact, the NCA's intervention here is crucial as there are already considerable failures in the market. Competition rules' advantage in these situations would be their flexibility and high adaptation on a case-specific basis. This approach may be traced from the reasoning provided by the CoS in its Eser Telecom[p] decision.

5 Assessment of the Turkish System and Suggested Solutions—The Filtering Method

In light of the foregoing explanations, all three of the methods assessed contribute to the harmony among the NCA and the NRAs in Turkey. On the other hand, the same assessments also demonstrate that none of the methods are fully capable of ensuring a smooth jurisdictional distinction that is applicable to different situations.

In light of the foregoing explanations, all three of the methods assessed contribute to the harmony among the NCA and the NRAs in Turkey. On the other hand, the same assessments also demonstrate that none of the methods are fully capable of ensuring a smooth jurisdictional distinction that is applicable to different situations.

Hence, we think it will be beneficial to briefly revert back to the development and characteristics of Turkish regulatory practice. This will help us understand the driving dynamics of the Turkish practice and gain insights that are necessary to come up with an alternative solution.

As we have explained, Turkey has a dynamic and developing regulatory practice. Even the competition rules, which are among the best examples of regulatory practice, have been in enforcement for around 20 years. Compared to the long-standing competition practice in the Anglo-American system or the accumulated enforcement experience of the EU institutions, Turkey has a young practice.

Considering that developing organic regulatory reflexes and crafting the system in accordance with economic and social dynamics takes years of enforcement, one could say that the Turkish system is yet to reach maturity. Therefore, the Turkish practice needs to rely on the experiences and insights collected in the global best practices, such as the EU and the US, to channel these into local enforcement. In so doing, the enforcers should also be mindful about the differentiating factors between Turkey and the foreign systems such as the community dimension in the EU or the federal government system in the US. Demographic factors or differences in the state of industry, importation/exportation and the value-added trade are also among the factors that may require fine tuning.

Accordingly, an important factor to keep in mind while assessing Turkish practice is the need for additional adjustments, while adopting best practices from abroad. By this token, it would be prudent to single out the young and dynamic structure of the Turkish practice as an extended point of exposure to jurisdictional overlaps. Considering that definition and depth of the differentiating factors may differ from one case to another, the solution for jurisdictional frictions should be structurally flexible. In other words, the flexibility required here should have a pattern of its own and should follow a certain structure to be effective against different kinds of jurisdictional overlaps.

When we act upon this need for structured flexibility, we believe that the filtering method may be a viable alternative for restoring cooperation between NRAs and the NCAs. It not only embraces and embodies the best practices we have highlighted in the sections above, but also brings a merit-based selection (filtering) process that aims to choose the best option for the given task with a pragmatic and practical efficiency.

In this regard, the filtering method suggests that the system described below would enable NCAs to interfere with undertakings that operate in markets with structural failures or in connected markets that are subject to some ex ante regulations (Ardıyok, 2019).

The first step is to assess whether the behavior in question violates any ex ante regulation in the relevant market. If not, it would be necessary (and sufficient) for the NCA to conduct a review based on the competition rules. If the relevant behavior is decided to be in violation of an ex ante regulation, we should assess whether such regulations are able to remedy the anti-competitive aspects of such behavior, not only on a theoretical level, but also taking into consideration the de facto application of the said regulations.

If the ex ante regulation turn out to be sufficiently remedying the antitrust concerns raised by the behavior in question, then we may conclude that the NRA's interference would be enough.

On the other hand, if we conclude that the relevant behavior violates both competition regulations and the sectoral regulations, and the ex ante regulations are not sufficient to remedy the competition law concerns in the market, then both agencies should take action based on their own legislations.

This method will optimize the resource utilization of administrative costs, while preventing the market failures from affecting social welfare in the relevant markets. In terms of execution possibilities, we think the regulatory framework we assessed in the foregoing sections contain the required level of flexibility and dynamism for adopting the filtering method.

By the same token, we also assess that the cooperation protocols that are in place between the regulatory agencies may also serve as functional tools to follow this method. In fact, the current state of the relevant texts already provides a parallel structure to what we have been describing under this section.

Lastly, we consider that the judicial jurisprudence we have assessed thus far also features a parallel approach with the filtering method. Considering the prudent and effective approach featured in the filtering method, we feel it would be beneficial for the judicial authorities to apply this method while reviewing the decisions of the NCAs and NRAs. To that end, we are awaiting the end result on the Enerjisa decision as one of the controversial cases in Turkey that is currently underjudicial review judicial review.

Notes

a. John Sherman on one side, J. D. Rockefeller and J. P. Morgan on the other. The so-called age of monopolies also featured many other business magnets such as Andrew Carnegie and Cornelius Vanderbilt.
b. Aalborg Portland and Others v. Commission, Joined Cases C-204/00 P, C-205/00 P, C-211/00 P, C-213/00 P, C-217/00 P, C-219/00 P (2004).
c. Published in the Official Gazette No. 29384 and dated 12 June 2015.
d. High Court of Appeals Assembly of Criminal Chambers decision dated June 2, 1986 and numbered T.1986/2-78, E.1986/324.
e. Decision of the Turkish Competition Board dated January 8, 2009 and numbered 09-01/2-2.
f. Verizon v. Trinko, LLP, 540 U.S. 398 (2004).
g. LinkLine Communications, Inc. v. SBC California, Inc., 503 F.3d 876 (9th Cir. 2007).
h. Deutsche Telekom v Commission, Case C-280/08 (ECR 2010).
i. Industriedes Poudres Sphériques SA v. Commission. Case T-5/97 (ECR 2000).
j. Napier Brown—British Sugar, Case IV/30.178 (1988), O.J. (L 284/41).
k. Commission Decision adopting interim measures concerning National Coal Board, National Smokeless Fuels Limited and National Carbonising Company Limited, (1976) O.J. (L 35/6).
l. Council of State decisions; "Borusan I", dated November 20, 2007 and numbered E.2006/2052-K.2007/7582; "Borusan II", dated February 13, 2012 and numbered E.2008/13184-K.2012/359; "Telkoder", dated May 8, 2012 and

numbered E.2008/14245-K.2012/960; "Kule A.Ş.", dated December 18, 2012 and numbered E.2009/2862-K.2012/73883.

m. Decision of the Turkish Competition Board dated September 11, 2008 and numbered 08-52/792-321.

n. Deutsche Telekom AG, Case COMP/C-1/37.451, 37.578, 37.579 (2003), OJ L 263.

o. Council of State Board of Administrative Chambers decision dated January 21, 2013 and numbered E.2008/1412.

p. Council of State Board of Administrative Chambers decision dated March 20, 2013 and numbered E.2008/3070-K.2013/982.

References

Akçollu, F. Y. (2003). Competition and regulation in the electricity sector (Elektrik Sektöründe Rekabet ve Regülasyon), Rekabet Kurumu Uzmanlık Tezleri. Ankara, p. 24.

Ardıyok, Ş. (2019). Regulation Law (Regülasyon Hukuku), On İki Levha Yayınları, İstanbul.

Ardıyok, Ş. (2002). Doğal Tekeller ve Düzenleyici Kurumlar, Türkiye için Düzenleyici Kurum Modeli, Rekabet Kurumu, Ankara.

Ardıyok, Ş., & Oğuz, F. (2010). Competition law and regulation in the Turkish telecommunications industry: Friends or foes? *Telecommunications Policy*, p. 234.

Ateş, M. (2008). An overview of EU competition law and policy in the light of recent regulations (Son Yapılan Düzenlemeler Işığında AB Rekabet Hukuku ve Politikasına Genel Bir Bakış). *Avrupa Çalışmaları Dergisi, 7*(2), 49.

Communication from the Commission. (2009/C 45/02). Guidance on the Commission's enforcement priorities in applying Article 82 of the EC Treaty to abusive exclusionary conduct by dominant undertakings (82. Madde Kılavuzu).

Dunne, M. (2015). *Competition law and economic regulation: Making and managing markets* (p. 35). Cambridge University Press.

EU Commission. (2014). Recommendation on relevant product and service markets within the electronic communications sector susceptible to ex ante regulation in accordance with Directive 2002/21/EC of the European Parliament and of the Council on a common regulatory framework for electronic communications networks and services.

Gürzumar, O. B. (2006). Obligation to contract based on the essential facility doctrine: Abuse of dominant position through refusal to contract with competitors (Zorunlu Unsur Doktrinine Dayalı Sözleşme Yapma Yükümlülüğü: Hakim Durumun Rakiple Anlaşma Yapmaktan Kaçınmak Suretiyle Kötüye Kullanılması, Seçkin Yayınları.

Hou, L. (2012). The Abusive Nature of Price Squeeze in the EU. Available at: http://papers.ssrn.com/sol3/papers.cfm?abstract_id=2185839

Karabel, G. (2015). The Principle of ne bis in idem in Competition Law (Rekabet Hukukunda ne bis in idem İlkesi), Rekabet Kurumu Uzmanlık Tezleri, Ankara, p. 23.

Law No. 5809, Electronic Communications published in the Official Gazette No. 27050 and dated 5 November 2008.

Law No. 4646, Law on the Natural Gas Market and Amending the Law on Electricity Market published in the Official Gazette No. 24390 and dated 18 April 2001.

OECD. (2006). The role of competition policy in regulatory reform.

Şahin, S. Y. (2010). The relationship between the competition authority and the sectoral regulatory authority in the turkish energy sector (Türk Enerji Sektöründe Rekabet Kurumu ile Sektörel Düzenleyici Kurum Arasındaki İlişki). Rekabet Dergisi (p. 25).

TÜSİAD. (2014). Competition law and competitiveness sector discussions: Energy sector (Rekabet Hukuku ve Rekabet Gücü Sektör Tartışmaları: Enerji Sektörü). İstanbul (p. 50).
Ulusoy, A. (2016). Allocation of authority between competition authorities and sectoral regulatory authorities and its legal consequences (Rekabet Otoriteleri ve Sektörel Regülasyon Otoriteleri Arasındaki Yetki Paylaşımı ve Hukuki Sonuçları). *Rekabet Forumu, 101*, 8–9.

Şahin Ardıyok is a partner at Balcıoğlu Selçuk Ardıyok Keki Attorney Partnership and the Head of Competition law, Public Policy and Regulation Team. He represents clients, government institutions, NGOs and multinational companies in anti-trust investigations and merger/acquisition notifications, regulatory authority filings, appeal cases, antidumping investigations (across Turkey, the EU, the US and Canada) and PPP projects. Before joining in the private practice, he worked at the Turkish Competition Authority for five years and conducted investigations in sectors including telecommunications, media and broadcasting, information technology and electronic appliances. He graduated from Business Administration Department of Turkish Army Academy in 1991 and Faculty of Law of Ankara University in 1997. He also holds an MBA from the Ankara University and an LL.M. from the University of Chicago. Previously, he taught the courses on "Telecommunications Law" and "Pharmaceutical Law" at Istanbul Bilgi University, and currently teaches courses on "Economic Regulation and the Law" (in English) and "Energy Law and Policy" at Bilkent University which is located in Ankara. Moreover, he is the Advisory Board Member of the Competition Law Center of Istanbul Bilgi University and he instructs various certificate programs at various universities, Bar Associations and other institutions. Recognized both nationally and internationally for his competition law work, he is an active delegate of the International Chamber of Commerce (ICC) National Committee in the ICC's Competition Commission and serves as an NGA for the various working groups of International Competition Network (ICN). He frequently speaks at conferences and symposia on competition law and regulated industries, he frequently writes on competition law, regulation and law and economics. His latest publications include a book on Regulation Law that provides an end-to-end coverage of all aspects in the field of regulatory practice. Legal500 and Chambers & Partners consistently rank him as a leading individual in the field of competition and compliance.

Armanç Canbeyli is an attorney-at-law focusing on competition law, mergers and acquisitions, international trade, compliance and energy. His work at Balcıoğlu Selçuk Ardıyok Keki Attorney Partnership includes representing domestic and international clients in anti-trust investigations, merger control filings, trade remedy investigations (across Turkey, the EU, the US and Canada), contract reviews, regulatory inspections, pricing regulations, compliance programs, anti-corruption and extraterritorial sanctions. He is particularly experienced in sectors including energy, automotive, digital platforms, FMCG, banking, cement and iron and steel. He has several publications in the field of competition and regulation law, compliance and international trade. His writings also focus on the regulation of electric cars and autonomous vehicles, smart mobility, alternative fuels and energy, sustainability, digital platforms and WTO regulations. He also speaks at conferences and post-graduate certificate programs including the Competition Law Center of Istanbul Bilgi University and the Turkish Bar Associations Education Programs. He holds an LLB from Bilkent University, Faculty of Law and an LLM from Galatasaray University, Faculty of Law. He is a member of the Istanbul Bar. In 2020 and 2021, Legal500 recognized him as Rising Star in the field of compliance in Turkey.

Index

M. Eroğlu and M. Finger (eds.), *The Regulation of Turkish Network Industries*,
https://doi.org/10.1007/978-3-030-81720-6

The manufacturer's authorised representative in the EU is Springer Nature Customer Service Centre GmbH, Europaplatz 3, 69115 Heidelberg, Germany. If you have any concerns regarding our products, please contact ProductSafety@springernature.com

Printed and bound by CPI Group (UK) Ltd, Croydon, CR0 4YY
30/09/2026
02225695-0019